CONCURRENT ENGINEERING FUNDAMENTALS

PRENTICE HALL INTERNATIONAL SERIES
IN INDUSTRIAL AND SYSTEMS ENGINEERING

W. J. Fabrycky and J. H. Mize, Editors

AMOS AND SARCHET *Industrial Safety and Health Management, 2/E*
AMOS AND SARCHET *Management for Engineers*
AMRINE, RITCHEY, MOODIE AND KMEC *Manufacturing Organization and Management, 6/E*
ASFAHL *Industrial Safety and Health Management, 3/E*
BABCOCK *Managing Engineering and Technology, 2/E*
BADIRU *Comprehensive Project Management*
BADIRU *Expert Systems Applications in Engineering and Manufacturing*
BANKS, CARSON AND NELSON *Discrete Event System Simulation, 2/E*
BLANCHARD *Logistics Engineering and Management, 4/E*
BLANCHARD AND FABRYCKY *Systems Engineering and Analysis, 2/E*
BROWN *Technimanagement: The Human Side of the Technical Organization*
BURTON AND MORAN *The Future Focused Organization*
BUSSEY AND ESCHENBACH *The Economic Analysis of Industrial Projects, 2/E*
BUZACOTT AND SHANTHIKUMAR *Stochastic Models of Manufacturing Systems*
CANADA AND SULLIVAN *Economic and Multi-Attribute Evaluation of Advanced Manufacturing Systems*
CANADA, SULLIVAN AND WHITE *Capital Investment Analysis for Engineering and Management, 2/E*
CHANG AND WYSK *An Introduction to Automated Process Planning Systems*
CHANG, WYSK AND WANG *Computer Aided Manufacturing*
DELAVIGNE AND ROBERTSON *Deming's Profound Changes*
EAGER *The Information Payoff: The Manager's Concise Guide to Making PC Communications Work*
EBERTS *User Interface Design*
EBERTS AND EBERTS *Myths of Japanese Quality*
ELSAYED AND BOUCHER *Analysis and Control of Production Systems, 2/E*
FABRYCKY AND BLANCHARD *Life-Cycle Cost and Economic Analysis*
FABRYCKY AND THUESEN *Economic Decision Analysis*
FISHWICK *Simulation Model Design and Execution: Building Digital Worlds*
FRANCIS, MCGINNIS AND WHITE *Facility Layout and Location: An Analytical Approach, 2/E*
GIBSON *Modern Management of the High-Technology Enterprise*
GORDON *Systematic Training Program Design*
GRAEDEL AND ALLENBY *Industrial Ecology*
HALL *Queuing Methods: For Services and Manufacturing*
HANSEN *Automating Business Process Reengineering*
HAMMER *Occupational Safety Management and Engineering, 4/E*
HAZELRIGG *Systems Engineering*
HUTCHINSON *An Integrated Approach to Logistics Management*
IGNIZIO *Linear Programming in Single- and Multiple-Objective Systems*
IGNIZIO AND CAVALIER *Linear Programming*
KROEMER, KROEMER AND KROEMER-ELBERT *Ergonomics: How to Design for Ease and Efficiency*
KUSIAK *Intelligent Manufacturing Systems*
LAMB *Availability Engineering and Management for Manufacuring Plant Performance*
LANDERS, BROWN, FANT, MALSTROM AND SCHMITT *Electronic Manufacturing Processes*
LEEMIS *Reliability: Probabilistic Models and Statistical Methods*
MICHAELS *Technical Risk Management*
MUNDEL AND DANNER *Motion and Time Study: Improving Productivity, 7/E*
OSTWALD *Engineering Cost Estimating, 3/E*
PINEDO *Scheduling: Theory, Algorithms, and Systems*
PRASAD *Concurrent Engineering Fundamentals, Vol. I: Integrated Product and Process Organization*
PULAT *Fundamentals of Industrial Ergonomics*
SHTUB, BARD AND GLOBERSON *Project Management: Engineering Technology and Implementation*
TAHA *Simulation Modeling and SIMNET*
THUESEN AND FABRYCKY *Engineering Economy, 8/E*
TURNER, MIZE, CASE AND NAZEMETZ *Introduction to Industrial and Systems Engineering, 3/E*
TURTLE *Implementing Concurrent Project Management*
WALESH *Engineering Your Future*
WOLFF *Stochastic Modeling and the Theory of Queues*

CONCURRENT ENGINEERING FUNDAMENTALS

Integrated Product and Process Organization

Biren Prasad

PRENTICE HALL INTERNATIONAL SERIES IN INDUSTRIAL AND SYSTEMS ENGINEERING

Prentice Hall PTR
Upper Saddle River, New Jersey 07458

Library of Congress Cataloging-in-Publication Data

Prasad, Biren

Concurrent engineering fundamentals: integrated product and process organization / Biren Prasad.

p. cm. — (Prentice-Hall international series in industrial and systems engineering)

Includes bibliographical references and index.

ISBN 0-13-147463-4

1. Production engineering. 2. Concurrent engineering. 3. Design, Industrial. I. Title. II. Series.

TS176.P694 1996

670.4—dc20 95-43132

CIP

Acquisitions editor: Bernard Goodwin
Cover designer: Jim Walsh
Cover design director: Jerry Votta
Manufacturing buyer: Alexis R. Heydt
Compositor/Production services: Pine Tree Composition, Inc.

Printed in the United States of America

10 9 8 7 6 5 4 3 2 1

ISBN: 0-13-147463-4

Prentice Hall International (UK) Limited, *London*
Prentice Hall of Australia Pty. Limited, *Sydney*
Prentice Hall Canada, Inc., *Toronto*
Prentice Hall Hispanoamericana, S.A., *Mexico*
Prentice Hall of India Private Limited, *New Delhi*
Prentice Hall of Japan, Inc., *Tokyo*
Simon & Schuster Asia Pte. Ltd., *Singapore*
Editora Prentice Hall do Brasil, Ltda., *Rio de Janeiro*

To Pushpa, Rosalie, Gunjan, and Palack,
for your patience and support.

Trademarks (TM)

Pro/Engineer^{TM}: Parametric Technology Corp., Waltham, MA.

I-DEAS Master Series^{TM}: SDRC, Milford, Ohio.

CADDS 5^{TM}: Computervision Corp., Bedford, MA.

Anvil 5000^{TM}: Manufacturing & Consulting Series, Scottsdale, AZ.

Catia^{TM} Solutions: Dassault Systems, North Hollywood, CA.

Unigraphics^{TM}: EDS Unigraphics, Maryland Heights, MO.

HP PE/SolidDesigner^{TM}: Hewlett-Packard, Ft. Collins, CO.

ICAD^{TM}, Icad Inc, Cambridge, MA.

I/EMS^{TM}: Intergraph Corp., Huntsville, AL.

CONTENTS

COMPUTER ACRONYMS

ANSI	American National Standards Institute
API	Application Programming Interface
ATIS	A Tools Integration Standards
CA	Computational Architecture
CAD	Computer-Aided Design
CAE	Computer-Aided Engineering
CALS	Computer-Aided Acquisition and Logistics Support
CAM	Computer-Aided Manufacturing
CAPP	Computer-Aided Process Planning
CASA	Computer-Aided Society of Manufacturing Engineers
CASE	Computer-Aided Process Engineering
CFD	Computational Fluid Dynamics
CFI	CAD Framework Initiative
CIM	Computer-Integrated Manufacturing
CPU	Central Processing Unit
DBMS	Database Management Systems
DDL	Dynamic Data Linking
DNS	Distributed Name Service
FEA	Finite Element Analysis
FEM	Finite Element Modeling
FMEA	Failure Mode and Effects Analysis
ICAM	Integrated Computer-Aided Manufacturing program
IGES	Initial Graphics Exchange Specification
IOS	Input/output Subsystems

ISO	International Standards Organization (ISO)
LAN	Local Area Network
MCAE	Mechanical Computer-Aided Engineering
MFLOPS	Million Instructions per Second
MIS	Mainframe Information System
MRP	Material Requirements Planning
NAS	Network Application Services
NC	Numerical Control
NCS	Network Computing System
NFS	Network File System
NURBS	Non-Uniform Rational B-Splines
PC	Personal Computer
PDES	Product Data Exchange using STEP
PHIGS	Programmers' Hierarchical Interactive Graphic Standard
PIM	Product Information Management
PPO	Product, Process, and Organization model
QFD	Quality Function Deployment
RISC	Reduced Instruction Set Computing
RPC	Remote Procedure Call
SLA	Stereolithography Apparatus
SME	Society of Manufacturing Engineers
SPEC	System Performance Evaluation Cooperative
SQL	Structured Query Language
SSD	Secondary Storage Device
STEP	Standard for the Exchange of Product Data
TCP/IP	Transmission Control Protocol/Internet Protocol
TQM	Total Quality Management
WAN	Wide Area Network

PREFACE

As the name implies, the book describes the fundamentals of Concurrent Engineering and explains the basic principles on which this very subject is founded. The most of the materials in the book are either original ideas or their extension to CE. Most is never reported elsewhere and is based on the author's successes while practicing CE on the job, and decades of his research and learning working with electronic, automotive, aerospace and railroad industries including Ford, General Motors, Electronic Data Systems, Association of American Railroads, NASA and numerous other places abroad. Concurrent Engineering approach to product design and development has two major themes. The first theme is establishing a ***concurrent product and process organization.*** This is referred to herein as ***"process taxonomy."*** The second theme is applying this process taxonomy (or methodology) to design and develop the total product system. This is referred to as ***integrated product development*** (IPD). Each theme is divided into several essential parts forming major chapters of this book.

The first volume called ***integrated product and process organization*** has nine chapters. The second volume named ***integrated product development*** has ten chapters. The materials in these two volumes are brought together to balance the interests of both the customers and the companies. The contents of Volume One are: Manufacturing Competitiveness: *Life-cycle Management, Process Reengineering, Concurrent Engineering Definitions, Cooperative Work Teams, System Engineering, Information Modeling, The Whole System,* and *Product Realization Taxonomy.* The contents of Volume Two are: *Total Value Management, CE Metrics and Measures, Concurrent Function Deployment, Integrated Product Development, Frameworks and Architectures, Decision Support System, Intelligent Information System, Capturing Life-Cycle Values, Life-Cycle Mechanization,* and *CE Implementation Guidelines.*

In Concurrent Engineering (CE) system, each modification of the product represents a transformation relationship between specifications (inputs, requirements and constraints), outputs, and the concept the modification represents. At the beginning of the transformation, the specifications are generally in abstract forms. As more and more of the specifications are satisfied, the product begins to take shape (begins to evolve into a physical form). To illustrate how a full CE system will work, and to show the inner-working of its elements, author defines this CE system as a set of two synchronized wheels. The representation is analogous to a set of ***synchronized wheels of a bicycle.*** Figure 1 shows this CE wheel set.

CONCURRENT ENGINEERING WHEELS

The first CE wheel represents the ***integrated product and process organization.*** The second CE wheel accomplishes the ***integrated product development.*** The two wheels together harmonize the interests of the customers and the fostering CE organization (frequently referred to as an enterprise). The contents of the first wheel are described in Volume One and the contents of second wheel are described in Volume Two of the ***CE fundamental*** books. Three concentric rings represent the three essential elements of a wheel. The middle ring represents the CE work-groups, which drive the customer and the enterprise like how a human drives a bike. The work-groups are divided into four quadrants representing the four so called CE teams. These teams are: *the personnel team, the technology team, the logical team, and the virtual team.* They are discussed in Chapter 5. The outer ring for each wheel is divided into eight parts. Each part represents a chapter of this book. The parts for the first wheel are discussed in Volume One. This volume explains how the CE design process (called herein CE process taxonomy) provides a stable, repeatable process through which increased accuracy is achieved. The book starts with an introductory chapter on manufacturing competitiveness reviewing the history and emerging trends. The remaining chapters of the book (Volume One) describe CE design process, explain how concurrent design process can create a competitive advantage, define CE process taxonomy, and address a number of major issues related to *product and process organization.* The parts of the second wheel are discussed in Volume Two. A separate chapter in the book is dedicated to discussing each part of the two CE wheels.

First CE Wheel: Integrated Product and Process Organization

The innermost ring of the first CE wheel is a hub. The layout of the hub is same for both wheels. The hub represents four supporting "M" elements: *models, methods, metrics and measures.* Models refer to information modeling. Methods refer to *product realization taxonomy.* They are discussed in Chapters 7 and 9 of Volume One, respectively. *CE Metrics and Measures* are discussed in Chapter 2 of Volume Two. The complexity of the product design and development (PDD) process differs depending upon the

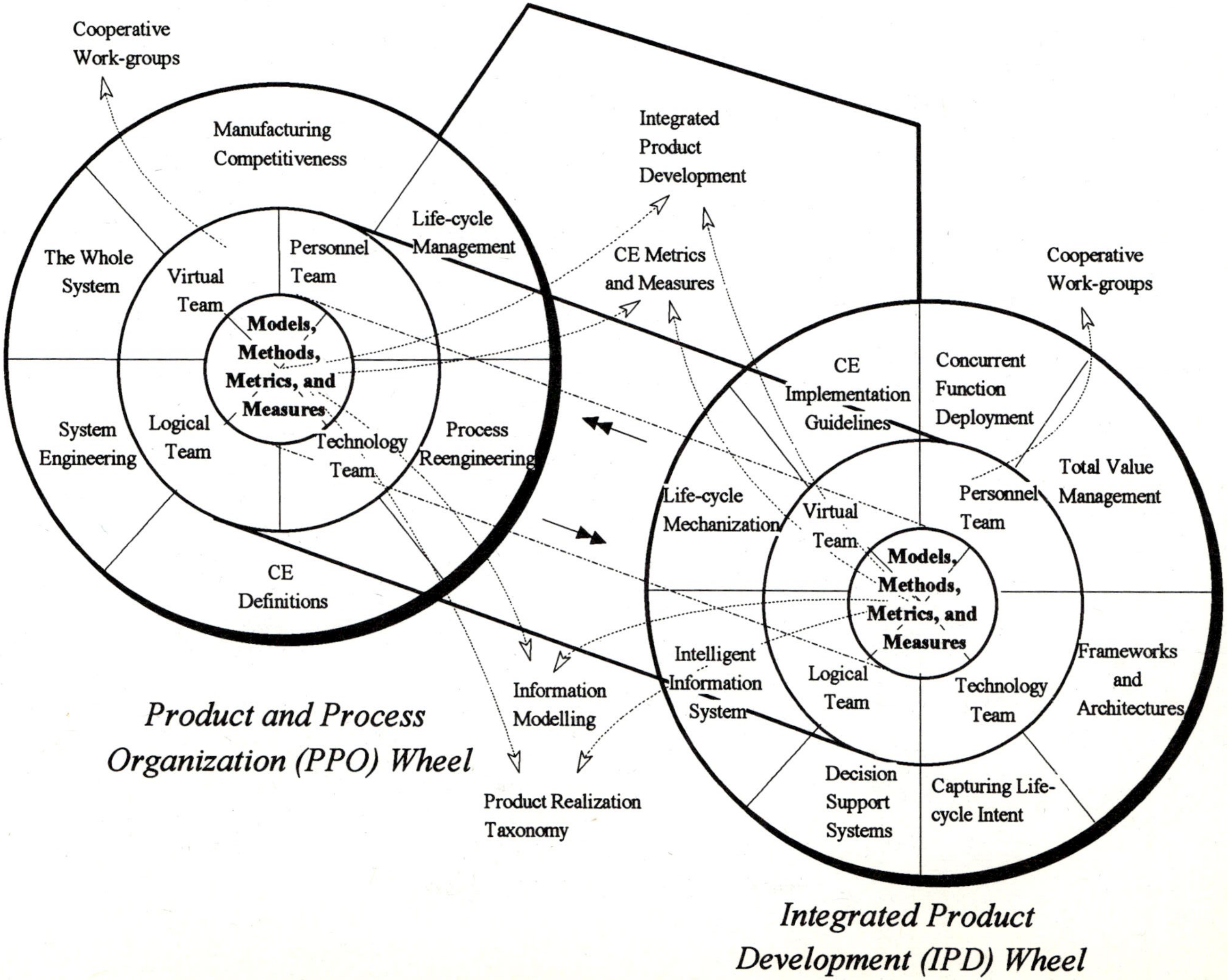

FIGURE 1 A Synchronized set of CE Wheels

(i) Types of information and sources
(ii) Complexity of tasks
(iii) Degree of their incompleteness or ambiguity

Other dimensions encountered during this PDD process that cannot be easily accommodated using traditional process (such as serial engineering) are:

(iv) Timing of decision making
(v) Order of decision making
(vi) Communication mechanism

The elements of the first CE wheel define a set of systems and processes that have the ability to handle all of the above six dimensions. In the following some salient points of the chapters are briefly highlighted:

- *Manufacturing Competitiveness:* The price of the product is dictated by world economy and not by one's own economy or a company's market edge alone. Those companies that can quickly change to world changing market place can position themselves to complete globally. This chapter outlines what is required to become a market leader and compete globally. Successful companies have been the ones who have gained a better focus on eliminating waste, normally sneaked into their products, by understanding what drives product and process costs and, how value can be added. They have focused on a product and process delivery-system—how to transform process innovations into technical success and how to leverage the implementation know-how into big commercial success. Many have chosen to emphasize high-quality flexible or agile production in product delivery rather than high-volume (mass) production.
- *Life-cycle Management:* Today, most companies are under extreme pressure to develop products within time periods that are rapidly shrinking. As the market changes so do the requirements. This has chilling effect in managing the complexity of such continuously varying product specifications and handling the changes within this shrinking time period. The ongoing success of an organization lies in its ability to: continue to evolve; quickly react to changing requirements; reinvent itself on a regular basis; and keep up with ever changing technology and innovation. Many companies are stepping up the pace of new product introduction, and are constantly learning new ways of engineering products more correctly the first time, and more often thereafter. This chapter outlines life-cycle management techniques, such as management change, and process improvement to remain globally competitive.
- *Process Reengineering:* The global marketplace of the 1990s has shown no sympathy to tradition. The reality is that if the products manufactured do not meet the market needs, demand declines and profits dwindle. Many companies are finding that true increase in productivity and efficiency begins with such factors as clean and efficient process, good communication infrastructure, teamwork, a constancy of shared vision and purpose. The challenge is simply not to crank up the speed of the

machines so that its outputs (per unit of time) are increased or doubled, but to change the basic machinery or process that produces the outputs. To accomplish the latter goals, this chapter describes several techniques to achieve competitive superiority such as benchmarking, CPI, organizational restructuring, renovation, process re-engineering, etc.

- *CE Definitions:* The changing market conditions and international competitiveness are making the time-to-market a fast shrinking target. Over the same period, diversity and complexity of the products have increased multi-folds. Concurrency is the major force of Concurrent Engineering. Paralleling describes a "time overlap" of one or more work-groups, activities, tasks, etc. This chapter describes seven CE principles to aim at: *Parallel Work-group, Parallel Product Decomposition; Concurrent Resource Scheduling; Concurrent Processing; Minimize Interfaces; Transparent Communication; and Quick Processing.* This chapter also describes the seven forces that influence the domain of CE (called here as agents or 7Ts) namely: *talents, tasks, teams, techniques, technology, time and tools.*
- *Cooperative Work-groups:* It has been the challenge for the design and manufacturing engineers to work together as teams to improve quality while reducing costs, weight, and lead-time. A single person, or a team of persons, is not enough to provide all the links between: human knowledge and skills; logical organization; technology; and a set of 7Cs coordination features. A number of supporting teams is required, some either virtual or at least virtually collocated. For the waltz of CE synthesis to succeed, CE teams need clear choreography. This chapter describes for the first time the four collaborative teams that are essential for managing a CE organization. Examples of collaborative features include capabilities of electronic meeting such as message-posting and interactions through voice, text, graphics and pictures.
- *System Engineering:* Most groups diligently work to optimize their sub-systems, but due to lack of incentives they tend to work independently of each other. This results in a product, which is often sub-optimized at each decomposed level. System engineering requires that product realization problem is viewed as a "system-centered" problem as opposed to "component-centered." *Systems Engineering* does not disagree with the idea of compartments or divisions of works, but it emphasizes that the interface requirements between the divisions (inter-divisional) and across the level should be adequately covered. That way, when the time comes to modernize other components of the system, an enterprise has the assurance that previously introduced technologies and processes will work logically in a fully integrated fashion, thereby increasing the net efficiency and profitability.
- *Information Modeling:* A successful integrated product development (IPD) requires a sufficient understanding of the product and process behaviors. One way to achieve this understanding is to use a series of reliable information models for planning, designing, optimizing and controlling each unit of the IPD process. The demands go beyond the 3-D CAD geometric modeling. The demands require schemes that can model all phases of a product's life-cycle from cradle to grave. The different aspects of product design (planning, feasibility, design, process-planning), process design (process-execution, production, manufacturing, product support), the human be-

havior in teamwork, and the organization or environment in which it will operate, all have to be taken into account. Five major classes of modeling schemata are discussed in this chapter. The are: (a) Product representation schemes and tools for capturing and describing the product development process and design of various interfaces, such as design-manufacturing interface; (b) Schemes for modeling physical processes, including simulation, as well as models useful for product assessments, such as DFA/DFX, manufacturability evaluation of in-progress designs; (c) Schemes for capturing (product, process, and organization structure) requirements or characteristics for setting strategic and business goals; (d) Schemes to model enterprise activities (data and work flow) in order to determine what types of functions best fit the desired profitability, responsiveness, quality and productivity goals; and (e) Schemes to model team behavior, because most effective manufacturing environments involve a carefully orchestrated interplay between teams and machines.

- *The Whole System:* Often while designing an artifact, work-groups forget that the product is a system. It consists of a number of subassemblies, each fulfilling a different but distinct function. A product is far more than the collection of components. Without a structure or "constancy-of-purpose" there is no system. The central difference between a CE transformation system and any other manufacturing system, such as serial engineering, is the manner in which the tasks' distribution is stated and requirements are accomplished. In a CE transformation system, the purpose of every process step of a manufacturing system is not just to achieve a transformation but to accomplish this in an optimal and concurrent way. This chapter proposes a system-based taxonomy, which is founded on parallel scheduling of tasks and a breakdown structures for product, process and work to realize a drastic reduction in time and cost in product and process realizations.
- *Product Realization Taxonomy:* This constitutes a "state of series of evolution or transformation" leading to a complete design maturity. Product Realization Taxonomy involves items related to design incompleteness, product development practices, readiness feasibility, and assessing goodness. In addition, CE requires these taxonomies to have a unified "product realization base." The enterprise integration metrics of the CE model should be well characterized and the modeling methodologies and/or associated ontology for developing them should be adequate for describing and integrating enterprise functions. The methodologies should have built-in product and service accelerators. Taxonomy comprises of the product, process descriptions, classification techniques, information concepts, representation, and transformation model for product realization. They are included as part of the taxonomy descriptions.

Second CE Wheel: Integrated Product Development

The second CE wheel defines the integrated product development (IPD). This is discussed in the second Volume. IPD in this context does not imply a step-by-step serial process. Indeed, the beauty of this wheel (integrated product development) is that it offers a framework for a concurrent product design and development. A framework within which, the CE teams have flexibility to move about, fitting together bits of the jigsaw as they come together. CE teams have opportunity to apply a variety of techniques contained in this volume (such as: *Concurrent Function Development, Total Value Management, Metrics and Measures,* etc.) and through their use have opportunity to achieve steady overall progress towards a finished product.

- *Concurrent Function Deployment:* The role of the organization and engineers is changing today, as is the method of doing business. Competition has driven organization to consider concepts such as time compression (fast-to-market), Concurrent Engineering, Design for X-ability, and Tools and Technology (such as Taguchi, Value Engineering) while designing and developing an artifact. Quality Function Deployment (QFD) addresses major aspects of "quality" with reference to the functions it performs, but this is one of the many functions that need to be deployed. With conventional deployment, it is difficult, however, to address all aspects of Total Values Management (TVM) such as *X-ability, Cost, Tools and Technology, Responsiveness and Organization* issues. It is not enough to deploy just the *"Quality"* into the product and expect the outcome to be the *World Class.* TVM efforts are vital in maintaining a competitive edge in today's world marketplace.
- *CE Merits and Measures:* Metrics are the basis of monitoring and measuring process improvement methodology and managing their effectiveness. Metric information assists in monitoring team progress, measuring quality of products produced, managing the effectiveness of the improved process, and providing related feedback. Individual assurances of DFX specifications (one at a time) do not capture the most important aspect of Concurrent Engineering—the system perspectives, or the trade-off across the different DFX principles. While satisfying these DFX principles in this isolated manner, only those which are not in conflict are usually met. Concurrent engineering views the design and evaluates the artifact as a system, which has a wider impact than just sub-optimizing the sub-systems within each domain.
- *Total Value Management:* The most acclaimed slogan for introducing a quality program in early corporate days simply was to provide the most value for the lowest cost. This changed as the competitiveness became more fierce. For example, during the introduction of traditional TQM program in 1990 *"getting a quality product to market for a fair price"* was the name of the game. The new paradigm for CE now is total values management (TVM): *"to provide the total value for the lowest cost in the least amount of time, which satisfies the customers the most and lets the company make a fair profit."* Here use of "value" is not just limited to *"quality."* To provide long lasting added value, companies must change their philosophy towards things like *x-ability, responsiveness, functionality, tools and technology, cost, architecture,* etc.
- *Integrated Product Development:* A systematic methodology is essential in order to be able to integrate: (a) teamwork; (b) information modeling; (c) product realization taxonomy; and (d) measures of merits (called CE metrics), and quantitatively assess the effectiveness of the transformation. This may involve identification of perfor-

mance metrics for measuring the product and process behaviors. Integrated product development methodology is geared to take advantage of the product realization taxonomy.

- *Frameworks & Architectures:* In order to adequately support the CE—4Ms (namely: modeling, methods, metrics and measurements), it is necessary to have an architecture that is openly accessible across different CE teams, information systems, platforms, and networks, Architecture consists of information contents, integrated data structures, knowledge-bases, behavior and rules. An architecture not only provides an information base for easy storage, retrieval, and tracking version control, but can also be accessed by different users simultaneously, under ramp-up scheduling of parallel tasks, and in synchronization. We also need a product management system containing work management capabilities integrated with the database. This is essential because in CE there exists a large degree of flexibility for parallelism that must be managed in conjunction with other routine file and data management tasks.
- *Capturing Life-cycle Intent:* Most CAD/CAM tools are not really "capture" tools. In static representation of CAD geometry, configuration changes cannot be handled easily, particularly when parts and dimensions are linked. This has resulted in loss of configuration control, proliferation of changes to fix the errors caused by other changes, and sometimes ambiguous designs. By capturing "design intent": as opposed to "static geometry," configuration changes could be made and controlled more effectively using the power of the computer than through traditional CAD attributes (such as lines and surfaces). The power of a "capture" tool comes from the methods used in capturing the "design intent" initially so that the required changes can be made easily and quickly, if needed. "Life-cycle capture" refers to the definition of the physical object and its environment in some generic form. "Life-cycle intent" means representing the "life-cycle capture" in a form, which can be modified and iterated until all the life-cycle specifications for the product are fully satisfied.
- *Decision Support System:* In CE, cooperation is required between CE teams, management, suppliers, and customers. A knowledge based support system will help the participating teams in decision making and to reflect balanced views. Tradeoffs between conflicting requirements can be made on the basis of information obtained from sensitivity, multi-criterion objectives, simulation, or feedback. The taxonomy can be made a part of decision support system (DSS) in supporting decisions about product decomposition by keeping track of what specifications are satisfied, in ensuring common visibility in the state of product realization, including dispatching and monitoring of tasks, structure, corporate design histories, etc.
- *Intelligent Information System* (IIS): Another major goal of CE is to handle information intelligently in multimedia—audio, video, text, graphics. Since IIS equals CIM plus CE, with IIS, many relevant CE demands can be addressed and quickly processed. Examples include: (a) over local or wide area networks, such as SQL, which connects remote, multiple databases and multimedia repositories; (b) any needed information, such as recorded product designers' design notes, figures, decisions, etc., can be made available on demand at the right place at the right time; (c)

any team can retrieve information in the right format and distribute it promptly to the other members of the CE teams.

- *Life-cycle Mechanization:* Life-cycle mechanization equals CIM + Automation + CE. Life-cycle mechanization is arranged under a familiar acronym: CAE, for CIM, Automation, and CE. Since CAE also equals IIS plus automation, the major benefits of mechanization in CAE come from removing or breaking barriers. The three common barriers are: (a) integration (this is a term taken from CIM), (b) automation, and (c) cooperation (which is a term taken from CE). CE provides the decision support element, and CIM provides the framework & architecture plus the information management elements. *Life-cycle Mechanization* refers to the automation of life-cycle functions or creation of computerized modules that are built from one another and share the information from one another. This includes integration and seamless transfer of data between commercial computer-based engineering tools and product-specific in-house applications. This tends to reduce the dependency of many CE teams on communication links and product realization strategies, such as decomposition and concatenation.
- *CE Implementation Guidelines:* The purpose of this chapter is to offer an implementation guideline for product redesign and development through its life-cycle functions. IPD implementation is a multi-track methodology. The tracks overlap, but still provide a structured approach to organizing product ideas and measures for concurrently performing the associated tasks. Concurrency is built in a number of ways (similar to what was discussed in Volume One), depending upon the complexity of the process or the system involved. This chapter proposes a set of *"Ten Commandments,"* which serves to guide the product and process iterative aspects of IPD rather than just the work-group collaborative aspects during the development cycle. The CE teamwork in the center of the wheel ensures that both local or zonal iterative refinements and collaborative refinements take place during each concurrent track.

A SYNCHRONIZED WHEEL SET FOR CE

All the above nineteen parts of CE put together creates a synchronized wheel set for CE, as shown in Figure 1. The teamwork, with four cooperating components (technological teams, logical teams, virtual teams, and personnel teams), is in the middle ring. The 4Ms (models, metrics, measurements and methodology) form the center of this wheel. The center has four parts to it: *Information Modeling; Product Realization Taxonomy; Measures of Merit; and Integrated Product Development.* The 4Ms are shown in the center because it provides the methodology for guiding the product realization process. The two inner rings, which are the same for both wheels, makes the wheels a synchronized set. The teams in the middle ring are the driving force of the methodology (4Ms listed in the center) and controller of the technologies (listed on the outer ring). *The emphasis of a team-centered wheel for CE is a departure from a conventional function-centered approach.* Outer rings of each wheel contain the remaining parts of "*integrated product and process organization* (first Volume) and *integrated product development* (second Volume), respectively. The idea of this middle ring is to provide a team-centered 7Cs (*Collaboration,*

Commitment, Communications, Compromise, Consensus, Continuous Improvement, and *Coordination*) interplay across layers of enabling technologies and methodologies. Everything is geared towards cutting and compressing the time needed to design, analyze, and manufacture marketable products. Along the way, costs are also reduced, product quality is improved and customer satisfaction is enhanced due to the synchronized process. There is, however, a finite window in which the benefits of time compression and cost cutting are available. As more manufacturers reduce lead time, what once represented a competitive advantage can become a weakening source. Fortunately, the CE wheel provides a continuum (dynamic) base through which new paradigms (process, tools, technology, and 7Ts) can be launched to remain globally competitive for a long haul.

Before we take a closer look at the different parts of this wheel as different chapters of this book, it is important to note that all the parts are not of the same kind. They emphasize different aspects of CE. The four major aspects are (see Figure 2):

- Philosophical aspect
- Methodological aspect
- Conceptual aspect
- Virtual aspect

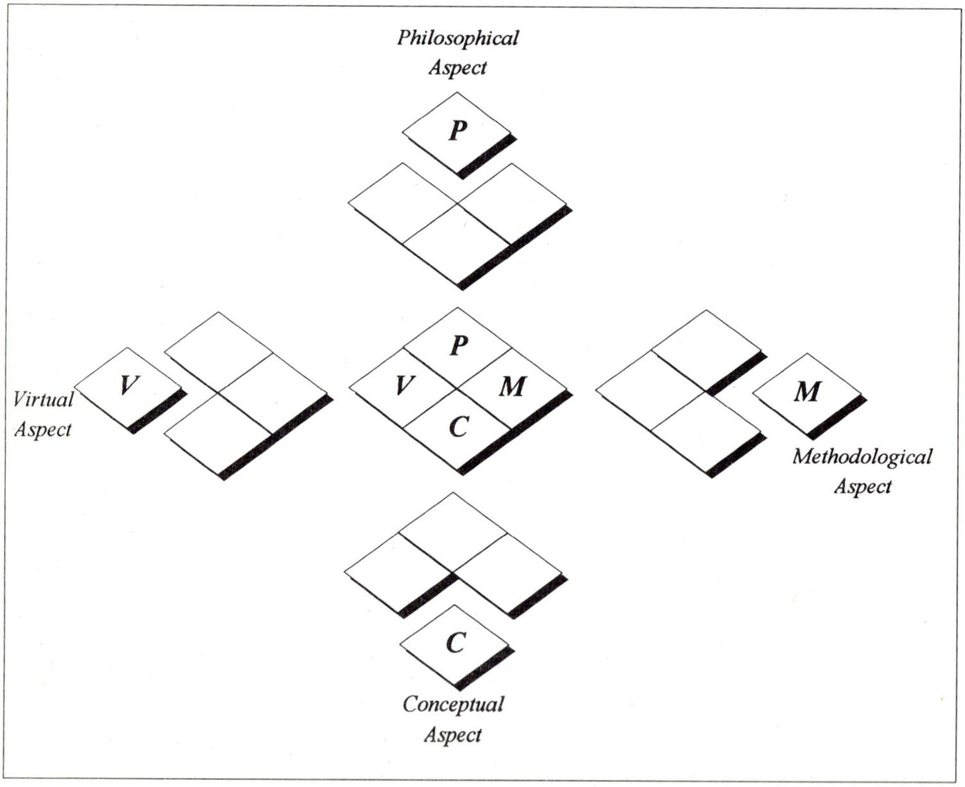

FIGURE 2 Four Aspects of CE

- *Philosophical Aspect:* Personnel CE team governs the philosophical aspects of CE. Philosophical aspect deals with the boundaries of the responsibility and the authority, culture, empowerment, teams make-up, program organization, supplier rationalization, management styles or philosophies, change management, workplace organization and visual control, physical proximity (collocation), management and reporting structure, etc. The chapters on *Cooperative Teamwork and Life-cycle Management* emphasize more of this aspect than others.
- *Methodological Aspect:* of CE is governed by technology team. Methodological aspect deals with system thinking, approaches to system complexity, system integration, transformation model of the manufacturing system, CE enterprise system taxonomy, integrated product and process development, transformation system for product realization, pull system for product realization, track and loop methodology, etc. The chapters on *Systems Engineering, the Whole System and Product Realization Taxonomy* emphasize more of this aspect than others.
- *Conceptual Aspect:* Logical CE team governs the conceptual aspect of CE. Conceptual aspect mostly deals with principles of CE, concurrency and simultaneity, modes of concurrency, modes of cooperation, understanding and managing change, reengineering approaches, work flow mapping, information flow charting, process improvement methodology, etc. The chapters on *CE Definitions and Process Reengineering* emphasize more of this aspect than others.
- *Virtual Aspect:* of CE is governed by a virtual CE team. Virtual aspect mostly deals with capturing life-cycle intent, information modeling, electronic capture of CE invariants. These CE invariants deal with product model class, process model class, specification models, cognitive models, communication through virtual proximity, agile virtual company, artifact intent definitions, etc. The chapters on *Information Modeling and Life-cycle Management* emphasize more of this aspect than others.

MAJOR FEATURES OF THIS BOOK

Whether you are a firm CE believer, or this is your first introduction to CE, this book provides a full view of CE from all of the above aspects and perspectives. The management perspective, which is a part of philosophical aspect, relates to organization and culture. Complete with a historical review and context, the author articulates these CE aspects by illustrating the differences between the best methodologies (or the best taxonomies) and what are currently being practiced in industries today.

Some examples of topics include:

- Japan Transformation from rigid culture-driven company to world leader in quality.
- Eight fundamental principles on which CE is founded
- Understanding the seven primary components of concurrency and simultaneity
- Modeling the CE environment and its five essential components
- Accounting for Seven C's to ensure cooperation among work-groups

- Seven primary influencing agents (called 7Ts) for achieving concurrency and simultaneity.
- Cooperative work-group environment spanned by four concurrent teams: (namely—logical team, personnel team, virtual team and technological team)

Integrated product and process organization (Volume One) deals with process taxonomy for CE. Process taxonomy is necessary to adequately classify, distribute and distinguish differences in behaviors of a complex enterprise from a traditional system. The innermost core of this process taxonomy is its foundation, which has four supporting M elements: models, methods, metrics and measures as mentioned earlier. The following table summarizes the major features of this book:

Features of this book	**How does this feature benefit readers?**	**What chapters or sections of the book contain these features or examples of them?**
This is the first CE book, which emphasized seven primary influencing agents (called 7Ts) for achieving concurrency and simultaneity.	This allows the readers to look for items that can significantly affect responsiveness and may be the root cause for cost, quality and productivity loss.	Chapter 4 (see Figure 4.1)
Book features manufacturing competitiveness, life-cycle management, process re-engineering, cooperative work-groups, systems engineering, information modeling, and PPO (product, process and organization) integration issues all described under a unified theme of "process taxonomy."	It allows the readers to consider a wider view meaning "integrating over the enterprise" while implementing CE. This eliminates the common problem of blindly automating design—that may mean repeating the same old mistakes but doing it more often and more quickly.	Chapters 1 through 8
Concurrent system tends to operate in one of the two modes. The book describes for the first time these two modes of concurrency. These modes are called in this book *overlapped pull system* and a *linked system.*	It allows the PDT groups to gradually built-up the right information and to link up the process activities with required skills so that project can be finished on time.	Chapter 4 (see Figure 4.11)
Organization of CE is approached by splitting the system level problem into its mutually separable transformation states, followed by modeling of each state, then the reconstruction of the system definition from the aggregation of the definitions of its constituent states.	This constitutes a "state of series of transformation" leading to an early completion of a mature design.	Chapter 9 (see Figure 9.10, Figure 9.16 and Figure 9.19)
No one has looked at the product realization process by decomposing its five components (work, product, process, system and enterprise) into their corresponding breakdown structures so as to exploit the inherent concurrency and independence of the decomposed parts.	The result is a virtual approach to defining multi-disciplinary problems and their solutions for improved productivity.	Chapter 8 (see Figure 8.1)

Features of this book	How does this feature benefit readers?	What chapters or sections of the book contain these features or examples of them?
The book for the first time views the integrated product development as a cooperative work-group environment spanned by four concurrent teams: (namely—logical team, personnel team, virtual team and technological team).	It allows the PDTs to come up with an effective team design. An effective team is like a peak-performing symphony orchestra: a group of specialists from these core CE teams creating an inspirational performance through mutual harnessing and cooperating process.	Chapter 5 (see Figures 5.7 through 5.9)
Cooperation is the key lynch-pin of achieving teamwork. The book describes for the first time the seven elements (called 7Cs) to bring in team cooperation philosophy.	Seven 7Cs help identify the extent to which the organizational culture of "self-interest" supports or detracts from achieving a unified product concept (or a common set of company goals).	Chapter 5 (see Figure 5.2)
Benefits of CE stem from a few basic principles. The book describes a set of eight fundamental principles on which CE is founded	Principles help the teams make timely decisions and early problem discovery. Approximately eighty percent of product's life-cycle development cost is driven by decisions made in the first twenty percent of the program effort.	Chapter 4 (see Section 4.1)
Concurrency is the major force of Concurrent Engineering. The book for the first time describes seven major components that assure concurrency and simultaneity	The rewards and satisfaction in CE come from the success of its components. Components yield a significant competitive advantage in product cost, quality, investments, and reduction in cycle time.	Chapter 4 (see Section 4.3)

At the end of each chapter, test problems are included. The instructor may choose a set of problems (ten or less) that he or she has covered in the class for that week from each chapter. Most test problems are based on the materials covered in the chapter itself. Some are based on materials covered in the earlier chapters thus stretching the student's grasp and understanding of the subject matters covered so far. Only a few test problems require stretching the students' imagination beyond what is discussed in this book. A rich reference section is provided for professors to reinforce the materials beyond what is discussed therein. The generous use of self-explanatory illustrations and bullets makes this book an easy and pleasant reading for everyone. Illustrations provide a quick visual grasp of the materials without the use of long and wordy sentences and paragraphs.

Biren Prasad
Electronic Data Systems
General Motors

ACKNOWLEDGMENT

Over the last several years, having associated with Concurrent Engineering: Research and Applications (CERA) Journal as a founding editor and having attended/organized numerous conferences dealing on this subject, I have steadily built up a massive collection of precious knowledge on concurrent engineering (CE). Many of the ideas set forth in this book are formulated based on the rigorous analysis of what has been reported in those journals and conferences, of what we found worked well in practice, and from our research of what we observed was essential and relevant for those that failed to be successful. In most cases, the materials in the book are actually built on trying these ideas on problems facing the automotive, electronic, and software industries (working with Ford Motor Company, General Motors, Electronic Data Systems, and other Delphi Automotive System customers). Many of the core CE concepts contained herein, therefore, are reported for the first time. The others are extensions to the ideas derived from various CE books, journal articles, and papers presented at various meetings but never published. Relevant references are contained at the end of each chapter. Many thanks to those, who supplied reprints of their articles and thesis, included therein. The author wishes to acknowledge the contributors of the CERA Journal and the members of its two editorial boards to whom the author corresponded on numerous occasions, which helped solidify many of the concepts reported in this book.

A very important aspect of almost any technical publication is the exposition of precise definitions of fundamental terms, and this book has a lot of them. In those areas, I can only take credit for bringing them together and packaging them in what is, one hopes, a convenient format. To this end, many thanks are due to my professional colleagues in a number of fields relevant to CE, TQM, quality circles, QFD, KBE, and product design and development.

The author wishes to acknowledge the assistance of General Motors, Electronic Data Systems, and Delphi Automotive Systems for providing the environments and opportunities, and assistance of our ACE colleagues and customers, whom I worked with. In addition, I thank the people who reviewed this manuscript and for the guidance they gave us in making this book more useful to our readers. A special thanks is extended to my spouse Pushpa and to my lovely daughters—Rosalie, Gunjan, and Palak—for their patience and understanding throughout these never-ending years while I was busy writing.

Biren Prasad

About the Author

Dr. Prasad is a senior consultant staff at Electronic Data Systems (EDS), a subsidiary of General Motors, where he is in charge of Automated Concurrent Engineering Group at Delphi Automotive Systems. Dr. Prasad's research interests include CAE, Structures, AI, Intelligent Vehicle/Highway Systems, Structural Optimization, Systems Identification, CAD/CAM Automation, and Simultaneous or Concurrent Engineering.

He has written or co-authored over eighty technical publications, including fifty archival papers and eight books. His most recent text is *Modern Manufacturing: Information Management and Control* (1994, Springer-Verlag). In 1989, he edited a set of three-volume books entitled *CAD/CAM, Robotics and Factories of the Future* (1989, Springer-Verlag). He has served as editor for several additional texts, monographs, and proceedings. He supports professional societies in several editorial and organizational roles. He served as the General Chairperson on two international conferences: ISPE's CARS & FOF'88 Conference held in Detroit, MI (1988) and 31st, AIAA/ASME/ASCE/AHS/ASC Structural, Structural Dynamics, and Materials Conference (SDM'90), held in Long Beach, CA (1990). He was the President of the International Society for Productivity Enhancement (ISPE). He has presented several short courses and has consulted frequently with external organizations. He has received three awards: AIAA's Survey Paper Citation Plaque & Award (1982), a NASA Award and a Certificate for Creative Development of a Technical Innovation on "PARS"—Programs for Analysis and Resizing of Structures (1981), and the ABI (American Biographical Institute) Commemorative Medal of Honor (1987).

Dr. Prasad was the Founding Editor-in-Chief for the *International J. of Systems Automation: Research & Applications (SARA).* This book on Concurrent Engineering Fundamentals is his latest book. He is the Managing Editor for the *International J. of Concurrent Engineering: Research & Applications (CERA),* which was founded in 1990.

Dr. Prasad earned his Ph.D. from Illinois Institute of Technology, Chicago, and a Degree D'Engineer from Stanford University, California. While completing studies for his D'Engineer, he worked at Xerox Research Center, and Failure Analysis Associates, both in Palo Alto, California.

Dr. Prasad first began his career in 1971 at ISRO (Indian Space & Research Organization) in India, where he worked as Senior Mechanical Engineer. After earning his Ph.D., he continued his career at Association of American Railroads (Chicago) as a Senior Research Engineer, where he participated in developing methods for the composite analysis of Railroad Track Support Systems (1977–1980). He then joined Ford Motor Company, Scientific Research Laboratory, where he researched structural optimization and developed techniques for automated design of large-scale vehicle structures (1980–82). During the latter years (1982–85) at Ford, he was responsible for developing design sensitivity and optimization procedures on EAL (Engineering Analysis Language). His work there supported a dozen prototype designs of the automobile energy absorbing structures for crush energy managements.

Dr. Prasad joined the new CAD/CAM department of EDS in 1985, and since then he has moved through several departments and ranks to this Senior Manufacturing Consultant position. Prior to this assignment, he was Special Projects Manager for Artificial Intelligence Services Group of EDS, where he was responsible for expert systems development in AI/Parametric technology areas. He presently consults on AI, parametric engineering, object-oriented tools, and expert systems. He also directs simultaneous engineering initiatives for automating product's design and engineering process. he also conducts applied research, primarily to study implementation issues in intelligent CAD design, capturing of design intent, concurrent engineering, and integration issues.

He is an adjunct professor of Mechanical & Aerospace Engineering at West Virginia University. He also teaches at Wayne State University and holds positions at several teaching institutions.

Dr. Prasad has severed on numerous committees for ASCE, ASME, and SAE. He is the Chairman of the SAE's Computer Readers' Committee. His professional honors include: Associate Fellow of AIAA, senior member of ASME, ASCE, SAE, AAAI, and fellow and life member of ISPE. He is listed in International Men of Achievement, International Biographical Center (IBC), England (1986), Dictionary of International Biography, IBC (1986), Who's Who in Frontier Science & Technology (1985), Who's Who in Technology Today (1985), Who's Who in Aviation & Aerospace (1983), International Directory of Engineering Analysts (IDEA) (1983), and Personalities of America (1986).

CONCURRENT ENGINEERING FUNDAMENTALS

CHAPTER 1

MANUFACTURING COMPETITIVENESS

INTRODUCTION

Manufacturing competitiveness and "world-class manufacturing" are often used interchangeably. Establishing manufacturing competitiveness is considered equivalent to being a worldleader in "world-class manufacturing." In other words, manufacturing competitiveness means sustained growth and earnings through building customer loyalty—by creating high value products—in very dynamic global markets. During the last decade, a number of modernization programs have been launched by many companies, such as FORD, GM, IBM, NEC, and Newport News Shipbuilding, dealing with how a company does business [Prasad, 1994]. These programs focussed on business processes and information usage in the manufacturing environment. Programs, such as enterprise reengineering, multi-dimensional integration of engineering, and manufacturing and logistics practices, have resulted in far reaching changes and profits. With changes, we are witnessing a movement from the "age of control" to the "age of flexibility." These changes describe a set of dominant trends in the business environment that have influenced the competitiveness of companies since 1980 [Goldhar, et al. 1991]. These changes could not have been possible without process tools like concurrent engineering (CE), and continuous acquisition and life-cycle support (CALS—earlier known as "computer-aided acquisition and logistics support") in action [Bauman, 1990]. Attributes such as empowerment, flexibility, total quality management, agility, fast-to-market, accountability, teamwork, and integration are inherent in the notion of CE.

Concurrent engineering is considered one of the key concepts that enables companies to attain world-class stature [Shina, 1991]. Today, it is an organizational keyword. The term refers to the same organizational process described as *simultaneous engineering, life-cycle engineering, parallel engineering,* and sometimes as the *multi-disciplinary team approach* or *integrated product and process development (IPPD)* [Hall, 1991; Ziemke and Spann, 1991]." Here, everyone contributing to the final product, from conceptual design to marketing teams, is required to participate in the project from its very inception. Some industrial analysts claim that this systematic approach is revolutionizing the manufacturing industries and could be just the boost that will enable firms to survive in the competitive global marketplace. Others argue that this practice is not really new and that it is very similar to the old fashioned "teamwork" that was prevalent a decade ago during product design and manufacturing [Dika and Begley, 1991]. A decade ago, the size of the project teams was small compared to today's standards (a couple of dozen people at most). It worked reasonably well, because small teams can work more closely than large teams. There are numerous anecdotes describing such small teams that expose challenging issues and resolve them by mutual understanding [Chisholm, 1973]. However, when project sizes became larger (involving hundreds of people), large teams created more chaos and anarchy. No matter how one looks at this, the global manufacturing paradigms have been in a dynamic state of flux for some time. In order to understand the changing pattern, let us review the situation by taking a closer look at the historical manufacturing processes and practices.

1.1 REVIEW OF HISTORICAL EVENTS

The term "manufacturing" is based on the Latin words for "hand" and "make." The roots of manufacturing date back as early as the nineteenth century when customers' orders were taken long before products were actually produced. Manufacturing was synonymous to individual handcrafting. The following are some of the historical highlights of manufacturing events (see Figure 1.1).

1.1.1 Pre-World War II Era

During the pre-World War II era, the business climate reflected a secured market environment. A secured market environment removes the competitive pressure that forces a manufacturer to make a quality product. The era was supported by isolationist politics and protectionism economic policies. There was no international competition. Even as late as the 1960s, high tariffs protected home industries from foreign competition leaving only domestic competitors. Other contributing factors were transportation costs and strong values of the US dollars. Most products were hand-crafted and orders were taken before products were delivered. In other cases, local markets were usually guaranteed. Manufacturers of consumer goods, such as watches, electronics, jewelry, appliances, and most other products (including automobiles) knew that due to fertile market conditions the products would sell well. For craft-based products, craftsmanship and perfectionism were the boons to success in business:

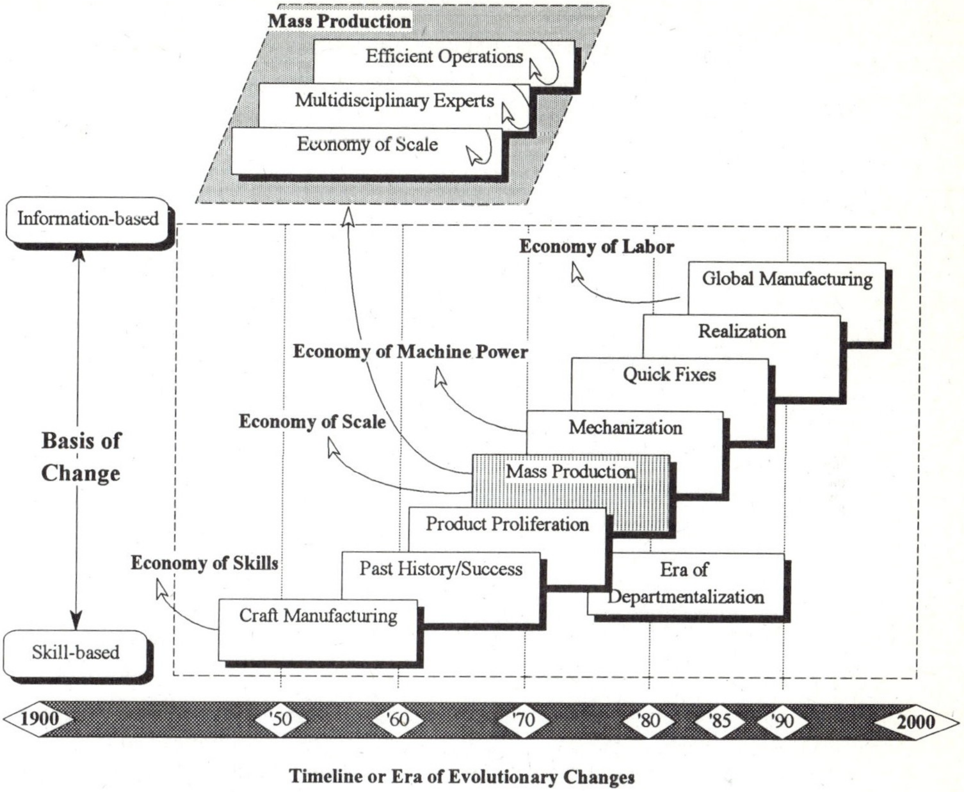

FIGURE 1.1 Major Historical Highlights of Manufacturing Events

- Products were designed and built to customer specifications.
- Each craftworker paid attention to details, quality was built-in by design, and in-process quality control was the responsibility of each worker.
- No in-process inventory was required, materials were purchased just-in-time, and finished products were delivered on schedule and on cost.
- Designers knew well in advance the manufacturing requirements—including the capabilities and the limitations of the production process; they had the power to modify in-process design (there was no need for empowerment) to take advantage of changing situations.
- Craftworkers participated in all decision making. They worked concurrently, switched roles, and helped each other.

In crafts manufacturing, since the product was crafted manually the following case occurred:

- A batch size, often a lot size of 1, was used.
- Productivity was limited to the capacity of a few craftworkers and their skill ability.

- Most costs were tied to production labors.
- Operations were sequential in nature.
- There were no major incentives to control costs as profitability was cost-centered.

Craft manufacturing provided a **first stretch (reach)** and a competitive edge to the manufacturing world based on the *"economics of skills."* This reach was based on leveraging human skills and dexterity to craft products that customers liked. In the post-World War II era (1950–60), early designs of products were obtained from incremental improvements. After a product was released for production, new designs were based on old design constructions. Creation of new designs was based on manual means like plotting of preliminary product boundaries or cut lines, and by creating models of geometrical kinematics of the parts. Manufacturers did not have many tools to aid their process. The product changes were introduced through a process that we refer to today as physical or manual aids. This usually consisted of either quick and dirty mockups (paper or cardboard models) or round-point pencil drawings. The drawings particularly dealt with large "tolerances" or envelopes, such as the "round-point pencil." Most companies had no large computers or sophisticated CAD (computer-aided design) terminals like those we see today. They managed to do their job without them. "Quality" meant compliance with specifications, and "inspection" was the main process used to monitor defective parts.

1.1.2 An Era of Product Proliferation: 1960+

During the early days of manufacturing, the same few individuals who designed the products, determined how they would be built. Ford's once popular automobile the Model T, was developed on a teamwork basis. Henry Ford himself sketched his ideas on a blackboard, Joseph Galamb and Spider Huff developed them into engineering specifications, while Charles Sorensen checked the manufacturing feasibility [Lacey and Little, 1986]. Henry Ford and other inventors of this time period even built the factories that manufactured their products. The existence of such practices was quite common even a couple of centuries ago. In 1712, when Thomas Newcomen, an English blacksmith, invented the first condensing steam engine, he did so all by himself. He conceived the design with a piston and other moving parts and purchased the required materials—irons and bricks. He handcrafted all the parts, assembled them at the mine site to pump water, and even operated and serviced it himself. It was the first design of its kind. Innovation was the prime motive here, as opposed to perfection in crafts manufacturing and functionality in automobile manufacturing. The need for cooperation was nonexistent because of the sheer size of the company and the familiarity with the products.

In the case of Henry Ford, the product designers, the process designers, and the sales and marketing staff were thoroughly familiar with the product. All were fully aware of the problems associated with getting the product to the market. Everyone knew the design and process requirements for the products as well as the manufacturing plant data and its limitations. During those days, planning for the major functions took place simultaneously or concurrently. There was no international competition and no great motivation to produce a world-class quality product. Whatever was produced sold well.

1.1.3 An Era of Mass Production: 1965+

During this period (1960–70), consumers demanded more and more functionality from products, especially durable goods. The concept of "mass production" took off. This evolution of competitive advantage via "*economy of scale*" was the **second reach.** Mechanization (i.e., use of machine power) was sought to efficiently build, fabricate, and assemble components (automation). The manufacturing process still consisted of discrete serial steps along the lines of *crafts manufacturing.* However, the *economics of skills* was no longer the equation for gaining competitive advantage in the manufacturing world. The skill sets for production of goods changed from a handful of people to a multidisciplinary team of experts. Developing a product required the assistance of a number of experts and other individuals experienced in the many aspects of the product's life-cycle, such as design, manufacturing, sales, product support, marketing, etc. Internal industrial emphasis changed from engineering and production to include marketing and finance. Interest in the marketplace and investment grew. Buying existing industries became more lucrative than improving efficiency of the internal operations or practices.

1.1.4 An Era of Mechanization: 1970+

During these ten years (1960–1970), as industry moved toward more complex products and the automated assembly line, new ways to get uniform results were sought. Standards were created for precise drawings, interchangeable parts, tolerances, and quality control. Skilled employees were hired to cope with these demands. During the latter part of 1970 a new emphasis was placed on mechanization—the ability to mass produce identical, interchangeable components. The "*economy of scale*" took the form of an "*economy of machine power.*" This became the **third reach.** This reach employed a process strategy based on "mechanization" for reaching an adequate "*economy of scale*" for equivalent mechanical products. Mechanization not only became popular to replace human or animal power, but also infiltrated the assembly lines. Use of robots were sought for the production of complex assemblies and repetitive chores. Mechanization drastically changed the manufacturing economy of scale. Manufacturers of products made from dissimilar parts soon found it difficult to compete. Due to mechanization, workers' skills became highly specialized and job scopes became highly focused.

1.1.5 An Era of Departmentalization: 1975+

As companies tried to foster greater revenue, organizations grew. It became imperative to divide the organization into subunits [Ziemke and Spann, 1991]. This was a beginning of the move toward "*departmentalization.*" Industrial engineers subdivided the complex tasks of creating a whole system into more manageable pieces on the basis of function, product features, and/or market orientations. The division of work into functions seemed to be the logical thing to do since it eliminated the multiplicity of responsibilities and central control. Some success in productivity and efficiency gain were achieved. However, the gain was short-lived. Over time it created more problems than solutions. The functional units acted like having a series of independent countries each with their own objectives and loyalties. While moving a product from design to marketing in several highly sequential phases, the designers, engineers, and manufacturing people tended to operate as independent entities. Members of subunits developed values consistent with their work practices and preferences. Engineering personnel, for example, valued building unique product design features, whereas management personnel valued cost-cutting measures. Divisions began competing with each other for customers.

Divisions could also differ in their need for time-to-completion. Assembly departments might have had a very short timeframe for assembling parts whereas design engineering departments might have had a much longer one. Based on their values and time requirements, divisions developed *divisional goals,* which might or might not have been consistent with overall *organizational goals.*

As organizational structures grew larger, productivity decreased and barriers developed. Interdivisional sharing of information became more difficult and bottlenecks formed. Efforts were skewed from making the "great product" to being the "best department." Each division over time built or acquired its own infrastructure of systems that were either redundant or were not compatible. Organizational goals were subverted by departmental values and time requirements. Government regulations and requirements grew. During the same period, organizational culture took deep root. Many divisions built their own turf and defended their past values and practices. Communication barriers developed between departments and became particularly stringent during bad economic times when organizational resources were limited.

1.1.6 An Era of Quick Fixes: 1980+

During the 1980s, as international pressure was mounting and American recession was taking its toll, researchers in organizational behavior proposed a series of quick fixes. They sought to understand and improve the ways in which a large, modern, industrial organization ought to structure and function. Unfortunately, untangling this web was a daunting challenge. Several quick fixes using techniques such as project evaluation and review technique (PERT), the matrix concept, value system, synchronous organization, and the use of value and nonvalue-added analysis and waste management systems were introduced to improve the functioning of organizations [Ziemke and Spann, 1991]. Management by meetings, committees, facts, and objectives became the new way of doing business. Vision became very short-ranged, with major emphasis on financial and bottom-line results. There was some interest in "quality," but quality was considered something to be assured through "*quality-by-inspection.*" The concept of designing for "customer-orientation" was recognized in some quarters, but a firmer commitment towards its implementation was missing. Quality was the province of a separate department, associated with, but playing an adversarial role to, production. The major role of quality departments was that of a police person—looking around the production floor for typical violations. Quality assurance techniques were known but widespread use was scarce. For violations, tickets were written and fines were assessed. Little effort was made to investigate the root cause of such problems or to take serious actions to prevent future recurrence. Quality prevention was followed in a "firefighting" sense. As a result, many defects were left undetected. As far as quality was concerned, only "symptoms" of production troubles were addressed as they popped up, too often inconsistently and with no "permanent" cure.

Employees and management wanted to operate in the "traditional" manner. Operational practice in most manufacturing industry, particularly automotive, followed a functional but "built-in control" approach. The tasks of product design and development were functionally distributed and performed by independent departments. However, to ascertain "adequate control" by each group, they followed a practice of what is today called as "sequential development process." Senior management conceived the need for a new product, outlined the concept, and "handed it over" to the design studio. There the product planners and stylists did their job and handed the theme package "over the wall" to engineering. The product engineers put details to the theme, established specifications, built and tested prototypes, created the release drawings, and handed that package over another "wall" to manufacturing. The back-and-forth throwing continued—manufacturing developed the processes, toolroom built and designed the tools, purchasing bought the parts, and construction plants assembled it. There were walls everywhere, some high and some not so high. Sequential operation, the "built-in control," "staged-release," and "over-the-wall" syndromes were widespread.

1.1.7 An Era of Realization: 1985+

During the latter part of the 1980s, an era of realization began. The industry started analyzing their situation more critically. They found that the employees had not changed very much, whereas the nature of business (products, process, and requirements) had changed significantly. Due to the complex nature of products that a company designs today, those same tasks, which used to be efficiently performed under individual departments, could not be performed independently anymore. It requires close interaction between members of various disciplines to design and develop a product. Sequential design and development processes, with very little information transfer and a lot of control, have resulted in extended time of bringing products to the marketplace.

During this period, although American, and most other Western producers did not lose teamwork, it became more difficult to cooperate (or function) in the realm of problems associated with the volumes of engineering changes and costly mistakes that ensue. The Japanese never strayed from the cooperative ideal, and succeeded in combining a teamwork-based approach to product development with another American concept—quality. The American and Western problems were the result of too much control without having the right information available at the right place, at the right time, and in the right format. During the last twenty-five years, not only have the products grown more "global" and "complex," but the use of computers has also increased. A large volume of information is constantly generated and has to be moved across "multidisciplinary" groups and departments. The need for information exchange has shown a manifold increase [Carey, 1992].

1.1.8 An Era of Global Manufacturing: 1990+

The manufacturing industry, in the 1990s, is entering an era of competition that is unprecedented. Just a few years ago, the automotive industry business in the USA, for example, was shielded by the big three—GM, Ford, Chrysler, and some foreign producers. The quality of their products was not so high. Today the picture has changed drastically. There are now about ten auto manufacturers in the USA and quality for each vehicle has improved significantly. Altogether, there are over thirty brands of vehicles available to the general public. There are many foreign and transplant operations that have cropped up recently. Manufacturing is becoming more and more global every day. The manufacturers are designing products at one place and building them at another. The "*economy of labor*"—the so-called **fourth reach**—is now a big factor in the competitiveness equation. Competition is fierce in spite of the declining value of currency imbalance (that is, yen to dollars), and the name of the game is the value (X-ability) of the product:

$$\begin{array}{c}\text{X-ability} \Leftrightarrow \{\text{ manufacturability, maintainability,} \\ \text{assemblability, serviceability, } \ldots \}\end{array} \tag{1.1}$$

This concept is elaborated in Chapter 8. In a 1987 survey, "durability and reliability" and "well-made car" were the top two selling point reasons for consumers buying cars. This has changed to "quality" as the top one reason in the 1990s [Carey, 1992]. Ten years ago, these reasons ranked nowhere. Today customers want value for their money (characterized in some way in the thing called "X-ability") and they will buy the product from anybody who meets their needs. This clearly shows a major shift in customer requirements. To meet customer satisfaction (and to win future business), a company must not only deliver what the customers have requested, but anticipate what the customers will need next (or in the near future) and package them into the products.

1.1.9 Why Change?

In the past, epochs were named on the basis of the predominant tools and materials of the time, e.g., the Stone Age, the Bronze Age, the Iron Age, the Crafts Age. Now, in the 1990s, we are living in the information-based automation age. Dramatic changes are underway as government and industries respond to:

1. Decreasing defense budgets
2. Evolving, highly competitive global marketplace

3. Technological advances
4. Closely aligned world economics and trade structures.

In fact, the sweeping changes in the trade and commerce policy of the world-body are threatening to "erode" the substantial progress that many companies have made over previous years. These challenges include the opportunity to be an "invisible company" with major functions globally or externally supported while still being capable of managing and controlling them effectively. Materials, tools, and processes (or methods) are still evolving. The role of the engineers and the methods of doing business are beginning to change. The people, the environments, and the cultures have all changed. The dominant material of society today is information, and the dominant process is mechanization. Almost everything is becoming mass produced or mechanized. Today a watch would cost thousands of dollars if it were built using the traditional crafts manufacturing technique. Now, with mass production, we can manufacture accurate digital watches for a price of $2 to $5 (United States currency). The technological needs are also demanding a more specialized group of people in the usual product development team (PDT). It is beyond the realm of a single person, a group, or even a department to optimize a new design with all its multidisciplinary requirements. There are too many factors or disciplinary skills involved for any one person, group, or department to handle. Besides the knowledge requirements in each disciplinary field, the depth and levels of skills required from these individuals are also very high. As a result, companies employ a large contingent of highly specialized knowledge workers and use various tools to develop new system or component designs. The methods and processes of doing the job are changing as a result of new shared ("work-group" centered) tools from the information age. Computer systems, intelligent machines, and "multimedia" software tools are as much in use today as were the stone, bronze, iron or craft tools in previous eras.

The marketplace is also rapidly changing (see Figure 1.2). During the last decade or so, 80 percent of the fortune 1000 companies have moved to client/server mode of computing. The following describes some trends:

- The shortage of trained and talented work force (including programmers) has increased.
- Information technology (IT) cost of hardware has declined.
- IT mix of computers and varieties (PCs, workstations, minis, micros, and brand names) has increased.
- The number of applications that are being run at the workplace has increased.
- The cost of the IT services has stayed fairly level.
- The global competition in the manufacturing sector has increased significantly.
- The computer speed and its power have increased.
- The integration cost of the software has steadily climbed.
- Time-to-market has been steadily declining.
- Product complexity has increased over the years.

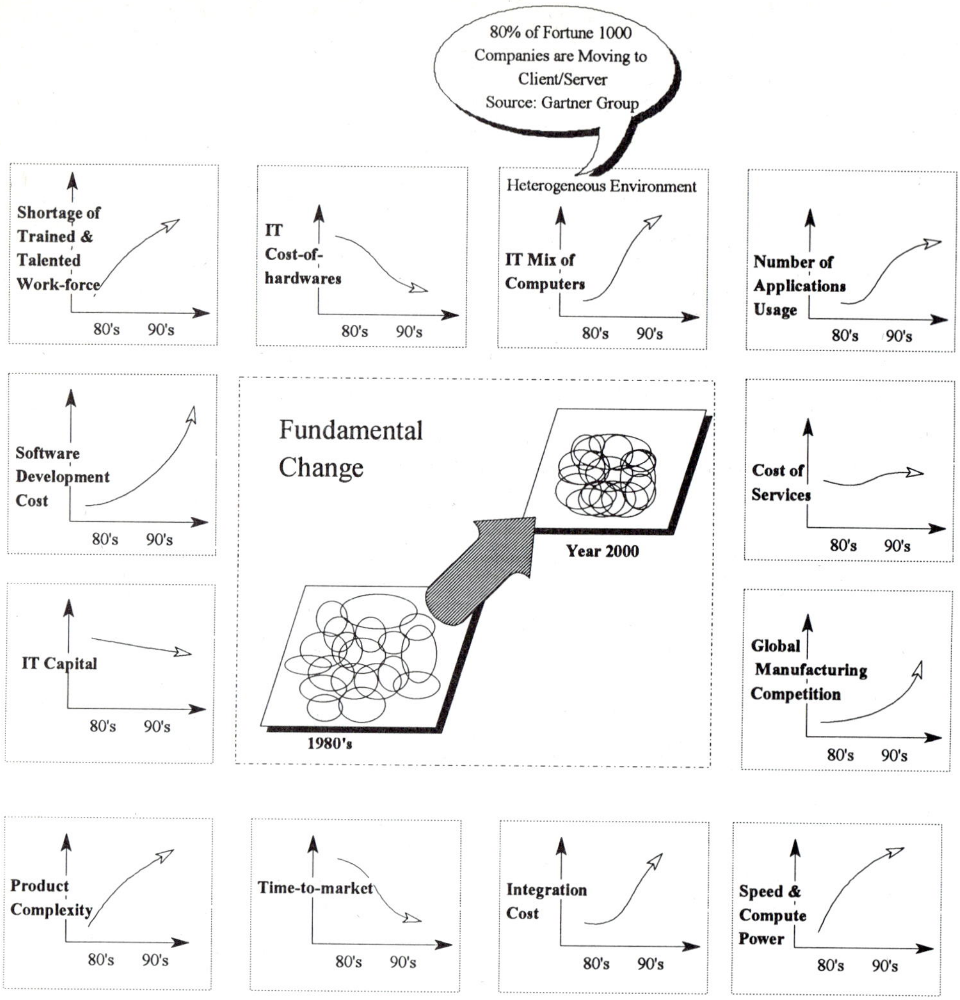

FIGURE 1.2 The Emerging Marketplace

- The IT capital cost has stayed fairly constant.
- The software development cost has been rising exponentially.

The factors that demand a significant change in our overall approach to design and manufacturing include the following:

- Rapid proliferation in the number and variety of products that we manufacture today
- Increasing dependency of product performance on manufacturing process capabilities
- Customer demands for improved quality and better product performance

- Dependence of cost on labor rates resulting from choice of plants to manufacture the product and their locations
- Demand for stronger and lighter materials
- Competitors steadily shortening their time to market
- Availability of world-class quality products at competitive prices in the world market
- Customer demand for producing the part right the first time and every time thereafter
- Declining cost of computing hardware versus the power it offers
- Rising costs of software development versus the hardware costs.

1.2 PUSH AND PULL FOR NEW PARADIGMS

In an effort to reduce time to market, foster teamwork, cut costs, and eliminate late engineering changes, managers in many organizations are trying to "redefine" the role of an employee in the context of a new paradigm. This new paradigm is characterized by recognizing the importance and real value of an employee as part of a cooperative work environment supporting concurrent functionality, and by horizontal and vertical integration within the enterprise [Prasad, 1994]. It is also characterized by "empowerment," or pushing decision making to the lowest in ranks, and emphasis on well-rounded experience and expertise. Other items include improving the "process" (e.g., benchmarking, continuous process improvements, process restructuring, process renovation, and process reengineering), considerations of both long-range and short-range goals, and a total customer focus [Ross, 1992]. Today, we are witnessing the demise of the "*control age*" and dawn of a "*flexibility era*"—translating into a new breed of customized products. Consumers want finished goods tailored to international (English versus metric measurement standards; *left-handed drive* versus *right-handed drive* vehicle regulations), national (emission requirements, etc.), regional, ethnic, and personal tastes. For manufacturers that means producing a great mix of product options at low volumes. The catch is that today's customers want the product features of skill-based manufacturing at the speed, quality, and cost of information-based mass production. Figure 1.3 describes some of the various ways in which new technology can impact an organization.

There are two aspects to new paradigms: push and pull. "Push" is where a new technology or a concept is substituted or inserted in a new or existing product. "Pull" is where customers demand an improved product or process irrespective of the current state of technology utilization. The triggering of a pull, from an improved product or process standpoint, happens only when the existing technology is not adequate or at its best. Not at its best means the technology, as it is, fails to provide the expected market share, quality targets, or profit margins. In the "pull" case, needed productivity can be gained through means other than switching to a new technology. Figure 1.3 summarizes these pull and push factors into four broad-based categories:

- Competitive pressure push
- Emerging technology push

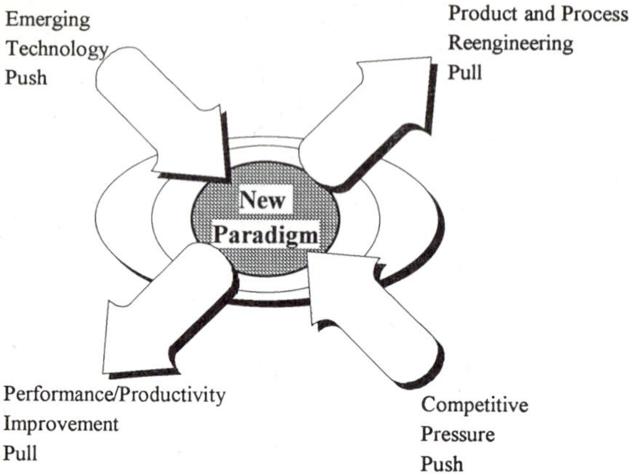

FIGURE 1.3 Push and Pull for a New Paradigm

- Productivity improvement pull
- Process reengineering pull

Push type of technology is easier to be copied by a competitor. There are two primary forces driving the technology push: the decrease in cost of automation tools; and increase in availability of automation technology through third parties. In American industries, during 1993–94, more than 30 million reported they use computers at work! About 5 years ago, the computer industry and technology were still expanding at a rapid rate. At that time the 286 CPU computer was considered the standard PC in the workplace and in schools. Everyone thought it was very reasonable and it was accepted widely. So that paradigm was set in the customers' mind. Anything less than Intel 286, such as Intel 186 or 88/87, was considered slow or undesirable. Since then, faster computers with new CPU have been introduced into the market every alternate year as shown in Table 1.1. Which means today customers have new paradigms. A 286/386 computer is considered obsolete and is no longer desirable by almost anyone. In the customer's eyes, the pentium computer is the new standard (paradigm).

1.2.1 Competitive Pressure Push

Organizational pressure is created by "globalization" of manufacturing and inadequate power of a company to provide a formidable competitive response. The competitive response is an organizational ability to produce a competitive product that ranks equal to or

TABLE 1.1 Paradigm Shifts due to New Technology

Paradigm Shift	A	B	C	D
Type of CPU/Chips	Intel 286	Intel 386	Intel 486	Pentium
Year of Introduction	1984–85	1988–89	1991–92	1994–95

surpasses quality, speed, cost, and in alike attributes, which are important to the customers or to the company. For a long time Switzerland was the world leader in precision watches. They have the best craftsworker of the watch industry and skilled employees for making mechanical watches. Switzerland had then owned most of the world market in precision watches. When the invention of the quartz watch came to being, the Swiss did not take it seriously—they more or less ignored this new paradigm (discovery and technology) and did not invest into it. The quartz technology was then brought over to Japan and the Japanese decided to invest into it. When the Japanese quartz watch came into the market and competed against the Swiss mechanical watch, the Swiss could not compete with the Japanese. Now, the quartz watch is much more popular everywhere and easily affordable—it costs less than five U.S. dollars. Japan still owns most of the world's market share in watches today.

Today, customers expect the same level of satisfaction and perfection that initially came from craft manufacturing but they want to pay a price achievable only by mass production. Customers want the best of both worlds—"features", as attention to details, and "quality" that come from craft manufacturing, and "low cost" and "consistency" that come from mass production. Though the underlying process and teamwork used in craft manufacturing represented a miniature replica of what was desirable for a customer-focused organization, it lacked the "recipe" for a successful company. The main question is how one would be able to provide customer satisfactions on a recurrent basis if the technology and organization keep on changing. Chain competition has driven engineers to include terms such as time compression, total quality management, teamwork, quality function deployment, and Taguchi into their vocabulary [Carey, 1992]. Concurrent engineering, coupled with automation efforts, is vital in maintaining a competitive posture in today's marketplace. Despite recent advances in computational and communication technology, it is still not an easy task to win competitiveness effectively. Despite all of the above efforts applied in the United States, foreign organizations appear to be gaining ground from the consumers' viewpoint. The "change" in external conditions adds new competitive pressure to further decrease the time and lower the costs. Today, companies are facing tremendous challenges of how to provide the agility that came from craft manufacturing with the cost benefits that were the results of mass production.

1.2.2 Emerging Technology Push

Technology is also changing; more and more vendors are opting for "open" architectures, "common" window-like user interfaces, and "plug-in" application interfaces. There is a great temptation to use advanced information technology to support widely "altered" organization forms. The truth is that technology only enables CE, it does not create it [Rasmus, 1993]. New technologies will continue to emerge. The changes in market conditions are driving the use of emerging technology, which in turn is driven by the changing processes it has to support. The two constantly interact, each pushing the other to an ever-higher limit. Nike used the athletic shoe pump technology to break into the medical equipment market. Group technology, for example, is forcing new ways of doing design engineering. In group technology, features common to many parts of a product are identified and classified in groups. This practice not only saves in process planning time for machining but a number of parts can be machined in a batch mode. Another case in point is that of an integrated system. In the mid-1980s, an integrated CAD/CAM/CAE system was considered to be an ideal choice, but not anymore. Many organizations, large and small, are adopting a "lease and integrate" (open system) philosophy rather than "buying a bundled (closed)" system. There are several other factors for this technology push.

1.2.2.1 Information Technology

The most changes that are occurring today at an accelerated pace is in the area of information technology [Morton, 1989]:

- ***Function Power:*** Desktop power is continuing its extremely rapid growth. The number of functions that computers can perform is rising. Manufacturers are now pointing toward the 3G machine that is several hundred times more powerful in processing, storage, graphics display, and multimedia than current 3M workstations.
- ***Compute Power:*** In the area of compute power, the capacity grows from the use of more powerful microchips. More MIPS (million of instruction per second) and MFLOPS (millions of floating point operations per seconds) computation power is becoming available on the desktop than ever before.
- ***Communication Power:*** In the communication area, technologies such as fiber optics are allowing for more new features to be added within the existing products. Communication power and implemented capacity is growing sharply. Current computer and telephone integration are allowing for instant voice, text, data, and picture transmissions.
- ***Price versus Performance:*** Price of hardware is declining while that of software is soaring day by day. More and more off-the-shelf software is meeting the requirements for a multifaceted organization today than in any time before. Not only is it economical to buy the hardware and software separately and plug them in, but this also gives organizations more control over the integration process.

1.2.2.2 Computer Technology Usage

The expanding technical roles and responsibilities of design professionals, as well as increasing accessibility of computational hardware have led to a proliferation of analytical CAD tools and user-friendly software. Most teams or groups have access to workstations or some type of visual-based (or interactive) computing.

- ***User-friendly:*** New and more powerful computers are showing up everywhere in the plant and in the office. Interactive usage is becoming more user-friendly, simple, and tailored for most applications, and is capable of being personalized according to the needs of an individual user.
- ***On-line, Real-time Access:*** Computer systems today are providing and managing information that is necessary to the organization's mission—that can help provide a

competitive edge in the marketplace. One of the most cited examples of such systems is the on-line airline reservation system. No major airline would even think of trying to compete effectively without a computer-based reservation system. Similarly, on-line ordering systems have given automobile dealers a new, competitive tool.

- ***Groupware Concepts:*** Applications emphasize "Groupware" concepts for cooperating work-group (*streams* of users). They are connected by a shared network through which each work-group can access these applications independently.
- ***Diverse but Integrated Solutions:*** Diverse applications for various engineering specialties are becoming more integrated. Unnecessary interactive activity is being automated or hidden (running in background), while simultaneously yielding better control in data management.
- ***Changing Computer Roles:*** The functions of computers are changing. More standard, open and common systems are being implemented and more comprehensive usage is becoming commonplace.
- ***Multimedia Access:*** Hypertext and multimedia applications now contain more useable knowledge representation. Graphics are becoming commonplace, giving a sense for photo-realism, animation, and real-time simulation on the desktop.

Computer Manufacturers or Vendors Perspectives Many vendors are developing software and hardware that follow certain industry specifications (conform to an open architecture and standards). Today, it is much easier to integrate in a "do-it-yourself" mode than it was a few years ago.

Management Perspectives This is directed toward protecting prior investments. Many organizations have already committed a large amount of resources toward their hardware and software needs and the associated training. It is not practical to start anew and do it all over again. Management often prefers a smooth transition to limit its impact on the business and possible productivity loss.

1.2.2.3 Design Technology

Recently, some of the newer generation CAD/CAM packages like Concentra/ICAD, Parametric Technology's Pro/Engineer, SDRC's I-DEAS Master Series, CADDS 5, I/EMS, Anvil-5000, and Intellicorp/Pro-Kappa, are adding some form of variable-driven modeling and design functions to their systems, as opposed to a pure explicit modeling system used in the past. For example, most of these systems now have the capacity to constrain the shape using parameters (or variable dimensions) or geometric relationships between the elements, such as parallelism, perpendicularity, or tangency. Some have the capability to record the interactive manual steps into a batch script file. Later, this script can be replayed when master values are changed. Many now have some high level programming capabilities to manipulate geometry. They do not just provide the ability to capture the geometry of the design as in conventional CAD/CAM systems, but also include various variable-driven features and positioning relationships of their parts to each other. Compared to a few years ago, interactive design technology is here now to stay, and the products we see in the marketplace have a lot to offer. The marketplace is getting more and more competitive. Popular CAD/CAM products such as EDS/Unigraphics, Catia Solutions, and HP PE/SolidDesigner, now have many built-in design linking and modification facilities which are changing the way CAD/CAMs are now applied to product design. Earlier, during modeling of parts, "explicit" values were independently assigned. Primitives were positioned using specific coordinate and spatial orientations. If product information was changed, users did not have much option except to recreate the design from the beginning. This is not true with some newer generation of CAD/CAM systems. Now we can link 2-D part's dimensions (coordinates and spatial orientations) with 3-D "master and slave" modes. Engineers could use simple shapes, developed with parametrically-driven 2-D profiles, to build 3-D solid primitives. Appropriate "mating" constraints and relationships can be defined as a part of the product data base. Three-dimensional solid parts can be recreated by altering the 2-D shapes through their 2-D profile parameters. This eliminates many time-consuming manual steps such as "record and play-back." Variable-modeling is quite useful if products have similar parts or features such as "family of parts." Besides product design, there are other areas where such "variable-driven" techniques are being applied:

- Assembly Engineering
- Cost Estimation
- Design/Build Automation
- Detail Design
- Engineering Modeling and Computation
- Finite Element Analysis
- Generic or Regenerative Modeling
- Library of Parts
- Market Trends Analysis
- Numerical Control
- Operation Analysis/Simulation
- Preliminary Design
- Process Planning
- Product and Part Geometry Creation
- Product and Process Costing Structure

The ability to network and gather information from around the globe has led to increasingly effective configuration management system (CMS). Due to many of the above reasons, version control or multiple teams or user's access is not considered a major chore. With careful variable-driven modeling and CMS, practically any user—in fact, anyone in the entire organization—can access data from any system, network, or application simultaneously. The result is the ability for the teams to make better and more time-critical decisions.

1.2.3 Productivity Improvement Pull

In most design and manufacturing organizations today, there is often a great disparity in the use of computer tools. The software and hardware they have acquired over the years can at best be described as firmwares—stand-alone compute islands. There are pockets in the organizations that are quite traditional—firmware-like—while some other parts are math-based. The levels of integration across departments are very poor. Usually the problem is more severe with bigger organizations, since, over time, the needs of the work-groups have been different. Considering the project needs and the work-group's familiarity with requisite tools, companies have acquired a multitude of "firmwares" that do not communicate with each other. Smaller organizations are lucky not to be constrained by such problems of size. Often they cannot afford a variety of software and hardware systems, and thus tend to be more integrated in their approach. Apart from the degree of integration, both types of organizations do, however, encounter the following basic challenges:

- ***Computers:*** Organizations have a multitude of workstations, personal computers, both mini- and mainframes, all of different makes and models acquired over time.
- ***Applications and Tools:*** They work fine when used alone but present a lot of problems when they are required to work together. Perhaps the biggest and most common problem someone experiences is in making an analysis program, such as Finite Element Analysis (FEA), an integral part of a CAD system. Most name-brand CAD systems can interface with an FEA program, but CAD and FEA models (geometry-wise) are not interchangeable. It thus presents a problem for the designers when they want to modify the geometry of the CAD design based on the results of FEA.
  - *Analysis Tools:* A variety of analysis tools exist—some are commercial tools, others are home grown (developed in-house).
  - *Synthesis Tools:* The workers have developed programs for solving specific product design problems, a large percentage of which are home-grown (in-house developed) and proprietary.
  - *CAD/CAM Applications:* CAD/CAM applications are vendor supported—bigger organizations have a variety of CAD/CAM systems depending on the particular inclinations of a group or a department.
  - *Office Automation Tools:* Office automation tools such as word processors and spreadsheets are most diverse and loose.
  - *Productivity Tools:* Tools such as those required to solve engineering and manufacturing problems, for example, equations solver, matrix analysis tools, etc., are group-dependent and are not well-maintained.
- ***Files and Data:*** Many kinds of files, such as geometric models, analysis results, NC tools, and process plans, generated by their parent applications are scattered

around on a number of computing platforms and are stored in a variety of formats: vector, raster, or text.

The technological race has created a dilemma for the users of such (computer-based) products. Due to a rapidly declining useful life of such products, technological obsolescence is becoming a major set-back. Moving to the new technological solution is not always the right answer. Though this may improve the potential loss in productivity of the work force, prior investments in technical memory and knowledge stored in old formats (legacy database) cannot be discarded completely.

1.2.4 Product and Process Reengineering Pull

In the midst of all of this, management focus and organization structure is changing. Many organizations are aligning themselves along the lines of strategic business units (SBUs). Many competitive, multitiered initiatives have rippled through the organization: just-in-time, quality function deployment, employee empowerment, ISO 9000, quality circles, six-sigma program, continuous process improvements, cross-functional teams, and process management and control, etc. [Carey, 1992]. The walls between engineering and manufacturing groups are crumbling. The computational tools that have been developed to perfection over the years work well with each specialized unit. This is because they are designed for independent departments. Coercing these tools to fit into the changing organizational structure does not meet enterprise needs. It creates a backlash of many problems, such as production delays and communication bottlenecks. There appears to be a growing technological imbalance among the activities of a production cycle.

In conjunction with technological push, many companies are introducing reengineering. Reengineering means evaluating a company's current product, process methods, and manufacturing practices, documenting what is successful, and recognizing wasteful and inefficient practices. It also means weeding out the nonvalued-added activities and pulling in the right things to do. Blindly following automation is not always the right way to enhance productivity. If we do not reengineer the process right, there is a danger that one might simply automate wasteful processes and make the same old mistakes only faster this time. Reengineering teams should be clearly represented in any PDTs just as sales, manufacturing, engineering, and others are.

1.3 AREAS OF MANUFACTURING COMPETITIVENESS

A basic premise of manufacturing refers to the best transformation of customer expectations and requirements into useful products and services (Figure 1.4). Alternately, the identification of the best manufacturing transformation process is that which produces satisfied customers recurrently. A large number of companies across Europe, America, and Japan were studied in [Ferdows and De Meyer, 1986], with the focus on manufacturing strategies and competitive priorities. It was observed that western countries lead the world in product innovation but do a poor job when it comes down to implementation. Manufac-

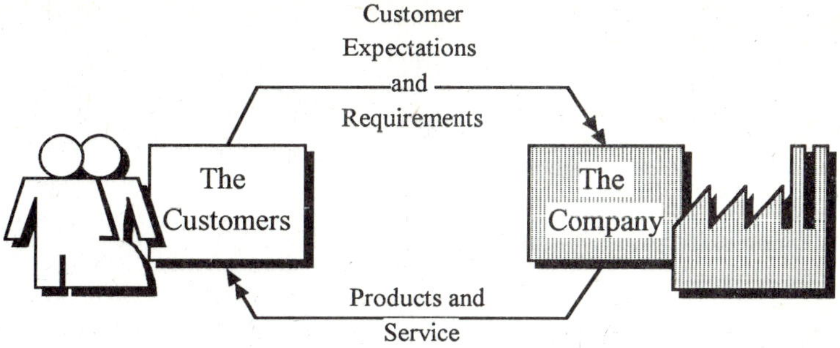

FIGURE 1.4 Basic Premise of Manufacturing

turers who at one time were able to differentiate themselves because of a lock on raw materials, technical knowledge, capital, process superiority, or innovation, have found that manufacturing is a vulnerable market. Technology by itself cannot create long-lasting competitive advantage. The relative affinities among push and pull elements are shown in Figure 1.5. If the push for emerging technology is high, the pull for product and process engineering is usually low. Similarly, high competitive pressure is created due to low performance or productivity improvement pull. If the performance or productivity level of a company is high, there is less of an urge toward an emerging technology push. However, this should not stop a manufacturer from striving to pull product and process reengineering even if push for competitive pressure is low. Manufacturers are constantly "watching" their competitors. When products come to market, forward thinking manufacturers always compare and frequently "benchmark" their products with other manufacturers to remain competitive. It is not very difficult to duplicate the "visible features" or even improve some of its salient characteristics. The competitive edge, in such cases, is usually short lived. For example, when IBM first introduced its 80286 chip based PC, in good faith, they openly published the interface specifications and directory. Soon, they were outstripped from market share by a multitude of clone manufacturers. Since IBM did not

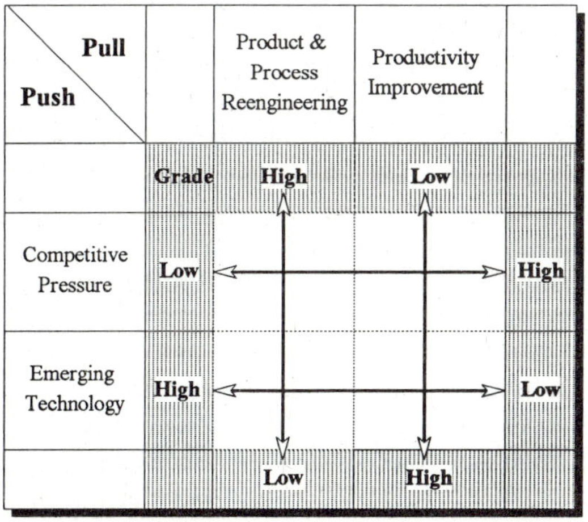

FIGURE 1.5 Relative Affinities among Push and Pull Elements

hold patent on the integral specifications or the architecture, they could not prevent its competitors from copying the functions of their PC ATs. IBM had to price their PCs higher than their competitors to cover their research & development costs. The clone market operated on narrow profit margins with low overhead and undercut IBM's price. In the absence of an exclusive right on the integral system-wide specifications of the PCs, functions were easily copied.

In principle, however, what is difficult to duplicate is how the technology is deployed into one's process. This may include deployment of product in terms of management techniques, organizational culture, and goals. Therefore manufacturers should continue to focus on the most efficient way of producing the product (achieving efficient product realization). The improvements made through deployment of technology and subsequent product and process reengineering can provide a real competitive advantage [Prasad, 1994]. The need for process reengineering is even more pronounced when different manufacturers produce similar products. Engineering schools and researchers tend to ignore the process reengineering factors and look exclusively at the visible technological solutions (for example, CAD/CAM, CAE, CIM, CAX, etc.). The industrial researchers, on the other hand, think of computers and all off-the-shelf tools as "commodities" that anyone can buy and use. Most truly successful companies (both in the United States and Japan) believe that process management techniques are the product of decades of "corporate learning" that others cannot easily buy or copy. The Japanese seem to be far ahead in mastering the technology and structuring it to fit their unique environments.

Two significant technological innovations are product innovation and process innovation. From 1955 to 1990, Japan's real gross domestic product increased almost nine times, the output of manufacturing in monetary value increased seventeen times, and the added value of this industrial sector increased twenty-one times [Hitomi, 1993]. Labor productivity in manufacturing increased during the same years at an average of 6.8 percent annually. Coming to terms with the Japanese market was one of the challenges Americans and Europeans had to meet to narrow the competitive gap. The only thing competitors cannot buy is someone's unique process or unique organizational culture. This can be a blessing in disguise or a curse, depending on how one looks at it. For most American automotive industries, the production process is deeply rooted in the way teams design and manufacture their products—thus inflexible—while the Japanese seem to have a better handle on process. With regard to culture, Americans seem to be more open minded than the Japanese—thus controversial—while the Japanese strong cultural ties facilitate better collaboration and teamwork. Americans, thus, seem to fall short on both ends. Japanese industry has practiced CE for a long time without calling it so. This was the conclusion of a recent study done in Europe that compared the time to market of Japanese with European's automotive manufacturers. Based on the eleven projects that they studied, Japanese companies could develop and introduce a new car to market twenty months faster than European [Hartley and Mortimer, 1991]. The two studies have also shown some subtle differences in the way an enterprise looks at its processes. Europeans still focus on quality improvements and operational efficiency during the process of manufacturing. Japanese seem to focus on flexibility while continuing improvements in quality, dependability, cost, and productivity. To the Japanese, flexibility in manufacturing means a rapid and efficient process of introducing changes in production volumes and product mix [Wheelright and Hayes, 1985]. In the field of manufacturing, the Japanese focus on a process of rapid development of new products is to become innovators of new process technologies. Locals of the Toyota factory in Japan can receive their car built to their specification within a few days of placing the order. Achieving perfection in process flexibility did not come without pain even for the Japanese. Such process flexibility cannot be attributed just to an edge in technologies. It has been observed [Putnum, 1985] that the success of the Japanese was largely due to practicing socially appropriate production, supremacy in process management, and continuous refinements. The Japanese seem to fully understand and internalize the tasks of transitioning product and process innovations into commercial success. They train their employees amply in techniques of implementing innovations, spend enough time early on, and work diligently on uncovering "how-to" problems of commercializing their technical success. Examples are the well-known Kanban system of production control, Kaizen (continuous improvements), Taguchi method of quality control, and market-oriented manufacturing, etc.

The recognized decline in the productivity of many U.S. companies has been a strong stimulant to search for ways to improve operational efficiency and become more competitive in the world marketplace. Many have changed their attitudes toward customers, their production processes, and their internal management approaches. Whereas, others still continue to search for the reasons of their demise. Successful companies have been the ones who have gained a better focus on eliminating waste, normally sneaked into their products, by understanding what drives product and process costs and how value can be added. Companies have focused on product and process delivery-system—how to transition process innovations into technical success and how to leverage the implementation know-hows into big commercial success. They have chosen to emphasize high-quality production in product delivery rather than high-volume production. With increasingly pervasive global competition, engineering excellence is becoming as fundamental a competitive weapon as manufacturing excellence. Significantly, what we are seeing is the completion of a definition of what it takes to be a world-class manufacturing company.

1.3.1 Measure of Competitiveness

Manufacturing competitiveness is a balancing act. There is no single solution—technical or nontechnical—that can easily be copied or bought from other successful companies. A well-orchestrated process, not just a series of programs, is required to achieve corporate goals and objectives [Rasmus, 1993]. The shorter lives of products today simply do not leave room to fix problems, correct design errors, iterate, or redesign products many times for lowering costs or improving quality [Prasad, 1994]. A company is considered to have reached a world-class manufacturing status if the goodness of products and services far outweighs the process and methodologies expended to produce it. Such a company gets the product right the first time. They measure the productivity not based on the inputs and output but in throughput and operating expenses. This is elaborated further in Section 1.5.

1.4 PRODUCT AND SERVICES

The nature of an enterprise business operation that distinguishes it from others is governed by four major controlling factors (see Figure 1.6):

- Tools and technology (hardware (h/w), software (s/w), applications, networks, etc.)
- Value system (people, business, customers, etc.)
- Human factors (cultural, social, knowledge, training, intelligence, etc.)
- Three Ps (policy, practices, procedures), business, management, cultural, technical, operational, etc.

Today, the nature of technology, people, and market are changing rapidly. The methods and practices of the past and, in particular, mass production of the Iron or Craft Age, are no longer valid. A recent trend is toward an "invisible enterprise" where key functions and services are all subcontracted to a set of selected suppliers. The time is not far when an enterprise will be totally virtual—merely serving as a glue tying all the required functions or processes back to the value system. There are many enterprises today in which at least one of the following life-cycle functions is outsourced (performed outside): (a) design; (b) manufacturing; (c) marketing; or (d) customer support. For example, an enterprise may choose to design its product in-house, manufacture it somewhere else, and rely on a third party (for instance, a value-added reseller—VAR) for marketing and servicing their products. The single rifle-shot approach of the past—that is trying to get everything right through marketing—has not worked and will not work. People gain a collective perception of a product by its actual field use, by its value to their lives, and by relative comparison to other similar products. Although vehicles may be produced in one

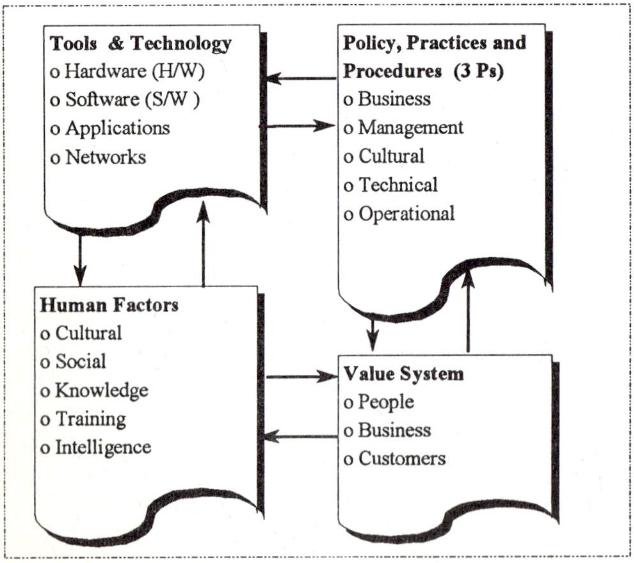

FIGURE 1.6 Nature of an Enterprise Business Operation

region and sold in another, the regional business unit is ultimately responsible for products sold in those regions in terms of their quality, cost, etc. Also, which company performs which function depends on several factors: currency exchange rates, facilities, company strengths and weaknesses, sheer size, company location, work force, etc. It is not just a matter of increasing quality in the sense of statistical conformance, but of creating products that are superior in both functional and cultural performance, and technological contents. Cultural performance denotes the customers' human aspects that make a product effective in the context of easing ones lives, profession, and work values.

In Figure 1.7, the items on the left side of the scale represent things that are visible to the customer. This side is mostly made out of weights belonging to products and services. Functions and features are two of the most visible items and therefore act as a primary weight. This is because if a product does not have the functions or features that a customer desires, it is difficult to maintain manufacturing competitiveness. Other primary weights belonging to this side are: cost, throughput, delivery and service, and agility. There is an invisible datum line passing through the support point or fulcrum in the center. This represents a datum level. The position of a lever-tip above this datum on either side denotes the company's weakness. In the same vane, the position of the lever-tips below the datum level indicates the strength. A desirable situation would be when the lever position is balanced, that is, weights on both sides are so placed that it deflects the beam large enough, on either side, so that the tip of the lever stays below the datum (the horizontal, invisible line as shown in Figure 1.7) level. Products and services are what the customer sees, touches, and feels. The weights that determine the items on the left hand side of the balance include the following:

- ***Functions and Features:*** The first step is to understand market needs and/or specific customer requests. Does the product serve the customer purpose? Does it have features that are versatile, simple, easy to use and handle? The product attributes that are important to the customer must be enumerated and translated into technical counterparts so that they can be measured by engineers. These product attributes should include the basic product functions, which the customer assumes, will be provided. They should include the essential level of product performance required by the marketplace, or the level necessary to lead technologically if this is a part of a competing strategy. They should also include those attributes that will attract and delight the customer, and differentiate this product from their competitors [Dika and Begley, 1991]. Manufacturing success requires the ability to produce top quality, often individually customized, products at highly competitive prices. Often they have to be mass produced or delivered to customers in batches as small as an unit of one with extremely short lead-times.
- ***High Quality:*** Quality is more than shiny paint—referring to an automobile for specificity. It ranges from the visual (such as good door fit, style, color, etc.) and nonvisual (such as reliability, producability, and scores of X-abilities), to customer perceived excellence (such as ride and handling, performance, drive-ability, noise-vibration and harshness, door closing efforts and sound, heating and air conditioning system). The myth that higher quality translates into higher costs is not always true. Many have found that overall production of higher quality products not only

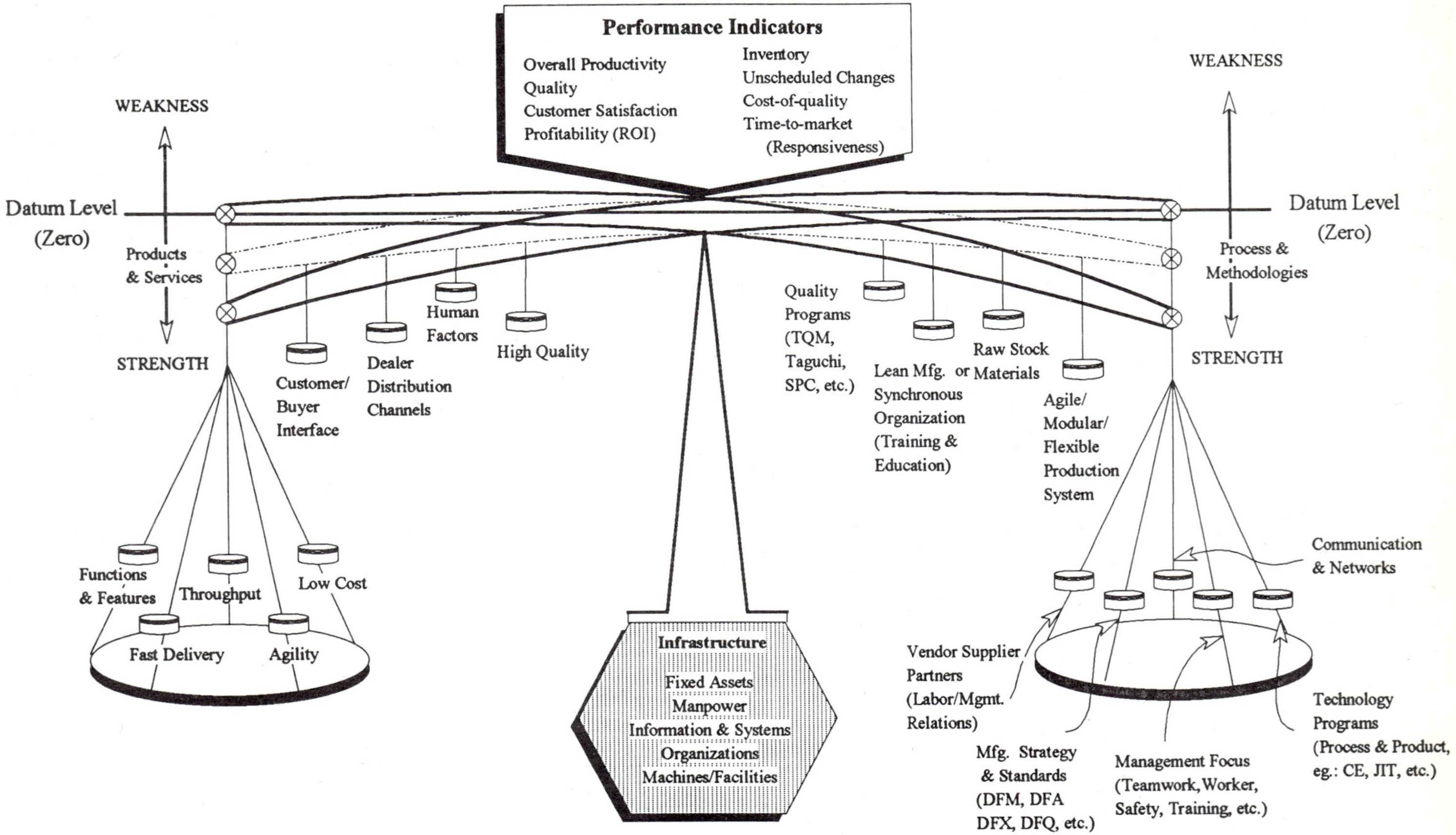

FIGURE 1.7 Balancing Elements for Manufacturing Competitiveness

costs less but also reduces cost overruns. In many cases, it has been shown that higher quality products yield—as much as 40 percent—higher return on investment than lower quality products do. In today's competitive world, quality is no longer an option; it is a necessity, which is to say, indirectly it is the way of doing business. It is imperative that each of the design alternatives, in consideration, meets the customer's quality requirements. Quality should not be an object when making trade-offs. If an alternative cannot satisfy the quality target, it should be removed from further consideration. On the same token, if the product functions exceed its quality expectations at a cost penalty, the customer will not consider it to be of great value. The product should not be over-designed in areas where the customer is insensitive [Dika and Begley, 1991].

- ***Low Cost:*** As quality is the price of admission in the competitive marketplace, cost is viewed as the "ticket for survival." Cost is influenced by many factors, and the relationship is more complex. For example, increase in quality, and for that matter any X-ability requirement, has an adverse effect on local cost. Cost reduction cannot be achieved by keeping status-quo. Cost reduction means looking at alternate concepts, materials, and process driven designs. Fixed costs (direct labor and materials) as well as variable costs (indirect labor and materials), and those due to waste in all forms—complicated manufacturing systems, inventory costs, equipment (single-purpose machines), facilities and layout—are all ingredients of "process-driven costs." Indirect costs are often the greatest contributor to increasing the final delivery costs. Elements of indirect costs include elements such as overhead expenses, supervision, material storage, rework, as well as hundreds of other operating expenses. Such costs can be three to four times more than the direct labor costs. Most designers focus on the direct, highly visible, and documented costs of labor and materials. They do not normally look at the indirect costs as elements they can possibly impact. Obviously, design by itself cannot eliminate a majority of the indirect costs. Though some indirect costs are not so bad, some indirect costs can be reduced or eliminated by strategic choice of tools, technology, and processes at an early consideration of designs. Reducing waste and process driven costs help in achieving higher profit margins and offer manufacturers opportunities to price their products more competitively.
- ***Fast Delivery (fast-to-market, responsiveness):*** In today's global marketplace, time is becoming a major competitive force. Quality used to take this place but not anymore. Today's customer sees quality and everything else as given (taken for granted). Every automotive company wants to get its new car models on the dealer's showroom floor in record time. Airlines spend billions of dollars in maintenance facilities and repairs to cut the nonflying downtime. Overnight delivery carriers are handling more and more packages with services faster than any time before. Even food chains (such as pizza parlors) are competing for home delivery on the basis of time rather than taste. The range for responsiveness and time-to-market is declining every time a new product is introduced. With the advent of newer and faster technologies, someday companies will perhaps be able to compress days into hours and minutes into seconds.

- ***Agility (economy of ease or flexibility):*** A prestigious study at the Iacocca Institute of Lehigh University has defined agility as the paradigm for manufacturing in the next century [Goldman and Priess, 1991]. Agility is defined as the ability to thrive and prosper in a competitive environment of continuous and unanticipated change, and to respond quickly to rapidly changing markets and customer demands. A key element of agility is the agile enterprise. Agile enterprises are totally integrated organizations. Information flows seamlessly among manufacturing, engineering, marketing, purchasing, finance, inventory, sales, and research [Dove, 1993]. Just as mass production in 1950 leveraged "machine intelligence" (economy of automation), and craft manufacturing leveraged people's skills and dexterity (economy of skills), agile manufacturing is the "*economy of ease*" with which a company can react to new opportunities. Becoming an agile, virtual company does not necessarily mean down-sizing. "*Economy of ease*" is taking the front seat as a **fifth reach.** It is not a technological solution, if plugged in, it will enable an enterprise to respond to change. It is merely a concept to be used while designing products and manufacturing processes so that when markets change or when customers need change, agility in the process can come to the rescue, allowing reconfiguration of the setup with calculated penalty and change in product offerings. Agility is part of the measure of merit that needs to be considered in product realization.
- ***Throughput:*** A global enterprise requires global or integrated thinking. The former CAD measures of productivity, such as "engineering CAD throughput" or the number of drawings created in a week are local (thus inadequate) throughput measures. It is not a true representation for maximizing the flow of work. Measures on "number of engineering changes per unit of time," or "ramp-up time-to-volume production" are closer to global measures. Such measures on work-flow provide a more accurate big picture. Throughput is the rate (per unit of time) at which the current factory process transforms the materials into finished "useable" products. Useable means the measure does not include defective or scrap parts (that is, those that can be converted into sales or profit dollars). Clearly, throughput is dependent on the amount of lead time required to finish each one of the activities. Throughput for serial engineering is lowest, since one starts when the other ends. Throughput for concurrent engineering is proportional to the degree of overlap or independence between the activities. If each of the activities is completely independent, then throughput will be proportional to the inverse of the maximum lead time that any one of the activities will take because when activities are independent all activities can be started at once.

$$\text{Throughput (for independent activities in SE)} \propto 1(\Sigma T_i) \tag{1.2}$$

$$\text{Throughput (for independent activities in CE)} \propto 1 T_{\text{maximum}} \tag{1.3}$$

Where

$$T_{\text{minimum}} < T_i < T_{\text{maximum}} \text{ for } i = 1, n \tag{1.4}$$

and

$$T_{\text{maximum}} = \text{Maximum lead time activity out of all the } n \text{ activities } (T_i) \text{ that could be performed in parallel}$$

$$T_{\text{minimum}} = \text{Minimum lead time activity out of all the } n \text{ activities } (T_i) \text{ that could be performed in parallel}$$

Besides, there are several secondary weights, such as human factors, dealer distribution channels, and customer and/or buyer interface, that can tip the balance to one side or the other depending on their locations on the lever.

1.5 PROCESS AND METHODOLOGIES

Process and methodologies form the "invisible" side of the balance. On this side, value is added to the raw materials through human skills. The general approach to management—the concepts used to market the product, the policy followed in making investment, the reward system for employees, and the importance assigned to customers—are all important methodologies.

The factors that are considered influential in tilting the balance on this side of the lever below the datum line include the following:

- ***Quality Programs*** (such as Total Quality Management (TQM), Taguchi [Eureka and Ryan, 1988], Statistical Process Control (SPC), Quality Circles, GM Delphi studies, Value Engineering, Six-Sigma Program, etc. [Carey, 1992]): One of the principles of SPC is that inspection, once a part is produced, even if it is done on a statistical basis, is wasteful. By the time a part comes off the end of a production line and is inspected, many bad parts will have been produced. It is likely that the common cause that produced the bad part will not be easy to identify. What needs to be done is to measure some key process parameters on a real-time basis, identify those which are sensitive to common cause of the variation and address the ones that are affected the most. Six-sigma program is named after the statistical figure of six times the standard deviation measurement. Like many earlier ones, quality programs require reengineering of the organization and how it operates. Quality programs should not only be directed toward minimizing defects in production, but also augmenting the capability of the plants to monitor and correct their own operations. Augmenting includes determining the level of variability in the process, on-line factory information, and performance feedback system to monitor the flow of the parts through manufacturing processes, and taking necessary corrective actions to eliminate those variations. It requires benchmark schemes, rationalized schema for products and processes, and a company-wide strategy for implementing team communication networks. The success will depend on the check and balances of the operations, for example, potential failures detected at any level need to be relayed back to team leaders. Through the chain of commands it ought to be conveyed to the design team for error-proofing the process so that failure does not occur again. Such quality programs will not only create good products but will also ensure continuous improvement guaranteed over the life of the product.

- ***Raw Stock or Materials:*** Raw materials are defined as a variable cost of producing a unit of a product. The advantage gained due to a lock on materials technology is considered part of this.
- ***Lean Manufacturing or Synchronous Organizations:*** Lean or synchronous, often used interchangeably, is an important and crucial manufacturing strategy. There are seventeen tactics that one can employ to be lean (see Figure 1.8). Synchronous organization improves efficiency through such things as quick set up, error-proofing, pull system, process control, and thirteen others [Hartley, 1992]. Lean manufacturing means developing an environment that is conducive to synchronous principles. Elimination of waste includes bottlenecks of information or material movements, smooth flow of work, committee structure rationalization, etc. Good lean manufacturing programs include training and education.

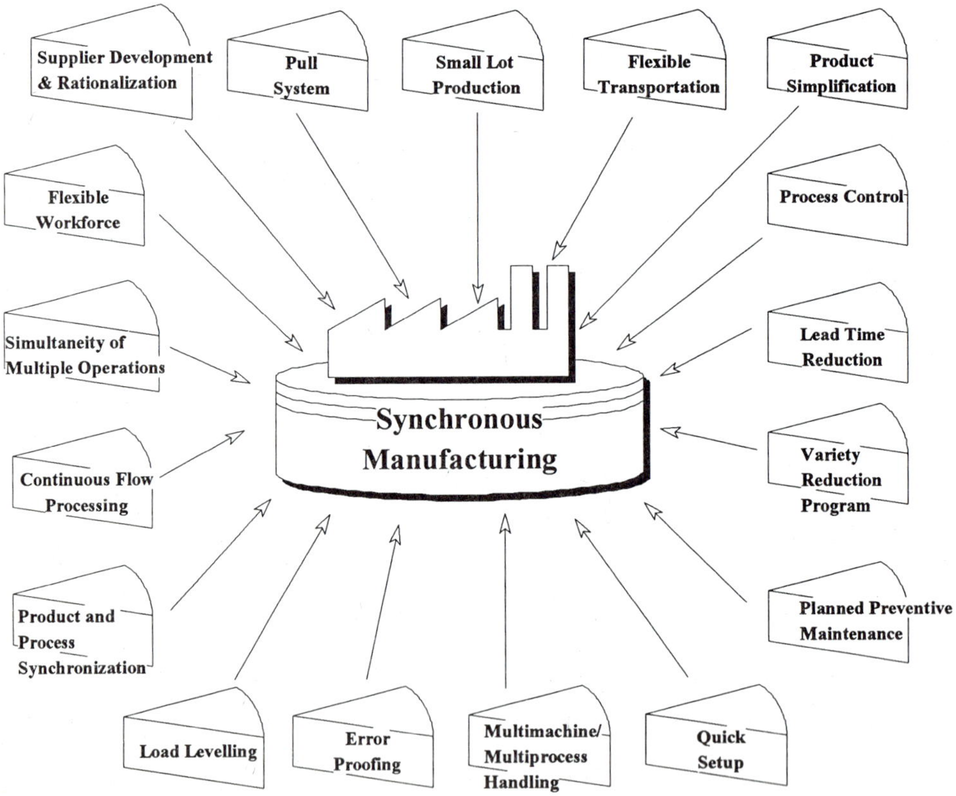

FIGURE 1.8 Seventeen Tactics for Lean (or Synchronous) Manufacturing

- ***Agile/Modular/Flexible Production System:*** Internal or invisible to the customer, is the ability of a manufacturer to be product-flexible, modular, and agile. Flexibility in manufacturing can reduce material costs, work force, inventory, idle facility or machine time and improve material handling. While these are invisible to the customer, they do affect the overall cost, quality, and timing of a product that is very visible to the customer. The tactics of flexible production system are very much company or process-focused. They are difficult to be duplicated by the competitors, since they often are not visible. These tactics are often directed toward: material cost improvements, reduced workforce through process-driven design concepts (such as flexible manufacturing), and inventory cost reduction through the combination of techniques (such as in-line sequencing, just-in-time delivery, defect-free supplier, and improved material handling).
- ***Technology Programs:*** This includes programs such as just-in-time (JIT) manufacturing, virtual reality, rapid prototyping, pull system, concurrent engineering, relational database systems. There is no single universal form that can fit all programs in a company. Each of these and other automation programs must be tailored to the needs of the individual company.
- ***Manufacturing Strategy and Standards:*** Strategies include measures such as DFM (design for manufacturability), DFA (assembly), DFQ (quality), DFC (cost), and other DFX measures. DFX stands for design for X-ability, where X-ability may symbolize any life-cycle concern such as quality, cost, assembly, manufacturability, serviceability, maintainability, etc. (See Figure 1.9). Kodak saved $60K in the redesign of a photocopier part by using rapid prototyping. Organizations must deploy common systems, standardized tools, methods, and practices to design and develop products, toolrooms, die design, fabrication and construction, and business support infrastructure. Polaroid deployed existing parts for 75 percent of its new ProCam, resulting in 35 percent savings in developmental costs.

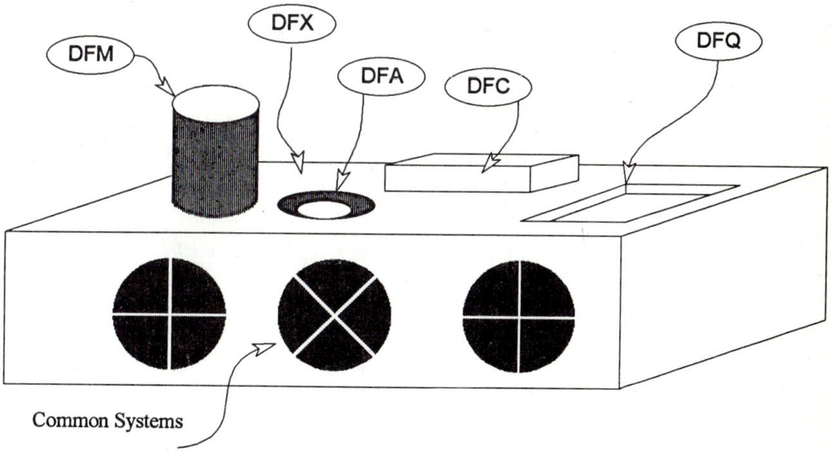

FIGURE 1.9 Examples of Competing Manufacturing Strategies

- *Use Common System:* Organizations must develop technologies that use common systems to produce a variety of products, each with its own unique features and distinctions. This means that processes and technologies are not developed with a single solution in mind; they are developed to be used in a variety of applications, such as 3-D CAD, solid modeling, etc. Technology in itself cannot be the end goal of an organization. Rather, an organizational objective should be to develop technologies that have broader applications throughout the SBU for solving existing and anticipated problems.

- ***Management Focus:*** Companies must adopt a management style that allows more leeway to workers (empowerment), pushes decision making to the lowest appropriate levels, and provides greater flexibility to managers. It must reduce the reliance on corporate headquarters and begin to establish intra- and intercorporation projects, with multidisciplinary focus teams and expertise. Emphasis should be on the process and continuous improvement, long-range thinking, and total customer orientation in the projects that are undertaken. There should be more and more decentralization of activities and better coordination between the groups involved.

- ***Vendor/Supplier Partners (including labor management relations):*** Manufacturing enterprises can, and most likely will, have borders going well beyond the confines of an individual company or its subsidiaries. This postulate is supported by the success of Japanese "Keiretsu" and is the key to the agile manufacturing concept [Goldman and Priess, 1991]. The latter is manifested in many recent international developments resulting in cooperative agreements such as U.S. congressional and presidential directives in the USA and Japan to mandate transfer of technology from federal research facilities to private sectors such as CRADA (Cooperative Research and Development Agreement) and U.S. National Science Foundation and Department of Commerce's strategic partnership initiatives.
  - *Strategic Sourcing:* Most companies are adopting policies to optimize outside supplier base, develop partnerships with strategic suppliers to develop mutual trust, and enter into various cooperative arrangements.

- ***Communication and Networks:*** The sleuth of electronic mail (e-mail) and Ethernet networks (such as local area networks (LAN), wide area network (WAN), and broad band) systems and a new generation of workstations are providing direct links between work-groups such as design, analysis, manufacturing, testing, and the CE team members. Improved communication features allow data to be transferred rapidly at a higher rate in one or two orders of magnitude faster than what was feasible a few years ago. Networks like LAN can quickly transmit huge data files almost instantaneously to a large number of interconnected users. Programs on these networks thus enable the rapid transmission of design information between work-groups and teams facilitating the CE approach. Ford boasts 2 million dollars a year saving as a result of their wide area multinational network.
  - *Electronic Data Interchange (EDI):* One of the biggest barriers to effective CE

programs has been the lack of a common graphics exchange standard. Much of the design generated by a CAD/CAM supplied by one vendor cannot be recognized directly by another CAD/CAM system. This has been eased at least partly by the introduction of exchange standards. Initial Graphics Exchange Specifications (IGES) is the most common approach that is used currently to allow dissimilar systems to talk to each other. In this approach, the transmitting system translates data into a second language—a so called IGES neutral file—that can be sent to different systems. Translators at the receiving end reformat the data into appropriate native forms. Today, many standards are being explored such as PDES and ISO STEP standards.

With such standards, the initial product definition data transfer capability of IGES has been extended to include feature-based and object-oriented representations. They are currently being expanded to many application domains, such as sheet metal forming, finite element, numerical control, process planning, etc. As this range of functionality and application interface expands further, it will be easier for engineers, designers, and manufacturing personnel to transmit data regardless of individual firmwares (brands of software and hardware systems) that may be in use.

- *Manufacturing Automation Protocol (MAP):* The proliferation of stand-alone manufacturing tools made it difficult, or impossible in some instances, for one machine to talk to another on the factory floor. MAP was subsequently developed by General Motors and its industry partners to facilitate and simplify communications between the growing number of machines and computers in today's automotive factories. MAP allows equipment to share data or "talk" to each other for improved efficiency in computer integrated manufacturing (CIM) factories.

Figure 1.10 shows an example of how process and methodologies will work in harmony to meet both the company and the customer interests. Three process examples are "manufacturing strategies," "agile/modular/flexible production system," and the "quality programs." As shown in Figure 1.10, the company interests are represented by one or more of the teams (multidisciplinary, multifunctional, or core competency). In the example of the manufacturing strategies case, teams are shown to take the initial step in determining the appropriate strategies, such as common systems, standards, DFM/DFA/DFQ/DFX, etc., which are suitable for the problem at hand. This means that they would be helpful in either enhancing the product values to the customer, or meeting the company interests (such as profitability). In the case of the quality programs example, the process is reverse. The customer (in terms of VOC) provides the basic needs and wants, which in turn are translated into appropriate methodology to be used in product design and development. Agile/modular/flexible production system on the other hand converts company interests into the customer interests. Many of the processes and methodologies described in this section will follow either a customer-to-company or a company-to-customer scenario. Like the examples in Figure 1.10, they will be leveraging the strengths of each other in protecting both the company and customer interests.

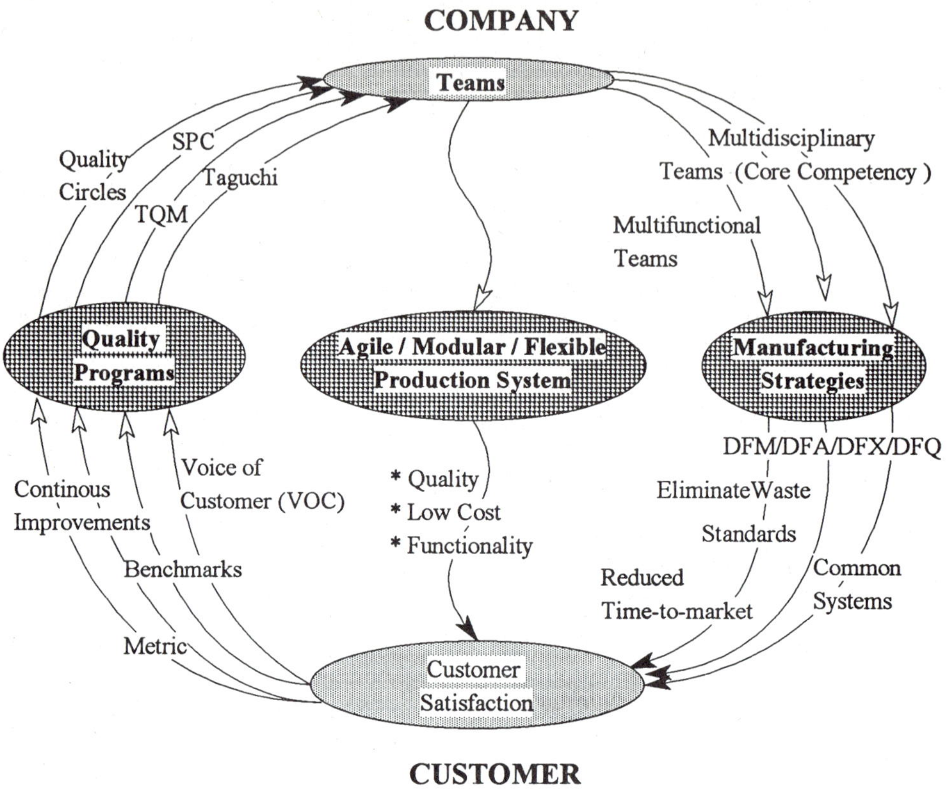

FIGURE 1.10 Examples of Process and Methodologies for Meeting the Company and Customer Interests

1.6 PERFORMANCE INDICATORS

Figure 1.11 shows a list of eight indicators that determine the performance of an enterprise competitiveness [Prasad, 1994]. Each indicator provides a measure of a company's efficiency in the world marketplace. Each indicator is shown by a directed radial line pointing away from the center of a unit circle. A point on the unit circle represents a world-class level for an indicator. Such points represent a normalized or scaled value of 1.0. A point at the center of the circle usually represents a value 100% out of range from the world-class. A point along a radial line inside the circle, thus, ranges from a value of 0 (at the center) to 1 (on the circle). A point outside the circle ranges from 1 (on the circle) to any positive number, depending on its distance away from the center. The desirable state depends on whether a performance indicator is to be maximized or minimized. The desirable state is outward of the circle (pointing away from the center), if a performance indicator is to be maximized. The desirable state is inward of the circle (pointing toward the center), if a performance indicator is to be minimized. For instance, a point 1 unit out from center may represent a level "twice" as good or bad from the world-class level. De-

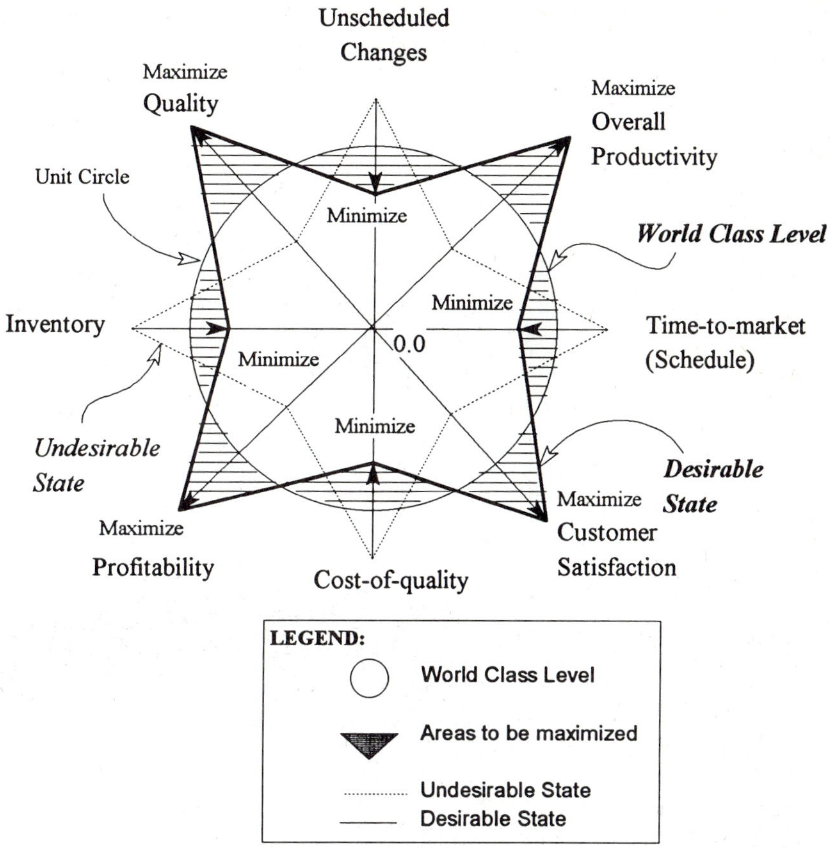

FIGURE 1.11 Performance Indicators for Measuring an Enterprise Competitiveness

pending on whether the performance is to be minimized or maximized the corresponding arrow is shown pointing inside or outside the circle. It may be noted that there are four indicators that need to be maximized and four that need to be minimized. They are placed alternatively around this unit circle. The solid lines show the current state (Figure 1.11). The hatched areas are formed due to the lines drawn connecting this max- and min- points and the unit circle representing the world-class. Clearly, the hatched areas represent the net contribution from each performance indicator. In order that the current state of the process must perform better than or equal to the world class, the hatched areas must be maximized.

$$\sum_{i=1}^{8} \text{Hatched Areas} \Rightarrow \text{Maximize} \tag{1.5}$$

The overall performance must show a net profit at the current conditions with or without the new product development or technology insertion. The objective is to mov the four indicators away from the center and four toward the center as much as possib

In other words, the objective is to maximize the hatched areas created due to intersection of the straight lines and the circle. Performance in this context represents system's performance. It is important to note that performance of an organizational unit is governed largely by the system in which it is contained. It could be a worthless exercise to improve the performance of a local unit without changing the entire system, if units were interdependent. New accounting measures (such as activity-based costing (ABC) and Goldratt's theory of constraints [1986]) are helpful in obtaining system's performance.

- ***Overall Productivity (gain or loss):*** Overall productivity means cumulative gain or loss. A higher level of productivity in one specific department or discipline is not a good measure. Productivity means creating concepts that positively impact the whole system—both the upstream and the downstream operations. The overall productivity is defined as the ratio of the throughput (T), to the operating expenses (OE). The point to note here, contrary to what is generally understood, is that productivity is not a simple ratio of the outputs to the inputs. Throughput in this context is defined as useful outputs (that customers can use)—end product or services completed in a given period of time. In other words, scrap or waste is not a measure of productivity.

$$\text{Productivity } (P) = T/OE \tag{1.6}$$

thus, productivity entails the effective measure of how inputs (people, materials, means, etc.) are utilized in a certain period (measured in terms of operating expenses), in order to realize useful outputs. All outputs are not throughput, some outputs (for example, scraps, defects, etc.) are waste. The throughput is defined as follows:

$$T = \sum_{i=1}^{N_o} \left[P_i * N_i * P_{v_i} \right] \tag{1.7}$$

where

P_i—the proportion of acceptable outputs (which are nondefective) of variant i
N_i—the total number of outputs produced of variant i
P_{v_i}—the production (or throughput) value per acceptable output i
N_o—the number of outputs (say number of assembly variants)

For convenience sake, defective outputs (or scrap assemblies) are assumed to have no production (or salvage) value since they cannot be sold to the market as is. Successful manufacturers are those who measure the difference between outputs and throughput, identify the possible source of such discrepancies, and take counter measures to prevent them at the source.

- ***Customer Satisfaction:*** One of the purposes of developing the product is to achieve satisfied customers recurrently. Customer satisfaction means meeting the customers' needs at the right time and in the quantity, price, and performance they want. The cornerstone of these performance measures is the customer. Of course, if the customer does not want to buy a product, improvements in cost, weight, and investment do not really matter. At the same time, if the customer becomes disappointed with the workmanship of the product or encounters problems over its life, he or she will not buy it again. The key to understanding customer satisfaction is the recognition that there are two basic types of activity: support and value-added. While support activities are necessary for internal planning and control, they consume the team's effort and time but they do not provide direct benefit to the ultimate customer. Value-added features or services are pleasant surprises to the customers.

- ***Unscheduled Changes:*** The success of rapid product realization depends on the team's ability to handle unscheduled changes. Unscheduled changes occur in many ways: Some are avoidable some are not. Avoidable changes are typical of products thrown over the wall before they were ready for manufacturing. Once the parts are sent back to the originating team, unscheduled changes have to be squeezed between work. Unavoidable changes occur when circumstances change, people move, and the steps are no longer valid. Unwanted changes are caused by changes in product lines, product functionality, technology, etc. Though a number is an important measure, unscheduled changes can be very serious. For example, if errors are detected late in the process (say during a downstream operation), it might be very costly to fix them.
- ***Inventory (I):*** Inventory includes all assets including property, plant, and equipment, but excluding value-added parts. The new definition, broadly stated, includes any item that the company could sell, not just the finished products. By including capital assets in the inventory category, teams are forced to focus on the way they are utilizing all of the investments under their control. The finished inventory is the amount the retailer must keep in stock. This amount is equal to the average demand over the order ship time plus a safety factor based on the standard deviation of demand over the order ship time.

  If n is the average demand for one day,

  Sigma (σ)—the standard deviation for a day's demand and,
  d—the order ship time in days,

  the required inventory is

$$\text{Inventory} = [\, n\, d + (3\sqrt{(d)}\;) * \sigma] \tag{1.8}$$

- ***Cost of Quality:*** Knowing how much quality costs and the way the cost is made up can provide a strong impetus for management to set off on the quality improvement trail. There are two contributory elements that affect the cost of quality: (a) cost to ensure quality (c-t-e-q) and (b) cost to correct quality (c-t-c-q). They are shown in Figure 1.12. Cost to ensure quality is the cost of doing things right (for example, choosing the right process), the cost of doing right things (for example, choosing right actions), and the cost of preventing mistakes (such as anticipating problems). Prevention costs are the expenditures on activities whose objective is to prevent the occurrence of failures. They are designed to ensure or build quality during designing, implementing, and manufacturing products and services. Typical examples include the cost of training, establishing procedures, insurance, preventive or contract maintenance, planning activities and analyses of performance data, surveillance, etc. The cost to correct quality is the cost incurred because of doing things wrong (for example, choosing the wrong process), the cost of doing wrong things (e.g.,

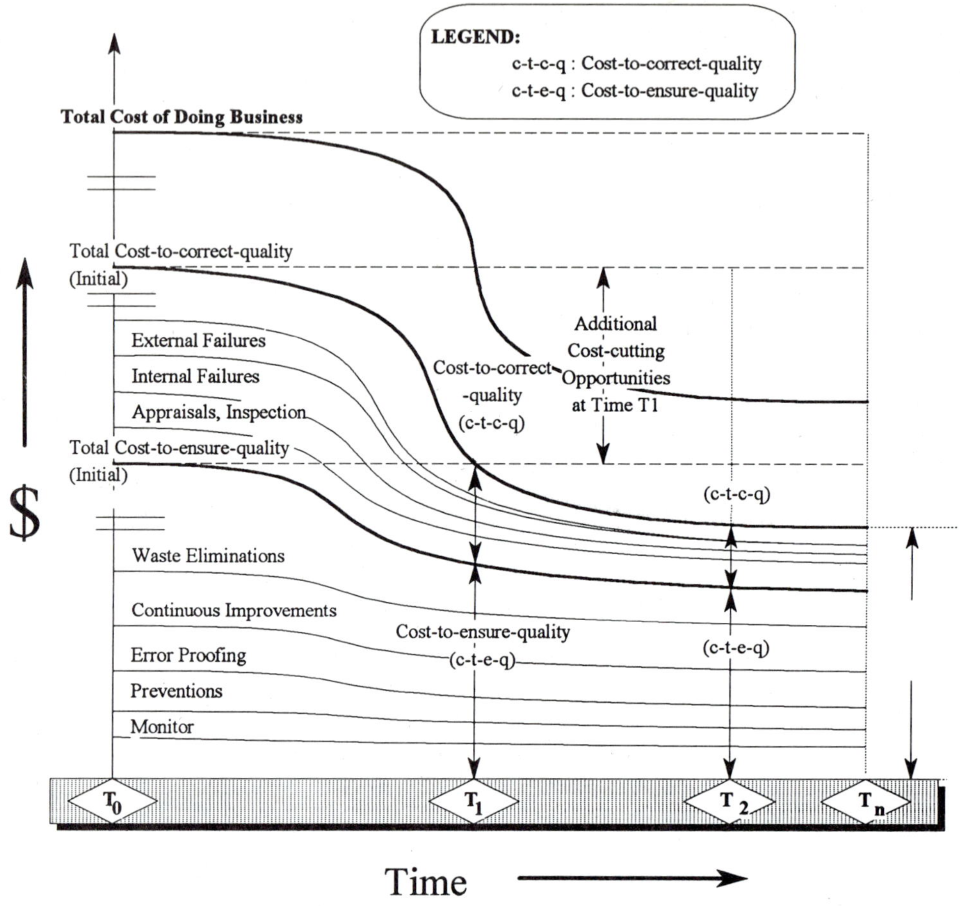

FIGURE 1.12 Cost-cutting Opportunities through Prevention (Measurement of Cost-to-Quality)

choosing wrong actions), and the cost of inspections to discover mistakes committed earlier. Cost-to-correct-quality falls into two categories:

- *Appraisals* are the costs associated with activities like checking, evaluating, inspecting, and measuring work, supplier monitoring, appraising performance, and conducting audits on work done to assure conformance to quality requirements. The conformance shows whether work has been performed according to the required specifications or standards. Other types of cost-to-correct-quality are internal and external failures.
- *Internal (or external) failures* are the costs incurred by failing to perform work right the first time. They are often associated with a product or service that does not meet the quality requirements (such as building codes) prior to transfer (or after transfer) to the customer. Costs of failures include the disposal or correction of incorrect work, scrap or excess stock, time spent on rework, bad debts, waiting for work, and dealing with complaints from customers.

Most cited product quality indicators attempt to measure the parts per million (PPM) level of conformance. This does not, however, account for criticality—for example a one dollar part failure may result in a thousand dollar part failure if one part is encapsulated into another. Another measure of overall effectiveness is to track cost-of-quality (c-t-q), both cost-to-correct-quality and cost-to-ensure-quality:

$$\text{C-t-q Effectiveness} \equiv [\ \{\text{cost-to-ensure-quality}\} \,/\, \{\text{cost-to-quality}\}\ ] * 100 \tag{1.9}$$

Where, cost-to-quality is the sum of two parts:

$$\text{Cost-of-quality} = \text{cost-to-correct-quality} + \text{cost-to-ensure-quality} \tag{1.10}$$

or

$$\text{c-t-q} = (\text{c-t-c-q}) + (\text{c-t-e-q})$$

If the c-t-q effectiveness number is close to 100, the company is doing things more right than wrong. The effectiveness number thus provides an analytical basis for decision making or tracking quality improvement opportunities.

- ***Profitability (ROI):*** The return on investment (ROI) is defined as the ratio of gain (G) minus the operating expenses (OE) to inventory costs (I), that is:

$$\text{ROI} = [\ \{G\text{-}OE\}/\{I\}\ ] \tag{1.11}$$

Where gain (G) is defined as

$$\text{Gain } (G) = \text{Net Sales} - \text{Cost of Raw Materials} \tag{1.12}$$

Where, net sales (or volumes) is defined as the irreversible transfer of product to the consumer. Such a definition of sales does not allow the transfer of goods in a consignment from a manufacturer to a dealer to be counted as a sale. *OE* is computed using all normal operating expenses plus direct labor and factory overhead. By grouping direct labor and factory overhead in an *OE* category, there is little reason for teams to over-build their inventory. Direct labor is recognized as a fixed cost.

- ***Time-to-market:*** This is a measure of time period required to design and develop a marketable product (from concept through rate production).

Some of these indicators might be contradictory. For example, quality-based focus drives costs down and time up, whereas time-based focus drives costs down and quality up. Additional performance indicators that are being used are in the areas of delivery, risk management, and teamwork communication. Figure 1.13 shows the productivity and quality rankings comparison for the years 1988 and 1989 for a number of American and Japanese automobile manufacturers. The comparison for the next 5 years (up to 1994) for only the Big Three manufacturers are contained in Table 1.2. The statistics are based on a recent competitive report by [McElroy, 1995].

The results are summarized into four categories:

- **Quality:** The quality rating is based on number of quality problems or defects per 100 vehicles. So the lower the number, the better is the performance. There has been

Company	Vehicles/ Employee Per Year*	Quality Problems Per 100 Cars**
Suzuki	70.4	--
Toyota	61.0	117
Daihatsu	57.0	--
Honda	56.2	112
Mitsubishi	50.4	--
Mazda	42.0	133
Isuzu	41.7	--
Nissan	39.5	111
Fuji (Suburu)	38.7	--
Ford	20.0	149
Chrysler	18.0	176
Peugeot	13.3	--
General Motors	12.5	169
Volkswagen	11.2	--

* Source: Automotive Industries, Vol. 169, No. 4 (April), 1989.

** Source: J.D. Power And Associates, 1989 New Car Initial Quality Survey

FIGURE 1.13 Productivity and Quality Ranking for Automobile Manufacturers

a significant gain in quality for american automobiles in the last few years. Ford's quality performance is very impressive. Not only did the company show an improvement in the last year, it was significant improvement. In fact, the company outperformed the industry average and broke through the 100 defects per 100 vehicles barrier. This is a first for any of the American Big Three [McElroy, 1994]. GM's quality rating deteriorated 3.7% and Chrysler quality rating deteriorated 7.5%.

TABLE 1.2 American Automobile Competitive Performance Rating

Category	Manufacturer	1990	1991	1992	1993	1994	Remarks
Quality	Chrysler Corp.	172.5	168	150	132	142	
	Ford Motor	147.6	132	125	101	91.3	Best quality performer
	General Motors	149	150.5	133	109	113.1	
Profit/Unit	Chrysler Corp.	$209	$427	$445	$1,709	$2,110	Best low cost producer in the world
	Ford Motor	$53	$700	$307	$240	$877	
	General Motors	$462	$883	$541	$208	$734	
Productivity	Chrysler Corp.	15.7	16.4	18.7	21.2	24.6	High Capacity Utilization
	Ford Motor	16.5	17.9	19.3	19.0	19.6	
	General Motors	12.0	11.8	12.5	14.1	16.1	
Market Share	Chrysler Corp.	13.2%	12.2%	12.2%	14.7%	14.6%	Minimum % Decline
	Ford Motor	23.8%	22.8%	24.6%	25.6%	25.3%	
	General Motors	35.5%	35.1%	33.9%	33.4%	33.1%	

- **Profit-per-unit:** The profit-per-unit shown in the Table 1.2 is based on dividing automotive operating income by total factory sales [McElroy, 1994]. Chrysler topped its best 1993 per-unit profit record again in 1994. Chrysler per-unit profit margin was the highest that ever had been achieved in the automotive industry in the world. However, the increase from its last year's performance level was not so good. It was the lowest per unit increase of the Big Three ($401/unit). Ford achieved a strong $637 improvement in its profit-per-unit in 1994 compared to a year ago, closely ahead of GM ($526/unit). GM improved is profitability on two fronts. First of all, it converted its net loss of $872 million in 1993 (corresponding to a profit rate of $208/unit) into $690 million net profit by raising the rate to $734/unit in 1994.
- **Productivity:** The productivity ranking is based on the number of vehicles produced per employee per year. Chrysler's productivity increased 16.3% in 1994, compared to 14.2% for GM, and 3.3% for Ford. In 1994, not only did Chrysler improve the most, it was already the best of the Big Three and was pulling away [McElroy, 1994]. It is hard to pinpoint the exact cause of this gain in productivity. One difference to note is that Chrysler is the least vertically integrated among the Big Three. GM improvement was quite impressive (compared to Ford), considering the number of problems GM had with the launches of its new vehicles in the US. Strangely, Ford's productivity improvements have stagnated. It may be that the major reorganization efforts that Ford completed during Spring 1995, called Ford 2000, will give Ford the right boost.
- **Market Share:** 1994 will long be remembered as the most profitable year in the history of the US auto industry and 1991 will be remembered as the worst year. All together the American Big Three posted $13.9 billions in profit—an all time record [McElroy, 1994]. Though there were many reasons, application of CE principles and practices were considered to be one of the major contributors in increasing pro-

ductivity, quality, and putting a stop to the fast rising imports and market share. Import brands altogether increased their share during 1994 only by a mere (1/10) of a percent increase to 27% total, while the Big Three ended up with a total of 73% of the US market.

Ford seems to do well in streamlining activities at manufacturing fronts and increasing efficiency [McElroy, 1994] when it comes down to applying structure techniques built around CE. Chrysler appears to gain the major benefits by applying it to the up-front processes: product development, capital investment, and marketing fronts. GM seems to be doing both.

1.7 INFRASTRUCTURE

The balancing act requires a foundation as a basis for running the operation.

1.7.1 Flexible Architecture

Flexible manufacturing architecture is the wave of the future. In short, it is not necessary to reinvent the wheel every time a challenge arises. Creating new solutions from existing systems is a more cost-effective and responsive approach. Flexible assembly operations promote the quick changeover of lines or entire plants to accommodate rapidly changing market demands. Technology elements in flexible architecture include reuseable processes and machines, versatile controls to empower operators, advanced sensors to let operators know in real-time what is going on, and advanced control algorithms to give the company a sustainable, proprietary competitive edge. Developing concepts of modular assembly systems and new techniques, such as the knitting cell, are essential components of concurrent engineering innovations.

- ***Information and Systems:*** A cooperative, seamless information and systems environment is a major component of concurrent engineering. Inherent to this environment, and directly stated in the Lehigh study report, is information sharing—within teams, within and among departments, and among suppliers and customers. It includes rationalizing hardware/software, CAD/CAM tools, and their deployment for improved communication and efficiency. Examples of strategic tools and technologies in an information system include the following:
  - Fourth generation languages (4GL)
  - Imaging systems
  - Enterprise data access and communications
  - Product and process data exchange
  - Manufacturing data collection and control
  - Engineering workstations (EWS)
  - Technical publications
  - Distributed computing
  - Machines and facilities

The strategic tools and technologies can be used to:

- Identify specific quality improvement opportunities; prioritize problem areas
- Assess the movement of quality costs from inspection to prevention
- Monitor reduction of appraisal costs, internal or external failure costs, and total cost
- Support waste elimination, continuous improvement, error-proofing, prevention, and monitoring efforts.

- ***Legislative and Regulatory Environment:*** This deals with changes to our social and legal structures allowing rapid formation of joint or cooperative ventures between the participating (at the national or international level) communities.

This book deals with product realization knowledge that is sufficiently general to provide at least partial intellectual support for a number of specific concurrent case histories, applications, and introduction of new products and systems. A number of fundamental techniques and methods for achieving product realization are extracted from the currently successful CE theory and practices. These are presented in two volumes and nineteen chapters of this book—*CE Fundamentals.*

REFERENCES

BAUMAN, R. 1990. "CALs and Concurrent Engineering: Business Strategy or Tool Survival?" *Aviation Week and Space Technology,* Vol. 133, (July), pp. s1–s12.

CAREY, W. R. 1992. *Tools for Today's Engineer Strategy for Achieving Engineering Excellence. Quality Function Deployment.* SP-913, SAE International Congress and Exposition, Detroit, Michigan, USA, February 24–28, SAE 920040.

CHISHOLM, A. W. J. 1973. "Design for Economic Manufacture," *Report of CIRP,* Vol. 22, No. 2 (2).

DIKA, R. J., and R. L. BEGLEY. 1991. "Concept Development Through Teamwork—Working for Quality, Cost, Weight, and Investment. SAE Paper # 910212, Proceedings of the Int'l Congress and Exposition, SAE. Feb. 25–March 1, Detroit, MI, pp. 1–12.

DOVE, R. 1993. Agility—The Key to Competitiveness in the 1990s. *Automotive Industries,* Vol. 173, No. 10, Oct., pp. A1–A15.

EUREKA, W., and N. RYAN. 1988. *Taguchi Methods and QFD: Hows and Whys for Management.* Dearborn, MI: American Supplier Institute (A.S.I.) Press.

FERDOWS, K., and A. DE MEYER. 1986. Towards an Understanding of Manufacturing Strategies in Europe—A Comparative Survey 1985–1986. Insead Research Paper.

GOLDHAR, et. al. 1991. "Flexibility and Competitive Advantage: Manufacturing Becomes a Service Business." *IJTM* (Special Issue of Manufacturing Strategy): pp. 243–59.

GOLDMAN, S., and K. PRIESS. ed. 1991. *21st Century Manufacturing Enterprise Strategy: An Industry-Led View.* Vol I. Iacocca Institute, Lehigh University, Nov.

GOLDRATT, E. M. and J. COX. 1986. *The Goal: A Process of Ongoing Improvement,* Croton-on-Hudson, NY: North River Press.

HALL, D. 1991. "Concurrent Engineering: Defining Terms and Techniques," *IEEE Spectrum,* Vol. 28, No. 7 (July), pp. 24–25.

HARTLEY, J. R. 1992. *Concurrent Engineering: Shortening Lead Times, Raising Quality, and Lowering Costs.* Cambridge, MA: Productivity Press.

HARTLEY, J. and J. MORTIMER. 1991. *Simultaneous Engineering—The Management Guide,* 2nd ed. Dunstable, UK: Industrial Newsletters Ltd.

HITOMI, K. 1993. "Manufacturing Technology in Japan." *Journal of Manufacturing Systems.* Vol. 12, No. 3, pp. 209–215.

LACEY, R., and T. LITTLE. 1986. *Ford—The Men and the Machine.* Brown & Company.

MCELROY, J. 1995. "Industry Report Card: Does it Get Any Better Than This." *Automotive Industries,* Vol. 175, No. 4, April, pp. 48–58.

MORTON, M.S.S. 1989. "The Impact of Information Technology on Organizations in 1990's," *Siemens Review,* Vol. 3, pp. 36–41.

PRASAD, B. 1994. "Modern Manufacturing." Chapter 1 in *Modern Manufacturing: Information Control & Technology.* ed. M. Zaremba and B. Prasad. Springer Verlag: London, pp. 1–15.

PUTNUM, A. O. 1985. "A Redesign for Engineering." *Harvard Business Review.* May–June, Vol. 63, No. 3, pp. 139–144.

RASMUS, D. 1993 "Learning the Waltz of Synthesis," Manufacturing Systems, Vol. 11, No. 6, pp. 16–23.

ROSS, E. M. 1992. CALS: Enabling the New Manufacturing Paradigm. *CALS Journal.* Winter, pp. 29–37.

SHINA, S. G. 1991. *"Concurrent Engineering: New Rules for World-class Companies," IEEE Spectrum,* Vol. 28, No. 7 (July), pp. 22–24.

WHEELRIGHT, S. C., and H. HAYES. 1985. "Competing Through Manufacturing." *Harvard Business Review.* Vol. 63, No. 1 (Jan–Feb), pp. 99–108.

ZIEMKE, M. C., and M. S. SPANN. 1991. "Warning: Don't be Half-hearted in Your Efforts to Employ Concurrent Engineering," *Industrial Engineering,* Vol. 23, No. 2, pp. 45–49.

TEST PROBLEMS: MANUFACTURING COMPETITIVENESS

1.1. Since the days of Henry Ford, manufacturing has responded to technological changes that are responsible for the various technical innovations. What factors are attributable for having brought on these changes?

1.2. Summarize the changing pattern in manufacturing with time. What were the five notable reaches that gave corporations significant competitive advantages? How does this relate to evolution of the engineers' function?

1.3. What are the major historical highlights that have occurred in manufacturing?

1.4. What is the significance of the new paradigm? Why is it so important in today's competitive market to shift paradigm?

1.5. What is the basic premise of manufacturing? Why it is important to consider case histories and their application in reaching consensus and launching new projects?

1.6. Cite examples where the balancing scale analogy (as shown Figure 1.7) may not be a good match for some elements of manufacturing competitiveness. What are those elements? Discuss why.

1.7. What are the independent measures of performance in a manufacturing unit? Outline an alternative approach to compare performance indicators than Figure 1.11? Identify the pros and cons of the two alternatives.

1.8. What is the concept of quality improvement? Why is one tempted to compare the best-in-class based on results of an international benchmark study?

1.9. How does push or pull affect the manufacturing environment? What are the differences between a "push" and a "pull" for a new paradigm? Describe their relative affinities.

1.10. Explain manufacturers, who are pushed to acquire emerging technology may not always be globally successful.

1.11. Explain the visible elements that constitute measures of competitiveness. Can some of them be invisible, if so when?

1.12. Explain the invisible elements that constitute the measures of competitiveness. Can some of them be visible, if so when?

1.13. Describe the emerging trends that you are seeing in the marketplace of the 1990s. Describe these changing patterns (in many economic measures or sectors) in a form of behavior curves with time.

1.14. Explain why a manufacturer should strive to push product and process reengineering even if competitiveness is low.

1.15. Why is a ratio of outputs to inputs not a suitable measure of productivity? What could be a suitable measure in its place? How do customer demands affect throughput time?

1.16. What was the true motivation behind the departmentalization of industry during the late 1970s?

1.17. How can a company get out of the rut of "quality-by-inspection" mode to some more effective alternatives: "built-in quality," TQM, etc.?

1.18. Does a consumer pull for a technology, functionality, or a features? What are the factors governing that pull?

1.19. Why was there no incentive to make a world-class quality product in 1960–1970? Why did the pre-World War II era have a secured market environment?

1.20. How would one go about measuring the level of competitiveness of a business unit/operation? What are the factors that need to be maximized and what needs to be minimized? How do you show them on the spider chart or a polar diagram?

1.21. Clarify differences between cost-to-correct-quality versus cost-to-ensure-quality.

1.22. Explain the concept of balancing in manufacturing. What do you understand by robust technology development?

1.23. How do you measure competitiveness? What are some of the examples of competing manufacturing strategies?

1.24. Describe some of the cost-cutting opportunities: (a) prevention type (b) inspection type.

1.25. How long will American manufacturers take to catch up to the Japanese car producers, assuming the rate of improvements in productivity and quality stays at the current pace (refer to Figure 1.13)?

1.26. What are the major factors that determine the nature of an enterprise business operation and distinguish them from their competitors?

1.27. Describe a set of seventeen lean production (or synchronous) tactics. How do these differ from what you perceive an ideal CE organization needs to have?

CHAPTER 2

LIFE-CYCLE MANAGEMENT

INTRODUCTION

The combination of new and old practices, such as old-fashioned habits, new life-cycle environment, changes, and mounting regulations, has increased the complexity of the product development efforts. The complexity results from five main sources: inherent product complexity, process complexity, team cooperation and communication complexity, computer and network complexity, and a maze of specifications including international regulations and safety. Over the past several years the diversity, variety, and complexity of *new product introduction (NPI)* has grown multifolds from *"very simple"* to *"very complex,"* while at the same time, the *time to market* aspect has shrunk [Prasad, 1994]. This is shown in Figure 2.1. The changing market conditions (such as global manufacturing, economy, and new innovation) and international competitiveness are making the *time-to-market* a fast shrinking target. Today, an automobile with complexity several times higher than before can be manufactured in less time (often less than three years). The same product, about half a decade ago, used to take over 5 years to bring into the marketplace. Whereas, its complexity 10 years ago, by today's standards, could be characterized only as *"very simple."* The workstation market is another good example. With new innovation in chip technology, workstation companies have continually shortened the time between new product introductions. In 1985, when a new central processing unit (CPU) was introduced, it was quite innovative—but was no where close to today's standard in complexity. Every eighteen months thereafter, new CPU, twice as complex was introduced at two times the performance at roughly half the price. In 1988, a four times as complex and four times faster CPU was introduced at quarter of the price in a twelve month period. In 1990, the development cycle for a new sixteen times faster CPU was in-

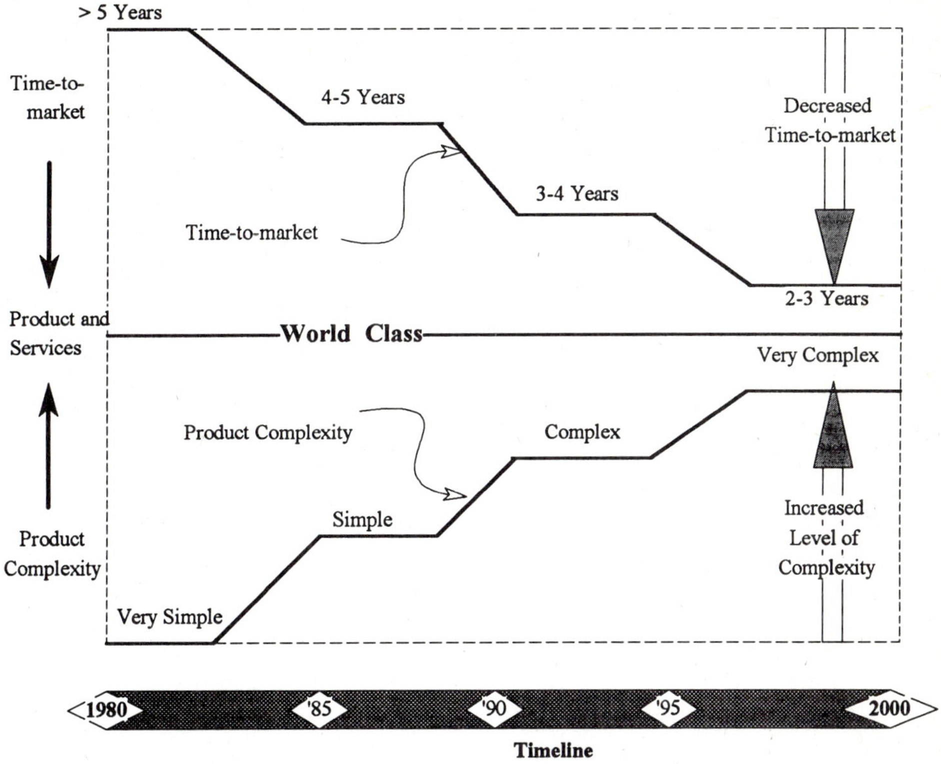

FIGURE 2.1 A Case of a Constantly Moving Target

troduced in only a six month time span nearly at 1/16 of its 1985 price. The trend goes on. The average development time for a compact disc (CD) player today is nine months, a PC is fourteen months, a knowledge-based engineering (software development) system ranges from two to three years. Among the web of such complexity, it is easy to overlook that requirements of the customers are also constantly changing. The customer is also becoming more sophisticated. Each time a company fulfills the customer's wants in a product, the level of customer expectation moves up a notch. They demand customized products more closely targeted to their personal, social, and cultural tastes. The same is true for the expectations of the performance indicators discussed in Section 1.6 and depicted in Figure 1.11. Products get old quickly—customer's excitement fades away, and demands decline. There is a great danger that a product introduced, after few years of its development, may not remain attractive for the market that existed at the launch time. Introducing new products at frequent intervals is also not a good business solution. New products require significant investments in redesign, retooling, and manufacturing costs. Development costs consist mostly of expenditures for staff and testing. These costs tend to increase proportionally with the overall time taken to complete the design. For this reason, most manufacturers have focused on shortening the time taken for new models to be designed and tested. Toyota, for example has set its sights on reducing the average development time of its automobiles from thirty months to eighteen months by 1996–97. US Department of Defense (DOD) Computer-Aided Acquisition and Logistics Support (CALS) initiative identifies CE as an enabling technology that can help potentially lower development and operational costs while appropriately managing the moving targets.

2.1 SHRINKING LIFE-CYCLE

There are many ways one can describe life-cycle effort. Terms related to time include useful-life span; lead time; art-to-part time; design and development time; launch-to-finish time, life-cycle time, etc. The literature has used these terms quite loosely [Thomas, 1990: Alting, 1993]. Figure 2.2 describes these terms so that their meanings are uniform and consistent throughout this book. The two terms that are widely used are lead time and life-cycle time. Lead time is the time required to finish one unit of a product, which could be an operation or a service. Since a product is made out of several units (subsystems, components, parts, etc.) there are many possible units of lead times. Figure 2.3 gives an example of a lead time for a machining operation. The machining operation is shown broken up further into smaller activities (for instance load/unload time, set-up time, process/operation time, move/transfer time, rework time, storage time, etc.) to facilitate accurate computation. Life-cycle time is the total elapsed time. It is the total time a product takes from cradle (the time when the initial idea was born) to grave (that is until parts of the product are recycled). The immediate effect of changing requirements and players keeping the process intact is to lengthen product development efforts. As the product design cycles stretch out, costs mushroom and quality suffers. The real pressure to reduce development costs and life-cycle time also comes from overseas competition. Not too long ago, mechanical typewriters had a 30-year useful life-span, and electromechanical typewriters over a 10-year lifespan. They were both quickly replaced by word-processors and personal computers. Development time and cost are becoming crucial in all engineering industries. It is becoming particularly serious in electronics industries where profits have been squeezed the most over the last decade. For example, the development life-cycle (when pay-off or returns-on-investment start coming in) of audio/video products, such as compact disc players and VCRs, is now less than a year (close to nine months) whereas the average useful life-span when someone replaces one, already in use or broken, has gone down to about 4–5 years. Figure 2.4 shows such trends (average) in useful-life span and development life-cycle time of products across a number of key competitive manufacturing industries.

$$\begin{aligned}\text{Elapsed Time} < \Sigma\ (&\text{load/unload time} + \text{set-up time} + \text{process/operation time}\\ &+ \text{move/transfer time} + \text{rework time} + \text{storage time}\\ &+ \text{delay/wait/idle time} + \text{certification/inspection time})\end{aligned} \quad (2.1)$$

The pay-off period kicks in when the product development life-cycle time ends. It continues until the product remains in use. The hatched area in Figure 2.4, thus, represents a time period during which the company reaps maximum profits. This is referred to as the "period of profitability". The period of profitability changes from industry to industry and

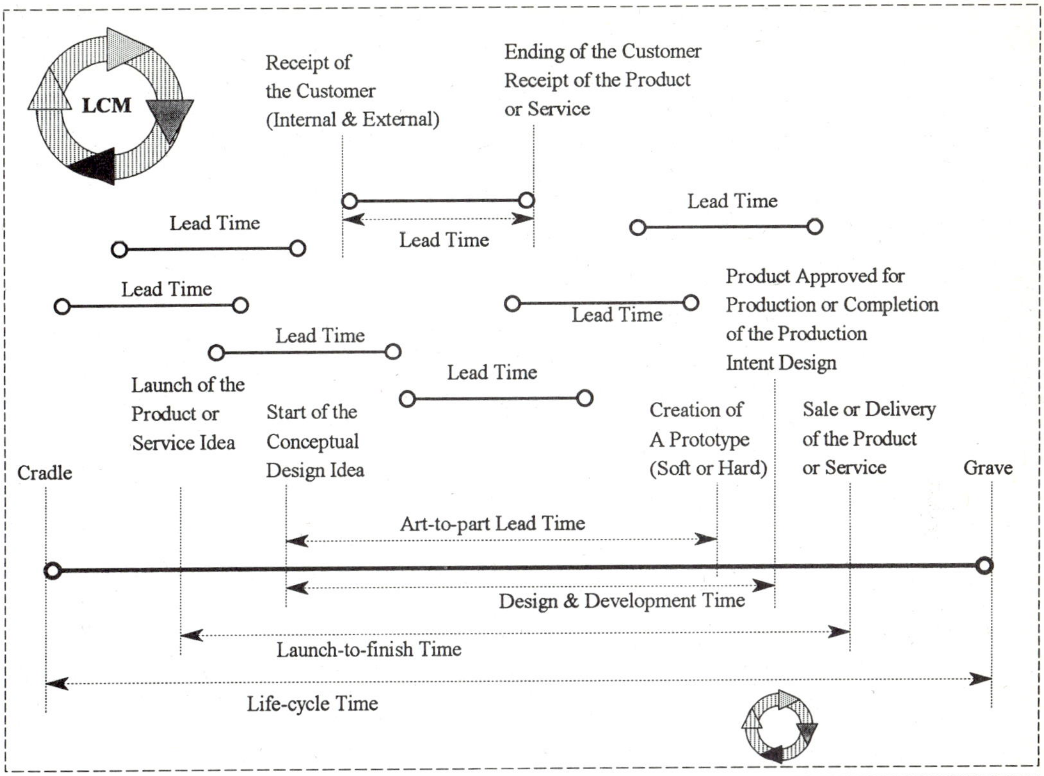

LEGEND

Useful-life:	Duration of time in which product is in use in the field by the customer
Lead Time:	Total time required to complete one unit of a product or service
Art-to-part Lead Time:	Design and development of a prototype part
Design and Development Time:	Time it takes from the start of the conceptual idea to the creation of the production intent design
Launch-to-finish Time:	Duration it takes from the time product is first launched until the sale or delivery of the product or service
Life-cycle Time:	Time it takes from cradle to grave -- until the part is recycled.

FIGURE 2.2 Terms Describing Life-cycle Efforts

from product to product. It is the lowest for consumer electronic industries and for computer products.

The global marketplace of the 1990s has shown no sympathy to tradition. The marketplace only recognizes results and is insensitive to efforts [Thomas, 1990]. Customers only appreciate what features they find useful in the products, they do not care how we got there. The reality is that if the products manufactured do not meet the market needs, demand declines and profit shrinks. As profit margins dwindle, so does the window of op-

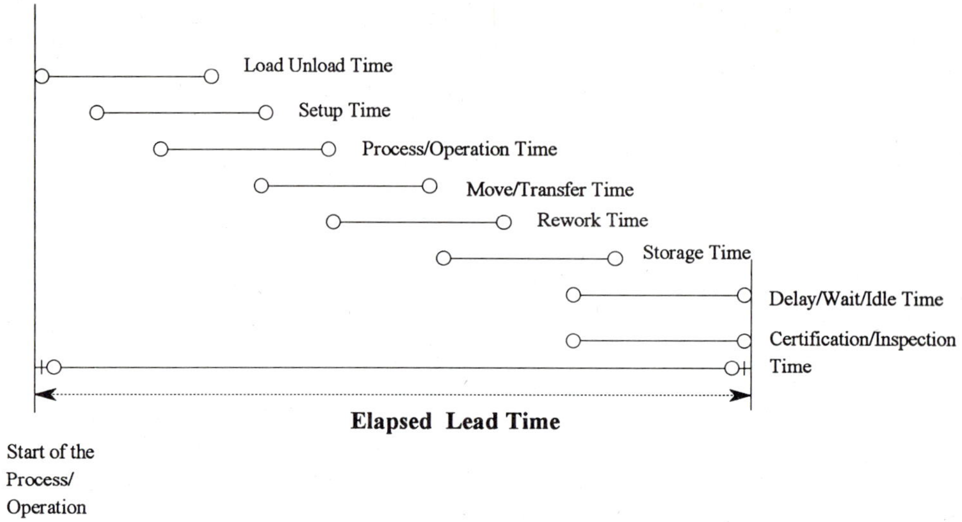

Elapsed Lead Time = < Σ (Load/Unload Time, Set-up Time,
Process/operation Time, Move Transfer Time, Rework time,
Storage Time, ...)

FIGURE 2.3 Unit Lead Time for a Machining Operation (an Example)

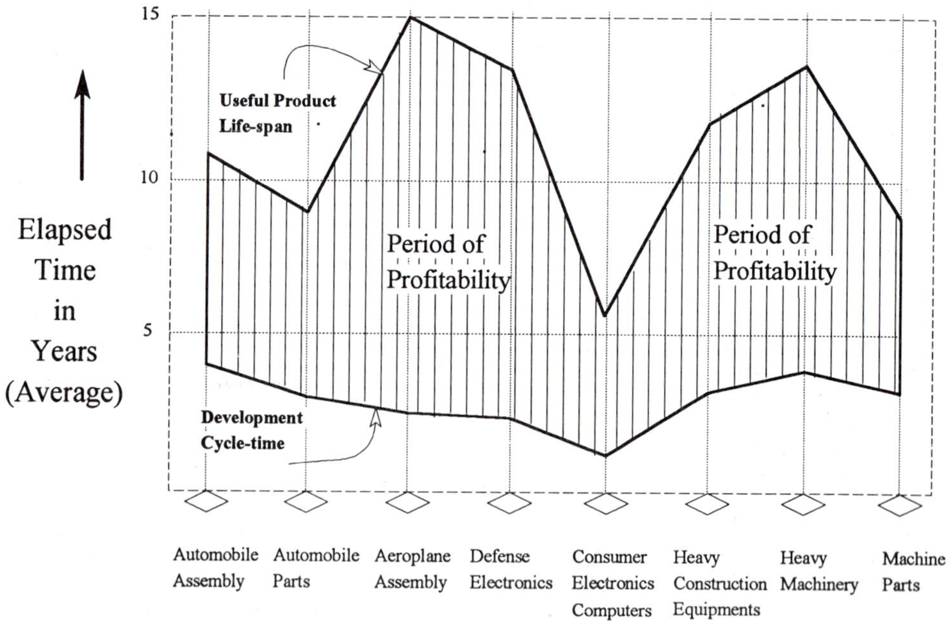

FIGURE 2.4 An Industry-wide Trend Showing Useful Product Life-span and Start of the Pay-off Period

portunity for the company to profitably change. Furthermore, suppliers, subcontractors, and partners feel the squeeze as their clients begin to cut costs and reduce time-to-market. The 1990's are not the first time the importance of time has been recognized. In the early eighties, manufacturers (predominately in Japan) had successfully developed a production technique for "assembly-oriented plants" to supply the component parts on a "just-in-time" basis. This technique was clearly one of the first to emphasize "time" in its orientation.

Today, most companies are under extreme pressure to develop products within time periods that are rapidly shrinking. As the market changes, so do the requirements. This is more pronounced if the products are consumer-based. For instance, the product that a consumer wants today, may not be liked when delivered three years from now. Associated with this are the urgencies and pressures on the manufacturers to modify their product characteristics based on the up-to-date requirements, while the product is still being developed. This has chilling effects in managing the complexity of such continuously varying product specifications and handling the ongoing changes. This is because it takes extensive time and effort to propagate the specifications throughout the product design and development (PDD) process and then turn them into opportunities for growth and profits. The ongoing success of an organization lies in its ability to do the following:

- Continue to evolve [Alting, 1993]
- Quickly react to changing requirements
- Reinvent itself on a regular basis
- Keep up with ever changing technology and innovation

Many companies are stepping up the pace of new product introduction, and are constantly learning and bracing for new ways of engineering products more correctly the first time, and more often thereafter.

2.1.1 Refocusing Product Development Efforts

A question often asked is "if the design process is to be changed, what would be an appropriate approach to product design?" The answer is not very difficult. Figure 2.5 illustrates the pattern of resources that were spent in a product's definition, design, and redesign phases for both British and Japanese companies [Wilson and Greaves, 1990]. The British pattern (mirrored in the USA) is one where meager resources are committed to the definition phase (17%), compared to what is ultimately spent in the redesign phase (50%). For example, some firms take people off projects/tasks that are just starting up and move them to projects/activities that are already in late. They feel that since the projects/tasks are just starting, taking resources away perhaps will not cause problems. They don't understand that the mess they were in could be the result of their own creations. The tasks were late because they made careless decisions in the starting stages. The result is that they are always in a fire-fighting mode; there is never time to do less important things because there is always something urgent occurring. To achieve a comparable level of quality, the

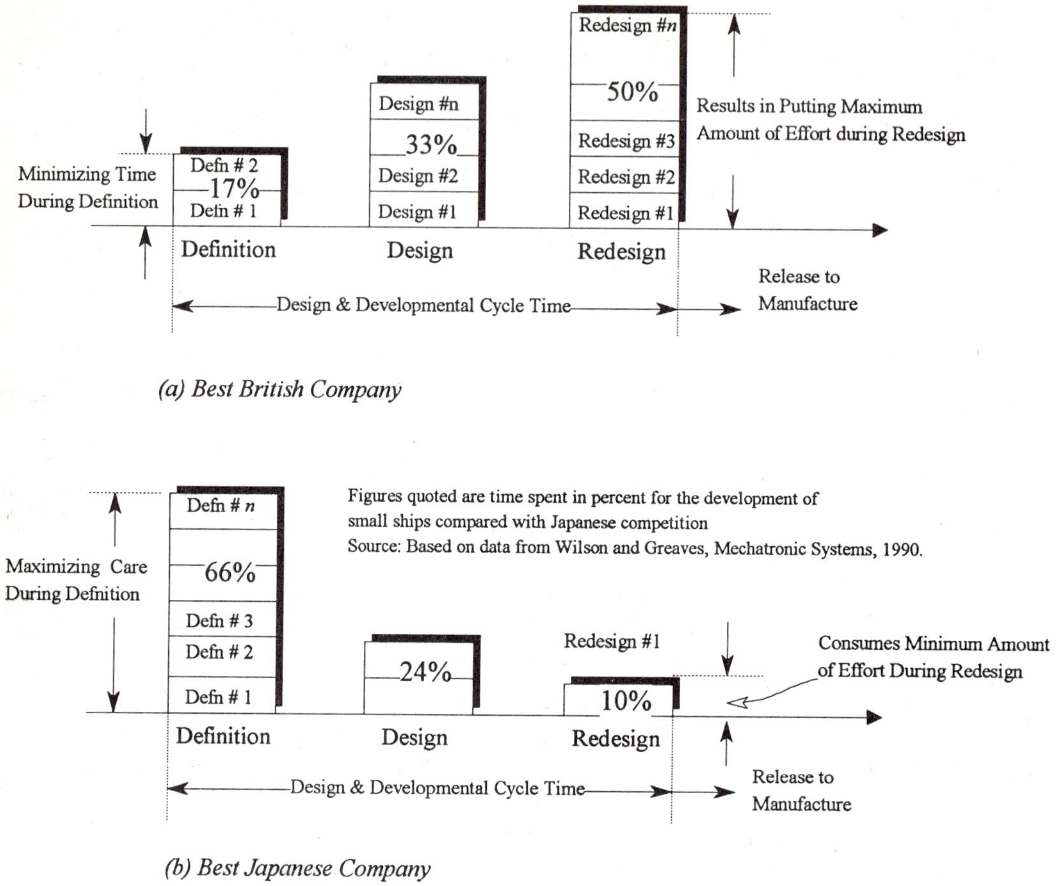

FIGURE 2.5 Distribution of Product Development Effort (a) Best British Company (b) Best Japanese Company

Japanese do things in reverse priority. They focus strongly on product definition supported by optimizing techniques during design (a hefty 66% of the effort is spent here). This results in getting the product design correct the first time, thus reducing the need for any extensive redesign. The average time spent by the Japanese in the redesign phase is relatively very small—somewhere in the 10% range.

In Figure 2.5 the percentage quoted in man-hours is for a British company designing small ships and that of a Japanese counterpart. In the British company case, the penalty was further compounded by the cost when another similar vessel was ordered. The incremental design cost for a second Japanese vessel was almost negligible (10%) but that for the British vessel was five times more (close to 50%) of the man-hours of the total effort [Wilson and Greaves, 1990]. Today, the relative gap, however, is closing. Some U.S. manufacturers have achieved impressive results. Chrysler brought the Dodge viper automobile to market in a 3-year development cycle time, significantly breaking previous 5-year standards.

The difference between the Japanese and British approach thus boils down to two main points:

1. Difference in life-cycle management methodology, which is discussed in this chapter.
2. Effectiveness with which life-cycle management is practiced; those who are able to make sound decisions during the early life-cycle will win the biggest competitiveness and profitability prize. This is discussed next.

2.1.2 Impact of Early Decision Making

The impact of early life-cycle decisions on product realization is far reaching. This is shown in Figure 2.6 by two exponential and two algorithmic cost functions: (a) relative cost of introducing a change, (b) degree of control to activate a change. The graph is based on data reported by the market research firm Dataquest Inc. The horizontal axis shows start-points of a seven-stage process. Depending on the product, the stages may be labeled differently and carried out either in a serial or a parallel mode. In Figure 2.6, for a concurrent process, only the portion of a stage (from the start-point of a current stage to a start-point of a next stage) which is not overlapped is shown. Notice how rapidly cost of

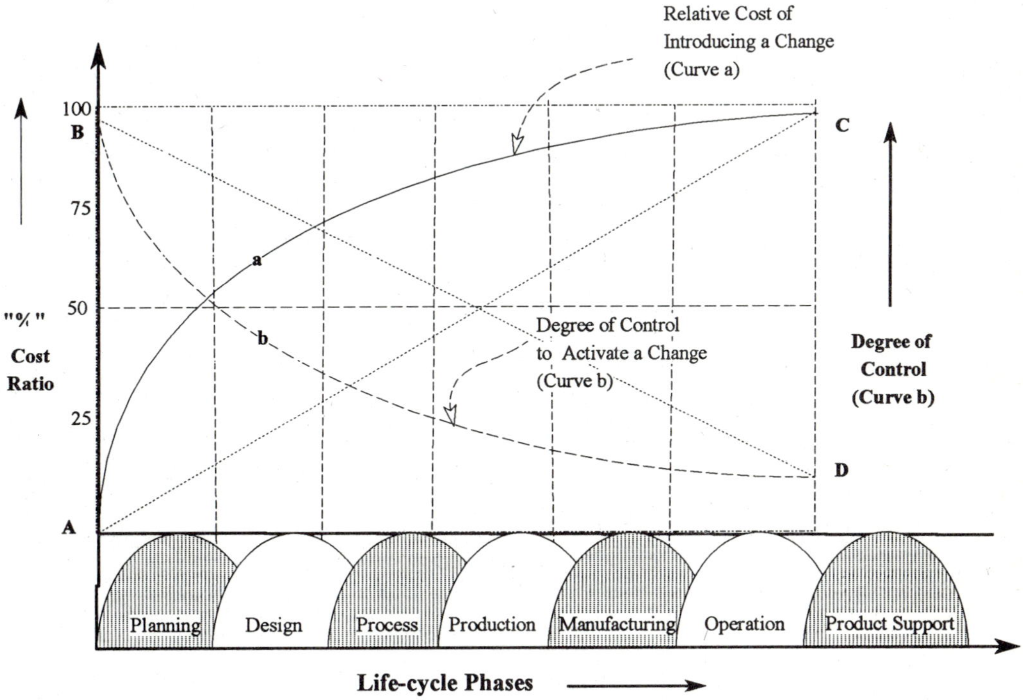

Source: Based on Dataquest Inc Data and an article published in. Business Week, New York: McGraw-Hill, April 30, 1990, p. 110.

FIGURE 2.6 (a) Relative cost of Introducing a Change (b) Degree of Control to Activate a Change

change accelerates if it is not made during initial stages. If "C" is the normalized cost, then the two cost functions are as follows:

$$C_a = \alpha_a + \beta_a \iota_n (\mu_a\, t/T + 1) \text{ (logarithmic curve a) } 0 \leq (t/T) \leq 1;\ \mu_a > 0 \tag{2.2}$$

or

$$C_a = \alpha_a + \beta_a\, e^{-(\mu_a\, t/T)} \qquad \text{(Exponential curve a) } 0 \leq (t/T) \leq 1;\ \mu_a > 0 \tag{2.3}$$

$$C_b = \alpha_b - \beta_b \iota_n (\mu_b\, t/T + 1) \text{ (Logarithmic curve b) } 0 \leq (t/T) \leq 1;\ \mu_b > 0 \tag{2.4}$$

and

$$C_b = \alpha_b - \beta_b\, e^{-(\mu_b\, t/T)} \qquad \text{(Exponential curve b) } 0 \leq (t/T) \leq 1;\ \mu_b > 0 \tag{2.5}$$

where

C_a—cost function is for curve *a* [Equations (2.1) and (2.2)]
C_b—cost function is for curve *b* [Equations (2.4) and (2.5)].
The coefficients α_a, β_a, and μ_a; and α_b, β_b, and μ_b are constants.
The time attribute under consideration ranges over the interval $0 < (t/T) < 1$.
t—the time lapsed after the launch of the product or service idea
T—the total life-cycle time interval.

In the above equations (2.2) through (2.5), μ (determines the shape of the cost function.

2.1.2.1 Exponential Cost Function (Curve a)

The constants α_a, β_a can be chosen to normalize the cost function in Equation 2.3, such that

$$C_a\,(t/T = 0) = 0 \text{ and } C_a\,(t/T = 1) = 100. \tag{2.6}$$

The normalization results in the constants α_a, β_a being selected so that

$$\alpha_a + \beta_a = 0 \tag{2.7}$$

$$\alpha_a + \beta_a\, e^{(-\mu_a)} = 100 \tag{2.8}$$

As a result,

$$\alpha_a = \frac{100}{[1 - e^{(-\mu_a)}]} \tag{2.9}$$

and

$$\beta_a = \frac{100}{[e^{(-\mu_a)} - 1]} \tag{2.10}$$

2.1.2.2 Exponential Cost Function (Curve b)

Similarly, the constants α_b, β_b can be chosen to normalize the cost function in Equation 2.5, such that

$$C_b\,(t/T = 0) = 100$$
$$C_b\,(t/T = 1) = 0 \tag{2.11}$$

The normalization results in the constants α_b, β_b being selected so that

$$\alpha_b - \beta_b = 100 \tag{2.12}$$

$$\alpha_b - \beta_b\, e^{(-\mu_b)} = 0 \tag{2.13}$$

As a result,

$$\alpha_b = \frac{100e^{(-\mu_b)}}{[e^{(-\mu_b)} - 1]} \tag{2.14}$$

and

$$\beta_b = \frac{100}{[e^{(-\mu_b)} - 1]} \tag{2.15}$$

2.1.2.3 Logarithmic Cost Function (Curve a)

The constants α_a, β_a can be chosen to normalize the cost function in Equation 2.2, such that

$$C_a\,(t/T = 0) = 0$$
$$C_a\,(t/T = 1) = 100 \tag{2.16}$$

The normalization results in the constants α_a, β_a being selected so that

$$\alpha_a = 0 \tag{2.17}$$

$$\alpha_a + \beta_a\, \iota_n\,(\mu_a + 1) = 100 \tag{2.18}$$

As a result

$$\alpha_a = 0$$

and

$$\beta_a = \frac{100}{[\iota_n(\mu_a + 1]} \tag{2.19}$$

TABLE 2.1 Table of Constants for Exponential Curves "a" through "f"

			Exponential Cost Function			
Constants	*Curve a*	*Curve b*	*Curve c*	*Curve d*	*Curve e*	*Curve f*
$\alpha =$	$\frac{100}{[1-e^{(-\mu_a)}]}$	$\frac{100\,e^{(-\mu_b)}}{[e^{(-\mu_b)}-1]}$	$\frac{100}{[1-e^{(\mu_c)}]}$	$\frac{100}{[1-e^{(-\mu_d)}]}$	Same as Curve a	Same as Curve b
$\beta =$	$\frac{100}{[e^{(-\mu_a)}-1]}$	$\frac{100}{[e^{(-\mu_b)}-1]}$	$\frac{100}{[e^{(\mu_c)}-1]}$	$\frac{100}{[e^{(-\mu_d)}-1]}$	Same as Curve a	Same as Curve b
$\mu =$	Shape Parameters (μ_a)	Shape Parameters (μ_b)	Shape Parameters (μ_c)	Shape Parameters (μ_d)	Shape Parameters (μ_e)	Shape Parameters (μ_f)

2.1.2.4 Logarithmic Cost Function (Curve b)

The constants α_b, β_b can be chosen to normalize the cost function in Equation 2.4, such that

$$C_b\ (t/T = 0) = 100$$
$$C_b\ (t/T = 1) = 0 \qquad (2.20)$$

The normalization results in the constants α_b, β_b being selected so that

$$\alpha_b = 100$$
$$\alpha_b - \beta_b\ [l_n\ (\mu_b + 1)] = 0 \qquad (2.21)$$

As a result

$$\alpha_b = 100 \qquad (2.22)$$

and

$$\beta_b = \frac{100}{[l_n(\mu_b + 1)]} \qquad (2.23)$$

If μ_a or μ_b tends to zero in the above equations, the slope changes to a straight line between AC and BD, respectively. Depending on the magnitude of the constants μ_a or μ_b, the shape of the two curves will vary.

Figure 2.7 illustrates the difference between the actual (incurred) cost and the future cost commitment (the locked in cost of future spending at that stage of development). Two curves are shown: curve "c" represents the actual costs incurred at any stage; curve "d" is the future cost that is committed at any stage due to decisions made prior to this stage. They are based on data reported by Computer-Aided Manufacturing (CAM) International Inc. In a separate investigation, Andreason, et. al. also reports a very similar distribution of the operating costs incurred by various departments [Andreason, Myrup, and Hein, 1987]. This is represented in Figure 2.8 by a pie chart. Clearly the design is a tiny piece of the development pie, but it locks in a bulk of the later (in downstream processes) spending. Notice the parabolic nature of the behavior:

$$C_c = \alpha_c + \beta_c * (t/T)^{(1/\mu_c)} \quad \text{(Parabolic Curve c)} \quad 0 \le (t/T) \le 1;\ \mu_c \ge 2 \qquad (2.24)$$

TABLE 2.2 Table of Constants for Logarithmic and Parabolic Curves "a" through "f"

	Logarithmic Cost Function		Parabolic Cost Function		Logarithmic Cost Function	
Constants	*Curve a*	*Curve b*	*Curve c*	*Curve d*	*Curve e*	*Curve f*
$\alpha =$	0	100	0	0	0	100
$\beta =$	$\frac{100}{[l_n(\mu_a+1)]}$	$\frac{100}{[l_n(\mu_b+1)]}$	100	100	$\frac{100}{[l_n(\mu_e+1)]}$	$\frac{100}{[l_n(\mu_f+1)]}$
$\mu =$	Shape Parameters (μ_a)	Shape Parameters (μ_b)	Shape Parameters (μ_c)	Shape Parameters (μ_d)	Shape Parameters (μ_e)	Shape Parameters (μ_f)

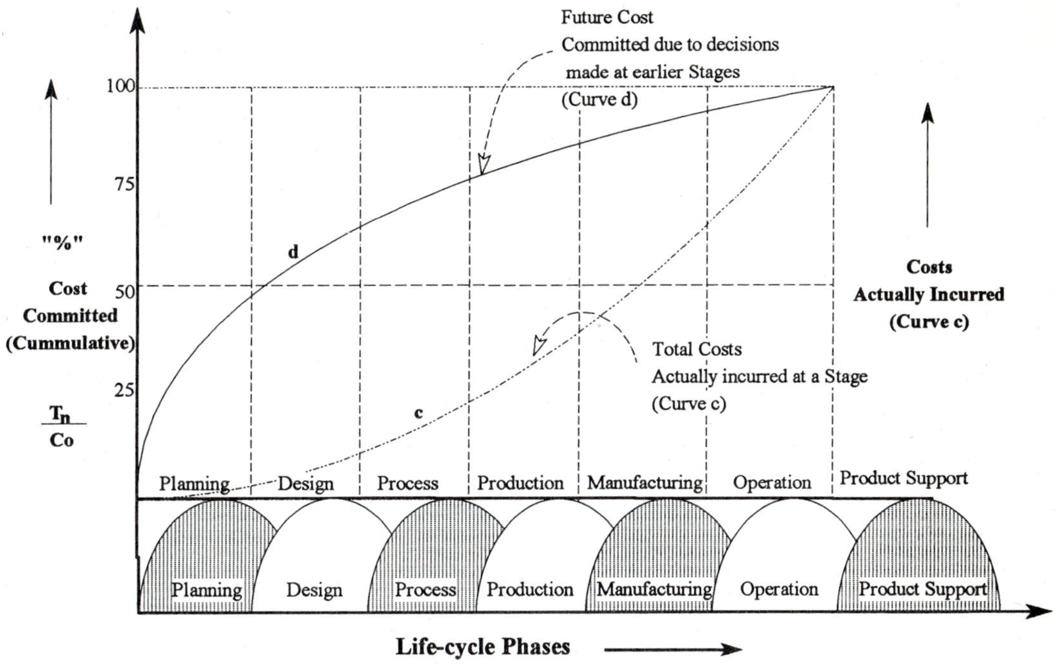

Source: Based on Computer-aided Manufacturing International Data and an article in Business Week, McGraw-Hill Publication, April 30, 1990, p. 110.

FIGURE 2.7 (c) Total Cost committed at any stage (d) Total Cost actually incurred at any stage

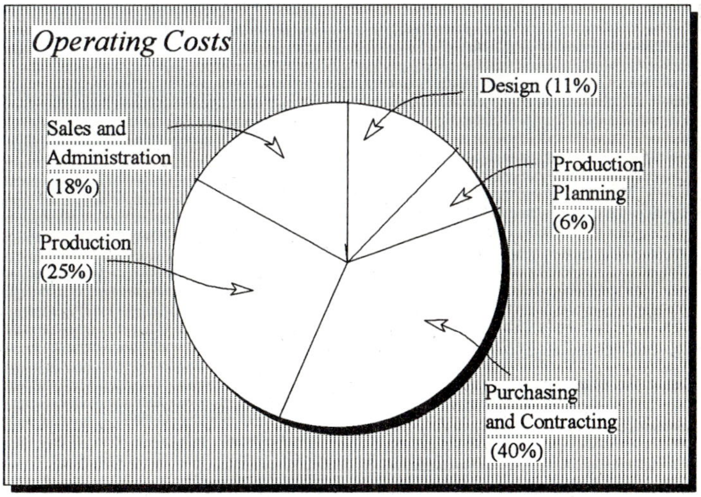

FIGURE 2.8 Percentage of Actual Operating Costs Incurred by Various Departments

or

$$C_c = \alpha_c + \beta_c\, e^{+(\mu_c\, t/T)} \qquad \text{(Exponential Curve c)} \quad 0 \le (t/T) \le 1;\ \mu_c > 0 \tag{2.25}$$

and

$$C_d = \alpha_d + \beta_d * (t/T)^{(\mu_d)} \qquad \text{(Parabolic Curve d)} \quad 0 \le (t/T) \le 1;\ \mu_d \ge 2 \tag{2.26}$$

or

$$C_d = \alpha_d + \beta_d\, e^{-(\mu_d\, t/T)} \qquad \text{(Exponential Curve d)} \quad 0 \le (t/T) \le 1;\ \mu_d > 0 \tag{2.27}$$

Where, constants α_c, α_d, and β_c, β_d can be chosen to normalize the cost functions (c) and (d) as follows:

$$C_c\,(t/T = 0) = 0$$

and

$$C_c\,(t/T = 1) = 100 \tag{2.28}$$

$$C_d\,(t/T = 0) = 0$$

and

$$C_d\,(t/T = 1) = 100 \tag{2.29}$$

2.1.2.5 Parabolic Cost Functions

The normalization results into the constants α_c, α_d, and β_c, β_d being selected so that

$$\alpha_c = 0$$

$$\beta_c = 100 \tag{2.30}$$

and

$$\alpha_d = 0$$

$$\beta_d = 100 \tag{2.31}$$

The user-defined variables μ_c and μ_d determine the shape (curvature) of the parabola. The two parabolic curves can be altered slightly by introducing a constant χ as follows:

$$C_c = 100 * (\chi t/T)^{(1/\mu_c)}$$

for $0.95 < \chi < 1.05$ (Parabolic Curve c) (2.32)

and

$$C_d = 100 * (\chi t/T)^{(\mu_d)}$$

for $0.95 < \chi < 1.05$ ((Parabolic Curve d) (2.33)

$\mu_c = 2$, $\mu_d = 2$, and $\chi = 1$ give a true parabolic curve, while any other value of μ_c, μ_d or χ (say $0.9 < \chi < 1.1$) gives a curve which is offset from the 100% datum line. One variation of the "curve c" is contended by CAM International, when the cost of any stage is a constant multiple of its preceding cost. If r is a constant ratio and C_0 is the cost of the initial stage (the zero stage), and C_n is the cost of the nth stage, then by definition;

$$r = \frac{C_1}{C_0} = \frac{C_2}{C_1} = \frac{C_3}{C_2} = \frac{C_4}{C_3} = \frac{C_5}{C_4} = \frac{C_6}{C_5} = \frac{\cdots}{\cdots} = \frac{\cdots}{\cdots} = \frac{C_n}{C_{n-1}} \tag{2.34}$$

The cost of the first, second, and third stage would be

$$C_1 = C_0\,(r)$$
$$C_2 = C_1\,(r) = C_0\,(r)^2$$
$$C_3 = C_2\,(r) = C_0\,(r)^3$$

The cost of the nth stage would be

$$C_n = C_0\,(r)^n \tag{2.35}$$

and the corresponding cost function $C_n\,(t/T)$ can be written as

$$C_n\,(t/T) = C_0\,(r)^{n(t/T)} \tag{2.36}$$

The total cost through the nth stage can be obtained by taking an integral of Equation (2.36) from

$$(t/T = 0) \text{ to } (t/T = 1)$$

In Figure 2.7, a sixth stage process is shown through a chain line.

The total cost incurred up to a nth stage (including the zero stage) for this process can be approximated using the Simpson's Rule for summation, that is

$$\Sigma_n = C_0\,[\,1/2 + r^1 + r^2 + r^3 + r^4 + r^5 + r^6 + \ldots + (1/2)r^n]/n \tag{2.37}$$

The total cost incurred up to a sixth stage (including the zero stage) is

$$\Sigma_6 = C_0\,[\,1/2 + r^1 + r^2 + r^3 + r^4 + r^5 + 1/2r^6]/6 \tag{2.38}$$

If a cost of a new stage is ten times more than its previous stage cost, that is when

$$r = 10 \tag{2.39}$$

$$\Sigma_6 = 1.019 * C_0 * 10^5 \tag{2.40}$$

Thus, the total cost of a sixth stage process would be hundred thousand times larger than the stage zero cost.

As shown by Equation (2.40), the cost is directly proportional to the initial (zero stage) cost. The first stage is an ideal place for new ideas, new concepts, new equipment, new processes, new technology, etc., to avoid possible changes in direction and back-tracking at later stages.

It has been reported [Patton, 1980] that 70% of the total cost of manufacturing a product is committed in the first few stages (conceptual formulation stage). Since the actual time and expense in product development during this initial stage is low (10%–20%), any changes introduced at this point cost very little but can greatly influence the subsequent costs of the production [Nevins and Whitney, 1989]. On the contrary, if the changes are made during the later stages, such as manufacturing planning of the part, only 10%–20% of the product cost is affected. Most people in many companies do not realize this fact. They start too late looking for the source of the problems and end up spending too much time and money in "fixing" the problems at a wrong place. In reality they end up fixing only the symptom of the problem. The "real" fix for a bad manufacturing process is not more of process control, SQC (quality control), or any similar controls on the factory floor, but the discovery and elimination of the source of the problems at the upstream stages, so that the redesigned process is insensitive to such variations.

It turns out that making few wrong decisions here and there at early stages, over a large number of tasks, may come out to be a more cost-effective way than being precise (correct all the times).

For example,

> sum of costs of canceling *N* tasks at 25% completion stage (if in doubt) plus associated penalty of making an unknowingly wrong decision (to cancel them), is expected to be
>
> <<< (far less than)
>
> the sum of costs of canceling those *N* tasks at 75% completion stage. This is true even when the original decision to continue at 25% point was clearly wrong.

where, *N* could be any number of tasks, usually more than one. The inequalities between the two cost scenarios are more pronounced when *N* is small. In general, the penalty for canceling even a few tasks at 75% point is normally so large that it does not make sense to wait for availability of precise information. In other words, it does not pay to make decisions late in the PDD process, even though most decisions at that point are likely to be the right decisions.

A similar trend occurs for the cost incurred in fixing a mistake and for the amount of control one has at any stage (see Figure 2.9). A mistake committed and discovered during the planning and design phase is comparatively inexpensive to fix. However, if it is overlooked and discovered later during process engineering, such a mistake can cost manufacturers several thousand times more. By the time a mistake comes to actual manufacturing, for example, it could cost millions more to fix compared to what it would have cost if detected earlier. After a few initial stages, changes are expensive because the CAD model; the prototypes, the intent-definition, the DFX-checks, analyses, documentation, and the processing have all been completed or begun and these must be redone or modified.

Since a cost function

$$C\,(t/T) \equiv e^{\,\mu(t/T)} \tag{2.41}$$

This means,

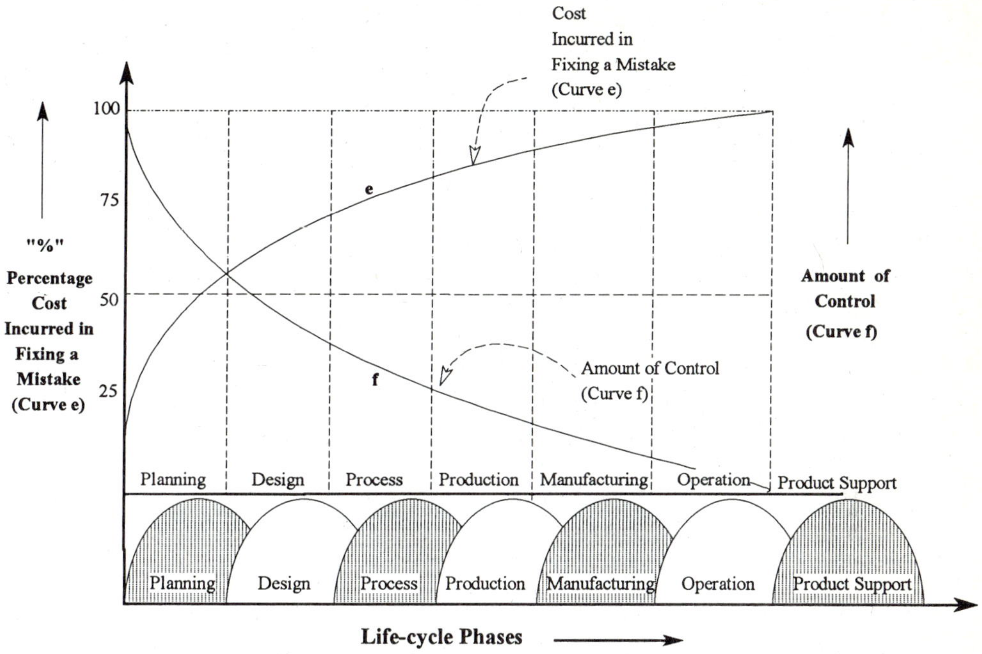

Source: Based on CAM International Data and an article published in. Business Week, New York: McGraw-Hill, April 30, 1990, p. 110.

FIGURE 2.9 (e) Cost Incurred in Fixing a Mistake and (f) the Amount of Control

$$\begin{aligned} C_{\text{planning}} &\propto e^{\mu(1/6)} \\ C_{\text{design}} &\propto e^{\mu(2/6)} \\ C_{\text{process}} &\propto e^{\mu(3/6)} \\ C_{\text{production}} &\propto e^{\mu(4/6)} \\ C_{\text{manufacturing}} &\propto e^{\mu(5/6)} \\ C_{\text{operation}} &\propto e^{\mu(6/6)} \end{aligned} \tag{2.42}$$

The above means that

$$\frac{C_{\text{operation}}}{C_{\text{manufacturing}}} = \frac{C_{\text{manufacturing}}}{C_{\text{production}}} = \frac{C_{\text{production}}}{C_{\text{process}}} = \frac{C_{\text{process}}}{C_{\text{design}}} = \frac{C_{\text{design}}}{C_{\text{planning}}} = \text{Constant} \tag{2.43}$$

Detecting and early fixing a design can save a lot of time which could otherwise result in material waste, additional planning time, design time, reprocessing, and lost time-to-market implied by the correction process. Normally, the actual cost of designing a typical product is a small percentage (10%–20%) of the total cost of the product. For heavy engineering products in the aerospace and defense industry, it could reach up to 40%–50%. Thus, there are dual disadvantages in delaying the decision making process—the cost becomes high and at the same time the degree of control is sharply reduced.

2.2 LIFE-CYCLE MANAGEMENT

Many progressive companies are interested in maintaining a competitive edge in the world market and in producing high quality products. They would like to maximize the life-cycle value of a product while containing costs and environmental burdens. These values, for example, include characteristics such as manufacturability, serviceability, recyclability, and other environmental issues. They would like to manufacture the product at a cost much lower than their competitors. Life-cycle management (LCM) is a process often used to accomplish these goals. LCM is actually a transformation process from a set of raw resources to a useful product, energy, or services that consumers want or intend to buy (see Figure 2.10). The resources may be present in the following forms:

- Material resources (such as water, wood, oil, minerals, etc.)
- Energy resources(such as chemical, nuclear, electrical, hydraulic etc.)
- In other logistic forms (such as capital, manpower, real estate, etc.)

There are three types of transformations that are usually present:

1. Product and process transformation that produces a useful product or unexpected scrap
2. Energy transformation that produces a useful energy and some unexpected energy waste
3. Seven Ts transformation that produces a value-added services and some wasted efforts.

LCM includes not only the effective conversion of the raw resources into useful outputs but also the management of the waste resulting from it. There are two types of waste: waste from the process of transformation and the consumers' waste that needs to be safely disposed or recycled.

To date, many companies view product realization as characterized by long lead times, a multitude of engineering changes, manufacturing complications, and ultimately heavy costs to satisfy the customer requirements. The number of engineering changes that occur in the best U.S. company is 40% to 60% more than the best Japanese company (see Figure 2.11). This is because in most U.S. companies, efficient decision making processes are lacking. They either limit the process to conventional "design review" or "red-team" meetings that inhibit free flow of information. These meetings serve no purpose but to postpone the decision from being made until after the meeting, or centralize the decision making authority in committees or hierarchical, tall-silos structure. For example, an engineer's choice of "design for X-ability" decision is often perceived as a functional service to be called upon periodically for incremental improvements in product quality, new product lead times, and costs. However, the perception is clearly different in successful engineering companies, where DFX is seen as a pervasive set of engineering activities that form the life blood of the CE cooperating teams. There, decision making steers the

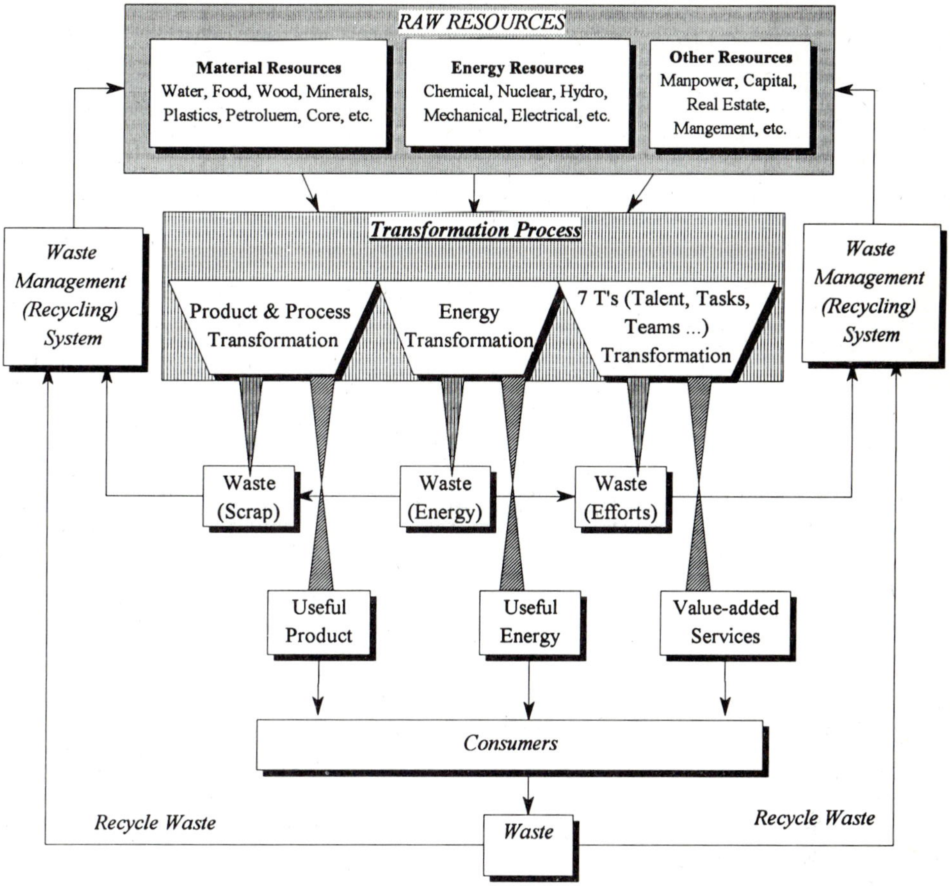

FIGURE 2.10 Scope of Life-cycle Management

PDD process. There, companies determine what subsystems, components, parts, . . . , to develop. Companies define a set of consistent product objectives with respect to company and customer goals, set priorities, and allocate resources.

Figure 2.12 compares the actual 1990 automobile production data between the United States and Japan. In all the five categories shown, U.S. production levels fall short. It takes 43% more design effort and about 21% more time to finish the design than it takes for the Japanese automobile company. At any time, the number of models in production for Japan is twice what the Americans seem to put through their own production system. The average replacement period per model for the Japanese is about 50% smaller, meaning they were able to replace two car lines in the same time period when Americans could do only one. The Japanese could manage to replace the car lines, even though their annual production was half of the American annual production volumes. This goes with what someone once said—"Japanese say adopt, then become adept, and only when there is nothing else to adopt, they adapt."

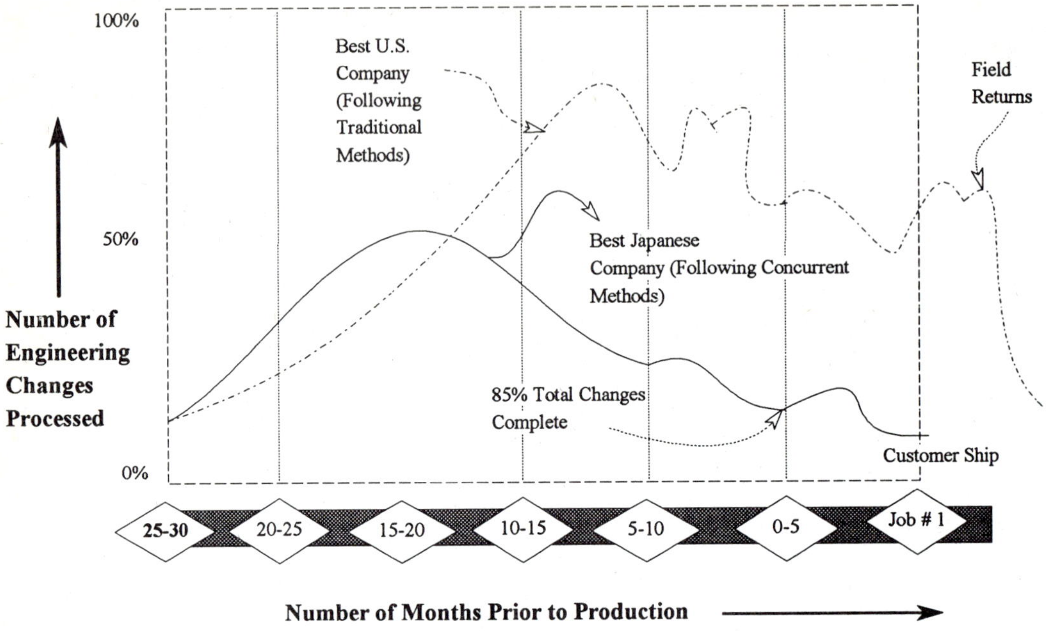

Source: Based on American Supplier Institute (L.P. Sullivan) Data

FIGURE 2.11 Number of Engineering Changes in Best US and Japanese Companies

Concept-to-delivery Life-cycle	**US (Japan) Data During 1990**	**Competitive Advantage Percentage**
Design Time per Model (in Months)	60 (47)	21.7%
Design Effort per Model (in Million Man Hours)	3.0 (1.7)	43.3%
Average Replacement Period per Model (in Years)	9.2 (4.2)	54.3%
Average Annual Production per Model (in Thousands)	230 (120)	47.8%
Model in Production (in Number)	36 (72)	50.0%

Source: Based on Data Published in The ECONOMIST, April 14, 1990.

FIGURE 2.12 US and Japanese Automobile Production Data

2.2.1 Risk Management

Development of an activity involves risk. As in serial engineering, collaborative concurrent design can also result in designs that have to be discarded. With the appropriate measures during product realization cycles, it is possible to minimize risks; whether they can be eliminated completely is debatable. If the problem is well understood, risk can be evaluated and managed. The old saying "no risk no gain" is very true in concurrent process as well. A low risk project is likely to succeed, but the associated rewards in terms of profitability are also usually very small. Rewards are proportional to the risk. The graph of rewards during various aspects of product development is shown in Figure 2.13. There are five segments to this curve corresponding to the challenges involved in bringing a competitive idea, a new feature, a function, or a product to the marketplace (Table 2.3).

There are "windows of rewards" for each of the above types of changes. The segment CDE window represents best returns on the investment but its associated risks are quite high (75% or higher). From Figure 2.13, it may be noted that the risks for copying an idea (in compliance with patent rights) from one's competitor product are higher than copying from one's own product and so are the rewards. Obviously, it is worth asking whether the return is large enough to take the associated risks in doing one or more of the

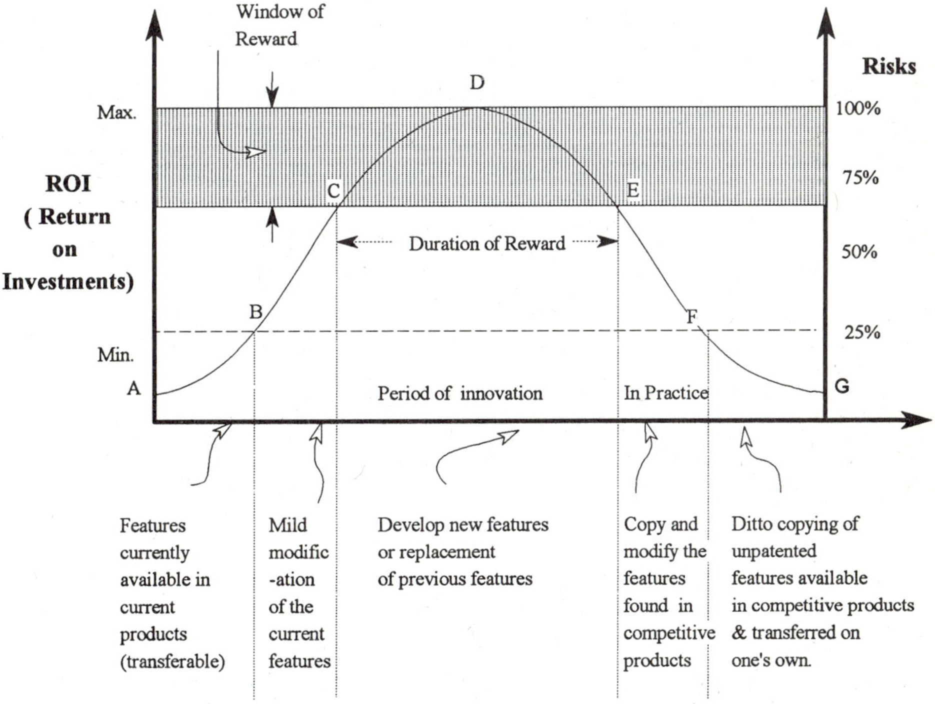

FIGURE 2.13 Risks and Rewards Associated with Various Aspects of Product Development

TABLE 2.3 Descriptions for the Segments of the Risks and Rewards Curve

Part of the Curve	Descriptions
Segment AB:	An idea or a feature that is currently available in the company's product. It is being transferred to a new product.
Segment BC:	An idea or a feature is not only transferred, it is being modified as well. Depending on the modifications, the level of risks can vary significantly.
Segment CDE:	The risks are highest when a modification takes the form of a new development such as the creation of a new idea or an equivalent feature, a functionality, or a replacement of an old idea by a completely new idea.
Segment EF:	This occurs when an idea or a feature that is found in a competitive product is copied and later modified to suit the situations. Deming points out that to copy an example of success, without understanding it with the aid of theory, may lead to eventual disaster—may be victim of an even larger risk [Deming, 1993].
Segment FG:	The risk is lowest when an unpatented idea is copied in its totality from a competitive product and transferred into one's own products.

following: develop a new idea, keep on modifying an old idea, or copy and transfer a competitor's idea into one's own products, etc.

2.2.2 Aspects of Life-cycle Management

Change happens all the time in all organizations. Most often, however, change is unplanned, unmanaged, and uncomfortable. Life-cycle management means learning to deal with new product introduction, changing technologies and systems, initiating quality leadership, process management, shaping direction for the change, taking control, and establishing the improvement process. Life-cycle management means management process for systematic incorporation of a new product family or a technology, handling continuity, and a revision-type product change.

There are four aspects of life-cycle management:

1. New product introduction
2. Strategic technology insertion
3. Managing Continuity
4. Managing Revision Change

With the advent of new process and design techniques, current processes may need to be restructured, reprocessed, or reengineered to exploit their maximum potential on product life-cycle. One of the major challenges in CE is to find an appropriate balance between continuity, revision, and re-engineering. Continuity is used here to indicate carrying the day-to-day operation, with the exceptions of "engineering change-order" procedures or minor alterations. "Major change" means changing the life-cycle methodology, process reengineering, or anything new that is introduced. Process reengineering is discussed in Chapter 3.

2.3 NEW PRODUCT INTRODUCTION

New product introduction (NPI) is similar to the improvement aspect of the change management process, which is described thoroughly in Chapter 3, Section 3.8. The product realization process in NPI involves iterative and multiple incorporation of changes across different elements of the end product, including all aspects of life-cycle considerations. These include simultaneous change management from its initial stage; life-cycle configuration management; and insertions of several "new tools and technology" along the way in product and process areas. The simultaneity reinforces adherence to TQM and other CPI change philosophies. Such adherence requires that a rationale of change be maintained to provide a basis for the product undergoing continuous change. A weakness of NPI is the lapsed time required to bring the product to market. Many manufacturing companies are losing the competitive race in this area to the speedy and effective execution process which other successful companies (for example, some Japanese electronic manufacturers) use. By introducing a product to market when demand or need for a product is at its peak, a company can lock-in a large sales volume.

2.3.1 Early Introduction

Concurrent engineering provides a way to bring the product into the marketplace early. Those companies that bring their product into the marketplace before their competitors, wrestle away a larger share of the marketplace. This is shown in Figure 2.14. Slow-to-market or slow responsiveness is due to inability to quickly react to changing market conditions. Such conditions force manufacturers inadvertently to loose market share to their competitors. As with the Big Three (Ford, GM, and Chrysler automobile manufacturers in USA), the mini-van and mid-size car truck market were heavily underestimated in the early 1990s. This caused consumers to wait for the mini-van, or to buy mini-vans and trucks from the competitors. Sports cars with convertible tops were also underestimated for 1994 sales of the FORD Mustang and GM Pontiac TransAm.

Let us denote the sales volume of two very similar companies as

$S_e(t)$ = Sales volume of company "E," who introduced their products early to the market place

$S_\iota(t)$ = Sales volume of company "L", who introduced their products late to the market place

"e" and "ι" prefixes in the above nomenclatures come from "early" and "late." T = Time when Company L introduced its product.

Company E had a 100% market share up to time T. After time T, Company L intro-

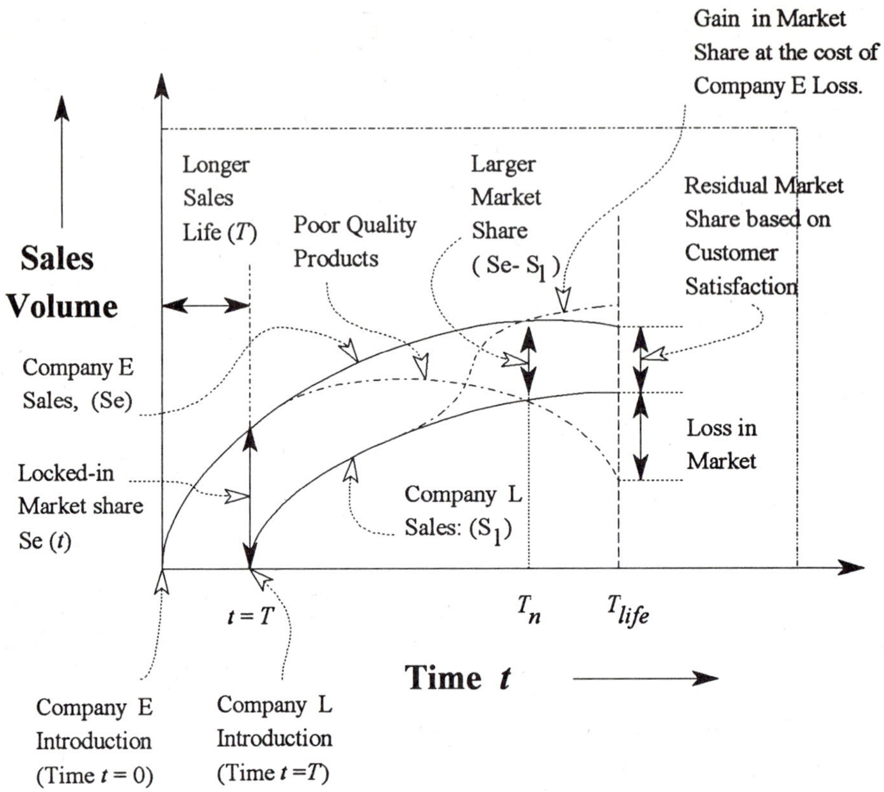

LEGEND:

S : Sales
E;e : Early Introduction
L; l : Late Introduction
t = Time elapsed after Initial Intoduction
T = Time when Company L Introduced its Product

FIGURE 2.14 Sales and Market Share Advantages of Early Product Introduction

duced its product and the market was shared between the two companies. If Company E were managed correctly, it could still enjoy higher market shares:

2.3.1.1 Market Share or Sales Advantage

$$
\begin{aligned}
&= [S_e(t) - S_l(t)]; \quad &\text{for } t > T; \\
&= S_e(t) \quad &\text{for } t < T
\end{aligned}
\tag{2.44}
$$

The earlier a product is introduced by a company, the better are its prospects for achieving and retaining a larger market share. Initial consumers or users tend to invest heavily at the introductory release of new or improved technology. Consumers usually make heavy capital investments and jump into a serious employee trainee programs. As these investments accumulate, consumers find it difficult, culturally and economically, to switch later to a new product and start over again. Consumers have realized significant productivity improvements with this early introduction. This, however, has a very positive effect on the sales-volume of Company E. By the time a new competitor's product was introduced, Company E had already captured and locked in a share of the market.

$$\text{Locked-in share at time } T = S_e\,(T) \tag{2.45}$$

$$\text{Longer sales life} = T \tag{2.46}$$

On the other hand, a one month slip in introduction (or development) is one less month of sales.

In some products, like software products and certain types of industries, when there is training and education involved, the potential locked-in opportunities for sales volume are quite substantial. Coincident with the initial introductions, buyers or users develop proficiency and commit resources: (a) locked in capital investments; (b) acquired training in the use of product; and (c) developed productivity short-cuts in the use of the product. It then becomes difficult for any company to switch to a newer product and start the whole process all over again.

The second benefit of early introduction is the price and cost advantages. Most of the profits from successful products are realized early after their introduction (see Figure 2.15).

Price, from the perspective of the seller, is viewed as the sum of the costs incurred in design and production plus a reasonable mark-up margin or profit. In general, the selling price of the product is several times its manufacturing cost.

$$\text{Price} = \text{Company cost} + \text{Mark-up} \tag{2.47}$$

Where company cost is the result of life-cycle costs:

Company Cost = Marketing Planning Cost + Design Development + Cost of Component Sub assemblies + Assembly Cost + QC/Inspection Cost + Investment Tooling Cost + Production Cost + Distribution Cost + Warranty Service Cost

Figure 2.15 shows conceptually the plots of the unit cost and market price of a product introduced by companies E and L drawn against time. For the purpose of discussion, let us denote

$P(t)$ = Market price of a product
$C_e\,(t)$ = Cost of developing a product by Company E
$C_l(t)$ = Cost of developing a product by Company L

Thus,

$$\text{Unit profit margin for Company E} = P(t) - C_e\,(t) \quad \text{for } t > 0 \tag{2.48}$$

$$\text{Unit profit margin for Company L} = P(t) - C_l(t) \quad \text{for } t > T \tag{2.49}$$

Cost advantage of early introduction with unit sale of the product is given by

$$[C_l(t) - C_e\,(t)] \tag{2.50}$$

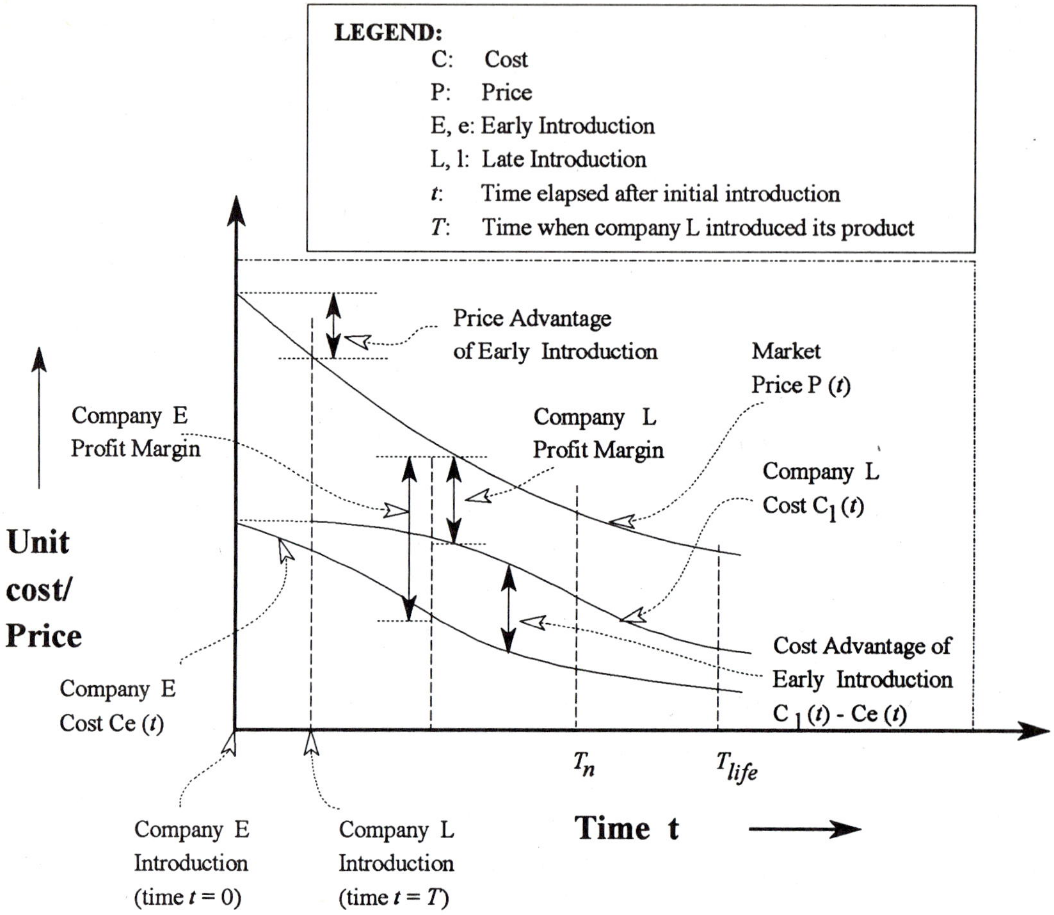

FIGURE 2.15 Cost and Price Advantages of Early Product Introduction

Total Cost of the products can be obtained by multiplying the sales volume and the unit product cost.

When a new product is introduced, there is usually no competition (that is during a period $0 < t < T$). The company has more pricing freedom and consequently has better profit margins (potential).

At time $t < T$; total profit margins (TPM) for company E:

$$\text{TPM}_e = [S_e(t)] * [P(t) - C_e(t)] \qquad t < T \tag{2.51}$$

TPM for Company L is 0, for $t < T$, because the product does not exist.

$$\text{TPM}_l = 0 \qquad t < T$$

During $T < t < T_{\text{life}}$, total profit margins for Company E are

$$\text{TPM}_e\,(t) = \{S_e\,(t) * [P(t) - C_e\,(t)]\,\} \quad t > T \tag{2.52}$$

and Total Profit Margins for company L at any time after $t > T$ are

$$\text{TPM}_l\,(t) = \{\,S_l\,(t) * [P(t) - C_l\,(t)]\,\} \quad t > T \tag{2.53}$$

Thus, as shown in Figure 2.15, total cost advantage (TCA) can be computed by subtracting the above two TPMs as follows:

$$\text{TCA}\,(t) \equiv \text{TPM}_e\,(t) - \text{TPM}_l\,(t) \tag{2.54}$$

$$\text{TCA}\,(t) \equiv \{\,S_e\,(t) * [P(t) - C_e\,(t)]\,\} \qquad \text{for } t < T \tag{2.55}$$

and

$$\text{TCA}\,(t) = \big[[S_e\,(t) * \{P(t) - C_e\,(t)\}] - [S_l\,(t) * \{P(t) - C_l\,(t)\}]\big];\ \text{for } t > T \tag{2.56}$$

If the sales volume are the same (say S), then Equation (2.56) gives

$$\text{TCA}\,(t) = S\,(C_l - C_e) \qquad \text{for } t > T \tag{2.57}$$

where, the figure in the bracket for Equation (2.57) represents the cost advantage of early introduction with a unit sale of the product.

As shown in Figure 2.15, market price of the product slowly decreases due to competition. After time T, the cost advantage gap does narrow down. However, there is still a cost advantage compared to the competition. Furthermore, there are secondary benefits of being first, such as early acquisition of the manufacturing competence, being ahead in the learning curve, and having time to react for market change or focus, etc.

For some products that have high switching cost, the benefits of early introduction are even larger. Besides the obvious cost advantage, it has the potential of gaining more customers who maintain their loyalty due to the heavy cost of switching to other products. Their loyalty creates a residual sale trail that is roughly constant. This is the minimum sales volumes if no new sales are incurred. If the residual sale trail is superimposed over the prior sales record, it gives the maximum sales volume that Company E can expect. Consequently an early introduction builds momentum which is carried through the later part of its life-cycle. What does this mean to the net profits? According to McKinsey & Co., [Musselwhilte, 1990], a high-tech product that reaches the market six months late, even on budget, will earn 1/3 less profit over 5 years. On the other hand, finishing on time, but 50% over budget, will reduce the company's profits by only 5%.

2.3.2 Increased Product Life

A by-product of early introduction—which is not so obvious from Figure 2.15—is that the product's sales life is extended. If a product is introduced earlier, it is seldom removed from the marketplace. It enjoys the same life as any new product introduced at a later time. Consequently, every month that is cut from its development cycle using CE or similar concepts, is added to the sales life. This means that all the revenues and profits generated during a period of T units are additions to the company profits.

2.3.2.1 Revenues and Profits (R & P):

The cumulative profit margin over a period of time ($t = T_n$) can be obtained by integrating the discussed curves over time t.

$$\text{Company E: (R \& P)} = \int_{t=o}^{t=T_n} S_e(t)[P(t) - C_e(t)]dt \qquad (2.58)$$

$$\text{Company L: (R \& P)} = \int_{t=o}^{t=T_n} S_l(t)[P(t) - C_l(t)]dt \qquad (2.59)$$

However, there are many situations which might affect the sales volume life-cycle curve. Economic factors, market trends, and product quality are some of the major examples. Economic factors such as recession, high inflation, or high interest rates can change (increase or decrease) the product's life-cycle in terms of sales volume. Market trends such as annual seasonal trends or fads could also lengthen or shorten the sales volume. An example is clothing styles, which are usually short-lived. Most damaging is the effect of quality or perceived quality of the product in determining its sales volume. If the quality of the product deteriorates or if the competition introduces a product which far exceeds the customer expectation, the product can lose the market share (as shown by dotted lines in Figure 2.14). A similar situation can happen in the case of a product's recall. Its effect on the sales volume however depends on the severity of publicity or the quality problem that the product has experienced. A highly publicized recall (such as the case in Ford's Pinto or General Motors' side-saddle gas tank pick-up trucks) can have a pronounced effect on the sales volume than a silent recall.

If such an adverse situation happens, it is in the best interest of the Company E to introduce a new product which exceeds or meets the customers' quality expectations. At time $t = T_n$, the sales volumes of products E and L are nearly equal,

$$[S_e(T_n)] = [S_l(T_n)] \qquad (2.60)$$

The decision to change the product must come quite ahead of this critical juncture. On the other hand, if the company continues its production to its full life-cycle, there is a danger of losing substantial market-shares as shown in Figure 2.14.

2.3.3 Calculation of Revenue Loss

The trend for revenue, as shown through the equations (2.58) and (2.59) usually follows a profile of an S-curve. This is shown in Figure 2.16 for both market growth and a market decline. Each follows an S-curve trend. Together they form a bell shape. In Figure 2.16b two bell curves are shown; one bell curve is for the revenue with an early (or on-time) market entry; and the other bell-curve is for a delayed market entry. In the growth S-curve case, the rate of growth is slow in the beginning, increases quite rapidly in the middle zone and then again slows down as it reaches the peak revenue. In the case of decline

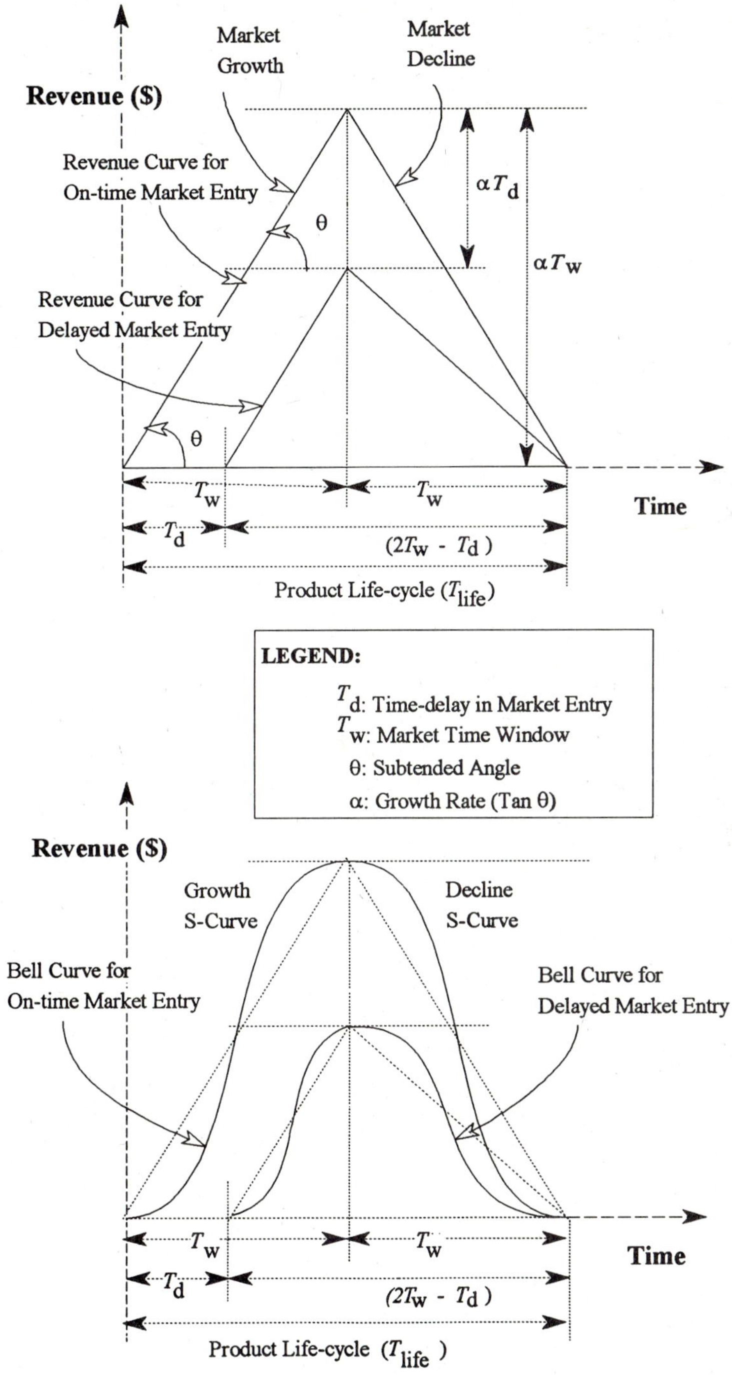

FIGURE 2.16 Computation of Revenue Loss due to Delayed Market Entry

S-curve, the rate is slow at the peak, sharply declines in the middle region, and again flattens up in the lowest region.

T_w = is the market window (time) when the growth revenue S-curve reaches its peak. If it is assumed that the time of decline is the same as the market window (time of growth), the two S-curves are symmetrical about the vertical axis. The total product life-cycle time can thus be computed as

$$T_{\text{life}} = 2T_w \tag{2.61}$$

The S-curves (or bell curves) are often very symmetrical about a straight line connecting a peak revenue point to a start-point. The bell curves of Figure 2.16b are approximated by a series of straight lines. Two straight lines are drawn—one for the market growth S-curve and another for the market decline S-curve. Another set of two lines are shown for delayed "market entry and the market decline." Carter and Baker also used a straight line to measure the impact of delays in launching a product [Carter and Baker, 1992]. As shown in Figure 2.16b, both these bell-curves are symmetrical about lines that form sides of a triangle. On the basis of this symmetry, the area under the bell-curve can easily be approximated by the area under this triangle.

If θ is the slope of the revenue S-curve and α is the rate of revenue-growth or revenue-decline, the peak revenue can be expressed as

$$\text{Peak-Revenue} = \alpha T_w \tag{2.62}$$

where

$$\alpha = \tan(\theta)$$

Revenue generated in the case of on-time market entry and delayed market entry can be computed by computing the area under the respective bell-curves.

Total revenue for early (or on-time market introduction) = Area under the larger triangle

$$R_{\text{early}} = (2T_w) * (\alpha T_w)/2 \tag{2.63}$$

or

$$R_{\text{early}} = \alpha (T_w)^2 \tag{2.64}$$

Total revenue generated when product introduction is delayed by T_d units

$$R_{\text{delayed}} = \text{Area under the smaller triangle}$$

or

$$R_{\text{delayed}} = (2T_w - T_d)\,(\alpha T_w - \alpha T_d)/2 \tag{2.65}$$

It is assumed that the growth rate for the delayed market entry is the same as the early or on-time market entry, meaning the rate remains constant.

If R_{loss} denotes a revenue loss term due to delay in introducing the new product, then

TABLE 2.4 Table for Revenue Loss Computation

Months Late (T_d)	Revenue Lost (R_{loss}) (%)	Remarks
1	12.2%	A 1 month market delay means 12.2% loss in total lifetime revenue.
2	23.6%	
3	34.4%	A 3 month market delay means 34.4% loss in total lifetime revenue.
4	44.4%	
5	53.8%	
6	62.5%	A 6 month market delay means 62.5% loss in total lifetime revenue.
7	70.5%	
8	77.8%	
9	84.4%	A 9 month market delay means 84.4% loss in total lifetime revenue.
12	100%	

$$R_{loss} = \frac{R_{early} - R_{delayed}}{R_{early}} \tag{2.66}$$

or,

$$R_{loss} = \frac{[(T_w)^2 - (2T_w - T_d)(T_w - T_d)/2]}{(T_w)^2} \tag{2.67}$$

or,

$$R_{loss} = \frac{[(3T_w - T_d)T_d]}{2(T_w)^2} \tag{2.68}$$

It is interesting to note that revenue loss is independent of the growth rate α.

The above Equation (2.68) represents an approximation of the actual revenue loss which must be computed using true S-curves or bell curves. This approximation can be used as a measure in calculating the impact of delays in launching a product. For example, considering a twelve months market window, $T_w = 12$, an impact of T_d months delay in launching a product can be computed using Table 2.4.

2.4 STRATEGIC TECHNOLOGY INSERTIONS

There is a point in the product's life-cycle when the benefit of a particular technology starts saturating [Prasad, 1994]. This is shown by an S-curve in Figure 2.17. There are three parts to it. The first part (part AB) is time for technology introduction, the second part (BD) deals with technology improvement and the third part (DE) is time for technology maintenance.

The simplest form of an S-curve is a cubic polynomial:

$$s(t/T) = a + b(t/T) + c(t/T)^2 + d(t/T)^3 \tag{2.69}$$

If we denote $\tau = t/T$

The points A and E on the curves are obtained by solving the slope equation, when

$$\frac{dS}{d\tau}\tau = b + 2c\tau + 3d\tau^2 = \qquad (2.70)$$

They correspond to point t_a and t_e given by

$$t_{a/T} \text{ or } t_{e/T} = -2c \pm \frac{\sqrt{(4c^2 - 12db)}}{6d} \qquad (2.71)$$

The point C is an inflection point on the S-curve. The point (tc/T) is given by

$$\frac{d^2S}{d\tau^2} = 2c + 6d\tau = 0 \qquad (2.72)$$

The constants a, b, c, and d define the shape of the S-curve. Their values are not independent. The length of the parts AB, BD, and DE depends on these constants.

S-curve can also be expressed in many other forms. The behavior of a trigonometric tangent function is very similar to S-curve, except it ranges from $-\infty$ to $+\infty$. If we can use a truncated portion of an inverse tangent function to define the shape of the S-curve, equation can also be expressed as:

$$S(\tau) = [(1./\Pi)\tan^{-1}\{\alpha(\tau + \beta)\}] + 0.5; \qquad -1.0 \le \tau \le 1.0 \qquad (2.73)$$

Where, α and β are constants and control the shape of the S-curve.

$$\alpha = \text{Slope parameters}$$

and

$$\beta = \text{Shift parameter}$$

In the above representation S-curve is normalized, that is, its value ranges from 0 to 1.0 at its two extreme points a and e. The smaller value of the slope parameter, α, gives the more gentle rising of the S-curve, while larger values of the parameter α gives a steeper rising of the S-curve. The shift parameter β gives the position of the inflection point, C relative to a normalized value of time, τ. This is the point when the satisfactory level is midway between the starting point and the saturation point. That is, shift parameter β gives the time τ when the magnitude of the S-curve is 0.5. The points a and b on the S-curve lies in between $\tau = -1$ and $\tau = \beta$. The point c is the inflection point when $\tau = \beta$. The points d and e lies anywhere between $\tau = \beta$ and $\tau = 1.0$.

The first and last part are more time intensive and costly than the middle part. Therefore,

$$[t_b - t_a] >>> [t_c - t_b]$$

also (2.74)

$$[t_e - t_d] >>> [t_d - t_c]$$

Usually, when a technology is first introduced, the startup cost is high. There is very little known about the impact of this technology on the production functions and features.

The learning that takes place in the early stage of insertion (part AB) is relatively costly compared to the demonstrated functionality it offers to the customers and the companies.

$$\text{Relative Cost (part AB)} \propto [t_b - t_a] \tag{2.75}$$

The second part (part BD) is when the company enjoys the benefits of early insertion of a useful technology. While the competition is playing a catching up game, the company enjoys a good rate of return without much additional expenses. The amount of time and effort required to provide a relatively steady level of performance through a number of easy improvements is small. Usually this part of the curve lasts only for a short while.

$$\text{Relative Cost (part BD)} \propto [t_d - t_b] \tag{2.76}$$

The third part (part DE) of the curve begins when the competition finds a better way of providing the same or similar functionality at a competitive price. Most companies usually fight back by aggressively fine-tuning the elements of technology that helped them on the first stage (part AB). Squeezing more improvements out of the same underlying technology costs more resources with very little improvements in functionality and other aspects. This is often an indication that the useful life of the technology has reached its saturation point. The company that keeps on improving without switching to a new technology or program often loses. Those who are visionaries invest time and resources early enough in the product cycle before they reach this stage.

$$\text{Relative Cost} \propto [t_e - t_d] \tag{2.77}$$

Three transition situations are shown schematically in Figure 2.17 by three S-curves. In the first S-curve (A'-E'), shown by a single dash line, an early transition from prior-technology to a new technology was made which was not necessary. The company did not utilize the full benefits of the previous technology. In the second case (S-curve A"-E"), shown by double-dash lines ("), transition to the new technology was rather late. There was an apparent transition gap. In Figure 2.17 the three S-curves for early transition, optimal transition, and late transition are shown carrying the same general shape. However, their shapes may vary. The shape controlling parameters are the constants (a,b,c,d) of the cubic polynomial, Equation (2.69).

$$\text{Transition Gap} = \text{Timing differences between points (C"-C"')}$$

and

$$\text{Cost Gap} \propto [t_{c\iota} - t_{co}] \tag{2.78}$$

Where, the first indices in $t_{c\iota}$ and/or t_{co} denote the point on the curve.

The second indices show the type of the curve ("*o*" stands for optimal and "ι" stands for late). The third S-curve (A"'-E"'), shown by triple-dash lines ("'), shows the correct timing of insertion. It shows an optimal transition point for switching from an old technology to a new technology. A far-sighted company starts work on a new S-curve much before a replaced S-curve has reached its maturity. They switch to the new curve at the end of the first stage so that for the remainder of the product life they can reap the optimum benefits. Transition to a new technology is used here to mean a number of things. For some, it may mean looking for fundamentally better ways of designing and developing products, such as new materials, a new concept, or a new process; for others it may mean searching for better productivity and efficiency means, high-potential techniques, or

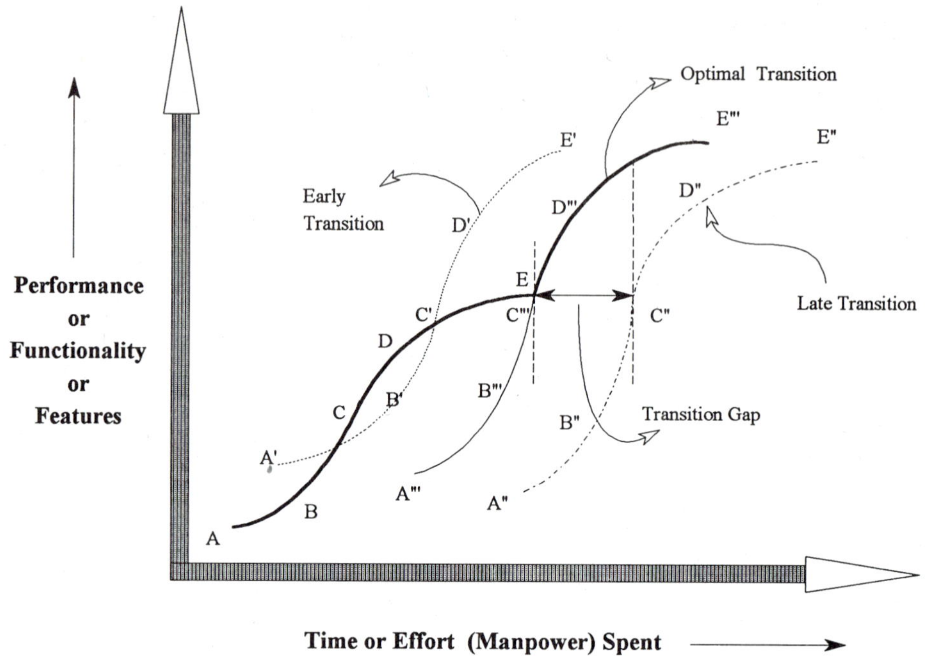

LEGEND:

A: Point of Technology Insertion
AE: Useful life
AB: Technology Introduction Period
BD: Technology Improvement Period
BC: Fine Tunning
CD: Modification
DE: Technology Maintenance
E: End of Technology Excitement

FIGURE 2.17 S-curves for Strategic Technology Insertions

alternate competitive advantages, such as labor management relations, multidisciplinary focus, empowerment, etc.

The above cost advantage relations (Equations (2.55) through (2.57)) and S-curves, (Equation (2.69)), can be used as measures of cost competitiveness. It is obvious from the relations that when sales volumes decrease and become smaller than the competition, the cost advantage too changes sign and becomes a cost disadvantage. The relations and S-curves are empirical tools for understanding the risks and benefits of "early introduction" of the product and "early insertion" of the technology. There is a great temptation on the part of the CE management to minimize the expenditure by "defending" the advantage already gained through prior insertions. However, this is a trap, and if one gets into it, it is often difficult to get out since there is a real-time lag between the point of "technology insertion" and the point of optimal "technology transition" (part AC).

2.5 MANAGING CONTINUITY

The continuity management process begins with the initial design of configuration management process. It allows the changes to take place without violating the outer boundaries around which configuration control is defined.

2.5.1 Configuration Management

Configuration management is a cross-functional process management framework for organizing various "as-designed" processes and product families [Prasad, 1995], where appropriate configuration control activities (where, how, and when changes) can take place. Because product structure drives the process, there is a need to create a configuration management system that can capture changes at the source. A comprehensive configuration management process is required earlier in the program. This involves shifting from a multidetail drawing-based system to a monodetail digital-based configuration (part and modeling structure) control system for product life-cycle management (see Figure 2.18). Product life-cycle configuration management (from inception to recycling) consists of three major stages: development, production, and service. The development stage starts

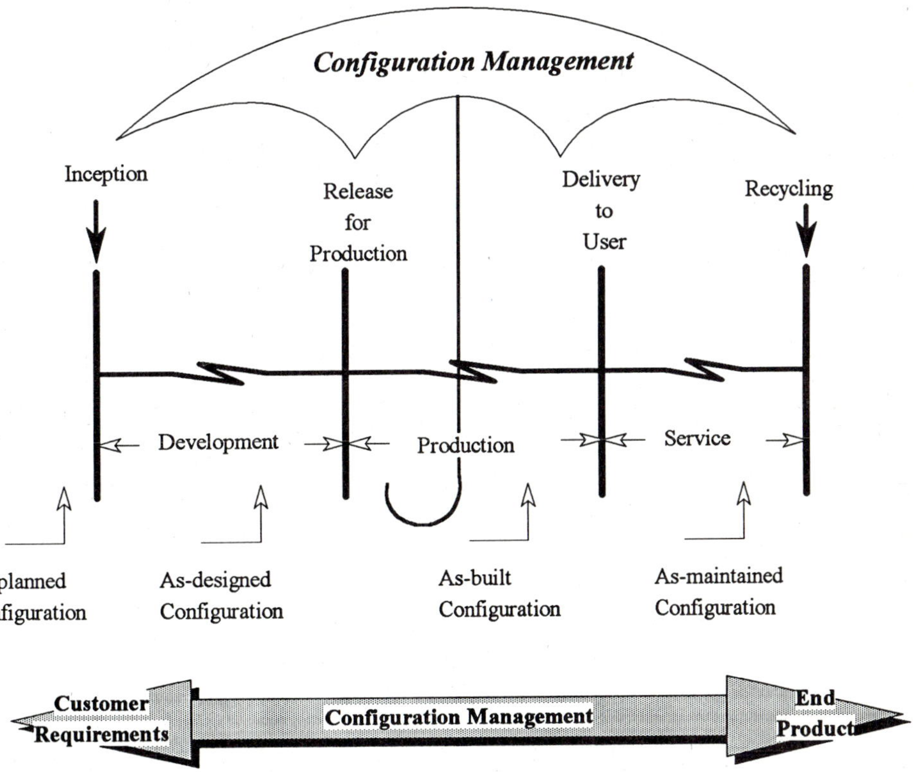

FIGURE 2.18 Product Life-cycle Configuration Management

with "as-planned" configuration and ends with "as-designed." On the development side, engineers work in an open-ended or evolving environment where product form, fit, and functions unfold creatively.

$$\text{Development Duration} = [t_{rfp} - t_i] \qquad (2.79)$$

Where, t_{rfp} and t_i are timings at "release for production" and "inception" points, respectively.

The production stage starts with as-designed configuration and ends with as-built configuration. Here, configuration management implies enforcing release procedures. Examples of such release procedures are only approved parts to be used, a top-level assembly can be promoted to a "release-level" if all of its components have also reached a "released-level" status, etc. At the production stage, plant engineers work in a stable environment so that they can operate the production at maximum capacity.

$$\text{Production Duration} = [t_{dtu} - t_{rfp}] \qquad (2.80)$$

Where t_{dtu} and t_{rfp} are timings at "delivery to user" and "release for production" points, respectively.

The service stage begins with an as-built configuration and ends with an as-maintained configuration. What is required for the configuration system to work efficiently are tools and interfaces that follow structured methodologies that are based on upcoming international standards.

$$\text{Service Duration} = [t_r - t_{dtu}] \qquad (2.81)$$

where, t_r and t_{dtu} are timings at recycling and delivery-to-user points, respectively.

Each stage methodology includes a "measurement of merits" of underlying processes in metric forms. Metrics include change incorporation control tied to designated product check-points, option incorporation, and family of product variation. This provides better cooperation, more flexibility, and allows for greater change during the development cycle.

2.6 MANAGING REVISION CHANGES

Because product specification process involves examination and changes by work-groups, configuration management process must be able to control and manage revisions. In addition, because work groups use dynamic processes, the change management process must provide control to manage change over time. Three basic elements that participate in providing inputs to change management are product specification process, product development process, and parts catalog. This is shown in Figure 2.19.

2.6.1 Incorporate Lessons Learned

Most companies gain a competitive advantage in the marketplace through a process of corporate learning that has gone into building their intellectual work-force. Though not obvious immediately, during this learning, a number of databases of experiences are created. If design programs are used, a trail of design history is inadvertently left behind.

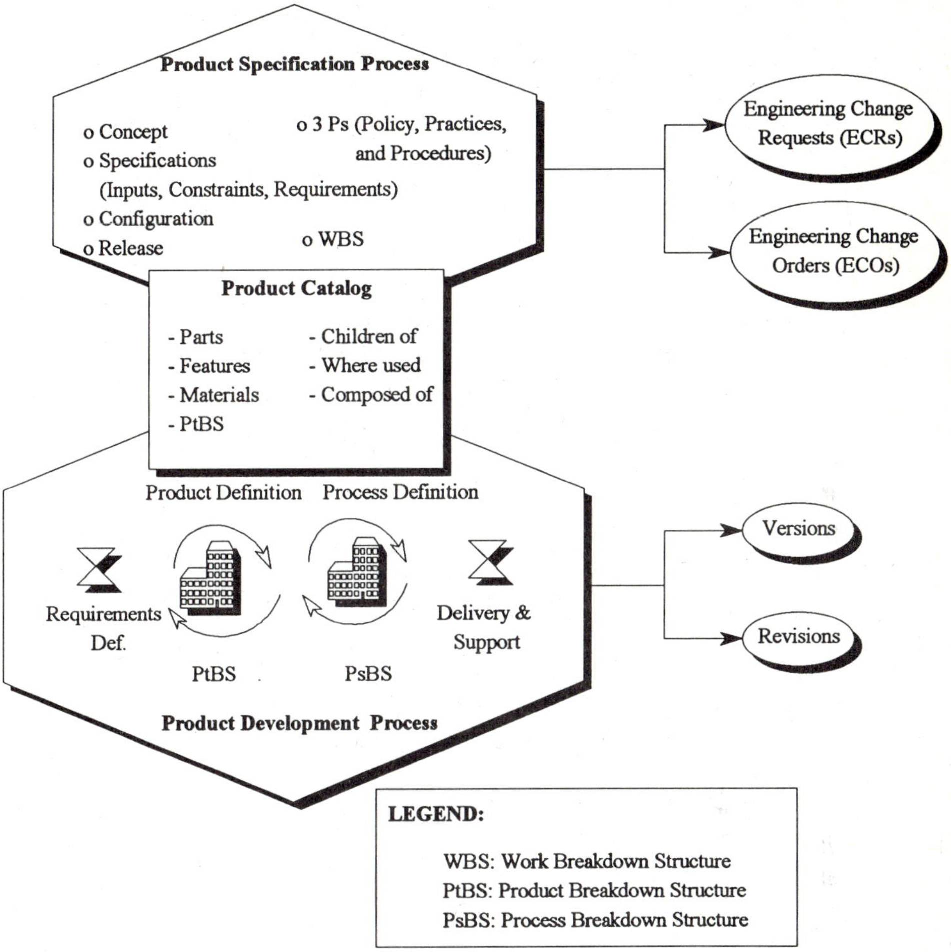

FIGURE 2.19 Participants of a Change Management Process

Lessons learned from previous programs and in-process developments can be immediately reflected if digital models or other forms of representations (such as knowledge base, database, etc.) are employed. This may be captured as a part of corporate technical memory containing both positive and negative past experiences. With proper keywords and on-line-search capability, knowledge-based teams can search for occurrences of similar situations and thus could capitalize on teams' past experiences.

2.7 LIFE-CYCLE COST DRIVERS

There are three main cost drivers during the entire product life from conception to grave: company costs, user costs, and society costs (see Figure 2.20). The goal of the life-cycle design is to maximize the value of a product, while containing its cost to the manufacturer, the user, and society.

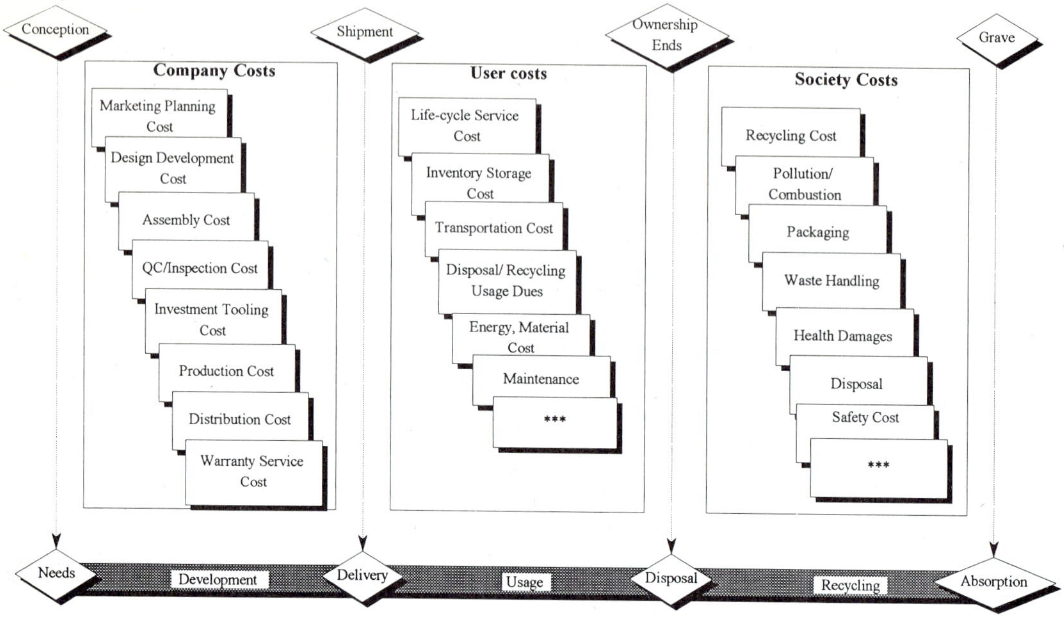

FIGURE 2.20 Life-cycle Cost Drivers

$$C_{\text{drivers}} = f\,[C_{\text{company}}, C_{\text{users}}, C_{\text{society}}] \tag{2.82}$$

where C_{drivers} stands for cost drivers; C_{company}, C_{users}, and C_{society} represent the cost contributions associated with company, user, and society.

Company costs are the costs of activities required in planning, design development, assembly, production, distribution, and servicing a quality product. It includes all costs that are incurred from needs to delivery until the product is shipped to the customer. Company costs are of two types: direct and indirect costs. Direct costs result from highly visible and documented costs of labor and material used on the factory floor. Indirect costs are everything other than labor and materials.

$$C_{\text{company}} \approx C_{\text{needs-to-delivery}} \tag{2.83}$$

User costs are the costs to the users of those activities that are performed by the user from the time when the product is delivered or shipped, to its disposal when the ownership ends. User costs will begin to include the costs for recycling or disposal. As an example, in some U.S. states, consumers will be asked to pay a disposal fee for old tires when new tires are purchased. Consumers may begin to pay penalties or taxes in terms of the use of freon in air conditioners, refrigerators, and freezers.

$$C_{\text{users}} \approx C_{\text{delivery-to-disposal}} \tag{2.84}$$

The society costs are the expenses that are inflicted on the society from the time the product is in user custody until it is disposed of or recycled safely. The ability to recycle the material, or its impact on the environment, is the major contributing cost to society.

These costs have already began to increase, even though many of these costs are intangible and unable to be measured or quantified accurately.

$$C_{\text{society}} \approx C_{\text{disposal-to-absorption}} \tag{2.85}$$

Most engineering and design teams do not go far enough in reducing life-cycle driver costs. Most focus on the company costs and in a narrow sense just concentrate on the direct costs. Very few teams attack the company's greatest cost challenge—the indirect costs. It is interesting to note that "indirect company costs" can reach four to five times larger than the direct (company) costs. In spite of this, only a very small portion of any engineering team considers reducing it during design and development phase. Most remaining product design teams believe indirect costs can only be attacked either at the management level or during production or the manufacturing assembly phase of its life cycle. The contention is certainly wrong.

2.8 LIFE-CYCLE MANAGEMENT TOOLS

Over the last decade or two, advances in product design and development, including manufacturing technologies, have been revolutionary. Computer-aided design (CAD), computer-aided manufacturing (CAM), computer-aided engineering (CAE), computer-aided process planning (CAPP), computer integrated manufacturing (CIM), and computer-aided testing (CAT) have helped cut down product development cycles and further reduce the variability inherent to the manufacturing processes. These and other advances in automation and system integration have helped reduce costs that were traditionally associated with mass-manufacturing.

$$\text{Computer tools} \quad \cup \quad [\text{CAD, CAM, CAE, CAPP, CIM, CAT, ***, } \ldots\ldots] \tag{2.86}$$

It is also necessary to understand and implement technological ideas such as rapid prototyping, parallel design philosophies, strategic forward design planning, rapid inch-up and scale-down, product covering, and product churning. That is,

$$\text{Technology tools} \quad \cup \quad [\text{rapid prototyping, forward engineering}\ldots\ldots] \tag{2.87}$$

These are not routine formulas to ensure concurrent engineering. One should not expect that simply by their use one can achieve competitiveness and industrial reliance. Computer tools and technology tools are like flexible tools in a toolbox. They have the potential to reduce cycle time and can be employed in a CE process. If a potential product niche is identified, the use of a right tool can only serve the benefits. Concurrent engineering teams must therefore keep abreast of these developments. They must have adequate tools to automate and effectively capture a CE process consistently. The major techniques that follow the concurrent engineering philosophy, or are easier to apply, are contained in this section. These are identified by their popular names and acronyms to provide a sense of the continuity of developments in this effort.

2.8.1 Systems Integration (SI)

This deals primarily with an information aspect that provides management and customers full access to technical information. However, full technical information integration in an environment other than small, simple, and relatively closed (and those, which are not based on object-oriented, Window, or UNIX-type operations) has generally proved to be extremely difficult. SI needs distributed databases along with global communication network.

2.8.2 Quality Function Deployment (QFD)

QFD is a market driven design and development methodology for products and services to meet or exceed customers' needs and expectations. The QFD technique originated in 1972 at Mitsubishi's Kobe shipyard site in Japan to build ocean transportation vessels. At first, the shipyard's management developed a form of upstream quality assurance to cope with a great deal of capital outlay. Market demanded quality products coupled with stringent government regulations. This management technique has since then come to be known as "Quality Function Deployment" (QFD) [Hauser and Clausing, 1988]. The Kobe engineers in 1972 developed a matrix relating "WHATs," such as government regulations, critical characteristics, and customer requirements with "HOWs" to control how the company would incorporate "WHATs" in their design. The matrix also showed the relative importance of each item, which ensured proper prioritization and allocation of available resources to the production planning. Other companies started to adopt the QFD methodology in the mid 1970s [Hales, Lyman, and Norman, 1990]. Toyota began to use QFD in the late 1970s and reported a 20 percent reduction in startup costs, when it introduced four new van-type vehicles, which increased progressively to 61 percent cumulative total by 1984 [ASQC/ASI, 1992]. Aisin Warner, which produced transmissions for Toyota, reported [ASQC/ASI, 1992] that QFD reduced the number of engineering changes by half resulting in significant improvement in quality. QFD has also been used [King, 1987] successfully by Japanese manufacturers of consumer electronics, home appliances, clothing, integrated circuits, synthetic rubber, construction equipment, and agricultural engines.

The application of QFD is a fairly recent innovation in the United States. Recent success can be attributed to industry interests in study missions to Japan and training in QFD offered by organizations like the American Supplier Institute [ASQC/ASI, 1992] and GOAL/QPC (Growth Opportunity Alliance of Lawrence, Massachusetts/Quality Productivity Center) [Akao, 1990; King, 1987]. The first recorded case studies in QFD were in 1986 [King, 1987]. Kelsey Hayes used it to develop a coolant sensor which fulfilled critical customer requirements like easy-to-add coolant, easy-to-identify unit, and cap removal instructions. A number of companies now use it, including Ford, General Motors, Chrysler, AT&T, Proctor and Gamble, Hewlett-Packard, Digital Equipment, ITT, and Baxter Healthcare. However, it has not yet found popularity as a design technique, though its use is appropriate for organizations of any size. Many companies are already experimenting with QFD ideas and applying them in diverse cross-functional team environments for problem solving [Bossert, 1990]. These companies have realized large benefits.

In the span of the first 7 years, between 1977 and 1984, the Toyota auto-body plant employed QFD and claimed that with its use the following has occurred:

- Manufacturing startup and preproduction costs had been reduced by 60%
- The product development cycle (i.e., time-to-market) had been reduced by 33% with a corresponding improvement in quality because of the reduction in the number of engineering changes [Wilson and Greaves 1990].

Use of QFD as a set of planning and communication routines by cross-functional teams is currently the most effective way known to cut through barriers to good design. QFD focuses and coordinates skills within an organization, first to design, and then to manufacture and market goods that customers want to buy. It is intended to better meet customer requirements, increase organizational capabilities, and at the same time, maximize company goals.

Figure 2.21 shows the effects of QFD on engineering changes schematically. With companies using QFD, engineering changes cut off early. Design is essentially frozen before the first product comes off the assembly line. Incorporating QFD in the design helps to identify the critical driving parameters in the design that should be addressed upfront where they benefit the evolution of the design the most. In companies not using QFD, changes continue to take precedence, requiring schedule and cost overrun. The upswings in the curves are the result of major changes. They may not be solely due to the number of changes taking place. Major changes have a way of propagating down to the detail level of design. If the changes occur at the subassembly or subsystem level, the number of re-

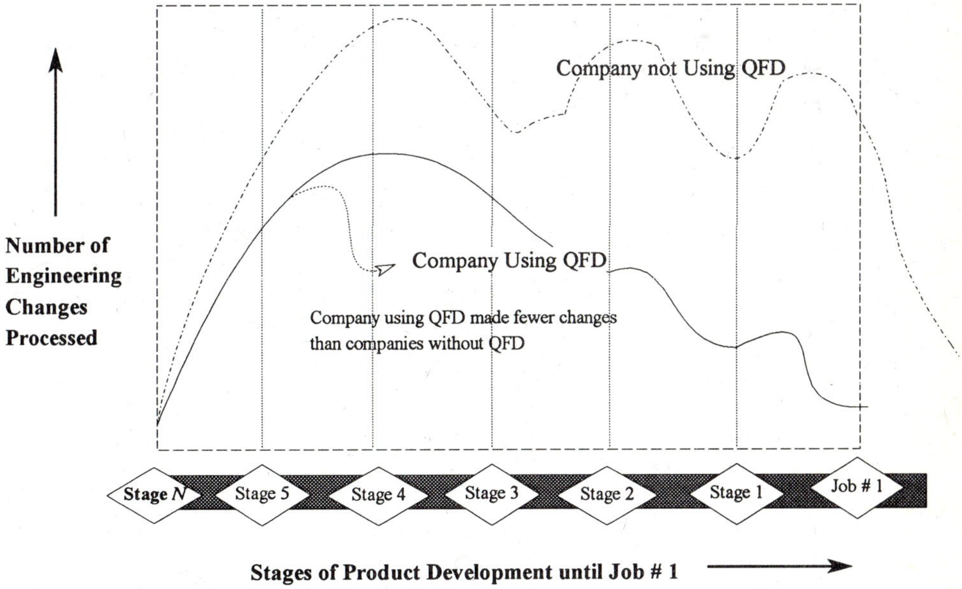

Source: Based on Data in "L.P. Sullivan, QFD, *Quality Progress*, June 1986."

FIGURE 2.21 Effect of QFD on Engineering Changes

sulting changes could be exponential in nature. Late fixes and design or process changes revamp even after reaching the final (Job #1) stage.

2.8.2.1 Focusing on Customer Satisfaction

The intent of QFD is to incorporate the voice of the customers into all phases of the product development cycle, through concept engineering and analysis, design, prototyping, production engineering and planning, management and control, manufacturing, and finally into delivery and support. In other words, QFD is customer driven product development methodology. The major goals for QFD are to incorporate customers' voice during product improvement and reduce product development time. The focus of QFD is on systematically capturing product information, such as market competitive analysis and customer satisfaction ratings, analyzing these ratings to improve product functionality, and then adding values that are important to the customers.

To understand what it takes to satisfy the customer, one must focus on the "voice of the customer" which is represented only in part by existing customers. Customers who do not buy a company's product also have a voice. The definition of quality, as stated earlier, relates to internal as well as external customers. Everyone wants the information—the voice of the customers—to be timely, accurate, and straightforward. Information is extracted (solicited) or comes from a variety of unsolicited means: Information can come in an active sense (solicited) because the company is looking for it, or because it is told to do so (unsolicited). It can be measurable (quantitative) or it can be subjective (qualitative). It can be organized/arranged in some order (structured) or it can be listed in a haphazard (random) way. No matter who is supplying the information, products, or service, the recipients rely on the information source (internal or external) for quality work. The internal requirements are as real as those of external customers—whether it is speed, accuracy, or measurement. Defining quality as "managing conformance to specification in order to achieve customer satisfaction" represents a minimum set of "do's" that a company has [Deming, 1986]. There are partners and internal customers who want their opinion to be sought and their issues to be addressed. The "voice of the customer" includes inputs from sources such as manufacturing, purchasing, field service, suppliers, etc. They represent a company's internal customers. Figure 2.22 lists various sources of data to develop customer requirements.

$$\text{Customer Requirements} \equiv f\,[D_{voc}, D_p, D_{wf}, D_{ma}] \tag{2.88}$$

where, $D_{voc}, D_p, D_{wf}, D_{ma}$ stands for voice of the customers, product, warranty or field, and market analysis data, respectively. The voice of customer data, D_{voc} data, in turn is a function of

$$D_{voc} \equiv f\,[D_{ic}, D_{ec}, D_{p\&f}] \tag{2.89}$$

where, $D_{ic}, D_{ec}, D_{p\&f}$ stand for internal customers, external customers, and past and future customers, respectively. Today, the customer focus is broader than just satisfying internal and external customers. Additionally, it includes making products that delight the user and cause a positive attitude toward their value, usage, and quality. As products become obsolete quickly in a highly dynamic marketplace, to many companies, maintaining value over the life of the product is becoming the most important quality dimension.

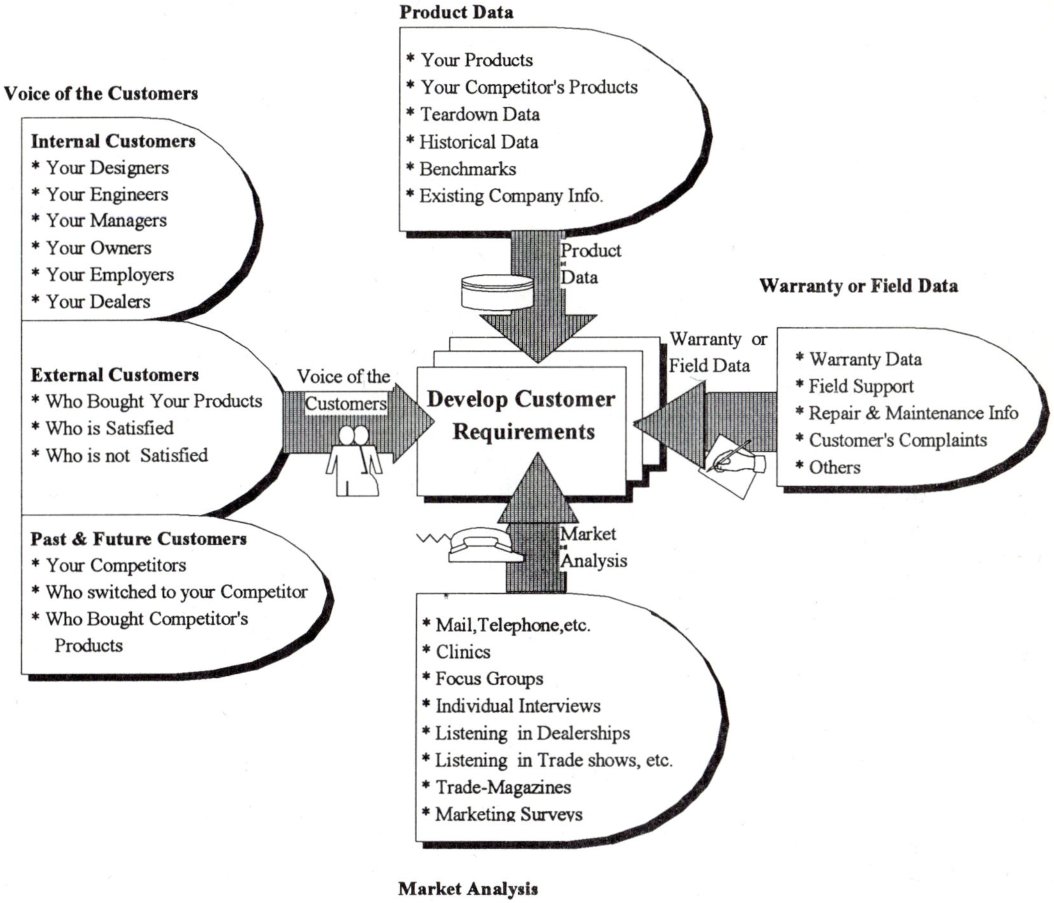

FIGURE 2.22 Sources of Data to Develop Customer Requirements

2.8.2.2 The QFD Technique

QFD is accomplished by multidisciplinary teams employing a series of matrix-based techniques to propagate critical customer wants throughout the product life-cycle. A three-legged structure of QFD: product planning, process planning and production planning are proposed. This is further discussed in later chapters. This is different from the American Supplier Institute's structure, which considers parts' deployment a separate phase—independent of the product deployment phase. The present deployment considers parts' deployment as a subphase of product planning. The three phases are as follows:

Phase I: Product Planning
Phase II: Process Planning
Phase III: Production Planning

Phase I can be further broken down into five subphases:

Subphase 1.1	Systems Planning
Subphase 1.2	Subsystems Planning
Subphase 1.3	Components Planning
Subphase 1.4	Parts Planning
Subphase 1.5	Materials Planning

These are discussed in detail in Chapter 3, Volume II.

2.8.2.3 House of Quality (HOQ)

The basic tool of QFD is the "relational matrix" concept. Matrices are schemata to define generically and relate directionally to multiple lists of identifiers, often referred as vectors [Prasad, 1993]. The basic matrix of QFD is the "house of quality," so named since the triangular matrix which forms its top structure—the roof—makes the diagram resemble a house. The quality matrix in QFD translates the voice of the customer into the engineering, design process, and production stages. The original ASQC/ASI version only had four rooms. Recent works have included "WHYs" in the matrix (see Figure 2.23). This way the definitions and manipulations of generic matrices can be combined into an expanded house of quality. Expanded QFD enables support for all prevailing QFD approaches that have recently emerged. Figure 2.23 is a schematic view of an expanded house of quality (HOQ) [Prasad, 1993]. This expanded house of quality has eight rooms. Four of the rooms form the basic perimeters of the house. These are lists of two row rooms: {WHATs} and {HOW MUCHes,} and two column rooms: {HOWs} and {WHYs.} Expanded HOQ also encompasses relationships between these lists resulting into four rela-

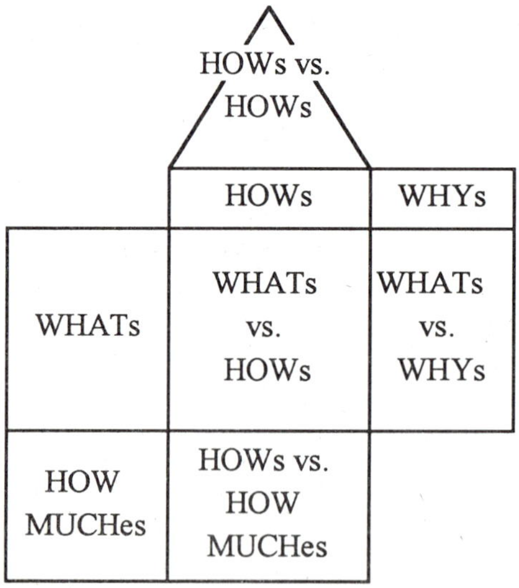

FIGURE 2.23 Expanded House of Quality (HOQ)

tional matrices: [HOWs vs. HOWs]; [WHATs vs. HOWs]; [HOWs vs. HOW MUCHes] and [WHATs vs. WHYs.]

QFD is a "market-in" approach instead of the traditional "product-out" approach. The voice of the customer is collected, analyzed, and translated into like attributes, such as {HOWs} (technical characteristics), functions, reliability, {WHATs} (objectives and goals), and product breakdown structure (PBS) such as systems, subsystems, components, parts, etc.

Figure 2.24 shows that in simple terms QFD relates a list of what the customer wants, for example, from a car door, easy to open and close, isolation, etc., to a list of attainable design requirements, for example, opening and closing effort, sealing insulation,

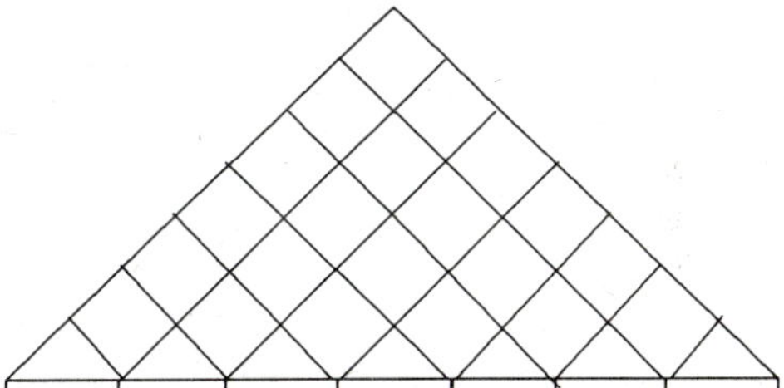

WHATs Vs. HOWs Legend

- Strong ●
- Moderate ○
- Weak △

		Open Close Effo-rts	Ener-gy to Close Doors	Check Force on Level Gro-und	Check Force on 10 Deg-ree Down-hill Slope	Seal-ing Insu-lation	Door Seal Resis-tance	Road Noise Reduc-tion	
		1	2	3	4	5	6	7	
Easy to Open and Close Door	1								1
Easy to Close from Outside	2		●	○	△			○	2
Stays Open on Hill	3		○		●		△		3
Leak Isolation	4								4
Doesn't Leak During Rain	5						●		5
Less Noise During Highway Driving	6							●	6
		1	2	3	4	5	6	7	

Source: Based on Houer and Clausing Data

FIGURE 2.24 Use of Quality Function Deployment: A Door Example

etc. To get the voice of the customer, the marketing people have to ask the right questions to the consumers. To respond to the question "how do I satisfy my customer?," the marketing people have to interrogate with the designers for the right answers. There is no point in the marketing people identifying something that is not feasible, or the designers identifying solutions that do not add value to the customer. Such issues should therefore be resolved early in the product definition stage.

In Figure 2.24 the central matrix is used to identify the strengths of relationships between the {WHATs} and the {HOWs}. For example, "door seal resistance" means that "rain does not leak in," but it will make it difficult to satisfy the "easier to close the door" criteria. The power of QFD is that it provides a logical way of focusing the analytical efforts of people (having different skills) and aligning them with selected actions (or indeed services—in Japan, banks use this technique). The simple matrix can be extended in several directions enabling customer-based assessments of current and competitive products. Critical parameters identified can be optimized in the design phase with Taguchi methods. QFD helps to deploy customer wants and needs to the appropriate design and delivery functions within one's organization.

2.8.2.4 Applications of QFD

QFD is a powerful technique which can be applied in the following manner:

- ***Management Tool:*** It serves as an organizational mechanism for launching new product development and making sure options and steps are not forgotten. QFD is a planning tool for translating customers' needs and expectations into product requirements. Thus, it reduces chances of downstream problems.
- ***Continuous Improvement Tool:*** A tool to improve current product functions based on customer desires. QFD is a systematic disciplined method that assists users in prioritizing resources, such as, when to focus time and where to concentrate on product improvements. It serves to identify nonvalue added activities by prioritizing what is not important to the customers, and streamlining what is important to the company.
- ***Requirement Adviser:*** A technique to organize the various levels of requirements as the product is decomposed from system, subsystem, components, parts to materials. It leads to more efficient, standardized, and consistent processes.
- ***Translation Tool:*** It transforms the requirements of one functional group into supporting requirements of another group. QFD is a type of conceptual roadmap that provides the means for interfunctional planning and communications. It streamlines communication about requirements among different functional groups, thus reducing chances of errors. It allows efficient coordination of activities with CE teams and timely completion.
- ***Mechanism:*** QFD is a mechanism to bring new products and improved existing products to market sooner with lower cost and higher quality. QFD ensures that product and processes will be designed right the first time and everything will fall in place.

Though QFD has been used in all of the aforementioned situations, the most common usage of QFD is during requirements analysis for product improvement. An example of this is when the design engineering changes are to be brought about in the next generation of the product that will incorporate a list of proposed customer desires for improvements. Usage of QFD for generating entirely new product ideas has been very limited. Clausing and Pugh [1991], have shown, however, that if the Stuart Pugh new concept selection method is coupled with QFD, its usage can be extended to *new product identification.* This enables teams to evaluate various concepts using the Pugh approach [Pugh, 1991] and later determine their importance ratings using QFD. Thus, coupling the two methods allows critical evaluation of a number of alternatives before a set of characteristics is finally chosen. The Pugh selection matrix becomes an input to the QFD process.

Similarly, Ross [1988] has shown that if the Taguchi method and design of experiments are coupled with QFD, its usage can be extended to process improvement. This enables teams to conduct a number of experiments to minimize the impact of variations in the process parameters using Taguchi method [Ross, 1988] and later determine their importance rating using QFD.

2.8.3 Pugh Concept Selection Matrix

The concept selection matrix initially proposed by Pugh is another matrix-based approach [Pugh, 1991] to measure a product concept. Like QFD, it is based on a list of product or customer requirements.

2.8.3.1 Steps in Pugh Methodology

The Pugh process uses the following seven basic steps (see Figure 2.25):

- ***Step 1: Generate an assortment of ideas or concepts.*** Ideas for the concepts are best generated by empowering the working team members or experts. Concept selection is best performed by teams of experts working together.
- ***Step 2: Prepare a criteria list.*** The team creates a formal customer or product needs/wants list called a criteria list. This information is based on voice of customers, product data, market analyses, and warranty and field data, as shown in Figure 2.22. The criteria list should be complete. If there are surprises or discoveries later, it may be possible to add them, but if they are overlooked altogether, it may result in the approval of a sub-optimum design. In general, each criteria should contain enough supporting features so that each, more or less, carries equal weight.
- ***Step 3: Pick a datum.*** Depending on the type of product being redesigned, a company may have a carry-over design, a proposed design, a supplier recommended design, a best-in-class competitive design, or a number of unsettled competitive designs. Out of the available alternatives (designs), one is selected as the datum, baseline, or standard for comparison. The team picks up its competitive features. The team creates a concept selection matrix—lists the design criteria with its features along the rows and the available design alternatives or concepts along the

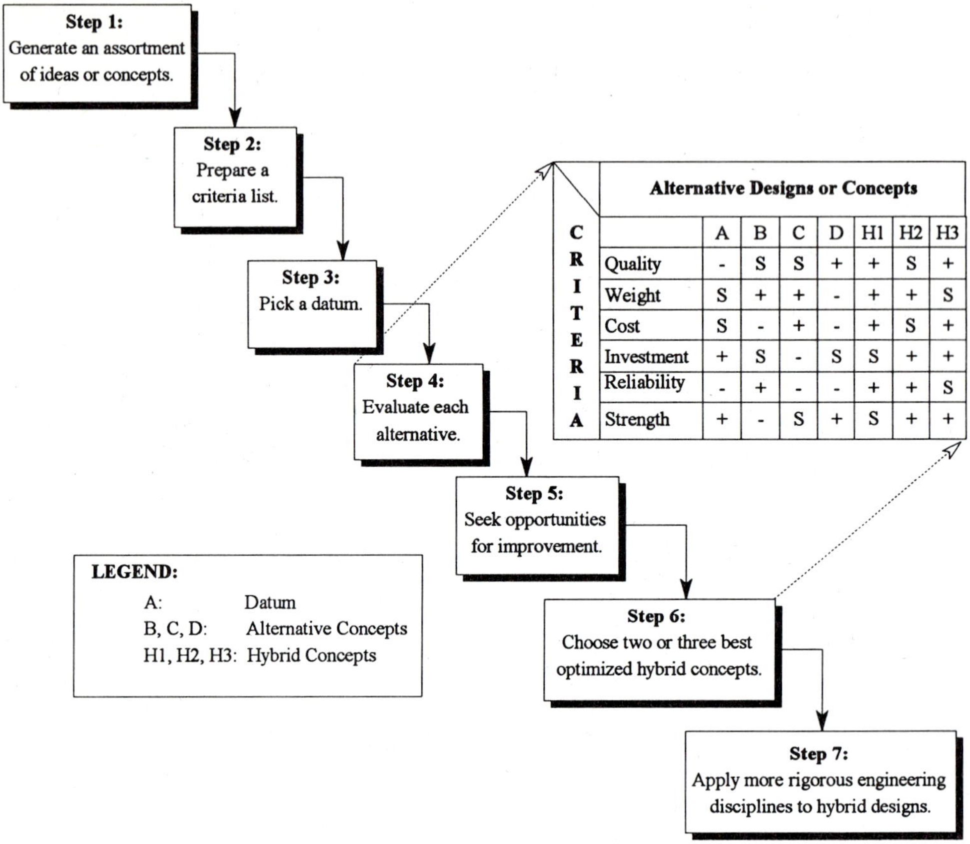

CRITERIA	A	B	C	D	H1	H2	H3
Quality	-	S	S	+	+	S	+
Weight	S	+	+	-	+	+	S
Cost	S	-	+	-	+	S	+
Investment	+	S	-	S	S	+	+
Reliability	-	+	-	-	+	+	S
Strength	+	-	S	+	S	+	+

FIGURE 2.25 Steps in Pugh Methodology

columns of this matrix. The column lists all of the design alternatives or concepts including datum.

- ***Step 4: Evaluate each alternative.*** The team evaluates each of the remaining concept or design alternatives, and compares their competitive features against the datum features for each of the criteria identified in the criteria list. The team then brainstorms and uses its collective experience to evaluate each alternative against each design criterion. The team determines the relative strengths and weaknesses and rates them using a simple comparison scale. A plus (+) is placed at the appropriate row and column to indicate that the particular alternative is inherently superior to the datum relative to that particular design criterion. A minus (–) is used to indicate that it is inherently inferior, and an "S" is used to indicate that it is essentially the same in terms of its ability to achieve those requirements or criteria.
- ***Step 5: Seek opportunities for improvement.*** Once this matrix is completed and the number of plusses and minuses are totaled, the selection matrix, depending on the outcome, gives a quick picture of the nature of each alternative design. Opportuni-

ties for improvement in the judgment of the team are then assessed. This consists of taking the good ideas and eliminating or reducing the weaknesses associated with those ideas to form a new concept or alternative design which is a hybrid of earlier alternatives or concepts.

- ***Step 6: Choose two or three best optimized hybrid concepts.*** The team can repeat steps 1 through 5 until the number of alternate designs or concepts are reduced into just two or three optimized hybrids. Here, an attempt is made to create hybrid concepts through redesign of the prior alternatives. The good points of the overall "bad alternative" are adopted into an overall "good alternative." The bad points of the overall "good alternative" are eliminated. This way the optimized hybrids will always turn out to be superior overall compared to any of the original proposals.
- ***Step 7: Apply more rigorous engineering disciplines to hybrid designs.*** More rigorous analysis such as FEA, FMEA, QFD, DOE, Taguchi, simulation, and prototyping can be applied at this stage to narrow down the hybrid selection to just one or two. A cost analysis is possible at this level of design. The detailed analytical techniques or testing helps to home in on the final decision or design alternative.

2.9 SEQUENTIAL VERSUS CONCURRENT ENGINEERING

At the beginning of the information age, we were wrongly led to believe that the control of information flow was the key component to success. We, thus, gave adequate power to the functional organization to control the flow of information. The organization common sense for this "era of control" then was hierarchical setup, that led to the creation of many tall "silos"—multilayered control structures. Product was task-focused and engineering was sequencing-based. During this age of control, linear product development appeared to be the right thing to do. The product design and development process was obtained in phases. The phases were essentially queued in consecutive order and segmented with some hard breaks between phases. Inputs from individual departments were sought in an ordered sequence as shown in Figure 2.26. Tasks were clearly labeled as requirements definition, product definition, process definition, and delivery and support. The tasks of the upstream phase were completed before tasks for a downstream phase were started. For example, requirements preceded product definition. Design engineers dominated the process (often drawing-based) in the earliest stages of conceptual design and preliminary development. Marketing experts give the needs to the designers, who determine product specifications.

In most sequential engineering processes, it was customary for the market research department to determine customer or user needs and throw its sales projection data over the wall to planning. The planning department developed the technical requirements for the product and threw its specifications over the wall to the product engineering group. This group then designed and developed the product on their own, in near-complete isolation from the production process. Later, the prototype was handed over to manufacturing so that their engineers could arrange to manufacture the product on a large scale. Seldom when a product develops this way will it go to production in one shot. Several engineering

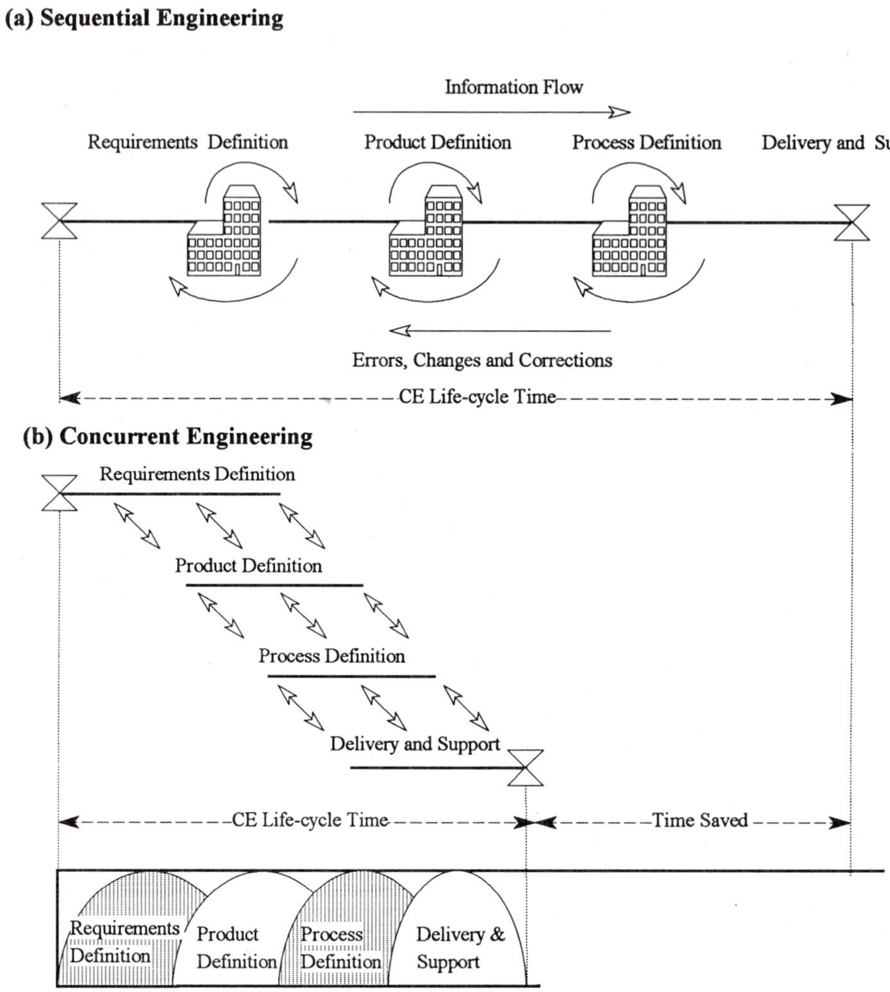

FIGURE 2.26 Sequential versus Concurrent Engineering

change orders are issued to fix problems, but due to time pressure, attention can only be given to fix the major show-stoppers (problems). Some of the major problems manufacturing engineers encountered are listed here:

- Unsuitable product design for production
- Unavailability of adequate manufacturing equipment
- Tight tolerance which could lead to extra work and high scrap generation
- Problem with parts assembly
- Inability to utilize the existing production equipment, tooling, automatic assembly, etc.

The feedback at each stage was in the form of errors, changes, and corrections. It took several rounds of negotiation and a long lead time before all pertinent design and manufacturing problems (errors, changes, and corrections) could be resolved. After a period of time, procurement experts must get involved to ensure that the necessary parts and materials are available for the assembly process. Next, marketing and sales personnel come in to introduce the product to its targeted market. The test analysis and fix (TAAF) process is often used to ensure component-level reliability. Analysis is conducted off-line and is treated as a way to fix design problems, not as a way to prevent them from happening. Should a design change be necessary, the control is returned to the front and the serial process is repeated. The difficulties occur as a result of what is termed the process *"serial engineering."* Here, each design phase starts mostly when the previous one is completed. During this process, an incomplete data is passed through a function or phase and augmented with new data only to be passed on to the next phase. The process is analogous to a relay race team that has to run x miles between the collective efforts of n runners (see Figure 2.27). Each runner is placed at a post equidistant (x/n miles) apart. Each runner is starting from its stationary position. The first runner takes off and passes the baton to the second runner at the end of x/n miles, the second to the third, third to the fourth, and so on. The last (nth) runner finishes the race.

If v_i is the maximum speed an ith runner can attain starting from its stationary position, then

$$\text{Average speed to reach its peak from its initial stationary position} \\ = v_i/2 \text{ miles/hour; for } i = 1, n$$

The time taken to finish the run can be computed as follows:

$$\text{Time it takes for an } i\text{th runner to traverse } (x/n) \text{ miles} = (x/n) * (2/v_i)$$

Total time it takes for all the runners to finish the relay race will be the sum of individual time. If we call this type of run as a serial relay run, the total time, T, is:

$$T\,(\text{serial-relay}) = \sum_{i=1}^{i=n} (x/n) * (2/v_i) \tag{2.90}$$

In the serial engineering process, the distribution is pretty much the same as the relay race, except the runner is replaced by a group of disciplinary experts (product development team); the distance to be traversed is replaced by the job to be finished; relay zones are replaced by the life-cycle phases for the product development; speed of the runner is replaced by the lead-time for completing a phase; mile post by the milestone set by the management or design reviews at the completion of each phase; and completion of the run is replaced by the design and development of a product. The total life-cycle time is usually a multiple of the sum of each of its individual phases. This type of approach is also known by other names, including *sequential engineering, linear* or *time-phased engineering,* and the *chimney method.* If we denote this process in short as SE, then:

$$\text{Total life-cycle time (SE)} = \\ [T_{\text{requirements}} + T_{\text{product def}} + T_{\text{process def}} + T_{\text{delivery and sup}}] * R_{\text{factor}} \tag{2.91}$$

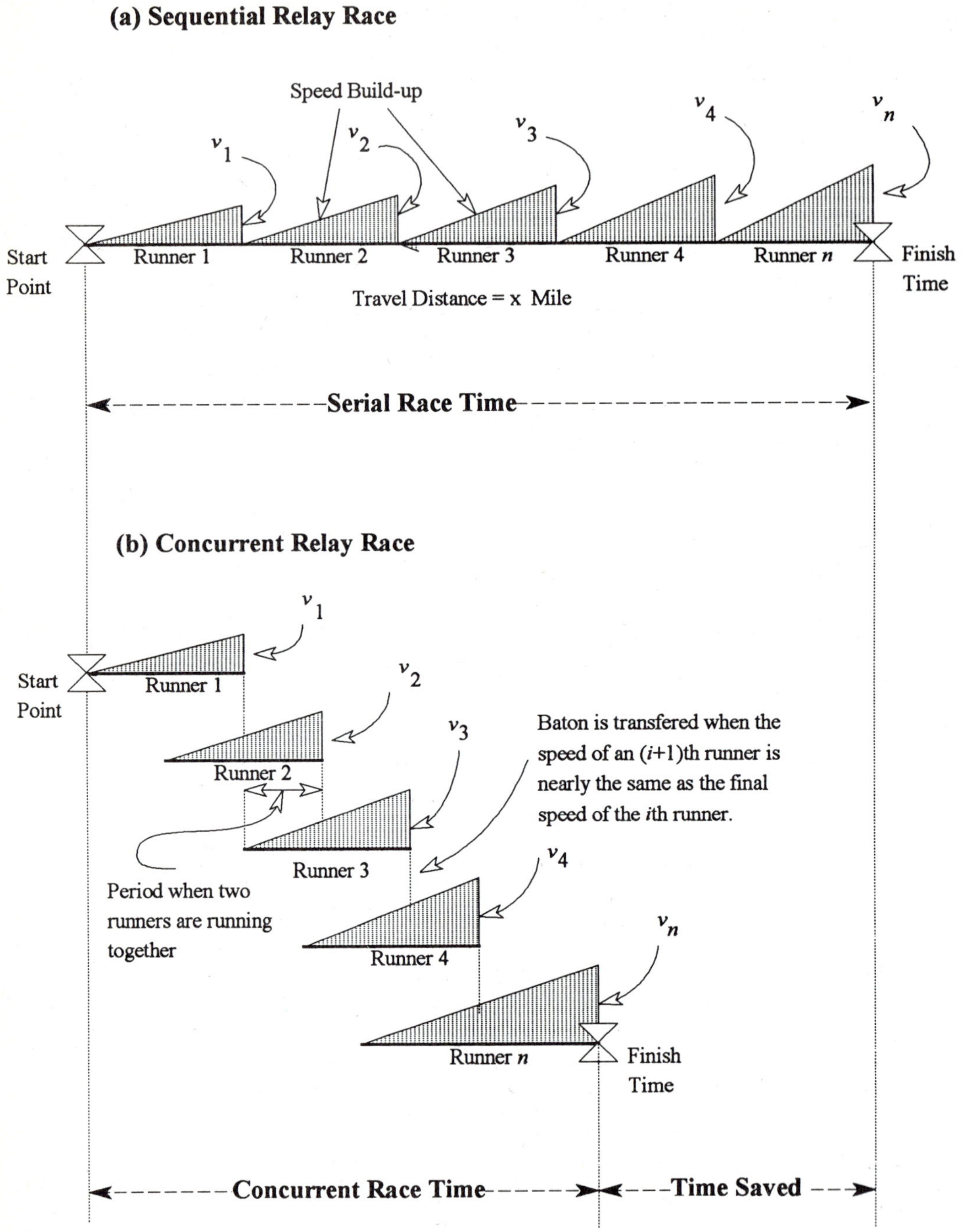

FIGURE 2.27 Analogy of CE to a Relay Race

Where, SE stands for Serial Engineering and $T_{requirements}$, $T_{product\ def}$, $T_{process\ def}$, and $T_{delivery\ and\ sup}$ represent the time it takes to finish a particular phase. R_{factor} is an equivalent repeat factor.

In SE, there is no consistent design, analysis, and documentation methodologies for conducting the business. Last minute engineering changes and higher product engineering costs result due to the lack of timely analysis of manufacturing considerations during early phases of product development. Preparation of data for analysis often takes too long, and by the time results are made available, the design often gets changed making the analysis useless. The result of such serial engineering has always been long design and build lead time, unnecessary complexity, and a scramble to obtain quality products. This shows in the high number of late-in-cycle engineering changes, design with excessive number of parts, retrofits, high customer complaints, manufacturing confusion, blaming mentality, high scrap and rework, increased field support and service, and only marginal customer appreciation or satisfaction compared to a competitor product. As a result, R_{factor} value for SE usually ranges above 2 or more.

$$R_{factor}\ \text{(for SE)} >>> 2.0 \qquad (2.92)$$

In recent years, computers have been extensively used in the product cycle. Product development cycle is now frequently carried out with the aids of computer tools such as CAD, CAM, CAPP, CII, etc. These tools automate the discrete manual process of product design without affecting the inherent linear (sequential) nature of the process. Although it was thought that this would decrease the development cycle time and increase the communication between design and manufacturing, this has not occurred. This is because, by automating the functions within a phase, it only reduces the time required for that phase, it does not affect the manual or serial process of passing the enriched information between the phases. A significant time is lost in maintaining a serial nature of the process and performing the manual control of the phase interfaces or intraphase data integration. This has created a proliferation of "islands" of departments due to the functional structure of the organization. Lack of management incentives for cooperation and inflexible culture of the organization to change have also impeded improvements in efficiencies and productivity. The impact of automation on the product development cycle has been, therefore, very little due to the unchanged inherent process.

The serial approach to product design, development, manufacturing, and marketing has several other shortcomings:

- It is based on the premise that a new phase cannot start until an old phase is completed and signed off. This usually means lengthening the product development cycle time.
- The linear input to product development implies that a significant portion (50% to 80%) of manufacturing cost may be committed before manufacturing engineers have a say in the product design.
- Due to the time loss, the final product may not remain suitable or viable for the market that was targeted initially at product launch.

In the 1990s, a new life-cycle management approach has emerged that focuses on time as critical force. It analyzes the PDD tasks' timing and duration across its entire organization with the goal of reducing tasks' time without any apparent loss in value to the finished product or service. This new approach focuses on tasks' time in all areas of product development that are value-added—from engineering to manufacturing, and from customer order to delivery. This management process is commonly referred to as *concurrent engineering (CE).* In CE, the tasks and phases run in parallel with feedback as and when needed, as shown in Figure 2.26. Each area of product development cycle has its own life-span that overlays the entire process. This series of smaller life-spans are shown in Figure 2.26 using "blinders". CE has brought these blinders closer together so they now overlap.

TABLE 2.5 A Comparison Table for Serial-relay and Concurrent-relay Race

Items	Serial-relay	Concurrent-relay	Concurrent Engineering	Remarks
Who are runners?	Individual runners.	A team of runners or partners.	Concurrent teams.	
Distance to be traversed	Mile-posts are placed at equidistant apart	Distance traveled by each runner is not necessarily the same.	Tasks or jobs to be done are not necessarily independent of each other.	
Relay-zones	Every runner starts from a stationary position. The mile-posts are placed at the start points. The end point of a previous runner and start point for a next runner is coincident.	The start point may be placed earlier from the mile-post. This allows the second runner time to build up speed before the baton is turned over. It means there are overlapped running zones.	The phases overlap. Two tasks are run simultaneously until the point when information build-up is adequate for the next task to proceed on its own.	
Speeds	Every runner starts from his/her stationary position (zero speed) at each mile-post.	Every runner starts from his/her stationary position but much before the finish point of a previous runner.	Information is built up to a point that can be useful for the early completion of the next task.	
Transfer-points	When a runner reaches the designated mile-post.	When the speed of the new runner equals or exceeds the previous runner speed.	The information build-up is sufficient to allow the next task to proceed independently.	
Mile-post/ Objectives	Minimize the time for traversing the distance between two consecutive mile-posts.	Minimize the total time for traversing the whole run.	Minimize the time it takes to do the entire job, not the completion of individual tasks.	
Completion Points	When the baton is finally passed to the last person in the race.	When the baton reaches the last post (how it gets there is immaterial).	When the whole job is successfully done (irrespective of order in which it is executed).	
Results/Outcome	All runners contribute to the best of their abilities.	Best of the individual and their combined team abilities/efforts.	Collaborative efforts.	

CE is analogous to a variation of the serial relay race described earlier (see Figure 2.27a). It is not essential that each runner be placed at an equal distance (for example, x/n miles) apart. Second, it is not essential that a runner has to start only after the previous runner has completed the x/n miles run. A runner can start earlier from his mile post, enough for him to catch up the speed and become equal to or greater than the speed of his partner, when the time comes to transfer the baton. The first runner stops when the second runner has achieved the first runner's speed. This way, the speed throughout the race remains fairly constant. If v_i is the maximum speed of each runner, time T for the serial-relay and concurrent-relay can be computed as follows:

The average speed for the race is equal to $(\Sigma v_i)/n$ and

$$\text{The time, the runners takes to traverse a total of } x \text{ miles} = \frac{(x * n)}{(\Sigma v_i)} \text{ unit of time} \quad (2.93)$$

If $v_i = v_o$, that is the same for all i's, $i = 1, n$; then it is not difficult to show that concurrent relay race would be 50% more efficient than the serial-relay race. If v_i is a linearly increasing speed from 1 to n miles/hour, that is ,

$$v_1 = 1; v_2 = 2; v_3 = 3; \ldots \text{ and } v_n = n$$

then

$$T \text{ serial-relay} = [\ (2x/n) * \Sigma(1/i)] \quad (2.94)$$

and

$$T_{\text{concurrent-relay}} = [\ 2x\ /\ (n+1)\ ] \quad (2.95)$$

Table 2.5 compares the differences between a serial and a concurrent relay race. It then lists the similarities between a concurrent relay race and CE process. Like in the concurrent relay race case, in the CE case, it is assumed that the concurrent work-groups can start working on their portion of the work, much before the previous work-group is finished. While the second work-group comes to a speed for understanding what the first group has done so far, real useful information can start flowing between the groups at frequent intervals to make both the units more productive. Instead of one time transfer of the baton in the relay race, the transfer between the consecutive work-groups are continuous and teamwork alike. The previous work-group stops their task, when no more input (by the incoming work-group) is required. The process continues when all the work-groups have made both their individual and team contributions. Individual contribution is when the teams are working alone (when there is no overlap) and team contribution is when they are working jointly as a team for a portion of the tasks which are overlapped. The passing of baton ceases when an artifact is finally manufactured and delivered to the customer.

It is usually necessary to use several computer-assisted programming tools to support this process. These tools should be capable of communicating with each other rather than creating isolated islands of automation. Not surprisingly, the cumulative time it takes to complete all the phases and tasks is much smaller than that of sequential engineering. R_{factor} usually ranges between 0.25 to 0.75:

$$R_{\text{factor}} \text{ (for CE)} \cong 1/2 \tag{2.96}$$

The word "engineering" in CE is used in a generalized sense. It is meant to include involvement of personnel from all required disciplines: engineers or nonengineers. The total process, from product development to mass production and sales, involves many nonengineers, including persons from the procurement, marketing, and sales departments. Hence, in concurrent engineering, all major parties involved in getting the product to market contribute to the development of the product. In contrast to the traditional sequential, linear, iterative (uncontrolled), and functional departmental practices, the concurrent engineering approach requires a parallel, interactive (managed), and cooperative multidisciplinary team approach to product and process development.

Just as craft manufacturing leveraged people's skill, muscles, and dexterity (*economy of skill*), concurrent engineering is about leveraging teamwork (*economy of cooperation*) to handle information and make informed decisions. Instead of using a human strength (physical power and muscles), CE is more directed toward utilizing the teams' intellectual power. These teams are not just the human teams (a term defined in Chapter 4, Volume I), but all the parties that are involved in the product realization process.

REFERENCES

AKAO Y. 1990. *Quality Function Deployment—Integrating Customer Requirements into Product Design.* Japanese Standards Association. Also in *Quality Deployment: A series of Articles.* ed. Yoji Akao and translated by Glen Mazur: Methuen, Mass: GOAL/QPC.

ALTING, L. 1993. "Design for a Life Time," *Manufacturing Breakthrough,* (May/June), pp. 29–33.

ANDREASON, M., MYRUP, and L. HEIN. 1987. *Integrated Product Development.* Springer, Berlin: IFS Publications Ltd.

ASQC/ASI. 1992. American Society for Quality Control, Automotive Division. American Supplier Institute, GOAL/QPC. Transactions from the fourth symposium on Quality Function Deployment (QFD), June 15–16, Nov., MI.

BOSSERT, J. L. 1990. *Quality Function Deployment.* Milwaukee, WI: ASQC Quality Press.

CARTER, D., and B. BAKER. 1992. *Concurrent Engineering: The Product Development Environment for the 1990s.* Massachusetts: Addison Wesley Publishing Company.

CLAUSING, D., and STUART PUGH. 1991. Enhanced QFD Workshop. Seminar Conducted at General Physics Corp., April 17–18, Troy, MI.

DEMING, W. E. 1986. *Out of Crisis,* 2nd ed., Cambridge, MA: MIT Center for Advanced Engineering Study.

DEMING, W. E. 1993. *The New Economics.* Cambridge, MA: MIT Center for Advanced Engineering Study.

HALES R., D. LYMAN, and R. NORMAN. 1990. Quality Function Deployment and the Expanded House of Quality International TechneGroup Inc. Ohio, USA, pp. 1–12.

HARTLEY, J. R. 1992. *Concurrent Engineering: Shortening Lead Times, Raising Quality, and Lowering Costs.* Cambridge, MA: Productivity Press.

HAUSER, R., and D. CLAUSING. 1988. The House of Quality. *Harvard Business Review.* Vol. 66, No. 3 (May–June), pp. 63–73.

KING B. 1987. *Better designs in Half the Time*—Implementing QFD (quality function deployment) in America. GOAL/QPC. Methnen, Massachusetts.

MUSSELWHILTE, C. 1990. Time-based Innovation: The New Competitive Advantage. *Training & Development Journal.* Jan., p. 55.

NEVINS, J. L., and D. E. WHITNEY, et. al. 1989. *Concurrent Design of Products and Processes.* New York: McGraw-Hill Publishing Co.

PARSAEI, H. R., and W. G. SULLIVAN (eds.). 1993. *Concurrent Engineering: Contemporary Issues and Modern Design Tools.* Boundary Row, London: Chapman and Hall.

PATTON, J. 1980. *Maintainability and Maintenance Management.* Research Triangle Park, N.C.: Instrument Society of America.

PRASAD, B. 1993. "Product Planning Optimization Using Quality Function Deployment", Chapter 5 in *AI in Optimal Design & Manufacturing.* ed. by Z. Dong, and series ed. Mo. Jamshidi. Englewood, NJ: Prentice Hall, pp. 117–152.

PRASAD, B. 1994. Competitiveness Analysis of Early Product Introduction and Technology Insertion. PED-Vol. 68–1, *Manufacturing Science and Engineering,* Volume I, ASME: pp. 121–134. *Proceeding of the 1994 Int'l Mechanical Engineering Congress and Exposition.* Nov. 1994, Chicago, IL.

PRASAD, B. 1995. "A Structured Approach to Product and Process Optimization for Manufacturing and Service Industries" *International Journal of Quality and Reliability Management,* special issue, Vol. 12, No. 9, 1995.

PUGH, STUART. 1991. "Total Design—Integrating Methods for Successful Product Engineering". Massachusetts: Addison Wesley Publishing Co. Inc.

ROSS, P. J. 1988. "The Role of Taguchi Methods and Design of Experiments in QFD," *Quality Program,* (June), pp. 41–47.

THOMAS, P. R. 1990. *Competitiveness through Total Cycle Time.* New York: McGraw-Hill Publishing Company.

WILSON, P. M., and J. G. GREAVES. 1990. "Forward Engineering—A Strategic Link between Design and Profit." *Mechatronic Systems Engineering.* Vol. 1, No. 1, Kluwer Academic Publishers, pp. 53–64.

TEST PROBLEMS: LIFE-CYCLE MANAGEMENT

2.1. How would you manage your company's product development efforts? What emphasis needs to be given to its subphases, if subphases are broadly categorized as definition, design, and redesign?

2.2. Discuss life-cycle management and explain its three main aspects. What are some of its strategies?

2.3. How does the complexity of a product affect life-cycle management?

2.4. Figure 2.13 details the risks and rewards associated with the various aspects of product development. One segment implies that there is a high ROI that can be associated with a high risk idea. Give examples where this will occur and where this may not be the case.

2.5. Give examples of components that fit each segment of Figure 2.13. Discuss the risks faced and the expected ROI.

2.6. Explain the time-to-market concept. Illustrate how to minimize it. What are some other factors on which the enterprise business operation is normally based?

2.7. Describe how the "future cost locked-in at any stage" and "total cost actually incurred" change if the CE tactics were applied from the onset of the program?

2.8. What are the two definitions of life-cycle time? What are the effects of long development time when the product is a consumer product and when it is a defense weapon?

2.9. Why with early product introduction, do companies quickly obtain a larger share of the marketplace? Give an example of an early product introduction that did not yield an advantage in market share. Why? What are some of the consequences of late introduction?

2.10. Explain why being second to market with a product similar to one previously introduced, with a cost less than the competitor and even slightly superior in performance is not good enough to win the lost customer?

2.11. Does CE provide a way to bring a product to market early? What are some other ways that companies can do the same? Are these techniques somehow related?

2.12. Can you approximately predict revenue loss due to a delayed market entry? What would be a loss to an automobile manufacturer if he/she chooses to defer a planned introduction of a three year automobile into the marketplace half a year late due to a major safety recall campaign?

2.13. What is the difference between Japanese and American companies in terms of operating practices. Briefly explain the differences. State some of the reasons for these differences and outline methods to overcome them.

2.14. What data would you focus on to ensure the right modification of your product development process?

2.15. How does a company decide what is the best time of introducing a new technology into an existing product?

2.16. If a company is six months late in switching to a new technology into a product, what impact would it have on the company's competitiveness, market share, and total profit margin (TPM). Assume both S-curves have 3 years useful life. Each S-curve has four segments with 30 degrees, 90 degrees, 60 degrees and 30 degrees slopes. Each segment is of equal length.

2.17. If for a company, it takes normally three years to bring a car to the marketplace, how much would a company save in percentage of the cost if the car was brought to fruition six months earlier (that is in thirty months)?

2.18. Draw a technology insertion S-curve, which has zero slope at its two tips. What does zero slope indicate?

2.19. If a car company expects 10% unit profit margin on each car it sells, what is the maximum company cost ($C(t)$) of the product if its market price ($P(t)$) is set at \$18,000/unit? How does the total profit margin (TPM) change if the company sells 200,000 cars rather than the 150,000 it originally projected?

2.20. Do the risks associated with new technology increase with shrinking life-cycle? If so, how would you go about managing such risks? Why should a company take any risks—calculated or otherwise?

2.21. Explain the concept of customer satisfaction. List the various sources of data for developing customer requirements in concurrent product development.

2.22. Going to market early has many advantages. In spite of those, explain how unexpected economic factors, market trends, and loss in product quality could quickly diminish the locked-in market share.

2.23. In terms of identified product's life-cycle cost drivers in the book, list some nontraditional costs attributed to these categories that can reduce profits or increase the price of the products to consumers. How much control does the company have in reducing the life-cycle cost drivers and who should be responsible to contain these costs?

2.24. How can one apply QFD for designing software applications? What are its benefits?

2.25. What part does QFD play in new product introduction and how can one use it? Describe other situations where QFD can be used.

2.26. What are the basic tools of QFD? How many rooms are there in expanded house of quality? What are the basic relational matrices of QFD?

2.27. To the best of your ability, speculate and compare the development time between a consumer product and a defense weaponry. Discuss the risk faced and associated ROI.

2.28. What aspect of the product design cycle is important? Illustrate your answer with Figure 2.5. How do such definitions reduce the number of engineering changes?

2.29. Assume that the sales curve for Company E follows a parabola and Company L stays fairly constant (follows a horizontal straight line). Company L however introduces the product one year late. The projected life expectancy of the product is 5 years. Assuming sales prices of the two products are identical; compute the total cost advantage of early product introduction.

2.30. Explain why it is more costly at the downstream stage to redesign the product even though we know how we have gotten there.

2.31. How do you determine the best-of-class product development practices?

2.32. What are the typical participants of a change management process work group?

2.33. Why is configuration management considered an essential part of the life-cycle management process? How does configuration management relate to CE?

2.34. Do you see a need for using Pugh's methodology during product development? Where is it more beneficial? Describe its seven major steps.

2.35. Are there any similarities between a relay race and a serial engineering process for a product development? Compare and contrast the differences between a concurrent and a serial engineering process in the context of a relay race.

CHAPTER 3

PROCESS REENGINEERING

INTRODUCTION

For many companies, to date, product development is characterized by long lead times, a large number of engineering changes, manufacturing complications, and, ultimately, excessive costs to satisfy the customer requirements. One area that is given more attention to in regaining the company's competitive position, is "product improvement". Product improvement in this case means improving product features—adding more competitive features (including bells and whistles) so that when the product comes out in the market it is attractive to the customer. The effectiveness and efficiency of the engineering, manufacturing, and/or business processes that support development and delivery of the products or services are given the back seat [Miller, 1993]. Lack of competitive situations in those companies is not due to product or technology related problems. It is the process—the way the companies carry out the design and development of a product and the way their teams spend their time. For example, technology might have gotten updated but engineers may not have changed corresponding design process or work habits accordingly. Design-process in this context means how a set of design-tasks is performed. Work-process means *how* a set of work tasks or job functions is performed. Most generally, a process is defined as a set of tasks arranged in a particular manner so as to transform a set of inputs into a specified set of outputs (goods or services). Since a task can be arranged in many different ways, there can be many process possibilities of performing a set of design-tasks, manufacturing-tasks or work-tasks. Some could be more efficient or effective than others.

Many progressive companies are interested in maintaining a competitive edge in the world market and in producing high quality products. They would like to do this at a lower net cost of production than their competitors. One easy way to increase one's productivity or efficiency is to squeeze more out of the current system. This often boils down to either working harder than before or automating some of the manual tasks rather than working differently. Automation of tasks to some may also mean repeating the same mistakes but doing so more often and more quickly. Many companies are finding that true increase in productivity and efficiency begins with such factors as clean and efficient process, good communication infrastructure, teamwork, and a constancy of shared vision and purpose [Prasad, 1995b]. The challenge is simply not to crank up the speed of the machines so that it outputs (per unit of time) are increased or doubled, but to change the basic machinery or the process that produces the outputs. To accomplish the latter goals, today many organizations are applying CE principles through benchmarking [Freeze and Aaron, 1990], CPI [Ezop, Jacoby and Leach, 1989], organizational restructuring [Juran and Gryna, 1993], "Ts" renovation [Prasad, 1995a; Perry, 1990; Carter and Baker, 1992] and process reengineering [Hammer, 1990; Hammer and Champy, 1993; Roberts, 1994]. The walls between various groups and departments, that existed a decade ago, have crumbled [Tutle, 1983]. Today, it is becoming more important to get inputs from all facets of organization, since no single group or department is expected to know or do everything. An organization is looking at how to run the business effectively and determine if it can be improved in some way. One item that is becoming important is that not only everyone in the organization should know what activity he or she is performing or engaged in, but the rest of the team should also know how their activities add to the big picture (constancy of shared vision and purpose). There are six parts (5W1H) to winning the competitiveness battle (See Figure 3.1):

- What? (tasks, object inputs, outputs, and process steps including measures, and decision points)
- Who? (talents, teamwork, subject customers, and supply chain)
- Why? (techniques, purpose, function, and rationale for decision making)
- When? (time, process order, and structure)
- Where? (technology gaps, space process relationship, and its context to the whole)
- How? (tools, method process boundaries, and process flow)

Knowing what information is required or task to perform is one sixth of this battle. How this information or task satisfies the corporate goals is a second part of the battle. The remaining pieces include the following:

- Who makes up the team?
- Who needs it?
- Why this technique or process will not work?
- Why is this information needed?
- Where will this information be used?
- When is the optimum time (to do it)?
- Where are the right places to use this technology?

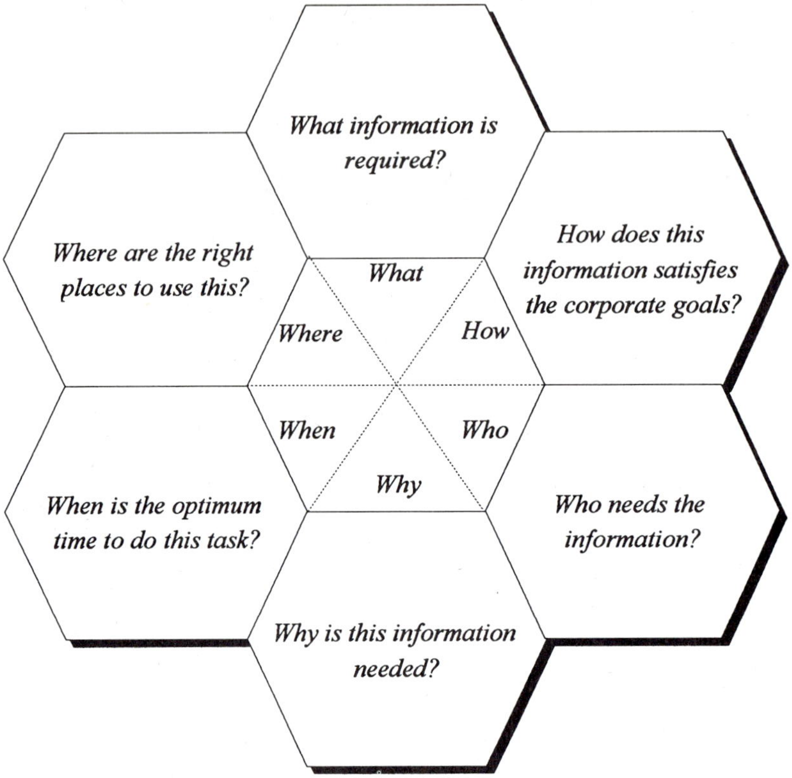

FIGURE 3.1 Questions Assessing the Improvement Needs

Though in Figure 3.1, parts are equally divided, in practice some pieces will be more important than others. "Who needs it" facilitates smooth communication and "why this information is needed" determines how valuable it is to a person, team, or the organization. "Where this information will be used" determines the right place, "when to do" denotes the right time and it is the contributing factor to meeting fast-to-market or concurrency goals. By knowing what we do today and how we do it, we will be in a better position to identify bottlenecks and barriers in the current system and possibly improve operations, if opportunities arise.

Process improvement is a concept often used to accomplish many lean production goals [Prasad, 1995b]. In some organizations, process improvement is often perceived as an after-thought—a functional service to be called upon periodically for productivity improvement. In such companies, process is viewed closely with work force productivity improvement (continuous process improvement) or organizational restructuring (reordering of tasks). Others who have paid little more attention have concerned themselves with process restructuring. Process restructuring is often targeted toward causing piece-wise or one-at-a-time improvements due to incremental or add-on approach of continuous improvement in manufacturing process, product quality, etc. However, the perception is clearly different in companies following lean production principles. In those companies, process improvement is seen as a pervasive set of renovation activities that form the life blood of the company's regenerating profit potential [Miller, 1993]. Process renovation is a modification approach that critically examines those 6 pieces of the battle, rethinks them through, and then redesigns the mission-critical products, processes, and services within an organization.

Process reengineering is used here to mean one or more of the above process modification strategies. In applying a process reengineering (PR) strategy, a team collectively comes up with a process that takes into consideration the needs of all individuals and work-groups, and above all, the needs of the company as a whole. The choice of the term "process" in PR speaks loudly that the focus is on processes as opposed to products. A process may refer to a design-process, a manufacturing-process or a work-process. The term "reengineering" implies that the change or effort is directed toward an array of process modification strategies. More frequently, reengineering implies either starting with a clean slate (a new process) or radically overhauling—meaning replacing the old processes with completely new ones. During an implementation of PR, the following is assumed:

1. An organization is in constant touch with new technological advances in all related fields such as engineering, process, computers, and systems.
2. The team regularly examines those latest advances in the fields, benchmarks and then reviews some key candidates for improvements in the product life cycle.
3. An enterprise employs those technological advances, if they sound appropriate or applicable, as a part of overhauling the process wagon to improve its efficiency, productivity, or performance. In actual practice however, these are difficult goals.

Reengineering generally provides companies with opportunities to do things differently and creatively and to distinguish themselves from their immediate rivals.

3.1 UNDERSTANDING AND MANAGING CHANGE

Changes are an essential part of any improvement. Whatever new steps or new tools we introduce require change. With the new tools, if we do not make the corresponding changes in the processes or work habits, we would be making the same mistakes but perhaps more often with the new tools. In other words, we may be computerizing or automating a bad process. So the question we must ask is how to successfully introduce and manage change.

- **The first step** before we introduce a change is to understand the change process. In CAD system for example, each change generates and stores a new version number of the design files, so that the designer can backtrack if required. Understanding the change process requires knowing

- When to change
- How to change
- How to cause change
- Where to change
- What to change
- How to promote change so that the improvement cycle repeats

The key factors that influence the change process are 7Ts (tasks, talents, teamwork, techniques, technology, time, and tools) [Prasad, 1995a]. Sometimes where to change and what to change has involved performing strategic review, which is an expansive term for a well-known competitiveness analysis process called SWOT. SWOT stands for *S*trengths, *W*eaknesses, *T*hreats, and *O*pportunities. Market analysis provides the points towards each of the four SWOT groups. They are listed in four quadrants of an axis-diagram as shown in Figure 3.2.

- **The second step** is to manage the change. Corresponding to the steps of understanding change there are also six steps to managing change:
  - Leading the change process
  - Setting the direction

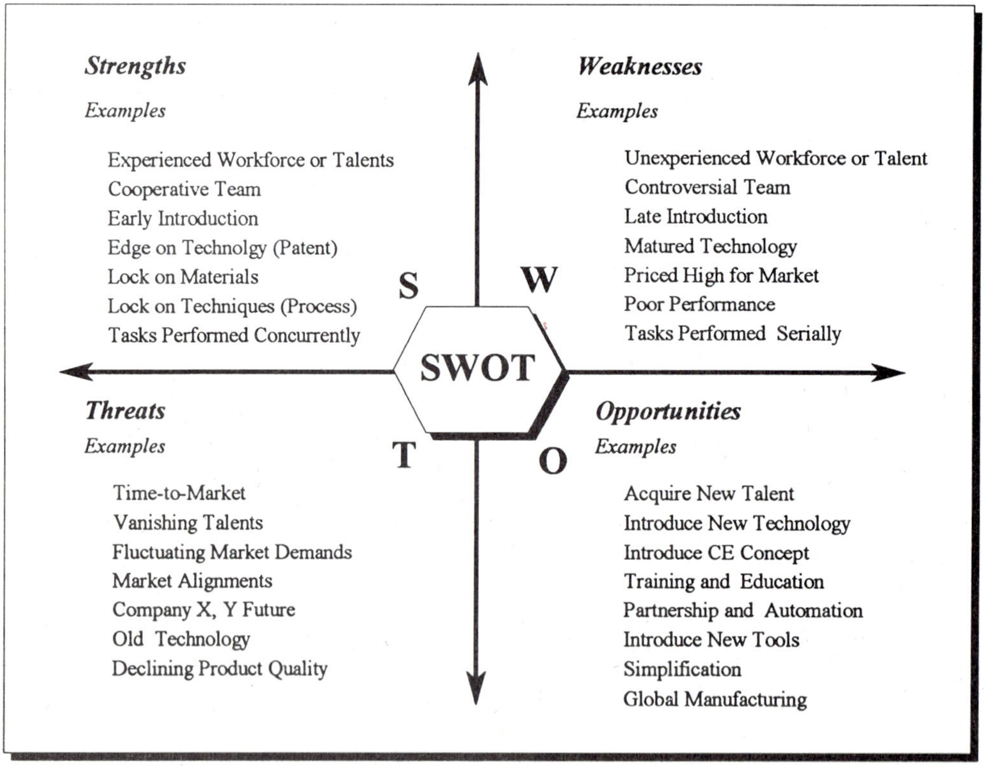

FIGURE 3.2 SWOT Analysis

- Creating the environment for change
- Challenging past practices and excuses
- Removing the barriers and roadblocks
- Rewarding the right things so that change continues to evolve.

The first and second steps are schematically shown in Figure 3.3 by two annular rings. Managing the change process is also referred to as $LSC^2 R^2$. Where the abbreviated word represents the first letter of the steps involved. Setting the direction involves method of data collection, determining the change frequency, and determining the change complexity. Collection of data that measures the number and magnitude of changes is important. The estimation of the frequency of changes (number of changes per unit of time) is usually made by monitoring the CAD files of the designers. The way to determine the change complexity is to examine the CAD file progression and to look for change notes, file numbers, dates, and times. Challenging past practices and excuses involves, to a large ex-

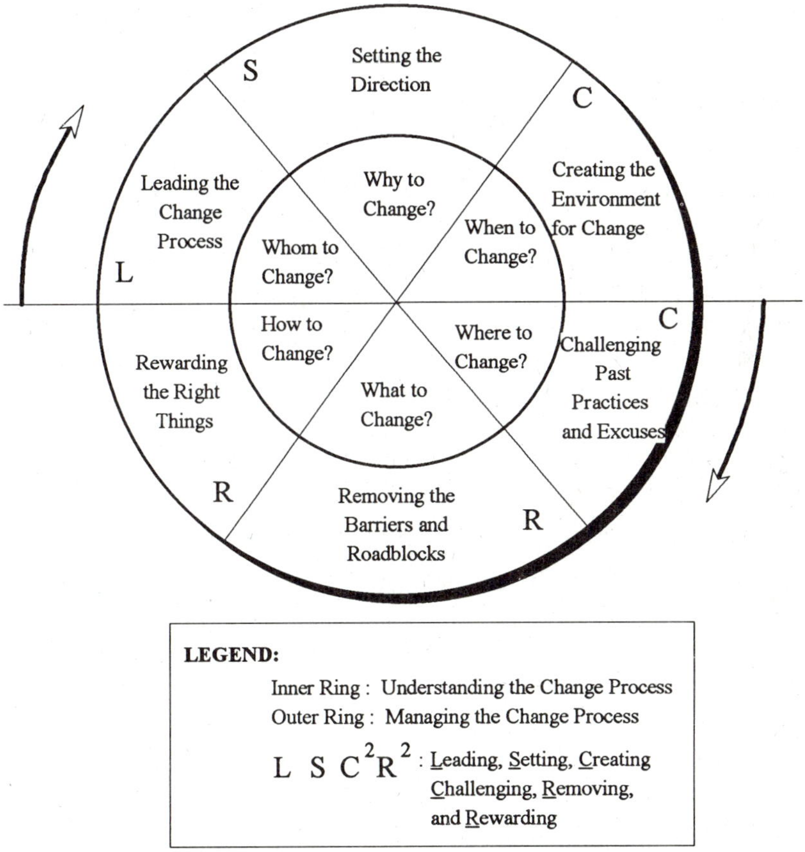

FIGURE 3.3 Understanding and Managing the Change Process

tent, understanding the sources of waste. Waste exists in all work activities, all process tasks, and at all levels in the organization. Taiichi Ohno has reported seven types of waste (muda) commonly found in a manufacturing work-site [Ohno, 1988]:

- Waste of overproduction
- Waste of correction
- Waste of material movement
- Waste of processing
- Waste of inventory
- Waste of waiting
- Waste of motion

The one type of waste (muda) that is missing from the Ohno's set is the waste of information movement [Prasad, 1995b]. Waste of information movement is concerned with unnecessary transfer of information between two or more dissimilar systems (computing systems or otherwise). Examples include conversion from one format to the other, upload and download of information, files retrieval and storage, unnecessary notification or notes, one-to-many communications instead of posting publicly (many-to-many), data security, etc. This completes the set of eight wastes shown in Figure 3.4

Removing the barriers and roadblocks to a large extent involves eliminating waste and the associated reworks. W. Edwards Deming has proposed fourteen things that companies, large or small, can do to ensure that change is an ongoing and positive experience [Deming, 1993]. These product improvement efforts can be categorized into four primary reengineering strategies:

1. A set of continuous process improvement (CPI) tactics (e.g., consistent or common environment)
2. A set of restructuring tactics (e.g., common best corporate system)
3. A set of organizational traits (e.g., agile and virtual organizational traits)
4. A set of renovation tactics (e.g., best industry practices, innovative and unmatched practices)

The key in managing change is, therefore, to establish an optimal balance between the types of process improvement strategies that were chosen and introduced from each category. Product and process reengineering follows its own life cycle. Figure 3.5 shows degree of severity in managing change. A soft common process improvement strategy is benchmarking. Benchmarking is a strategy that is common to all four reengineering strategies. Benchmarking is an aboveboard and perfectly legal way of finding out how others are doing compared to one's own system. One could also learn from this exercise and imitate what is identified as better in the competitors' product or perhaps improve upon their approach. Other soft strategies are "as-is" flow-charting, value engineering, value analysis, etc. Figure 3.5 illustrates four approaches of reengineering. This section begins with CPI strategies and traits, to effectively manage the changes.

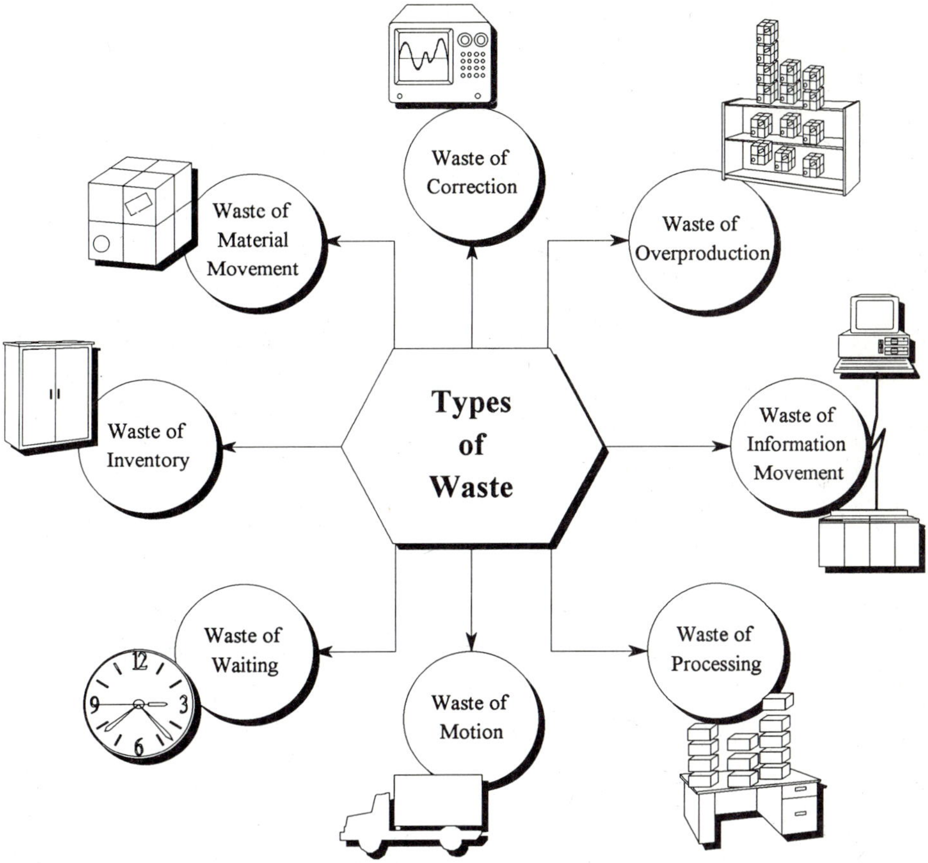

FIGURE 3.4 Eight Types of Waste

3.1.1 CPI Traits

Continuous improvement is the basic trait for causing change. If the process is stable, CPI allows pace with the known common changes. In CPI, some prevalent tactics used are to identify and eliminate wastes and to identify and eliminate rework. Some popular CPI traits include the following:

- Motivation and reward system
- Identification of key elements of scope and applications
- Employee participation
- Management commitments

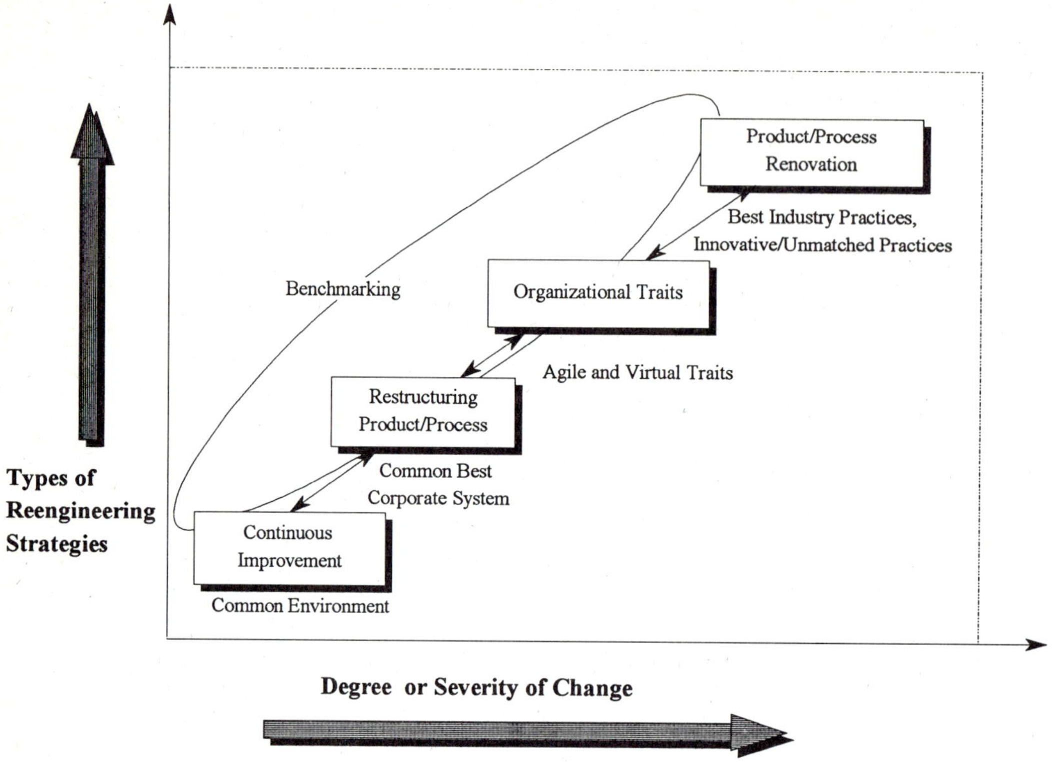

FIGURE 3.5 Degree of Severity in Managing Change

3.1.2 Restructuring Traits

Restructuring is the next level of process improvement strategy. Restructuring means transforming from old ways of conducting business to a new way:

- Use the same level of abstraction in product, process, enterprise, and behavioral modeling
- At a minimum, maintain the systems' performance (product functionality and semantics) and efficiency at the same level as the old

Some examples of restructuring include the following:

- Refocusing the efforts in the definition phase, so that the product is done right the first time it is released for design intent
- Prioritizing the tasks with the customer in mind
- Defining a common best corporate system for CE

Some examples of restructuring traits include the following:

- Partnering with supply base, collaboration, supply chain
- Cross-functional integration
- 3Ps (policies, practices, and procedures)
- Empowerment

3.1.3 Organizational Traits

Over the last few years there has been a flurry of activity in the study of manufacturing systems, and, more particularly, the mechanical design process. There is no consensus about what the term "design" means, nor is there an agreed upon description of a process improvement methodology. Often, process improvement has been wrongly construed as simply an organizational restructuring. By carefully restructuring departments into a modern multi-functional setup, an enterprise cannot expect to reap all the desired productivity gains. Though organizational restructuring has the potential of breaking down cultural barriers, the conventional product realization process, such as serial engineering, remains intact.

One of the important elements of "managing change" is the organizational traits. There are two elements of organizational traits: agile and virtual. They form the left arm of reengineering strategies. The right arm consists of CPI (continuous process improvement) and restructuring. These two arms are sandwiched between renovation traits at the top and CE infrastructure at the bottom (see Figure 3.6). There are four agile organizational traits supporting the business goals:

- Lean manufacturing seventeen tactics
- Reconfiguration
- Responsive
- Plug compatibility

Likewise, there are four virtual organizational traits supporting the business objectives:

- Decision making
- Team cooperation
- Virtual links
- Communication networks

Chapter 6, Section 6.6 discusses these traits in much greater depth.

3.1.4 Renovation Traits

Renovation is the highest level of strategy for managing change that commonly cannot be handled by the continuous improvement, restructuring methods, or organizational traits. In the words of Hammer this type of, "reengineering strives to break away from the old rules about how we organize and conduct business . . . (by) recognizing and rejecting

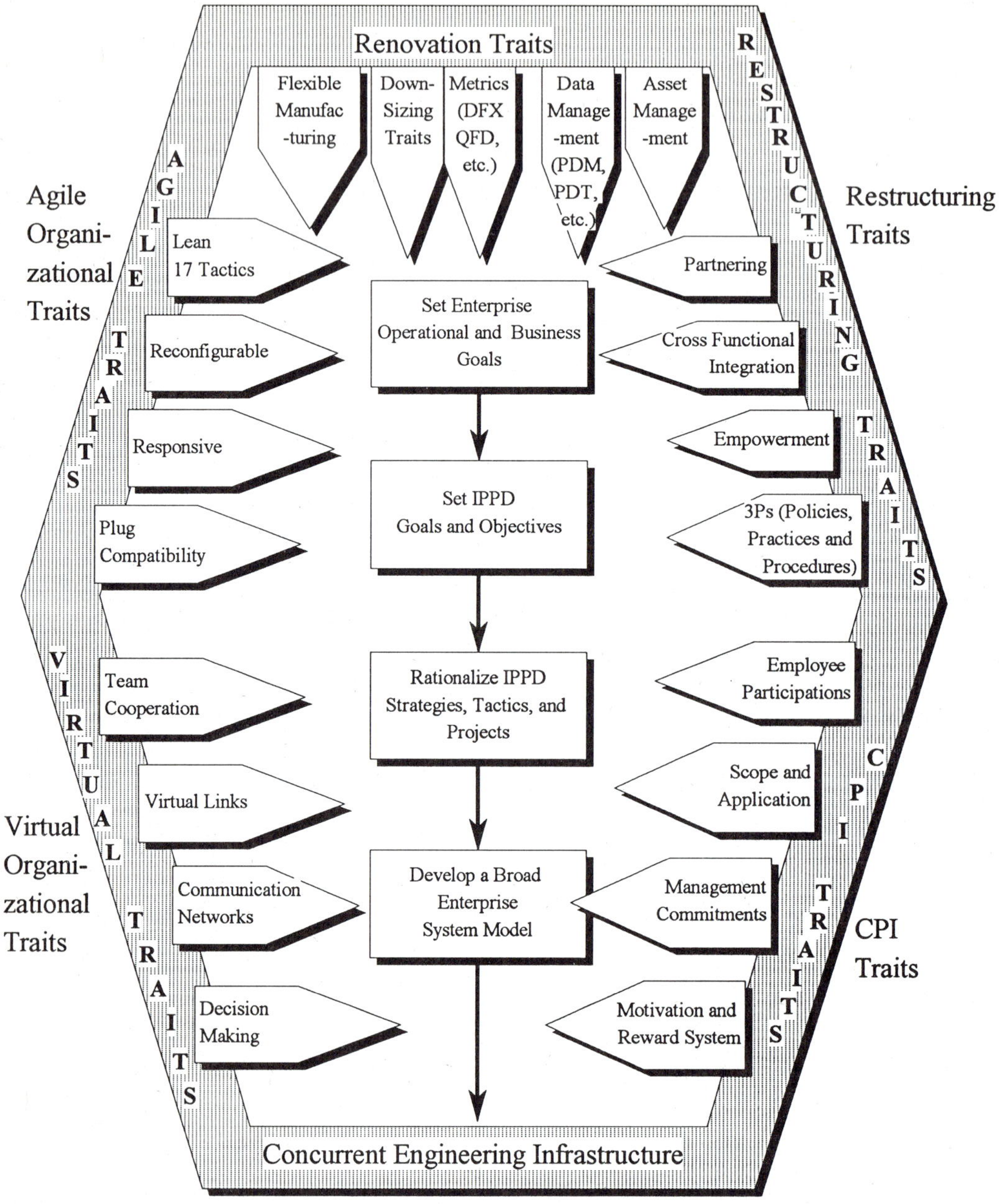

FIGURE 3.6 Reengineering Strategies and Business Goals

some of them and then finding imaginative or innovative new ways to accomplish work" [Hammer, 1990]. In essence, the three are complimentary rather than opposing approaches to improving processes. Some examples of renovation traits include the following:

- Flexible manufacturing
- Down-sizing traits
- Metrics (DFX, QFD, CAE, etc.)
- Data management (PDM, PDT)
- Asset management (see Figure 3.6)

3.2 REENGINEERING APPROACHES

Reengineering plays a pivotal role in the CE process. The key to any (manufacturing company) competitive posture lies in its ability to reengineer a business for agility—both physically and logically. Factories, systems, and organizations must differentiate and remove nonvalue added functions from the chain of design-tasks, manufacturing-tasks and work-tasks, and foster open lines of communications. Reengineering helps define strategies for bringing manufacturers, suppliers, and customers closer together.

Reengineering means taking steps to redesign and simplify business systems and processes, search out best industry practices (3Ps) to develop a more competitive work force, and to explore new process methods (see Figure 3.5). It fosters out-of-comfort-zone thinking, relies on value-added benefits to both the customer and the business, and focuses heavily on 7Ts (talents, tasks, teams, techniques, technology, time, tools, etc.). Reengineering requires follow through until the new process is firmly entrenched. CPI is one of the simplest reengineering strategy. Work-groups empowered and charged with CPI typically lack the authority, perspective, and/or capability to implement radical changes that cut across functional lines. It is not true for restructuring. The restructuring team involves multidisciplinary leaders from each facet of the organization. Such technical leaders charged for restructuring have the needed authority and perspective to implement radical changes that cut across functional lines. Unlike restructuring, renovation involves alterations in the level of abstraction to reconfigure the subject system. It may involve reconstituting this subject system into a new form or to a new level of abstract descriptions, and a prior implementation of the altered form. In renovation a majority of the experts in the team come from outside the immediate area of investigation. This way, the changes proposed can impact other processes or otherwise impact the interest of the organization as a whole rather than a few specific units. An example of renovation is the reorganization of a company by a "product line." In this case, each area of the organization, which is involved in the particular product, must work together with common organizational goals and objectives. Incorporation of best industry practices or innovation, and unmatched practices are some examples of renovation efforts.

There are many ways to accomplish reengineering: top-down, bottom-up, or incremental. The top-down, and bottom-up approaches are discussed in Section 3.6.2. In the case of incremental reengineering, new tools and systems can be introduced one batch at a time. As the team gets familiar or trained in one set of systems or its use, other sets are introduced progressively. The deployment is done on an incremental, *"pay-as-you-go"* basis to ensure a managed impact.

Reengineering can also be achieved by combing a couple of process improvement strategies discussed in Section 3.1. The two most popular reengineering strategies are listed as follows:

- ***Combining CPI with Restructuring:*** Some processes do not need to be redesigned at frequent intervals, since most of the original processes remain intact. It might be just enough to first design the system for optimum performance and then incrementally improve it (the system) over time as the needs for fine tuning become evident. Reengineering in this case consists of a one-time design for optimum performance and a series of CPI, followed by restructuring fired at regular intervals as shown in Figure 3.7a.
- ***Combining Restructuring with Organizational Traits:*** Some processes tend to have a useful life of their own. Other times, due to changing external environment, market conditions, or changing technology, these original processes no longer remain an effective solution/option. In such cases, these processes have to be restructured a number of times throughout the product's life cycle. Applying agile or virtual organizational traits at frequent intervals may very well be in the best interest of the corporation to stay healthy for a long haul. Reengineering in this case consists of a one-time design for optimum performance and a series of restructuring followed by the application of organizational traits at regular intervals as shown in Figure 3.7b.

3.3 TENETS OF PROCESS IMPROVEMENT (PI)

Regardless of how a process ultimately becomes implemented or accomplished, process improvement (PI) is always concerned with defining a broad range of metrics: some that are productivity related (how efficiently are the resources being utilized) and some that are performance related (how effective are the results or the returns). Productivity measures involve anticipated or measured levels of activity, such as the number of parts manufactured per unit of manpower, number of lines of coding per unit of time, number of reports done, etc. Performance measures involve anticipated or measured level of outputs relative to a specified set of goals—for example, expected return on investment, customer satisfaction, expected or measured profitability, market share, etc. PI metrics are necessary for both program planning and program evaluation. Program planning relies on past performance and judgments whereas program evaluation needs measurement information combined with judgments in computing either the efficiency or the effectiveness factor. In this context the following definitions apply [Roberts, 1994]:

- ***Process efficiency*** is concerned with degree of economy—how well a process uses available 7Ts (talent, tasks, teams, techniques, technology, time, and tools) to achieve

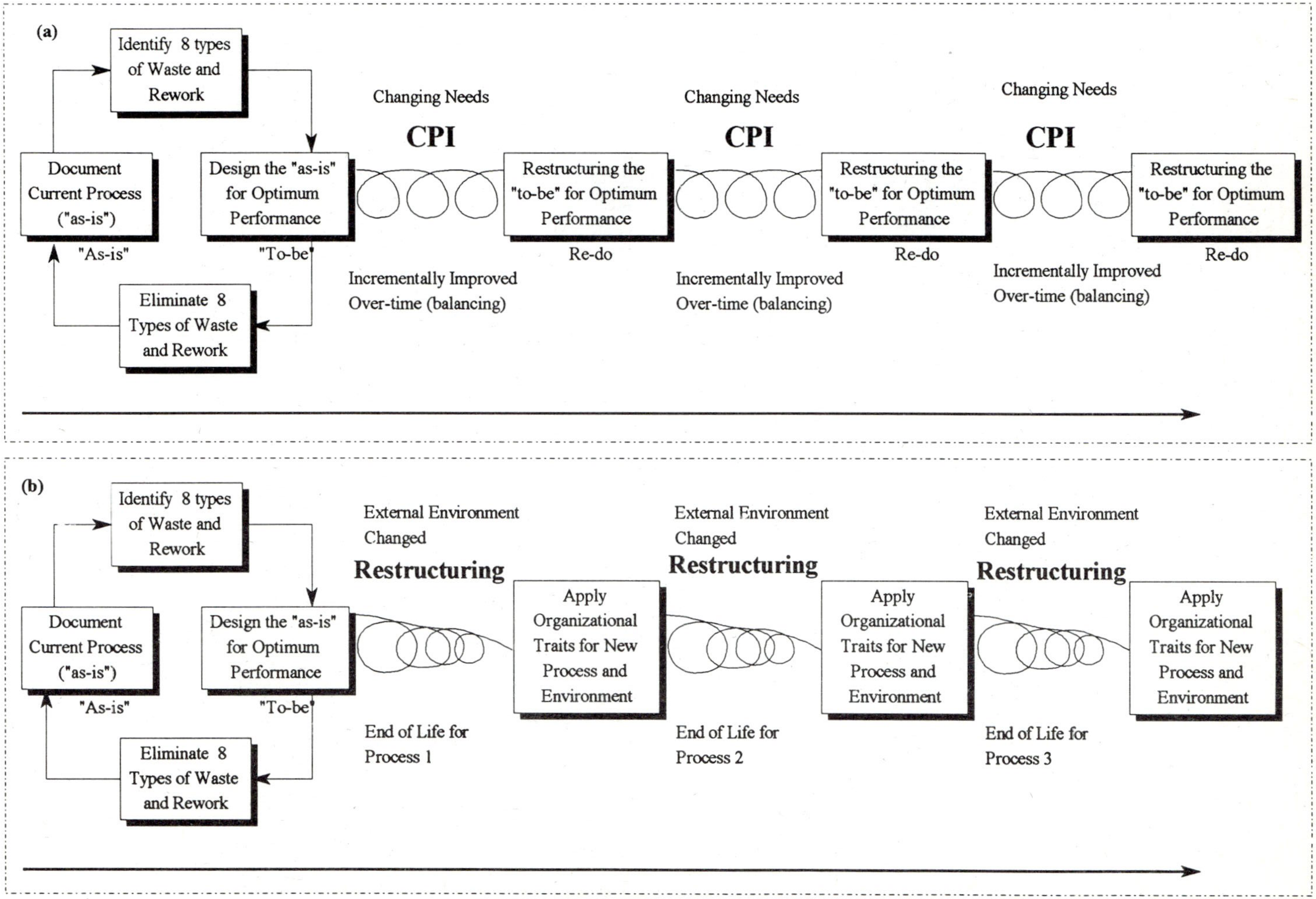

FIGURE 3.7 Combined Reengineering Approaches

the desired results. Improvement in process efficiency is concerned with eliminating process waste and reworks. *"Waste of processing"* is the unnecessary process-related efforts, which add no value to the output (product or service) (see also Figure 3.4 for other types of waste). Examples of "*waste of processing*" include expensive machines or a process to produce the same quality part, lack of clear customer specifications, enhancement that is transparent to the user, process bottlenecks, etc. Examples of process rework include endless refinements, repeated functions, unnecessary iterations, redundant approvals, etc. Waste of processing is of two types:

1. Waste caused by the individual slack conditions
2. Waste caused by the lack of integration or coordination

During waste of processing slack condition, one or more of the following six elements are active at the work-site: machine, management, manpower, materials, method, and money. They are analogous to the input drums of a friction powered system as schematically shown in Figure 3.8. Central drum is an output drum, which is being powered by a set of six surrounding friction drums. Friction is the driving force in transmuting power. The central drum denotes the return on investments (ROI) or profitability. The six rotating friction drums represent the individual contributions to the ROI. If there is no slip between the friction drums and the center drum, the system will operate at its full efficiency. The no-slip conditions are given by:

$$\omega_i * d_i = \text{Constant} \qquad \text{for } \forall\ i;\ i = 1, 6 \tag{3.1}$$

Where ω_i is the angular velocity of an ith friction drum and d_i is the corresponding diameter of the ith friction drum. If all of the six drums are not synchronized, slip will occur at the interfaces. A nonsynchronized condition is given by:

$$\omega_i * d_i \neq \omega_0 * d_0 \qquad \text{for } \forall\ i;\ i = 1, 6 \tag{3.2}$$

Where ω_0 is the angular velocity of the central drum and d_0 is the corresponding diameter of the central drum. The symbol ≠ denotes "not equal to." The above two equations are true for all values of i.

Efficiency is the result of the integration of 6Ms (machine, management, manpower, materials, methods, and money) at the work-site. Some common factors that can cause slips (or waste in the manufacturing setup) are unevenness of the mating surface, overburden (due to heated or loaded conditions), and the process method used. The process method is the associated mechanism to transfer energy from one drum to the other. In the case of the drum example, friction represents the process method. Slip will deteriorate the individual effectiveness of the friction drums. Similar to slips—which is due the individual contribution of the 6Ms—overburden, unevenness, and process methods are also caused by lack of integration or lack of synchronization between 6Ms and the outputs.

The lack of efficiency has an effect in reducing the product's life-cycle time. There exists an inverse relationship between process efficiency and the life-cycle lead time. Figure 3.9 shows the relationship between process efficiency, η, and the lead time, T. Figure 3.9a shows the trend when the process is at an original un-

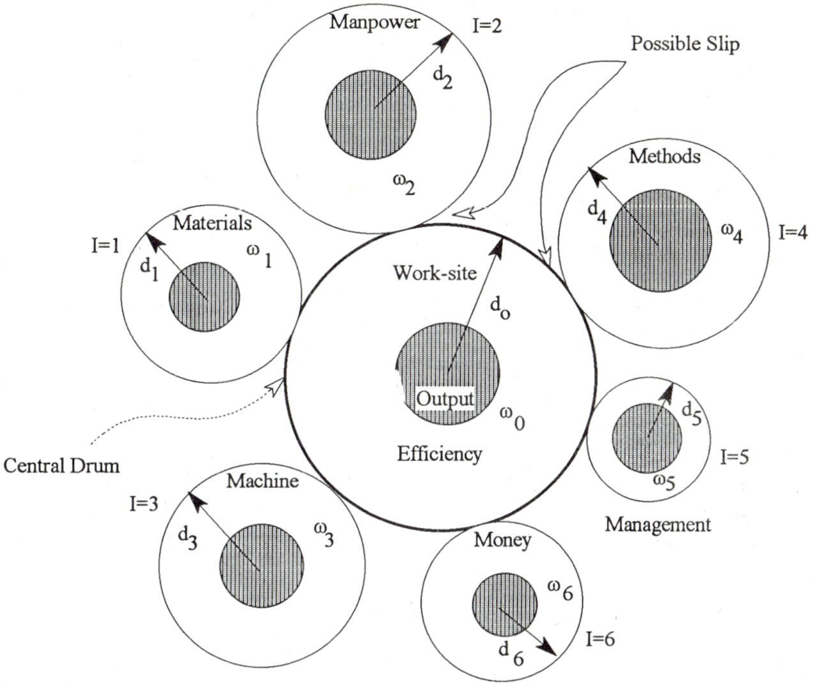

LEGEND:
I = An Identification for a Rotating Friction Drum
d = Diameter of the Friction Drum
ω = Angular Velocity of the Friction Drum

FIGURE 3.8 Elements of Process Efficiency at the Work Site

changed state. The dashed lines show how the lead time decreases from T_1 to T_2 when the efficiency is improved from η_1 to η_2. Figure 3.9b shows how conditions might change if the original process is altered or improved. Three curves are shown (in Figure 3.9(b)) corresponding to the following three situations: When the initial state (given process configuration) is altered (a) using CPI, (b) using restructuring techniques, or (c) using renovation strategies. In each case, at the start of the improvement process, the rate of decrease or drop is obviously quite steep indicating that a small increase in efficiency has a large impact on the lead time. This trend slows down as incremental improvements continue to be made in the efficiency of the original state of the process. At a later point in improvement, it is more difficult to squeeze in additional decrease in cycle lead time. A lot more effort (7Ts) is required to achieve meaningful gain in lead time. This means the efficiency has reached its peak for a given state of process configuration. In Figure 3.9(b), therefore, three peaks are shown corresponding to these three process improvement situations. Clearly the peak efficiency is a function of the (i) state of the current

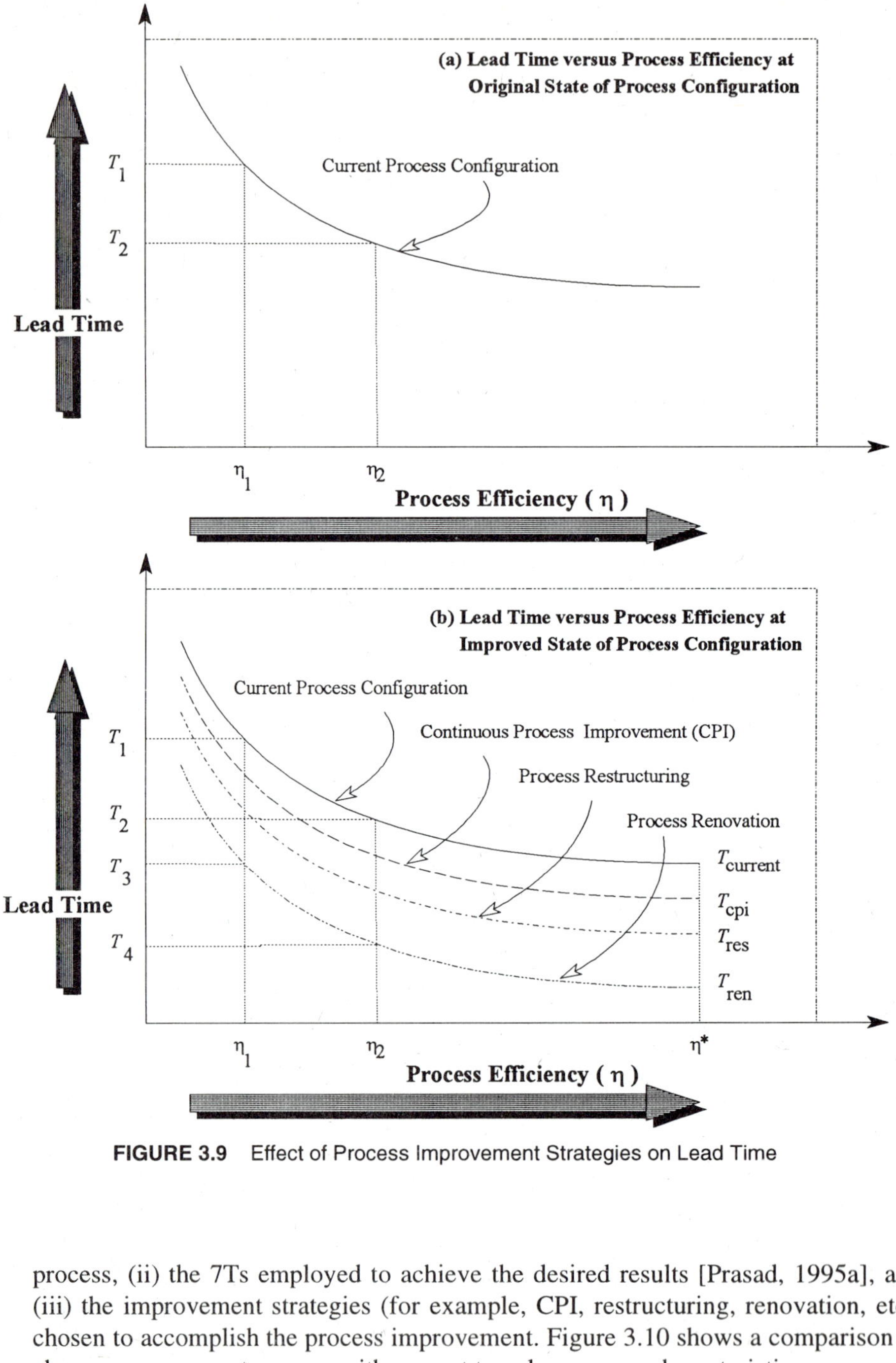

FIGURE 3.9 Effect of Process Improvement Strategies on Lead Time

process, (ii) the 7Ts employed to achieve the desired results [Prasad, 1995a], and (iii) the improvement strategies (for example, CPI, restructuring, renovation, etc.) chosen to accomplish the process improvement. Figure 3.10 shows a comparison of change management process with respect to a dozen or so characteristics.

Elements	Continuous Improvement	Restructuring	Renovation
Requirements	Little investment, but collective effort and time put in are large	Little investment, but more management effort is required	Large investment, but little effort to maintain
Approach	Collective input, every vote counts	Selective input, key groups consensus driven	Rugged individualism, results driven
Basis	Built on conventional know-how and attitude	Built on existing trained staff, leaders, new training, and reorientation	Built on technology transfer, in-house innovation, or technological breakthrough
Orientation	People/work-groups	Organization, leaders, management staff	Technology, process, methods, etc.
Involvement	Everybody	Leaders and management staff	Selective, few "champions"
Focus	Geared to eliminate waste, defects, error, or to make process improvement	Geared to change the pace, manpower rotation, new product orientation	Geared to make quantum jump in technology use or break traditions
Time-frame	Continuous and ongoing	Incremental and periodic	Intermittent and fragmental
Mode	Maintenance and Improvement	Reorgaize and Restructure	Scrap and Rebuild
Change	Gradual and Components-wide	Incremental but system-wide	Radical and sometimes volatile
Pace	Slow, smaller steps	Medium and large steps	Giant leaps
Effect	Long-term, permanent, and unnoticeable	Short-term, permanent, and sustainable	Short-term, dramatic, and noticeable
Constraints	Time	Skills, employees' talents, ability to manage	Technology feasibility, applicability, or cost
Return on Investment (ROI)	Mild	High	Dramatic

FIGURE 3.10 Comparison of Change Management Process

- ***Process effectiveness*** is concerned with how well the process actually accomplishes its stated goals, here again from the "constancy-of-purpose" perspectives. Managers of concurrent engineering teams face a demanding complexity of unconnected functions and processes as they attempt to improve enterprise operational efficiency. There is a tendency in most teams to operate out of their own self-interest, and within their own comfort zone. They tend to create goals and objectives that primarily address their own unit's contribution without much regard to their impact on the rest of the organization or teams. This has been a part of our behavior psychology and cultural heritage for generations. The problems arise when such goals and objectives are in conflict with other teams' goals. When conflicts arise, the overall quality of the product suffers, affecting time-to-market and customer satisfaction targets. With greater emphasis on autonomy within CE teams than any time before, it is more likely that some teams, if not careful, may create inconsistent or even unrealistic goals. Serious difficulties arise when such goals start competing with the corporate ideology or the so-called enterprise value system. The interest of the company suffers, which may result in serious business loss—team dissatisfaction, loss of productivity, profitability, etc. In CE, therefore, a "system-based improvement metric," and not just a team-based (or functional-based) improvement program, is required to achieve enterprise goals and objectives. QFD can be used to define this system-based improvement metric. This can be accomplished by identifying the common areas of interest (rather than the differences), that tend to align a company's separate functions toward a constancy-of-purpose goal. Integrated product and process development (IPPD) is the term often used to describe this improvement process. IPPD process, in the context of CE, is meant to include enterprise-wide operation, not just the product subsystems' integration.

Current product design and development is a complex undertaking with a multitude of external and internal inputs, outputs, and relationships. All goals and objectives at each level should be in agreement with and supportive of the corporate level vision and mission statements with constancy-of-purpose [Dika and Begley, 1991]. This is shown in Figure 3.6. An integrated approach to manufacturing entails the following:

- Set enterprise operational and business goals and objectives so that the management and strategic teams
  1. Understand the needs for "constancy-of-purpose" between themselves and downstream product development teams (PDTs)
  2. Know who their customers and their competitors are and develop a well thought out strategy for making and selling the product
  3. Communicate to PDTs design and manufacture goals of the product (such as sales, marketing, finance, and manufacturing) so as to be competitive in that business.
- Set IPPD goals and objectives according to "constancy-of-purpose" so that the product development teams (PDTs)
  1. Understand their roles in this strategy and know the process and tools for designing and developing the products
  2. Develop clearly defined product objectives, including well-defined strategies for product functions in support of specific strategy for making and selling the product; most commonly used examples of product objectives are planning for design, process planning, production planning, fabrication, or assembly, etc.
  3. Integrate product goals and objectives in support of the management goals and objectives for making and selling the product; the integration of the different strategies and functions is critical to success.
- Rationalize IPPD strategies, tactics, and projects to
  1. Maintain a constancy-of-purpose between the various teams
  2. Understand and analyze PDT functions, fabrications, and assembly processes so that enterprise operations can be performed with consistency and quality

3. Establish criteria for measuring the IPPD performance in conjunction with enterprise goals and objectives
4. Optimize the IPPD performance first by redefining internal and external interfaces and relationships, and then optimize its individual functions; it is important to optimize the overall performance rather than attempting to suboptimize each function individually.

- Develop a broad enterprise system model that
  1. Can establish a manufacturing system design coordinated with product design
  2. Does early economic analysis of design-process and manufacturing-process alternatives to permit rational choices
  3. Is capable of rationalizing the system under the combined influence of product, process, and production requirements.

3.4 WORK FLOW MAPPING

Mapping is the activity of creating a detailed flow-chart of a work-group process or function using graphics or visual icons to show its inputs, requirements, constraints, and outputs, in an ordered sequence. A work-group process or function (used interchangeably) is actually made out of a series of steps, tasks, or activities and has a beginning and an end. Mapping is used to lay out the details of this process for converting the inputs into an output. Mapping should provide answers to a broad range of questions related to work, process, or methods used to perform an activity. The questions include the following:

- Who is doing what?
- What steps are involved in executing an activity?
- What are the inputs and the outputs?
- How long does it take to execute a particular activity?
- Where is the decision making?
- What is the order in which activities are performed?

A process or function in this context means either a design-process, a manufacturing-process, or a work-process. Something that adds value to the inputs by changing them or using them to produce something new—either a tangible product or an intangible service as its output. Mapping is merely an enabler—a means to a more illustrative or perhaps to a more meaningful end. A generic mapping process in shown in Figure 3.11.

3.4.1 Process Map

Figure 3.11 shows major steps involved in creating a process map or making process improvement. It is divided into three primary steps:

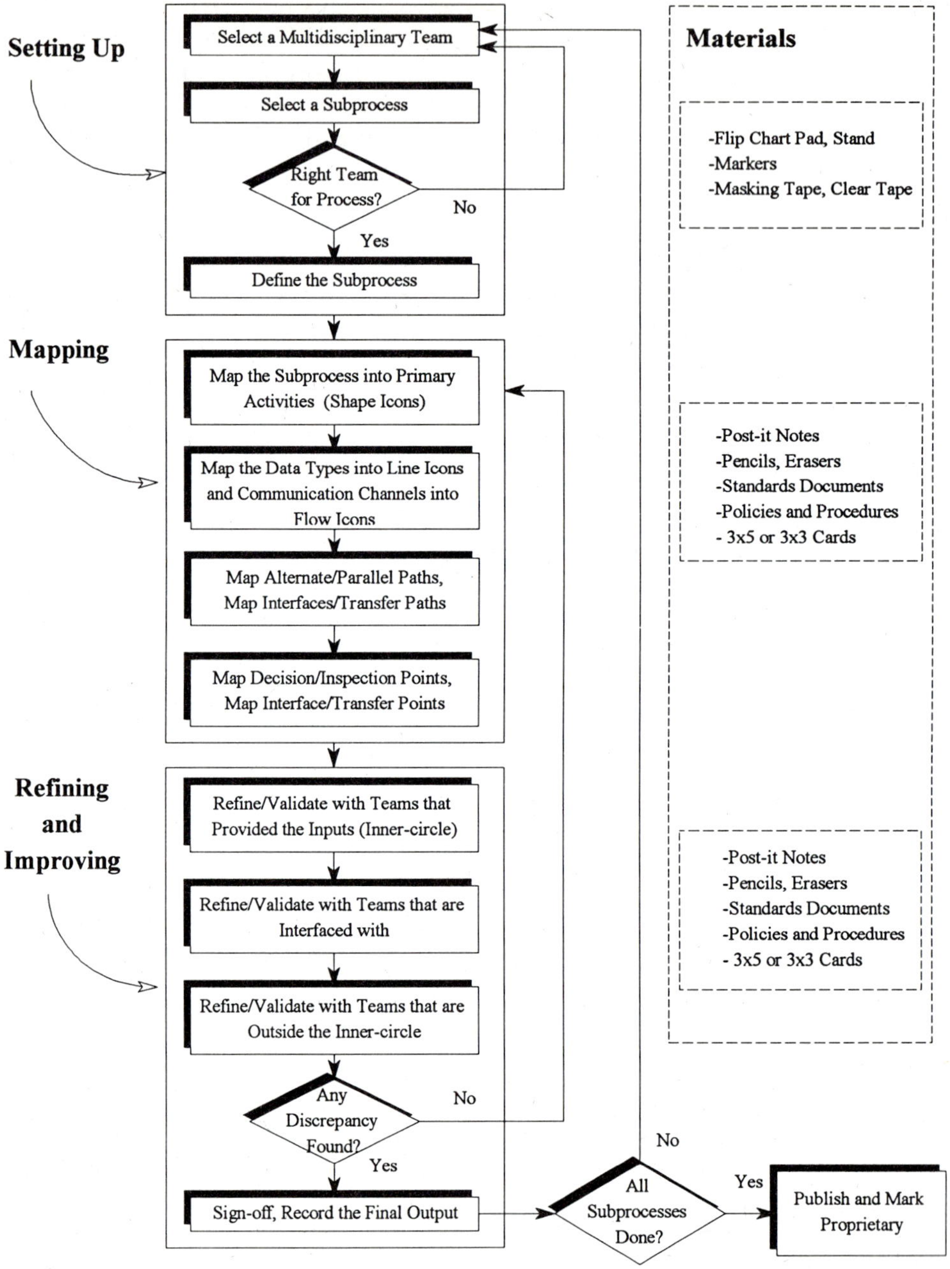

FIGURE 3.11 Major Steps for Creating a Process Map and Improvement

- **Setting Up:** The first step is broken up into four substeps. To select a multidisciplinary team means to select a group of team-members who are familiar with the current process and are asked to create a process map. Besides the members who do the actual work, the team consists of members from three other groups: those, who give the requirements, those who benefit from the work (or process), and those who manage the resources. Process is considered as decomposed into several subprocesses, which are large enough to mandate their own mapping. The order in which a process is mapped is based on the chronology of events. If the initial team is not a right team for the process selected then the team can be reorganized. The last substep of "setting up" is to define the subprocess. This entails breaking the subprocess into activities, tasks, etc.
- **Mapping:** The next step is mapping the subprocess itself. The map is the paper trail of the subprocess that exists now with all its flaws and inefficiencies. The map captures through proper icons (shape icons, flow icons and line icons)—what activities are performed under the heading of this subprocess. At this stage, generally, we do not bother to ask "why we are doing it" and "who is doing it." Each function or action is represented using inputs, requirement, constraint, and output variables. Icons are used to differentiate between various activities (work functions), communication channels, and data connections. Activities are represented by shape icons, communication channels by flow icons, and data connections by line icons. The primary subprocess with its decision points, alternate paths, and loops is likely to be general. It shows the major critical steps on a macro-level.
- **Refining and Improvement:** This is the last step. This consists of validating the results of mapping with teams of experts from the people who are doing the work, from those who have interfaced at one point, and from those who are managing the resources. Most, if not all, of the steps of primary activities can be broken down further into secondary activities. If the results of validation show any discrepancy, mapping is repeated, otherwise the procedure may be repeated for the remaining subprocesses. Each step in the subprocess can be further broken down into subelements and represented as a sub-subprocess. If wanted, it is possible to take a sub-subprocess and break it into its sub-sub-subprocess.

The items in the right column in Figure 3.11 (contained within dotted rectangular boxes) represent examples of resources required to complete each mapping step.

3.5 INFORMATION FLOW-CHARTING

A flow-chart or process map, is a graphical representation of a process, showing a sequence of steps, tasks, or activities (used interchangeably here) in terms of some standard flow-charting symbols or icons. It is a picture of how a team of employees or work-group does their job. A road network is an example of a map that shows many possible roadways from a chosen starting point to a desired destination. Process maps are similar to road networks in that different teams choose different routes to the same destination for different reasons. A good process map is the one that leads to the outputs in some value-added fashion (such as smallest span of time or minimum cost). It meets all the constraints along the way and leads the organization to a world-class manufacturing or service provider. (See Figure 3.12). A process map is a vehicle to express and release the knowledge, creativity, and energy that lies within every work-group, regardless of its ver-

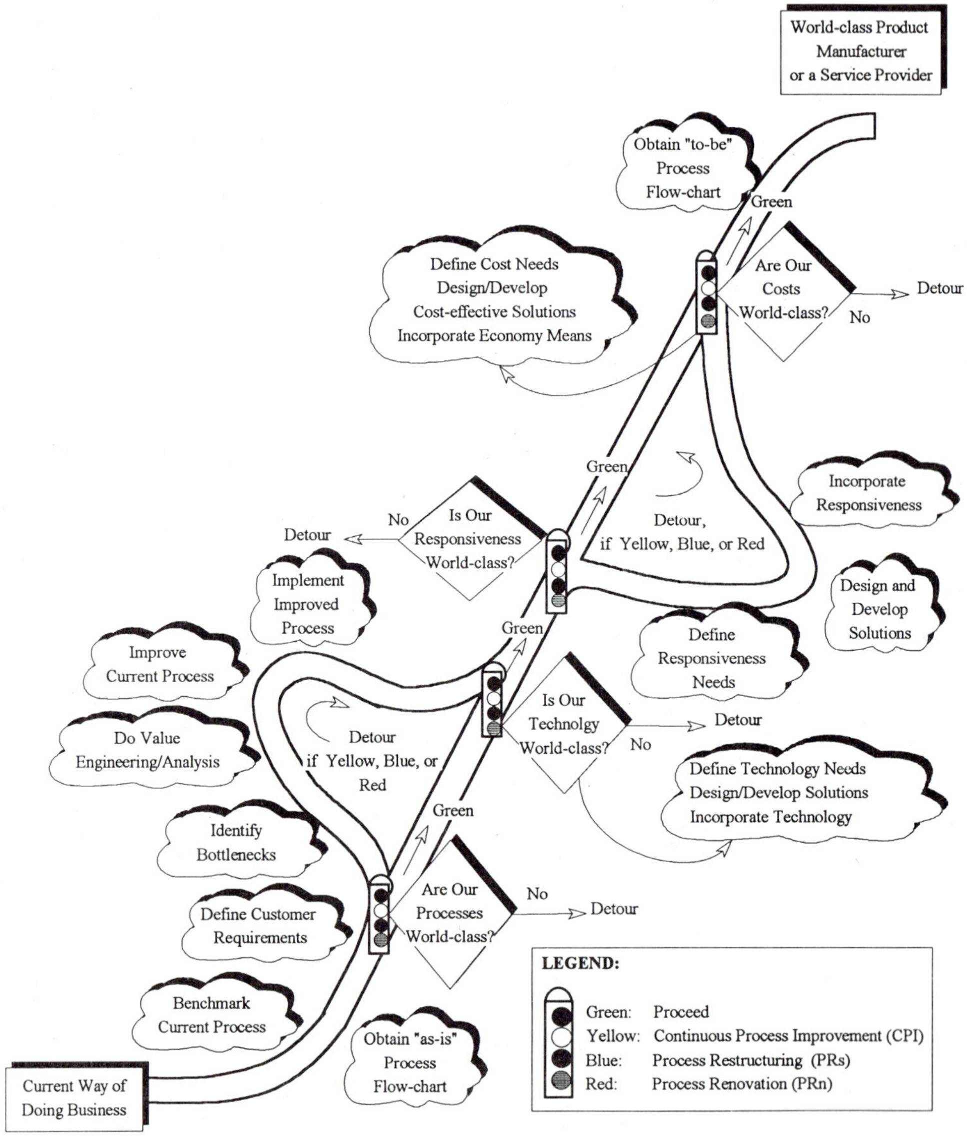

FIGURE 3.12 Roadmap to World-class Manufacturing or World-class Service Provider

tical or horizontal levels within an enterprise. A process map lets you detour when certain world class objectives are not met. It incorporates automatic signals for checking the state of the progress:

- How do our processes compare to world-class?
- Is our technology close to world-class?
- Does our responsiveness compare well with world-class level?
- Is our product cost competitive?

If the results are satisfactory, it lets you proceed to satisfy the next set of requirements for world-class. This corresponds to a green signal. A good process map, like a road signal, provides green, yellow, blue, and red signals, and detour conditions. If the current state does not meet the world-class requirements, the process map points to a detour condition. There are three independent detours one can take, depending on the degree of mismatch with respect to the world-class. If the current state is slightly off the world-class level, this signifies a warning (or yellow) condition. A process improvement strategy based on CPI might just do the trick. If the difference indicates a blue condition, process restructuring might be needed. The most unmatched situation is reflected by a red condition. It means the current state of conditions is quite far off the world-class level. In that case, major reengineering strategies such as "process renovation" are usually employed. A good process map is capable of redirecting to the right path—letting you choose right process improvement strategies depending on the differences in the current state of the product features compared to the world-class level. A good process map is the foundation for continuous process improvement efforts. A process route is considered good when teams analyze the "initial conditions" and agree on the routes that are favored by most people. Use of flow-charting provides a visual mechanism, a method for continual improvement of product or service, and continual learning [Deming, 1993].

Process improvement is evolutionary, requiring a systematic approach—capturing the "as-is" process in enough detail with relevant activities, data flow, and their connectivity. The capture and categorization of activities within the process flow are important to determine their relative importance and values. Similarly, types of data flow and data connectivity are useful to serve as an aid in identifying information flow's bottlenecks. Not obvious to persons actually performing the work, bottlenecks do exit in most processes. They can cause delay in information movement, deterioration in the data accuracy, or problems with communications. A formal mechanism is required to identify these specific types of bottlenecks from the current process and provide solutions to overcome them.

One of the major roadblocks in implementing process reengineering is that there is no formal methodology of visually monitoring and applying this concept on a regular basis. During process design, teams need to determine at a conceptual level how the product and the related process will function when a particular technology or process is changed. There is a need to capture schematically the function and behavior of the system without any physical model or prototype so that the relationship can be studied at a conceptual level [Prasad and Strand, 1993]. One way is to capture pictorial (graphic) representations of work-flow, information-flow, and data connectivity on the same chart. Such a conceptual model is often referred to as an information flow-chart.

In addition, the methodology used to create the information flow-chart must be capable of the following:

- Providing "a simple mechanism" towards easy understanding of the current work, methods, and practices
- Encouraging an open forum of discussion among inside and outside groups to promote a free flow of information and activities.

On a stand-alone basis, the methodology must provide visual means to improve the current process in the following manner:

- Seeking peer inputs from the work-groups who were not involved in the interview process in the first place when the information was initially captured
- Seeking inputs from management, customers, and plant personnel, and getting their feedback on the effectiveness of the process being used.

Thus, the methodology must provide natural means of accomplishing continuous improvements without interfering with the day-to-day business. Organizations, which are interested in applying the principles of synchronous manufacturing, and the Japanese Product Development Team (PDT) approach, must find this methodology to be useful both as a catalyst as well as the means for the evolution of a more efficient process [Carey, 1992; King, 1987; Oakland, 1989; Eureka and Ryan, 1988].

3.5.1 Forms of Representations

There are many ways the information about a process can be captured or represented. Some modeling schema provides better flexibility than others. An integrated definition (IDEF) model fits the definition of a typical process reengineering tool [U.S. Air Force, WPAFB, 1981; Yang, 1989] on which more efficient system can be planned. Information or business leaders use such models to begin the transition from planning to implementation. Other modeling tools employed during process reengineering are activity rule, business rule, or information flow-charts. Information flow-charts can be captured in many ways. Chapter 7 discusses the use of block diagrams, context diagrams, fishbone diagrams, bond graphs, and schematic diagrams for information modeling.

3.5.1.1 Common Methodologies

In this section four key methods for capturing the process information are described. Petri Nets and System Specification Language are based on Venugopal and Huang [1989]. IDEF is one of the most widely used methodologies for conceptual modeling. The Air Force's Integrated Computer Aided Manufacturing Program (ICAM) initially developed this methodology for describing complex business rules in their definition program [U.S.

Air Force, WPAFB, 1981]. It has been formalized in the Federal Information Processing Standards declared by the Commerce Department's National Institute of Standards and Technology. There are two versions of IDEF methodology, namely IDEF0 and IDEF1X. The configuration parts of IDEF0 and IDEF1X are shown in Figures 3.13a and 3.13b, respectively. IDEF has been found to be most successful in capturing a business process. However, its use as a tool to bring about the necessary continuous improvement in engineering and manufacturing process management has been very limited.

- ***IDEF0:*** IDEF0 is mostly used for functional representations of activities during conceptual modeling. The IDEF0 functional model, as shown in Figure 3.13a, is composed of ICOMs (Inputs, Controls, Outputs, and Mechanisms), arrows, and boxes. Each activity or function is conceptually represented by a rectangular box with three inward arrows and an outward one. Arrows represent interfaces (ICOMs) among activity boxes. Each activity can be decomposed into many levels, each following the same conventions. An entire IDEF0 model is a hierarchical representation of a process composed of activities or functions at various levels. Lower-level activities make up an upper level function. A solid line arrow indicates the flow of ICOMS at each level. The arrows do not cross between levels. The interaction of activities is managed through the boxes representing the activities. Because of the uniformity and limited interaction facilities, IDEF0 has been widely used for business modeling. Its application to modeling of engineering and manufacturing processes has encountered some difficulties since it does not allow iterations between the levels.

$$\text{IDEF0} \Leftrightarrow \text{ICOMs } (\textit{I}\text{nputs, } \textit{C}\text{ontrols, } \textit{O}\text{utputs, and } \textit{M}\text{echanisms}) \tag{3.3}$$

- ***IDEF1X:*** IDEF1X is best suited for relationship representations between the activities. In IDEF1X, a piece of information is represented as an ECAR (an Entity or object, Cardinality, an Attribute to an entity, and a Relationship) [U.S. Air Force, 1985]. In place of attributes, some types of assertions can also be used. Figure 3.13b shows common IDEF1X syntax and conventions where the parent-child relationship is shown using ECARs. Entities or objects are represented by rectangles and arrows. IDEF1X follows a top-down structure, where the top entities are composed of bottom entities. Some entities are shown with a dashed rectangular outline, either to indicate incompleteness, to designate an open relationship, or to clarify a relationship.

$$\begin{array}{c}\text{IDEF1X} \Leftrightarrow \text{ECARs (an } \textit{E}\text{ntity or object, } \textit{C}\text{ardinality,} \\ \text{an } \textit{A}\text{ttribute to an entity, and a } \textit{R}\text{elationship)}\end{array} \tag{3.4}$$

- ***Petri Nets:*** A Petri net is a mathematical representation of the system. Like IDEF0, the four major elements of a Petri net are PTIO (a set of Place, a set of Transitions, an Input function, and an Output function). A Petri net is often denoted as PN:

$$\begin{array}{c}\text{PN} \Leftrightarrow \text{PTIOs (a set of } \textit{P}\text{lace, a set of } \textit{T}\text{ransitions,} \\ \text{an } \textit{I}\text{nput function and an } \textit{O}\text{utput function)}\end{array} \tag{3.5}$$

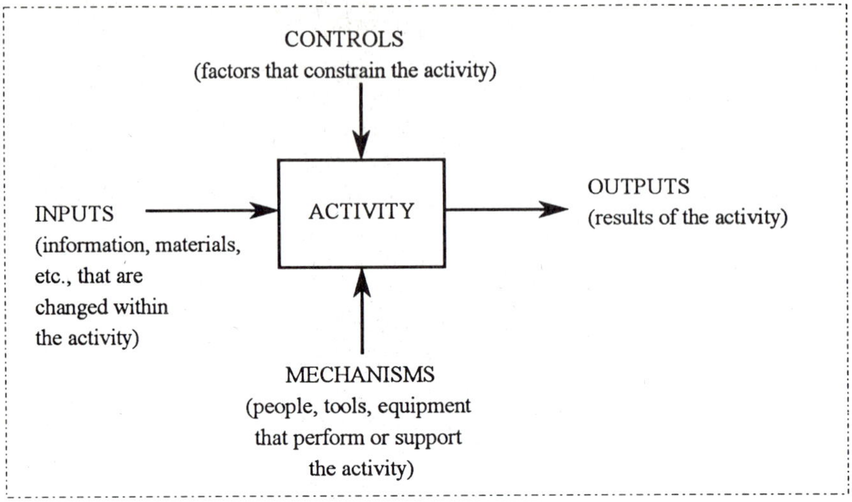

(a) IDEF0 Methodology Diagram

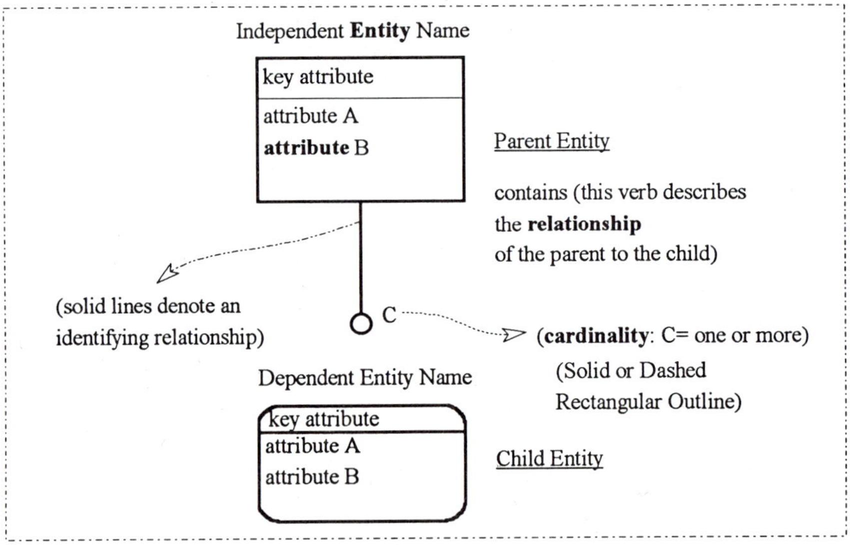

(b) IDEF1X Methodology Diagram

FIGURE 3.13 Information Modeling Methodology (a) IDEF0 (b) IDEF1X

The input and output functions relate places and transitions. Analyses of Petri net models show both the structure and dynamic behavior of the modeled system. Petri net models capture the control associated with the process. They do not capture the information flow as IDEF0 does. Thus, mathematically, Petri nets are more rigorous than IDEF0, but are not very useful for change analysis from a product realization viewpoint [Peterson, 1981].

- ***System Specification Language (SSL):*** SSL is another methodology to capture the behavior of the processes [Alford, 1986]. Behavior is conditional sequence and concurrence of functions transforming inputs into outputs and can describe the actions of a black box [Venugopal and Huang, 1989]. It is considered a part of DCDS (distributed computing design system). It consists of F-nets and I-nets. F-nets are structures containing time activities, while I-nets are structures containing data items.

3.5.2 Modeling Conventions

During information modeling, the capture and categorization of product or process flow functions, called "activities," are important to determine their relative values and weight. Different types of connectivity are useful as an aid in identifying the bottlenecks of the information flow that exist in most manufacturing processes. Similarly, different types of communication channels or data flows are useful as an aid in tracking the communication barriers. Problems with data flow and data connectivity can cause delay in information movement, deterioration in the data accuracy, or problems with communications. A conceptual modeling framework is required to identify these specific types of bottlenecks in the current process and provide means to overcome them.

- IDEF has no formal mechanism to indicate iterations among a group of activities.
- IDEF has no inherent mechanism to represent various forms of data communication such as manual, oral, or verbal (voice-based), or paper-based, computer-based, or physical-aid based. IDEF has a single representation for an activity, a rectangular block type and a single representation for connectivity—a solid line type.

This book has followed a convention similar to IDEF, which overcomes the above limitations. This methodology captures the variations in product and process information (activities, information-flow, and data connectivity), along with their functional characteristics. Functional variations of activities are captured through designated graphic icons and connectors. Consequently, this methodology is called information flow-based (IFLO) methodology [Prasad and Strand, 1993]. IFLO associates each activity with a functional type and each connectivity with a data type that links the process information. Thus, IFLO has multiple representations of activity, information-flow, and data connectivity types.

3.5.2.1 Functional Activity, Flow, and Data Connectivity

Icons and blocks are usual means to reflect the various types of activity, types of information flow lines, and data connectivity. Flow lines represent communication channels or flow directions. Figure 3.14 shows eight shape icons, three line icons and five flow icons.

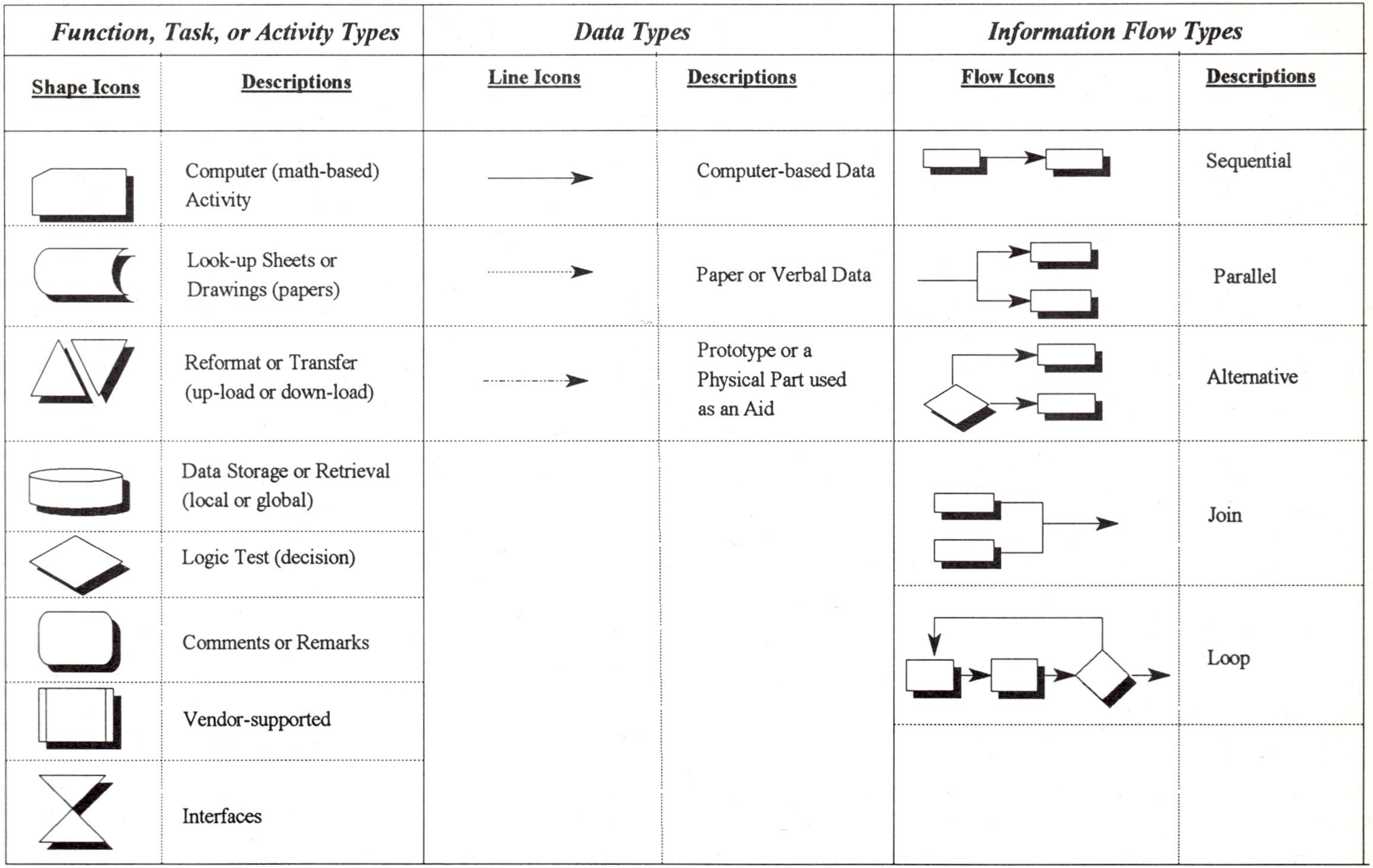

FIGURE 3.14 Shape Icons, Line Icons, and Flow Icons Employed during Enterprise Modeling

Line icons identify the type of data connections. Flow icons join the activity elements using one or more of the data icons. They are employed in the modeling enterprise process. The shape-icons in Table 3.1 correspond to a set of eight basic functions, enabling any complex enterprise process to be modeled (or decomposed into basic functional sets).

Data elements are represented by a line with an occasional arrowhead. The arrowhead indicates the movement of data. There are many ways data connections can occur. Data type is the type of data connections linking two or more shape icons. Line icons are the graphical representation of such data connectors. Table 3.2 displays the types of line-icons.

The flow icons connect the shape icons with line icons and set a flow direction. There are many ways such direction setting can occur. Flow type is the type of information flow between two or more shape icons. Such direction setting options are called connecting flow operators. A flow icon is the graphical representation of such flow operators.

TABLE 3.1 Types of Shape-icons

Shape-icons	Graphical Representation	Remarks
Computer (math-based) Activity	See Figure 3.14	This shape icon is a computer card, indicating that a process or an activity is performed using a computer, or is CAD- or math-based.
Look-up Sheets or Drawings (paper form)	See Figure 3.14	This shape is represented using a document icon indicating that a document is being used in the process.
Reformat or Transfer (up-load or down-load)	See Figure 3.14	This shape is represented using an up-triangle or a down-triangle icon. It shows activity that pertains to up-loading and down-loading the information or to a transfer (when information moves from one media to another).
Logic Test (decision making)	See Figure 3.14	Decision making is identified by a decision symbol (diamond shape icon). Following this shape-icon, the process would branch into two or more directions depending on the decision possibilities. For example, the decision might be between a pair of outcomes: "Yes" or "No"; "Go" or "No Go," etc.
Data Storage or Retrieval (local or global)	See Figure 3.14	This activity (shown by a cylinder or a drum) pertains to file management or data management, which may be locally accessed or is available over the network.
Comments or Remarks	See Figure 3.14	This shape icon is a rounded rectangle. It identifies a comment or remark that elaborates or describes an action, an activity or a data flow-line.
Vendor-supported (activity performed by an outside source)	See Figure 3.14	In a global manufacturing setup, vendors perform functions just-in-time to complement production activities. A partitioned rectangle icon identifies the vendor-supported activity.
Interfaces (sources or origins of information)	See Figure 3.14	In CE, teams work in parallel, with shared tools, and with outside vendors, suppliers, and partners. A cross-hair icon identifies such interfaces.

TABLE 3.2 Types of Line-icons

Line-icons	Graphical Representation	Remarks
Dotted Line	- - - - - - - - - - - - - -	Indicates that manual or verbal types of data are involved in the process.
Solid Line	————————	Indicates that the data is computer generated or computer-based.
Chain Line	-·-·-·-·-·-·-·-·-·-·-·-·-	Indicates that the data is present in some physical form such as cardboard mock-ups, hardware prototypes, or some visual aid-based form.

Five types of such flow operators are usually present in most modeling situations: They are listed in Table 3.3.

Eight shape icons, five flow icons, and three line icons are used to represent the various types of activities, information flow directions, and data connections, respectively. Although, each graphic icon is designed to represent a specific work-group function—activity, information flow, or data connection, each can be used in combination to get deeper insights into a particular product or a process situation.

Studies indicate that 40%–60% of the typical engineer's or designer's time is spent on nonvalue-added activities. These may represent tasks such as operating system functions (integrating multiple platforms, multiple operating systems), business interfacing (data transfer, file management, backup), communication (fixing network, protocols), and others. Such activities represent time that is not well spent in core business—which are to design and develop the end products (see Figure 3.15a). Some industry observers think that, by careful planning, a well-orchestrated enterprise could boost an internal teams' value-added activity (in percentage of their time) to the 75%–80% range (see Figure 3.15b). "Well-orchestrated" is used here to indicate a flexible enterprise where tools, concepts, devices, interfaces, control, and communication links are built-in at all levels of the enterprise's business. To achieve this, firms need a logical view of "how they do the business now, and how they want to do the business in the future." One work-group needs to understand, in detail, the work and needs of the other work-group that follows. This is called enterprise modeling—a work flow-charting of the organization.

TABLE 3.3 Types of Flow-icons

Flow-icons	Graphical Representation	Remarks
Sequential	See Figure 3.14	The two activities or functions are executed in series.
Parallel	See Figure 3.14	Two or more activities or functions are placed parallel to each other meaning either they lag behind slightly or are executed simultaneously
Alternative	See Figure 3.14	This is a decision operator. Depending on the decision outcome, this operator is used when controls need to be transferred to one or the other shape functions.
Join	See Figure 3.14	This operator is used when outcome of two activities is to be joined to constitute one larger activity.
Loop	See Figure 3.14	This operator is used when there is an iteration involved. This usually occurs when a decision icon is present as shown in Figure 3.14.

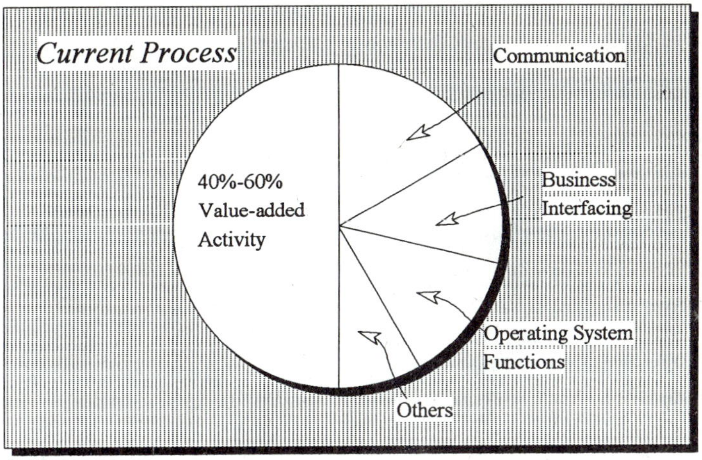

(a) Percentage of Team's Time Spent in the "As-is"

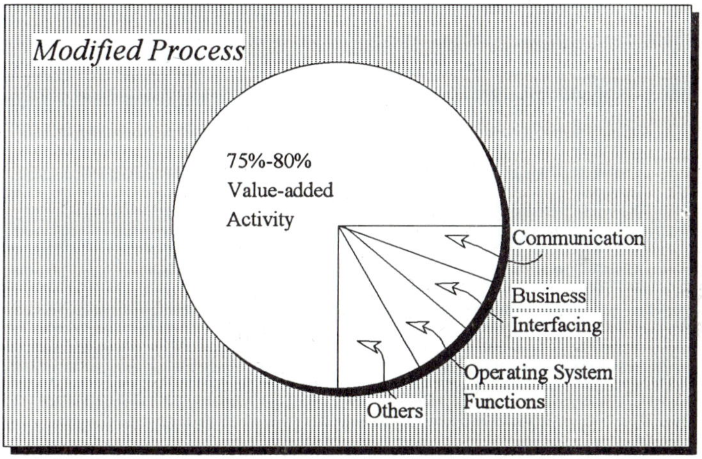

(b) Percentage of Team's Time Projected in the "To-be" Process

FIGURE 3.15 Typical Team's Time Distribution Chart (a) Time Spent in the "As-is" Process (b) Time Projected in the "To-be" Process

3.6 ENTERPRISE MODELS

Most firms review their enterprise operations at least once a year. They logically look at a conceptual level, how they conduct their business (the current practices), and how they would like to do the business in the future. They view the enterprise conceptually through some paper trail of its operations. This is often called enterprise modeling. It is a set of models that provides a precise and logical description of the types of activity and flow of information in an enterprise. These models capture the flow of information through an organization unit (say through a strategic business unit (SBU)) that is responsible for the product manufacturing. Most models of such types reflect the information needs of both the manual and computerized aspects of the manufacturing operation:

- The **first phase** of enterprise modeling is to set up a quality taskforce in each production area and to capture the business and life-cycle processes. By capturing the information exchanges and the interactions, the impact of changes on various performance indicators of an enterprise (as listed in Section 1.6, Volume I) can be studied. The interactions may take place between business operations (such as marketing and purchasing) or between the life-cycle operations (such as product definitions and process definitions). Similarly, by capturing the information exchanges and interactions with the external operations (such as relationships with vendors, suppliers, and customers), the effect of changes in external variables (such as market share fluctuations, consumer preferences and demands, and material cost) can be studied. The first phase provides a framework for taking an integrated look at the organization structure, processes, and tools of the entire enterprise.
- The **second phase** involves flow-charting the operations within a manufacturing enterprise in some conceptual fashion. Flow-charting, with its large visual graphical format for shape icons, flow icons, and line icons, seems to hold the most promise (see Figure 3.16). There are nine steps to flow-charting. The modeling steps and the corresponding graphical representations are shown in Figure 3.16.
- The **third phase** embarks on a detailed process analysis to examine the product/process development activities in enough detail and to identify redundant procedures and bottlenecks.
- The **fourth phase** involves cross-training the PDTs (product development teams). This requires institutionalizing a cross-training program, which breaks down the common barriers between the work-groups and the departments.

There are two types of enterprise models that are often invoked: The "as-is" and "to-be" models. An objective of the "as-is" model is to identify the current process and flow of information including timing aspects associated with designing a product. The idea of a "to-be" process is to define an improved process (improved in relation to "as-is")—detailing how the SBUs should conduct their operations in the near future, say three years from now. "To-be" mapping provides answers to a broad range of questions related to the organization including efficiency and effectiveness of the process, and of the tasks performed. These questions include:

- How well have we performed in these categories?
- Have we received adequate value for our investment?
- Is the program structured efficiently?
- What are the bottlenecks in the "as-is" process?
- How are we going to eliminate those bottlenecks?
- How well are we likely to do?
- Are we likely to receive adequate value for our investment?

Sr. No	Modeling Steps	Graphical Representation
1.	Identify the subprocess to be mapped.	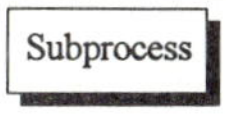
2.	Identify the input and output functions that this subprocess performs. An input or output function may itself be a subprocess that needs to be remapped later. They can be represented by one of the shape icons.	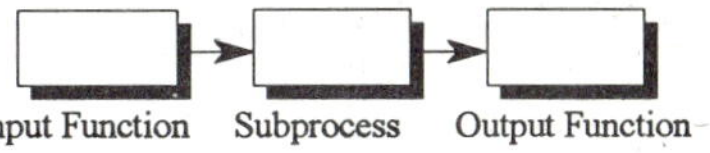
3.	Identify the type of the shape icons for those functions (inputs, subprocess, and outputs)	See Figure 3.14 for the graphical representation
4.	Arrange the functions in the order of work assignments or chronological processing steps. They look like a function chain with a catalogue of actions and methods.	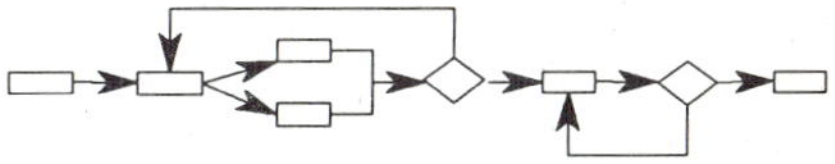
5.	Identify information flow-types. Connect the functions identified by appropriate flow icons. There are five types of operators (flow icons): o Sequential	
	o Parallel	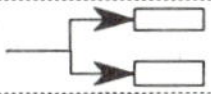
	o Alternative	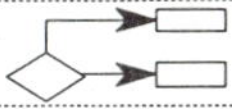
	o Join	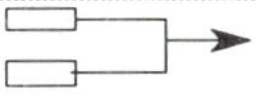
	o Loop	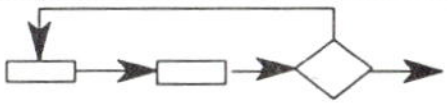
6.	Identify type of data (line icons) connecting the functions (electronic data, voice/manual/ paper, or physical part, etc.) They are represented by solid, dotted, or chain line, respectively.	Computer Data Manual/Voice/Paper data Physical part
7.	Describe the complete specifications associated with each identified function. This means identify requirement and constraint functions (besides inputs and outputs identified earlier).	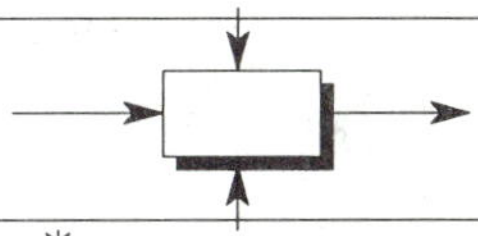
8.	Merge the arrows (line icons) with the corressponding flow icons as shown in the diagram.	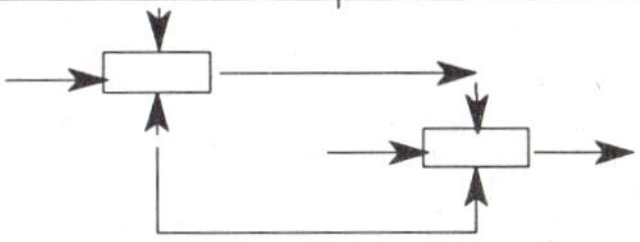
9.	Further, decompose the subprocess. Repeat the steps, if necessary.	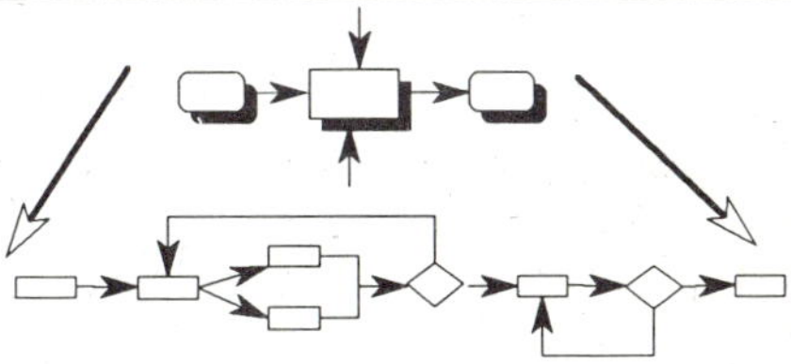

FIGURE 3.16 Modeling with Predefined Shape Icons, Flow Icons and Line Icons

Coming up with a "to-be" model from the "as-is" model is a daunting task. It is important to identify the parts and processes, and the human teams that are essential. This may require pulling all relevant core and support teams, including suppliers, into a multidisciplinary coherent work force. Teams form the "bricks and mortar" of a process reengineering workforce.

The idea of "as-is" is to determine the following characteristics:

- ***Flow and Timing:*** Identify not only the current flow of information and material through them, but also the actual flow of information and the timing associated with designing and developing the product.
- ***What and How:*** Identify what to measure in the process and how to measure it.
- ***Relationship:*** Assist and predict what components of the system will be affected and by how much.

Through "as-is" models, people within the organization can gain valuable insight into the process details. It also leads to opportunities for continuous process improvement by providing teams with the ability to predict how changes in the "as-is" structure or changes within the steps of "as-is" will affect the performance indicators. Section 1.6 lists a set of eight performance indicators.

The idea of "to-be" is to determine how we should conduct the business in the future. To come up with a "to-be model," one usually starts with the "as-is" model, keeping the essential parts, key processes, and the human teams. The human team forms the "bricks and mortar" of the new product realization process. The act of flow-charting in the "to-be" process determines where value-added measures can be put in and where waste can be eliminated from the process—both for manufacturing and service areas. The cycle of enterprise modeling—starting with "as-is" and coming up with "to-be"—may be an ongoing process. Answering these questions requires a broad range of metrics, including some that are based on efficiency (for example, is the program structured efficiently?) and some that are based on effectiveness (for example, how has the performance changed?).

3.6.1 "As-is" Model

The "as-is" model defines the steps in the current process. It involves identifying types of activities performed, types of information flow, and types of data connections across activities that may, in any way, influence the parts and processes. The steps involved in an "as-is" process are (see Figure 3.16) as follows:

- ***Step 1: Empower a knowledge acquisition team to interview and identify the "as-is" process.***
  - The members of the knowledge acquisition team generally are outsiders (preferably not currently be a part of the group being interviewed). This is because it is easy to overlook many of the simple steps by those who are close to the job. People on the job including managers and supervisors should preferably be a part of

a validation team not such an acquisition team. The three key members of an acquisition team are facilitator, scribe, and a timekeeper. The facilitator leads the meeting. He or she ensures that discussions stay on track. The scribe records ideas on the flip chart or white board and makes sure ideas are not lost. The timekeeper keeps track of the time spent and time left.

- The sponsoring organization should inform the appropriate managers and supervisors of the real purpose of conducting the "as-is" modeling. They must understand that the purpose is not to identify who is doing what but rather to find out what process is in place—what are the activities and how are they being performed. This clarification is important to eliminate the fear among employees. It is important to emphasize that this knowledge gathering is not a part of performance appraisal or in any way related to it. It is a mere "documentation" of how the business is done and the procedure in place of achieving that.

- ***Step 2: Identify the information or activities performed and categorize them by types of activities.*** Some of the activities are bidirectional, such as Data translation or Interfaces (see Figure 3.14). The most commonly performed functions in "as-is" modeling are as follows:
  - *Manual or manual-aided:* This designation is used when someone performs an activity manually. It is also used when an activity is performed by an individual for the sole purpose of producing intermediate outputs, drawings, soft-prototypes, physical prototypes, etc.
  - *Computer-based or computer-aided:* This designation is used when a computer program or a software is used to do a required step or to support an activity.
  - *Look-up sheets or drawings:* Modes of communications other than electronics, such as paper, or verbal are used.
  - *Format or Data Transfer:* This designation is used when an electronic data is converted from one format to another such as IGES, PDES, one CAD system to other, CAD to paper drawing or vice versa.
  - *Interfaces:* Interfaces are the groups other than the primary group from which information is sought to complete an activity or to whom the information is being sent.
  - *Decision logic:* This designation is used to identify conditional logic when a process splits into multiple branches or parallel activities.
  - *Vendor-supported function:* This icon is used for services or activities that are performed externally; for example, services that are contracted. Suppliers and vendors in many companies perform a number of supporting roles.
  - *Comments and remarks:* Often this designation is employed to identify checkpoints or milestones in the process. It has little relevance to the flow of information or data connection and is meant to explain circumstances or group involvement.

- ***Step 3: Identify the types of data that flow through these activities.*** This is usually identified by icons (Figure 3.14):
  - *Computer-based data:* This is represented through a solid line. This designation is used to connect functions or activities, which are computer-based.

- *Paper or verbal data:* This designation is used when the information relay between two adjoining functions is carried via paper or through verbal communication.
- *Prototype or physical data:* This designation is used when data that is passed to an adjoining function is not computer based, or in a form of paper or verbal. It is employed when data takes the form of a prototype, which aids in the realization of the desired product.

- ***Step 4: Map these functions and data steps into flow-charts:*** An icon-based flow-charting program may be used for this purpose. ABC flow-charter was used by Prasad and Strand [1993]. However, any suitable program could have been used in its place.
- ***Step 5: Create back-up sheets:*** This contains additional information, such as statement of work, who is doing it, how long it takes, and additional inputs and outputs requirements. The information is captured in a standard form developed for this purpose. Information is entered in designated fields so that later, if the needs arise, they can be processed using some type of spreadsheet or database programs. A sample example of a back-up sheet form is shown in Figure 3.17.

3.6.2 "To-Be" Model

The "as-is" model provides a good basis to study the current process and to identify the redundant functions that are either repetitive or present little or no value. The idea of a "to-be" model is to reorganize and streamline the "as-is" activities or even combine or eliminate the tasks so as to control the waste. In order to lead a company to the twenty-first century, one must increase the throughput and improve productivity. This means "doing it right the first time," and eliminating time-consuming and costly engineering change requests. The delineated work units are identified here, barriers are broken down, and an alternate work-flow is determined. Following the principles of TQM, the team members assume responsibility for several elements of the work units rather than just one functional area or department. It is constructed on a broader range of principles, such as in-house team knowledge, observations through hands-on use of the product, and increased flexibility. There are two ways of achieving a "to-be" model: top-down approach and bottom-up approach (see Figure 3.18).

3.6.2.1 Top-down Approach

In this approach the management of the company appoints a team of experts from various disciplines and empowers them for coming up with a vision of how the future operations should be run. Usually, with such objectives little or no constraints on the existing systems, such as cultures, cost investments, automation, etc., are imposed.

In the top-down approach, top executives provide the mission statements and a bird's view of the enterprise on which to base the future. The team looks over the "as-is" process and asks a series of "whys." Figure 3.19 illustrates this process. One does not stop until the root cause of the initial problem is found. Examples of first level "whys" are also contained in Figure 3.19. The group then collectively comes up with alternate ideas about

Computer Entry Form for Tasks Details

Field	Prompt
Source:	*What's the name of the person being interviewed?*
Date:	*When was the person interviewed?*
Data Flow Task Number:	*Choose an Identification Serial Number*
Task Name:	*Can you give a short name to this task? It is better if you choose a name that signifies the task being performed.*
Statements of Work:	*What does this task entail? Explain what this means in technical, business, and operational terms. What are the reasons for doing this task? How is this task related to the other tasks that are being performed?*
Task Responsibility:	*Are you responsible for doing the task? If not who is (name of the person and group)?*
What Triggers the Task:	*Who requests you to do this task? What information is required to do your job?*
Task Duration:	*How long does the job take to finish?*
No. of People:	*How many people are working?*
Deliverables:	*Why you do this? What comes out of this? What are the tangible or intangible outputs? What do you want to accomplish?*
Customer for Deliverables:	*Who asks for it? Who are your internal customers?*
Inputs Required:	*What inputs do you require to do your job? Who do you depnd on? What information does he or she provide?*
Souce of Inputs:	*What is the source of your input? Is the information available in electronic form? If so, fill in the next four blocks.*
Computer System Requirements:	*Inputs (Software):* / *Inputs (Hardware):* / *Outputs (Software):* / *Outputs (Hardware):*
Disconnects:	*Identify what you would change, if you have the freedom of changing the proces. How do you see the job (to be done)? What benefits may it bring if you make this change? Who will be affected by this change?*

FIGURE 3.17 A Sample Back-up Sheet for Capturing Other Information

a possible product or service—what the customers need that would alleviate this root cause. The team flow-charts the ideas and determines a process that meets the customer needs and the company's vision. The team finally comes up with a sound reengineered process for executing the tasks. The process for achieving this should embody the principles discussed in the reengineering strategies Section 3.2. This results in a prescriptive model of an enterprise called "to-be." Clearly, the first step in getting a "to-be" model is

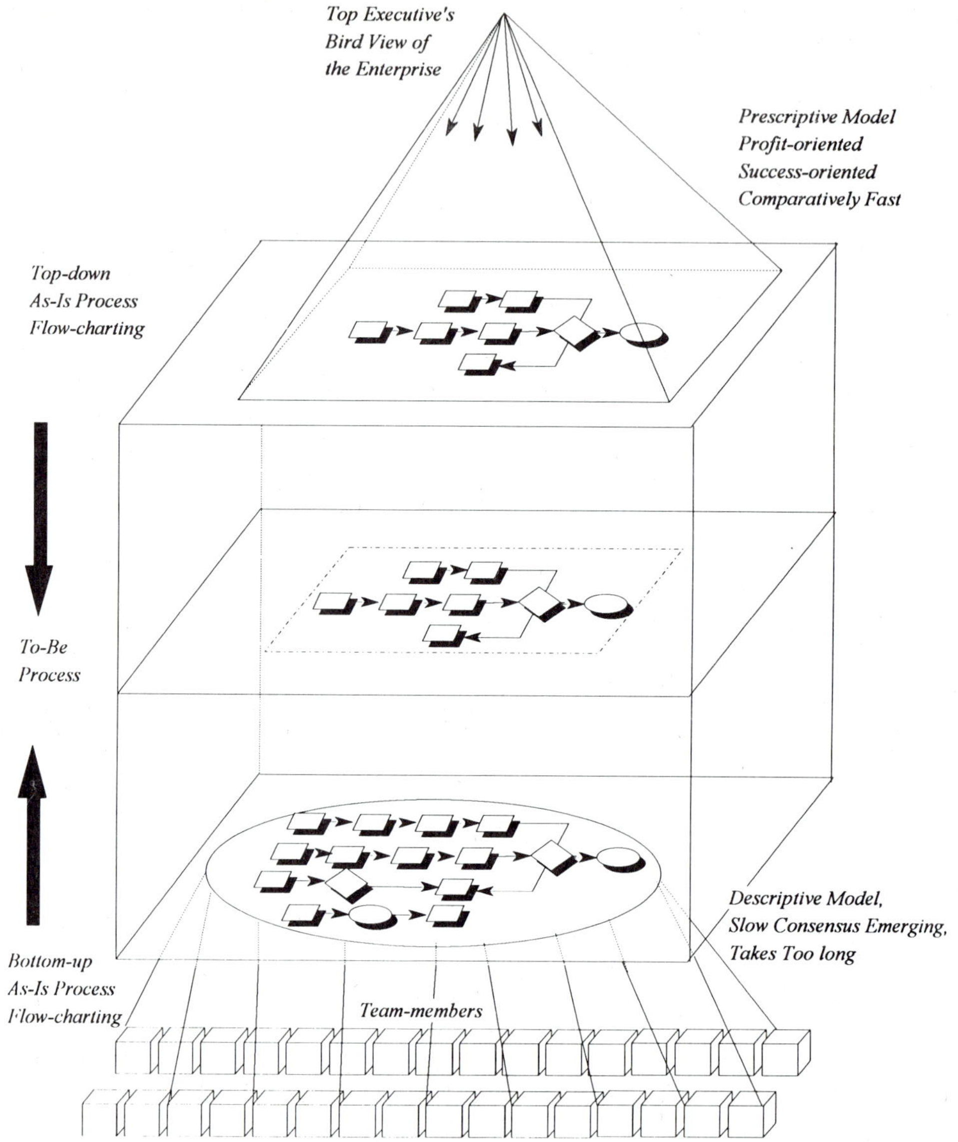

FIGURE 3.18 Two Ways to Achieve "To-Be" Model

to study the "as-is" process and identify the redundant functions that are either repetitive or nonvalue-added. The main idea of studying is to eliminate, reorganize, and streamline the activities. Sometimes the aim is to combine the tasks so as to contain waste, increase the throughput, and *do it right the first time* to eliminate time-consuming and costly engineering change requests.

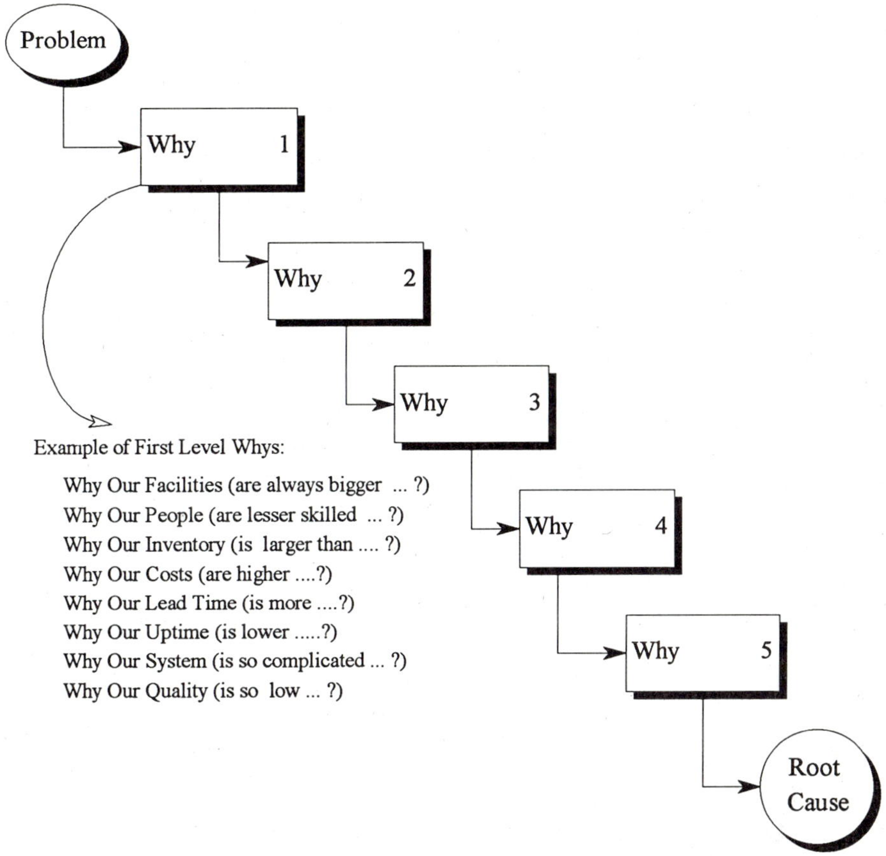

FIGURE 3.19 How to Get to the Root Cause?

The steps of a top-down process, which leads to a "to-be" model are as follows:

- ***Step 1: Strategic Vision:*** The team members understand their group's vision in relation to the company's overall vision. They either know or learn of the direction the company is heading and how the units will be supporting it. They also study the lines of products they will continue to support, what is being refined, and their global needs.
- ***Step 2: Global Participation:*** Begin to conceive a concept where there exist no barriers of disciplines or group participation. A global manufacturing philosophy, where geographical location has no barrier, is followed. Global means conceiving a product in one place, getting the design done by another group or in other country, and later manufacturing the product in another place or plant.
- ***Step 3: Common Objectives:*** Consider a set of common objectives, such as cost and

quality, customer satisfaction, etc., that is uniform across various departments and related units. Use this as a basis for step 4. Each activity should be analyzed from its own right and how it impacts customers, suppliers, engineering teams, product teams, manufacturing teams, plants, etc.

- ***Step 4: Simplify/Reorganize:*** Reorganize the activities to do one or more of the following: eliminate disconnects between data changeover, for example, manual to computer, computer to paper, etc; eliminate duplications, rework, repetitions, and waste. Waste is anything that does not add value to a product or service. Most design teams see activities but have difficulties identifying the types of data needed to execute them.

3.6.2.2 Bottom-up Approach

Here, the "to-be" process is evolved from the "as-is" model as opposed to being defined a priori. A process called "synchronous manufacturing" or "value analysis/engineering" is employed to transform the "as-is" model into a "to-be" model. Multidisciplinary teams that are formed at the early concept stage are given far greater responsibility. They meet frequently and review the process in detail for all sorts of bottlenecks. Thus, many such decisions are made lower in the ranks. Management still retains the right of "review and reject," but teams shape the vision, streamline parts, and process work flow, recommend solutions, and implement improvements. This results in a descriptive model of an enterprise called "to-be." The root of the model is based on the "empowerment" philosophy. The philosophy is based on the belief that employees on-the-job are close to the problems, know a lot about them, and, therefore, should be capable of solving them as well. The approach encourages teamwork and enforces faith and confidence. The only problem with this approach is that it could be time-consuming. The process of change could be slow. It may take a long time to reach a consensus. In today's rapidly changing market conditions, one may not have as much time. Often, far reaching and bold decisions are needed to change the course quickly. A combination of both, starting with a bottom-up proposal, then a top-down "to-be" and a bottom-up follow-up, may be the right prescription.

3.7 PROCESS IMPROVEMENT METHODOLOGY

In the following a PI methodology is proposed which is based on combining both top-down and bottom-up approaches with value engineering. In this methodology, first, one starts with a bottom-up proposal, then a top-down scenario, followed by a bottom-up revision. The trio combination, proposal-scenario-revision, often leads to a more balanced result. The following are the major steps of the combined process:

- **Plan for "as-is" activity charting:** Activity charting is an important step of value engineering for building quality into manufacturing. The initial steps involve
  - Identify who and what has been done so far
  - Review and prioritize the work to be done

- Put together a policy document outlining purpose, scope, and deliverables
- Review with the management and get their approval.

It focuses on the entire manufacturing process and their interfaces, rather than its subset. The normal procedure is to start with one key area and use this as a team exercise.

- **Identify the function of the product or service:** The next major task is to review the current process for this area and to get all the relevant facts. For example, the team gets (a) information about specialty products, materials, processes, vendors; (b) talks to company and industry experts to understand the functions they are performing.

1. **Begin work flow modeling:** This step includes drawing a process flow-chart (following the standards for flow-charting) in minute detail identifying what is planned to be done. Besides documentation, time-based examination of work flow is detailed. A structured format is used to emphasize the impact of sources of variations on the process. The emphasis is on procedure *source* not its agent, or the person who is running the procedure. In addition, the tasks involved in this step are listed here:
   - ***Define a candidate process model or value tree:*** The result of this charting is the development of specifics—important data and objects—called a "candidate" process model.
   - ***Draw a value graph—process description sheets:*** The team should be able to break down the flow into smaller chunks in a form of a process description sheet so that each chunk accurately describes the associated manufacturing method.
   - ***Identify the sequence:*** It identifies sequence, the main assignments, part flow, timings, etc.
   - ***Identify indirect expense and control parameters*** (such as statistical quality control (SQC), statistical process control (SPC), and scheme associated with just-in-time manufacturing).
   - ***Identify process parameters:*** During this step, process assumptions are identified, machines and people are sketched, critical process characteristics are confirmed.
   - ***Identify new investment:*** New machinery or equipment cost is also identified and estimated, and noted against the appropriate entry in the bill-of-material.
2. **Develop performance metrics:** These are criteria by which organizations measure efficiency and improvement. Building on IFLO [Prasad and Strand, 1993], managers can perform activity-based analysis for measuring process performance while identifying critical performance drivers. Better performance metrics are critical to the success of PRs. It gives managers some clues to measure the value of the system.
3. **Establish a value for that function or service:** The process begins with looking at the "candidate" process model of an enterprise. One then performs functional economical analysis in support of process redesign and new investment justification, if any. The teams model the process using icons and blocks. In addition they do the following:

- ***List sources of variations:*** The team lists all sources of variation potentially affecting each operation.
- ***Identify value-added activities:*** Brainstorming and team dynamics are tools often used to identify value-added activities from nonvalue-added activities in the candidate process. We may never be sure, but we can break down the activities in the following three categories that are helpful in making this assessment [Roberts, 1994]:
  - *Real value-added (RVA)*—refers to the activities that are essential to the process in order to meet the essential customer expectations.
  - *Business value-added (BVA)*—refers to activities essential to conducting business, such as employee happiness, benefits (vacation, sick leave, etc.), policy and regulation compliance, that add cost or time to the process but do not add value from the customer's perspective.
  - *No value-added (NVA)*—refers to activities that neither add value to the process from the customers' perspective nor are required to conduct the essential business. Examples of NVA include unnecessary indirect costs (such as storage, movement, multiple measurements, end-of-the line monitoring, tracking, all forms of duplication, rework, long approval cycle, etc.) Departments collect these types of costs under the heading of "overhead" or "burden." Some indirect costs are not so "bad," others do not add much value to the corporation. Most reengineering teams are not trained or motivated to make indirect costs as part of their process reengineering equation, despite the fact that this is perhaps the best place to begin attacking them.
- ***Develop fishbone diagrams:*** The completion of this phase is typical of most brainstorming sessions and is equivalent to developing a fishbone diagram for each operation.
- ***Brainstorm to identify the new "improvement" process:*** Continue elimination of the most obvious barrier each time in the current process.
- ***Create and refine:*** Based on the consensus, the concepts are redefined, combined, or subdivided in order to provide a clear identification of activities that add values from those that do not. A graphic tool for placing the values can be used to aid this process.

4. **Seek creative ideas:** The creative ideas and the graphic pictures obtained during the analysis phase is evaluated here. Here, teams think creatively and analyze costs, (such as assigning cost ($) equivalent of tolerances), apply judgment, and timings. The teams also look at the affected dimensions or changed characteristics at each step in the process. Such activities can be categorized to fall into three definite ranks:
   - ***Analyze the information and make changes:*** This includes identifying gaps or the activities that are impacted by gaps.
   - ***Initiate a control plan for each operation:*** Each affected dimension or changed characteristic and its associated description are recorded. If "gauges" are available for measuring process results, they are recorded on the control plan. Only "perform" activities imply a transformation process from inputs to

outputs that adds value. Others are means to get to the end but are generally not an essential part of the process.

- ***Evaluate alternatives:*** Brainstorming and other tools are used to provide alternate methods of doing the same work. The team looks at each activity one at a time and studies the measurement system variability. Participants are encouraged to suggest ideas no matter how difficult or costly they may be. The objectives are to overcome gaps by use of standards, use of specialty products, or by using one's own judgment. Inappropriate ideas may lead to good ideas.

5. **Redesign an improved process:** The redesign phase aims at settling into an improved process that leads to error-free designs with embedded performance monitoring and provisions for self-correcting. The steps are as follows:
   (a) Mark the process flow activities that are impacted.
   (b) Invite experts to critique the charts or seek inputs.
   (c) Revisit and look at every detail of how the business is run and determine if it can be improved every step of the way.

   During this phase, the team overcomes roadblocks and constructs a characteristic matrix, which displays relationships between product requirements and operations. The matrix helps to evaluate the importance of characteristics (value-added) while showing important upstream relationships. Based on this, an improved process is redesigned. This process uses standards, effective employee time, and eliminates wastes or redundancies. It contains a significant portion of value-added activities in its flow ("perform" type activities). Outputs of one activity normally are made inputs for a subsequent activity in a pipeline mode. The preparations and testing time are kept at a minimum. However meeting this threshold consistently at a lower cost is often difficult. Complexity, process variation, and uncontrolled situation (such as dealing with consequences instead of outputs) are some of the elements creating such difficulty.
6. **Do a global gap category ranking as a group.** First, identify the system level gaps, timings, bottlenecks, and major disconnects. Then, identify the system's level impact on the "to-be" process and the enterprise's functions. Then, invite experts to critique the numbers and the gaps.
7. **Define ideal process:** The goal is to provide the services at the lowest possible cost, in the least possible time, without diminishing previously set performance or affecting quality. The possible steps are listed here:
   - ***Think creatively:*** This can be a paradigm shift in terms of how we do business today. Though the goal is to find the best possible solution within the limits of organizational and cost barriers, it is possible, however, to cross them if the plan provides extraordinary savings and adds values to the company.
   - ***Identify top level NVA, gaps/waste, and create gap categories.***
   - ***Eliminate waste, NVA and gaps:*** Eliminate all the NVA, waste/gaps inherent in the process.
   - ***Define the ideal situation:*** The team looks at all the possible alternatives and recommends the ideal situation that ranks well. Document the improved process timings and ratings.

8. **Repeat the steps (1) through (7) for each functional area:** Invite the experts and repeat this step as often as necessary

Process improvement (PI) methodology provides a wide range of tools for reengineering. Such tools take the form of software-based system for drawing work flow diagrams and identifying value management tasks. Most software products for PI contain modules for IDEF-like charting, work flow, and cost analysis. Some are equipped for collaborating, benchmarking, monitoring, applying change-management principles, implementing just-in-time delivery, doing process and route mapping, reducing cycle times, and other PI tasks.

3.7.1 Cumulative Weight Analysis

Value analysis is most often the process of identifying or ranking critical items or functions. The value analysis begins with the "as-is" process flow-charts. One looks for items for improvements and a number of areas, where some local improvements can be possibly made, are identified. However, it is usually not clear, looking at the charts, which of these activities ought to be changed, modified, or expunged and why. One does not readily know what basis to use in setting up the change priorities. What is lacking is a rational basis for decision making—what is important and what is not. Though value analysis provides one such basis, many other criteria do exist [Prasad and Strand, 1993]:

- *Critical Index:* This is a measure of criticality. This index is assigned a value of 1 if an activity falls on the critical path and 0 if not. Thus critical index takes a logical value of 0 or 1.
- *Time Index:* This is a measure of how long it takes to perform an activity.
- *Cost Index:* Some activity cannot be measured by time but requires capital investment. Cost index is a measure of the cost associated with performing an activity.
- *Issues Index:* Many times, some activities are subject to a lot of management and implementation issues not related to either time or cost. Issues index provides such measures.
- *Priority Index:* As defined earlier, it measures the priority of performing one activity in relation to others.

It is difficult to decide which one of these indexes should be used as a basis for determining the overall importance of an activity. If one is considered more important than others, it might be true for one set of activities but may not be true for the others. We needed a schema for arriving at a combined rating from the individual ratings. This section describes such a schema. Figure 3.20 shows a partial result of combining the individual ratings. The principles, from which the combined rating is obtained, are derived on the basis that each index is equally important from process improvement considerations. If this is not true, weighting factors can be assigned to the indexes to balance the results out. The following discussion describes the approach.

	A	B	C	D
1	DIE-PROCESSING DATA SYNTHESIS MATRIX			
2				
3			Individual Basis of Ratings &	
4	Activity No.	Activity Description	Does it fall on the cri	How long it takes to
5	Type		Critical	Time
6			Index	Index
7	Weighting Factor		20%	30%
8	Minimum Value		0	0
9	Maximum Value		1	4
10	Normalized Value		4.58	8.74
11	Standard Deviation		0.49	1.23
12	S1	Advanced prints (if any)	0	0
13	S2	Plant information (press information)	0	0
14	S3	Receive work order from SME for die processing	1	1
15	S4	Input work order and assign job to die processor	1	1
16	S6	Start of die processing or die pre-processing phase	0	0
17	S7	Review work order and advanced prints if available	0	4
18	S8	Upload latest product data from CAD System	1	1
19	S9	Is this a new part?	1	0
20	S10	Product released databank (corporate)	0	0
21	S11	Is the released data same as preprocessed?	1	0.5
22	S12	Are there any major changes since last process?	1	0.25
23	S13	Input from formability platform, Senior Manufacturing Engineer	0	0.5
24	S14	Tip part (based on user experience)	1	2
25	S16	Determine die type and number of die operations	1	4
26	S17	Versatec tipped plots (full-size or scaled down)	0	0.25
27	S18	Is any product change required?	1	0.25
28	S19	Change request signed off by Senior Manufacturing Engineer	0	0.25
29	S20	DES user files (mainframe)	0	0
30	S21	Is this acceptable by SME and design staff?	1	0.5
31	S22	Write die line-up (minus blank die) information	1	3
32	S23	Preliminary die line-up review	1	1
33	S24	Preliminary line-up given to layout and design group	1	0
34	S26	Update line-up as required	1	2
35	S27	Die design change request input if any from die design	1	1
36	S28	Is this a minor FCO change ?	1	0
37	S29	Die line-up review with plant and Senior Manufacturing Engineer	1	4
38	S30	Die processing tracking database (for scheduling)	0	0
39	S31	Make any required change, if needed	1	2
40	S32	Sign-off	0	0.5
41	S33	Is this a pre-die processing phase?	1	0
42	S34	Provide final die line-ups and MM drawings to layout	1	0
43	S36	Ship line-ups and MM drawings to car platform group	1	1
44	S37	Update tracking database (a PC-based system)	0	0.25
45				

FIGURE 3.20 Relative Ranking of the Activities

	E	F	G	H	I	J	K
1	DIE-PROCESSING DATA SYNTHESIS MATRIX						
2							
3	Specifications			Overall Ratings			
4	Whats' equivalent di	Influence of the acti	Does this activity ad	Relative Severity Index Calculations			
5	Cost	Issues	Priority	Weighted	Weighted	Average	Average
6	Index	Index	Index	Value	Rank	Value	Rank
7	20%	20%	10%	1.00		1.00	
8	1	0	0	0.03		0.03	
9	1	1	1	0.29		0.28	
10	5.74	1.60	3.91	0.78		0.77	
11	0.00	0.27	0.47	0.07		0.06	
12	1	0	1	0.06	1	0.09	2
13	1	0	1	0.06	1	0.09	2
14	1	1	0	0.24	8	0.23	8
15	1	0	1	0.14	4	0.15	5
16	1	0	0	0.03	0	0.03	0
17	1	0	0.5	0.18	6	0.15	5
18	1	0	1	0.14	4	0.15	5
19	1	0	0	0.08	2	0.08	2
20	1	0	0	0.03	0	0.03	0
21	1	0	0	0.10	2	0.09	2
22	1	1/3	0	0.13	4	0.13	4
23	1	0	1	0.08	2	0.10	3
24	1	0	1	0.17	5	0.18	6
25	1	0	1	0.24	8	0.22	8
26	1	0	0	0.04	0	0.04	0
27	1	0	0.5	0.10	3	0.11	3
28	1	0	0	0.04	0	0.04	0
29	1	0	1	0.06	1	0.09	2
30	1	1	0	0.22	7	0.22	7
31	1	2/3	1	0.29	10	0.28	10
32	1	0	0.5	0.13	4	0.13	4
33	1	0	1	0.10	3	0.13	4
34	1	0	0	0.15	4	0.12	4
35	1	0	1	0.14	4	0.15	5
36	1	0	0	0.08	2	0.08	2
37	1	0	0	0.22	7	0.17	5
38	1	0	0.5	0.05	1	0.06	1
39	1	0	0.5	0.16	5	0.15	5
40	1	0	0	0.05	1	0.05	0
41	1	0	0	0.08	2	0.08	2
42	1	0	1	0.10	3	0.13	4
43	1	0	1	0.14	4	0.15	5
44	1	0	1	0.07	1	0.09	2
45							

FIGURE 3.20 *Continued*

Let us assume S_{ij} is a measure of jth index for an ith activity. Then the combined ratings can be formed as follows:

$$\text{Normalized Index } N_{ij} = S_{ij}/L_j; \tag{3.6}$$

Where

$$L_j = \sqrt{\sum_{j=1}^{j=5} [S_{ij}^2]} \tag{3.7}$$

If w_j is the weighting factors associated with the indexes, where

$$0 < w_j < 1.0 \qquad \text{for } j = 1, 5 \tag{3.8}$$

and

$$\sum_{j=1}^{j=5} w_j = 1.0 \tag{3.9}$$

The combined ratings can be obtained as:

$$\text{Combined Ratings } R_i = \sum_{j=1}^{j=5} \{[N_{ij}] * w_j\} \qquad \text{for } i = 1, n \tag{3.10}$$

The normalized ranking of R_i can be obtained by:

$$\text{Normalized Cumulative Effectiveness Factor (CEFs)} = R_i * 100 / \sqrt{\sum_{i=1}^{n} [R_i^2]} \tag{3.11}$$

It may be noted that $j = 1$, indicates the critical index, $j = 2, 3, 4$, and 5 denote time, cost, issues, and value indexes, respectively. n denotes the maximum number of activities.

The method outlines how to obtain an effective combined rating from a set of individual indexes. The activities, that are closer to 1.0, mean that those are quite important. The activities, that are closer to 0 and less than 0.5, represent weak links. They represent candidates for possible modifications or elimination. One can combine this approach with the normalized relative importance rating obtained from Quality Function Deployment. Using these two ratings, one can also compute a value index as follows:

$$\text{Value Index} = \frac{\begin{array}{c}\text{Normalized Technical Importance Rating (TIR)}\\ \text{obtained from QFD}\end{array}}{\begin{array}{c}\text{Normalized Cumulative Effectivess Factor (CEFs)}\\ \text{computed from Equation (3.11)}\end{array}} = \frac{\text{TIR}_i}{\text{CEF}_i} \tag{3.12}$$

Value index is defined here as the ratio of the normalized technical importance rating (TIRs) computed using QFD and the Normalized Cumulative Effectiveness factor

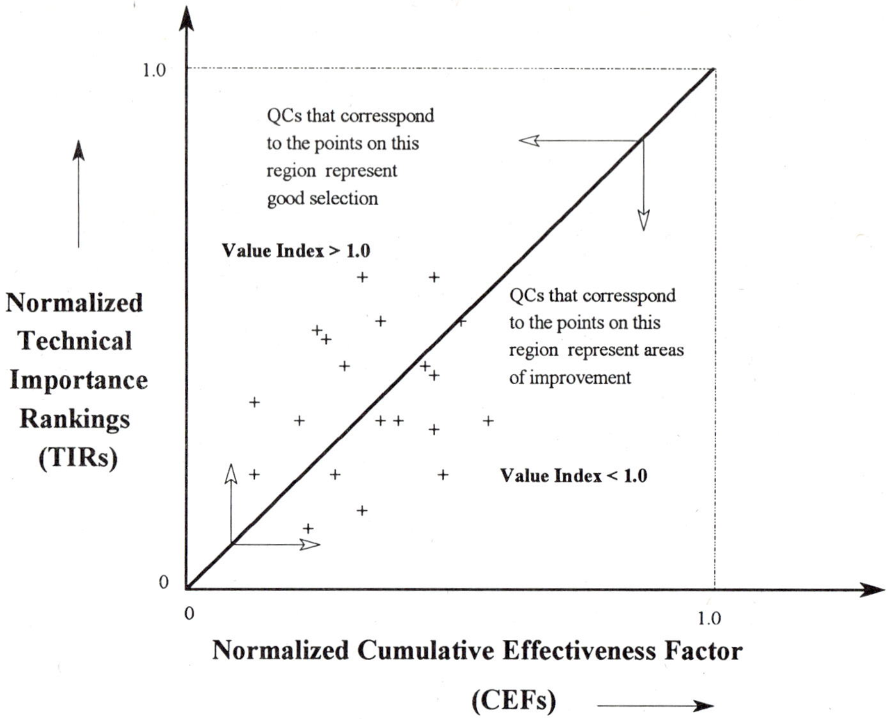

FIGURE 3.21 Value Graph of Quality Characteristics

(CEFs) computed from Equation (3.11). Since each QC has a corresponding TIR value, there will be as many value-indexes as TIRs. These indexes can be plotted on a value graph (see Figure 3.21). The points on the value graph represent the value index for each of the QCs. The value points that fall on the diagonal line (slope = 1) represent a break even point. The points that fall below the unit slope line represent the areas of possible improvements in performance or efficiency. The points that fall above the unit slope line represent the good points.

3.8 CHANGE MANAGEMENT METHODOLOGY

Understanding where we are starting from is an essential prerequisite for establishing the need for a change. Methodology for change management (CM) has four main elements [Prasad, 1995c] as shown in Figure 3.22:

1. **Quality Leadership Process (P_{ql}):** The quality leadership process is focused on determining the sources (many known and unknown factors) such as regularity mandates, customer preferences, technological breakthroughs, and world of busi-

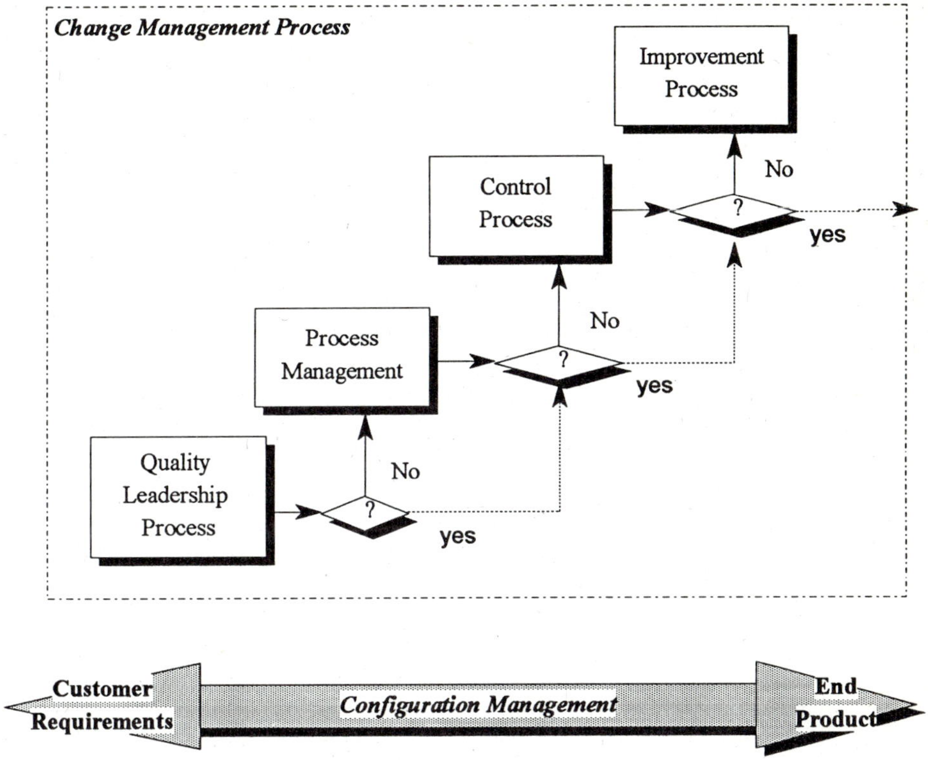

FIGURE 3.22 Elements of a Change Management Process and their Relationship to Configuration Management

ness realities that may legitimize the change from both "must-have" and "want-to-have" perspectives.

2. **Process Management (P_{pm}):** This element involves validating the changes—teams analyzing and evaluating the various processes and types of changes under consideration.
3. **Change Control Process (P_{cc}):** The third element is a change control process. There are many special causes of variations, such as improper setup, operator errors, defective gauging, broken tools, and materials that are under the direct control of the operator (production team). Through preventive action and continuous elimination of their sources of variation, teams can turn an unstable and unpredictable process into a well-behaved process that has a predictable outcome. This process utilizes a routing and queuing model to simulate the impacts of potential change.
4. **Improvement Process (P_{ip}):** The final element is bringing about the needed improvement. This is very similar to (continuous process improvement, CPI) and is discussed later in this chapter.

In other words,

$$CM \equiv f\,(P_{ql}, P_{pm}, P_{cc}, P_{ip}) \tag{3.13}$$

Where CM stands for change management and *f* stands for function.

3.8.1 Quality Leadership Process

In the area of change management, it is important to facilitate and provide a well-balanced quality leadership, P_{ql}, as unobtrusively as possible. All aspects of a company's operations should be continuously scrutinized, reexamined, and questioned. Total Quality Management (TQM) is a formalized initiative aimed at improving the industrial productivity and quality of products produced by the global manufacturing industry. The Big Three automobile manufacturers have changed their quality standards, which the suppliers must meet to be viable suppliers. Ford Motor went from Q101 to Q1 and now Total Quality Excellence [Fauteck and Studzinski, 1994]. General Motors changed from SPEAR 1 to Targets of Excellence and Chrysler Motors moved from Penta Star to Excellence. Organizations need to keep a constant watch on new technologies as they appear, utilizing those that are appropriate for TQM missions:

$$P_{ql} \equiv f\,[P_{tqm}, P_{qs}] \tag{3.14}$$

Where, P_{tqm} denotes total quality management and P_{qs} denotes quality standards, respectively.

3.8.2 Process Management

The second element of change management is understanding the change process. Improvements may result from product change, process change, cultural (human factors) change, requirement change, or enterprise operations change. Juran and Gryna [1993] mention that a change has a cultural bearing to one's formed comfort zone. When the level of change exceeds this zone, management often encounters some degree of resistance. The effective management of a process necessitates a structured approach that takes into account the "who," the "how," and the "what" of the changes. The process management (P_{pm}) can be viewed as a five-stage process that can determine company progress. The progress depends on the following measures (see Figure 3.23):

1. Current status (where are we now?)—P_{current}
2. Future status (where would we like to be?)—P_{future}
3. Analysis of what is lacking in the current process (review of who, how, and what)—P_{analysis}
4. How to accomplish the transition (getting there)—$P_{\text{transition}}$
5. Rational criteria or a basis for measuring the company progress

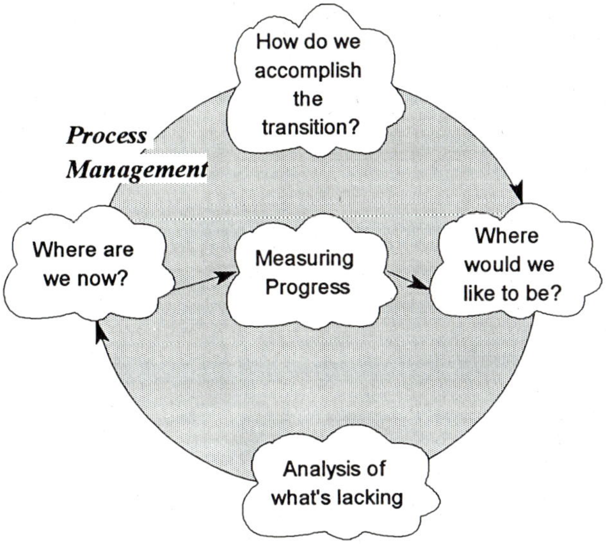

FIGURE 3.23 Elements of Process Management

The process management is thus a function of the first four measures. This can be expressed as

$$P_{pm} \equiv f\,[P_{\text{current}}, P_{\text{analysis}}, P_{\text{future}}, P_{\text{transition}}] \tag{3.15}$$

The measurement process is a function of first two measures. This can be expressed mathematically as:

$$\text{Measurement Progress} \equiv \frac{\Delta[P_{\text{future}} - P_{\text{current}}]}{[P_{\text{current}}]} \tag{3.16}$$

Carrying out a measurement at the starting point is necessary for the success of the whole change management process. Measurement serves as a mechanism to further systematically refine the process, and if needed, to improve and optimize the overall output of the company.

3.8.3 Change Control

The third element is change control process. There are two aspects of change control P_{cc}: (a) minimizing change variety; and (b) adherence to change specifications.

$$P_{cc} \equiv f\,[P_{cv}, P_{cs}] \tag{3.17}$$

Where, P_{cv} stands for change variety and P_{cs} stands for change specifications.

Both P_{cv} and P_{cs} reinforce the need for a good configuration management system having a built-in change control process. The latter ensures an orderly set of practices that could recognize and fix errors, could promote CPI and TQM, and could add discipline to decision making.

3.8.4 Total Quality Management Process Map

TQM is an iterative approach to optimize results and is an extension to what is commonly referred to as Continuous Process Improvement (CPI). The development and implementation of a CE process into a manufacturing enterprise is also a classic example of an iterative approach [Hartley, 1992].

Continuous improvement is a cyclic process of product and process optimization over a product life-cycle. Optimization implies that an organization is keeping in constant touch with new technological advances and frequently employs them to improve an existing product. Cycling means that an organization is continually exploring new frontiers in manufacturing technologies. The latest advances in related fields such as computers and systems are often reviewed regularly for possible inclusion in the product development cycle. The appropriate technology is captured and utilized, if it is proven to be the cost effective thing to do. The right decisions are made at the right time while sorting out an array of issues and choices. Successful implementation requires that teams work together, monitor quality, identify controlling factors affecting them, and find the appropriate technologies. Use of a continual refining process such as QFD is important. Many companies are regularly using such programs in quality and continuous improvement to maintain a world-class competitive position.

One of the hallmarks of the Toyota Production system is Kaizen. The Japanese word "Kaizen" may have no clear English translation, but the old-fashioned American phrase, "every day in every way I am getting better and better" seems a practical way of describing it. Here, the Japanese tend to study and improve processes manually for years before computerizing it. This is useful for processes that are subject to frequent changes. For processes that are fairly static, Kaizen may appear to be simply a mechanism for obtaining consensus among workers before automation. The Japanese perspective, however, is very different. Their objective (through Kaizen) is to permanently create a culture within a growing organization. This would either (a) help workers prevent the development of mistakes in the designed product or manufacturing processes, or (b) help prevent errors being passed on to the next operation in the process. This culture includes the utilization of a structured problem-solving process integrated with a mistake-proofing methodology that focuses on product design and manufacturing, both new and existing. Both Deming [1993] and Feigenbaum [1990] emphasize the importance of continuous improvement to assure a firm's long-term viability.

A structured approach to problem solving in a customer-driven mode is TQM. A process map for TQM is shown in Figure 3.24. The map provides a visual guide to help improve quality of the activities, services, and products. It consists of three connected balloons. It shows CPI as one of the balloons but product improvement is a three-way process. The problem prevention starts with CPI suggesting initial areas and motivation for change. The project management methods interact with QFD to determine an im-

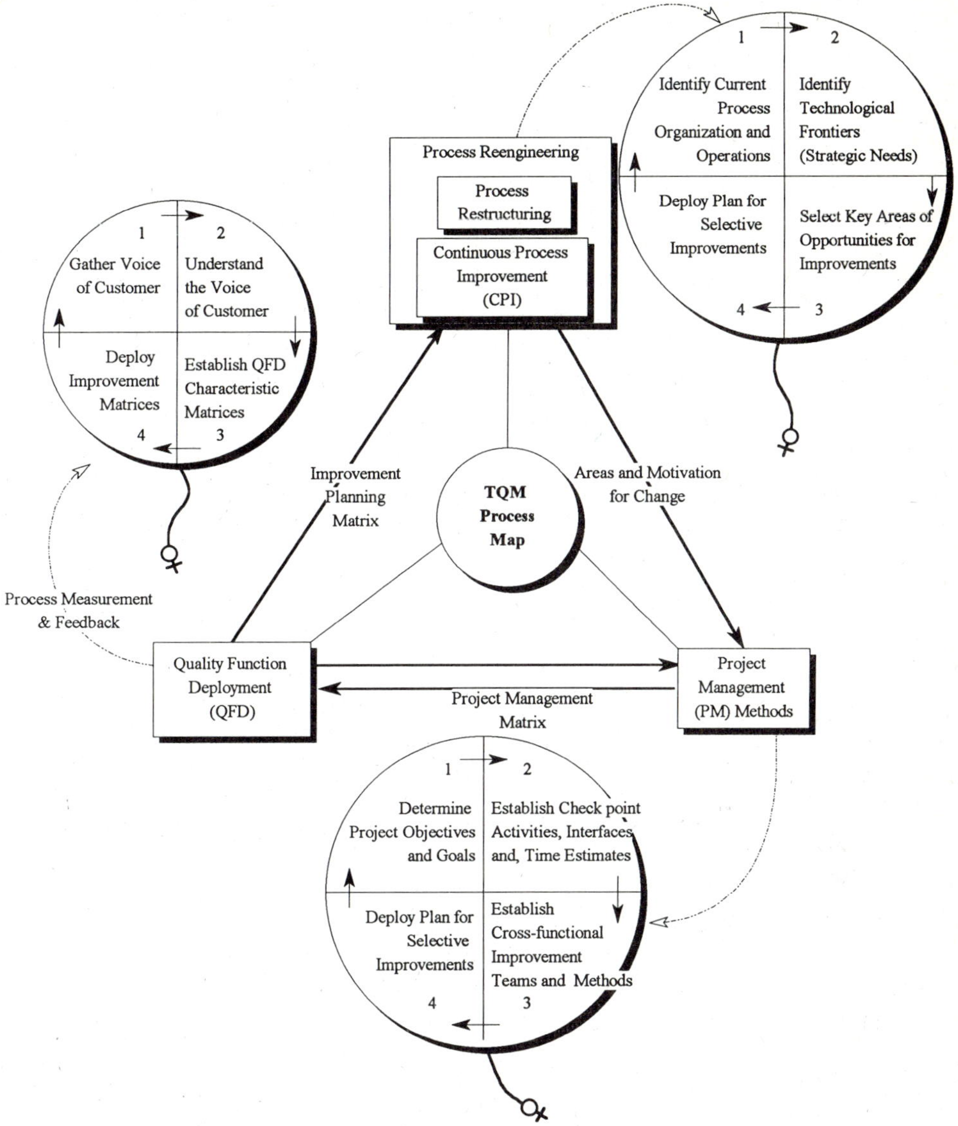

FIGURE 3.24 Elements of a Total Quality Management (TQM) Process Map

provement planning matrix. In the beginning, project management method suggests a trial project management matrix. QFD evaluates the matrix and provides process measurement and feedback. If the improvement planning matrix is satisfactory, it is fed back to CPI for further cycling. Total Quality Management (TQM) for product, process, or service improvement is a function of

$$TQM \equiv f\,[CPI, QFD, PM] \tag{3.18}$$

Where, in addition to CPI, QFD and project management (PM) are required to control each product development project. The inclusions of quality function deployment and project management methods add additional power to classical CPI approaches. The latter is indicative of a loop of a product realization process discussed in Chapter 7. Besides the three-way interactions, CPI, QFD, and PM each performs its own independent four-part cyclical loops. Each loop cycles the information clockwise as shown in Figure 3.24. The steps in CPI loop are as follows:

- ***Identify current process organization and operations:*** It may consist of selecting missions/goals/strategic issues, which are critical to organization or the operations.
- ***Identify technological frontiers:*** This may consist of identifying strategic needs and documenting performance, measures, and key processes.
- ***Choose key areas of opportunities for improvements:*** This may consist of documenting opportunities for product, process, or service improvements.
- ***Deploy plan for selective improvements:*** This includes developing and documenting action plans. Implement, monitor, and recycle the results for the next improvement cycle.

With TQM process maps, companies have been able to steadily decrease lead time and cost of the product while satisfying the customer with a robust design [Taguchi, 1986]. In the foreseeable future, there is a need for radical changes in the way engineering is managed and executed, just to stay in the race. In the future, design is more likely to be characterized by [Wilson and Greaves, 1990] the following developments:

- Zero distinction between researchers, designers, and production engineers
- No annotation of detail until manufacturing intent is assured
- No manufacturing engineering sign-off because multidisciplinary teams solve conflicts early
- One hundred percent predictable and controllable quality by design, not by inspection
- Total replacement of classical PERT project management techniques with goal-oriented approaches.
- Research, design, development, production engineering, and manufacturing engineering activities carried out in parallel and not in series.

3.9 CONCURRENT PROCESS REENGINEERING

In one sense, the whole idea of using process reengineering traits seems simplistic and easy, but its implementation in our concurrent product and process environment can be extremely difficult. Using a reengineering trait as an organizational strategy in CE is a long term commitment. It requires reviewing the 8 types of waste and reworks in each new process, each time a new change is introduced. The reengineering process, the way it is done traditionally, takes a long time.

A typical approach that one can take in implementing reengineering in a CE environment, where product and process design are performed concurrently, is to begin with a multi-functional team analyzing the current process. The team identifies wastes and reworks and recommends a modified design of the new process that eliminates these wastes. Translating the "as-is" process into an optimum performing design can be done using any one of the process reengineering traits: CPI, restructuring or renovation—as described in this chapter. Reengineering is a cyclic process as indicated in Figure 3.25. Identification of waste components in some processes may be taking place simultaneously while elimination of waste components in other processes is in progress. Once an optimum design of the "as-is" process is set, the product and process design conceptualization begins concurrently. Three parallel tracks of process reengineering are shown being performed concurrently in Figure 3.25 corresponding to process 1, process 2 and process 3, respectively. All these processes are assumed independent and are required for the product design. This parallel concept is called here "concurrent process reengineering." During each track, three steps are shown to be performed: a restructuring of the current process, an evaluation of the alternate processes and their impact on the design, and a reengineering of this particular process. A process could be a work-process, a design process (such as a reliability design, finite element analysis, crashworthiness analysis, etc.), or a manufacturing process (such as stamping, machining, forging, etc.) In the latter case, process 1 could be checks on manufacturability (such as cutting or machining), process 2 could be the checks on formability (such as stamping), and process 3 could be the checks on moldability (such as injection molding). The examples of checks could include limits on processing costs, accuracy, finish, risks, process capability, manufacturing practices, etc. During the second step of a track, alternate processes are evaluated and their significance, in eliminating or reducing the undesirable behaviors, are flagged. During the third step of a track, these flagged processes are used as discussion points (by the multifunctional team) during the reengineering step of this track. As critical or undesirable process behaviors are resolved by the teams, reengineering of the combined process then takes place. It consists of checking now the combined process and its environment against the combined system functionality. Since basic requirements for manufacturing and all other "ilities" have been concurrently checked, and since critical process decisions have been made by the multifunctional team, catastrophic process errors that could later come back to haunt the team will be eliminated upfront during the least costly phase of the product development. The above steps outline during concurrent process reengineering approach are meant to be time reducing, error-proofing, and simplifying techniques. Many of the routine, tedious process steps are concurrently dealt with. This allows the process teams to concentrate their expertise on the more complex, critical process as well as the creative aspects of the process design. The use of concurrent process reengineering in an integrated product development environment fosters engineering creativity while promoting higher grade process designs engineered in less time and lower cost.

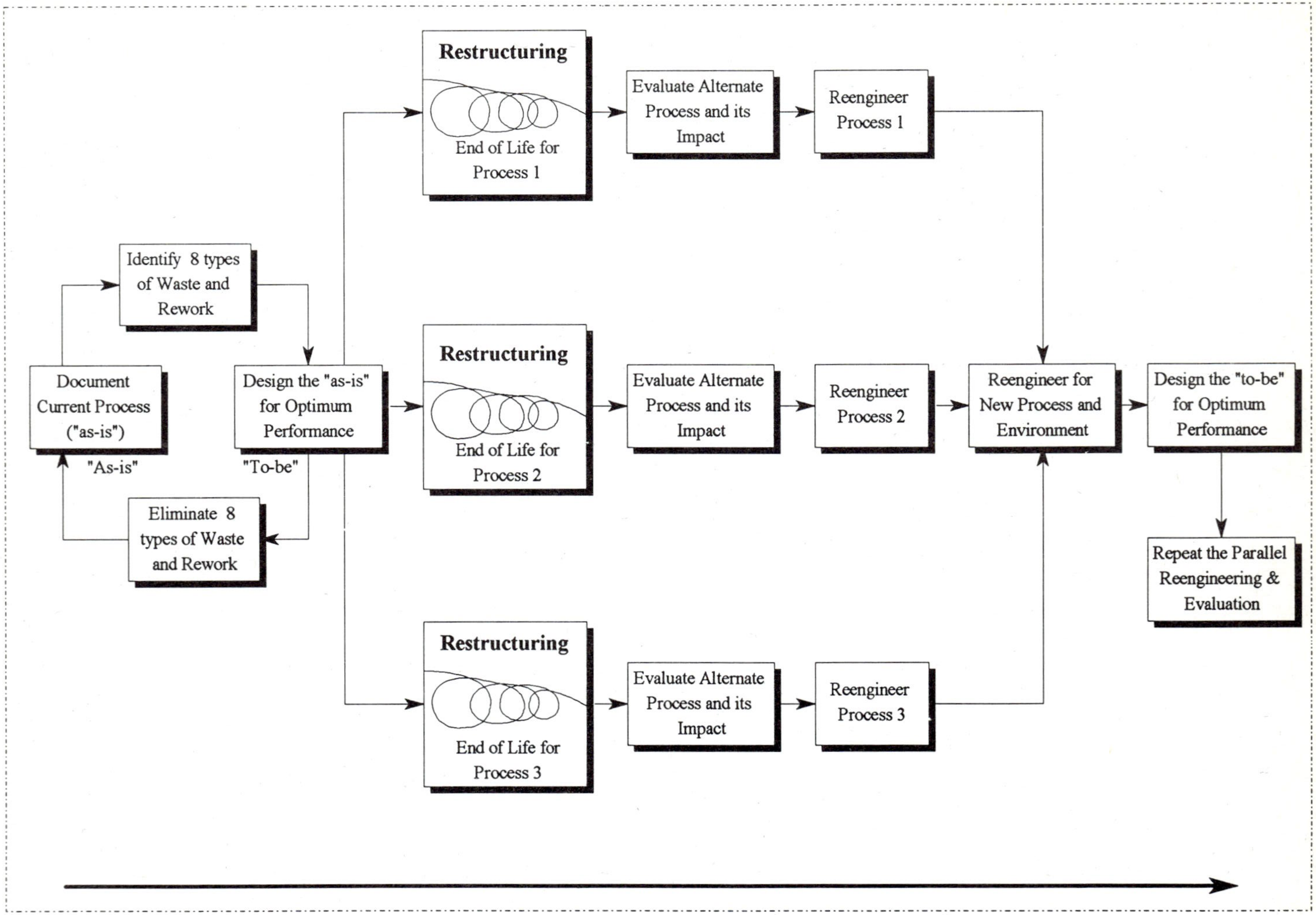

FIGURE 3.25 Concurrent Process Reengineering

3.10 CONCLUDING REMARKS

Reengineering strategies combined with flow-charting methodology show the power of "visual" techniques to process improvements. The different types of shape icons for the activities, line icons for the data, and flow icons for the connectivity help CE teams to capture the current (as-is) process very easily. It also allows teams to communicate the information through simple iconic means rather than through plain words. By depicting the activities with suitable icons, team members can easily visualize the entire process without much need for explanations and tutorials. The charts are easy to follow and self-explanatory. This strategy also provides a medium of discussion for process improvements in many areas of productivity concerns. The implementation of the change process as a result of those improvements overcome possible resistance from employees. The barriers become vividly clear when the entire process is flow-charted, and at that point members seem forthcoming in suggesting new improvements in their own areas and compromising on other issues.

REFERENCES

ALFORD M. 1986. "SREM at the Age of Eight; The Distributed Computing Design System." IEEE Computer Society, *Computer.* April 1986.

BEBB, B. 1992. "Learning to Thrive on Change" *Design Management Journal,* Spring 1992, pp. 10–19.

CAREY, W. R. 1992. *Tools for Today's Engineer—Strategy for Achieving Engineering Excellence.* Section 1: Quality Function Deployment. SP-913, SAE 920040, SAE International Congress and Exposition. February 24–28, Detroit, Michigan.

CARTER, D., and B. BAKER. 1992. *Concurrent Engineering: The Product Development Environment for the 1990s.* Massachusetts: Addison Wesley Publishing Company.

DEMING, W. E. 1993. *The New Economics.* Cambridge, MA: MIT Center for Advanced Engineering Study.

DIKA, R. J., and R. L. BEGLEY. 1991. "Concept Development Through Teamwork—Working for Quality, Cost, Weight and Investment." SAE Paper #910212. Proceedings of the International Congress and Exposition, SAE. Feb. 25–March 1, Detroit, MI.

EUREKA, W. and N. RYAN. 1988. *Taguchi Methods and QFD: Hows and Whys for Management.* Dearborn, MI: American Supplier Institute, A.S.I. Press.

EZOP, E., T. JACOBY, and L. LEACH. 1989. "Continuous Improvement With Quality Function Development." SME-IQ890405, SME Technical Papers. Dearborn, MI.

FAUTECK, R. D. and G. C. STUDZINSKI. 1994. "Supplier Improvement Process Plan." SAE Paper No. 940661, Quality and Design Issues in Automotive Simultaneous Engineering. SAE SP 1035. pp. 25–34.

FEIGENBAUM, A. V. 1990. "America on the Threshold of Quality." *Quality,* Jan. 1990, pp. 16–18.

FREEZE, D. E., and H. B. AARON. 1990. "Customer Requirements Planning Process [CRPII] (Beyond QFD)." SME Technical Papers. SME-MS90003, Dearborn, MI.

HAMMER, M. 1990. "Reengineering Work: Don't Automate, Obliterate." *Harvard Business Review.* Vol. 68, No. 4, (July–Aug.), pp. 104.

HAMMER, M. and J. CHAMPY. 1993. *Reengineering the Corporation: A Manifesto for Business Revolution.* New York: Harper Collins.

HARTLEY, J. R. 1992. *Concurrent Engineering: Shortening Lead Times, Raising Quality, and Lowering Costs.* Cambridge, MA: Productivity Press.

JURAN J. M. and F. M. GRYNA. 1993. *The Quality Planning & Analysis.* 3rd ed. New York: McGraw Hill, Inc., pages 158–160.

KING, B. 1987. *Better Designs in Half the Time—Implementing QFD Quality Function Deployment in America.* GOAL/QPC. Methnen, Massachusetts.

MILLER, L.C.G. 1993. *Concurrent Engineering Design: Integrating the Best Practice for Process Improvement,* Dearborn, MI. Society of Manufacturing Engineers (SME).

OAKLAND, J. S. 1989. *International Conference on Total Quality Management* (2nd: June 14–15 1989: London), Total Quality Management-Proceedings, New York: Springer-Verlag.

OHNO, T. 1988. *Toyota Production System: Beyond Large-Scale Production.* Cambridge, MA: Productivity Press, Inc.

PERRY, T. S. 1990. "Teamwork Plus Technology Cuts Development Time," *IEEE Spectrum,* Vol. 27, No. 10 (Oct.) pp. 61–67.

PETERSON, J. L. 1981. *Net Theory and Modeling of Systems.* Englewood Cliffs, NJ: Prentice-Hall Inc.

PRASAD, B., and N. STRAND. 1993. "A Flow-chart-based Methodology for Process Improvement." *Quality Concepts '93 World Class Manufacturing.* Detroit, Michigan: Engineering Society of Detroit, pp. 53–74.

PRASAD, B., 1995a. "On Influencing Agents of CE," *Concurrent Engineering: Research & Applications—An International J.,* Vol. 3, No. 2 (June), pp. 78–80.

PRASAD, B. 1995b. "Is Planning a Strategic Requirement for Concurrent Engineering Success?" *Int. Journal of Concurrent Engineering: Research and Applications,* Vol. 3, No. 3, pp. 164–166.

PRASAD, B. 1995c. "A Structured Approach to Product and Process optimization for Manufacturing and Service Industries," *International Journal of Quality and Reliability Management,* Vol. 12, No. 9, (October).

ROBERTS, L. 1994. *Process Reengineering: The Key to achieving Breakthrough Success.* Milwaukee, Wisconsin: ASQC Quality Press.

TAGUCHI, G. 1986. *Introduction to Quality Engineering.* Tokyo, Japan. Asian Productivity Organization.

TUTLE, H. C. 1983. "Breaking the Wall Between Design and Manufacturing," *Production,* Vol. 91, No. 5 (May) pp. 63–68.

U.S. Air Force. 1981. Integrated Computer Aided Manufacture (ICAM) Functional Modelling Manual (IDEF0). Air Force Materials Laboratory, Wright-Patterson AFB, Ohio, Report # AFWAL-TR-81-4023, (June, 1981).

U.S. Air Force. 1985. Integrated Computer Aided Manufacture (ICAM) Architecture Part II Information Modeling Manual (IDEF 1X), Air Force Materials Laboratory, Wright-Patterson AFB, Ohio, Report # AFWAL-TR-85-4023.

VENUGOPAL, R., and S. S. HUANG. 1989. "Framework for Determining the Effects of Change in Product Realization Processes. *Concurrent Product and Process Design.* DE-Vol. 21, PED-Vol.

36, Proceedings of the Winter Annual Meeting of the ASME, (edited by Chao and Lu), San Francisco, CA., (Dec. 10–15), pp. 41–46.

YANG, Y. 1989. IDEF1X Style Guide For PDES Users, Draft, (August 31).

WILSON, P. M., and GREAVES, J. G. 1990. "Forward Engineering—A Strategic Link between Design and Profit," *Mechatronic Systems Engineering,* Vol. 1, No. 1, pp. 53–64.

TEST PROBLEMS: PROCESS REENGINEERING

3.1. Which are the barriers of winning a competitiveness battle? What is the SWOT analysis? What benefits does it offer?

3.2. Why worry about introducing a change? How does a company successfully introduce and manage change? Give examples to illustrate "managing the change" process.

3.3. Describe categories of major strategies that companies, large or small, can adopt to ensure that change is an ongoing positive experience. Does the adoption of a particular strategy depend on the degree of severity for managing change?

3.4. What does reengineering encompass? From an organizational standpoint, should reengineering be the responsibility of an existing group/department or should a separate organization be created to perform this task? How can reengineering solutions be implemented effectively throughout a company?

3.5. Illustrate a set of reengineering strategies and how they would be used to meet basic business goals? Identify situations where one is more appropriate than another. Under what circumstances would renovation strategy be a better option than restructuring?

3.6. What are some of the common methodologies for capturing process information? What must this be capable of?

3.7. Identify factors that demand significant change in our approach to product design and development. Provide examples to support your reasoning.

3.8. Identify a list of quality characteristics (QCs) that you manage at work every day. Utilizing these QCs, draw a work-flow diagram. Then sort them into the following categories: prevention, appraisal, internal failure, and external failure.

3.9. Highlight some major quality concepts that have been utilized in the early 1980s from a total product development viewpoint.

3.10. Explain the major points in the concept of reengineering strategies. Compare this to the concept of departmentalization, which has occurred in the 1980s.

3.11. What changes usually take place in product life-cycle? How could you manage them effectively?

3.12. Describe the "Do it Right the First Time" philosophy. Does CE account for this or not? Explain.

3.13. Discuss company interests versus self-interests? Why is it essential to consider a "system-based" improvement matrix, not just "team-based"?

3.14. Can reengineering be achieved by combining some of the popular strategies? Give two examples to support your arguments.

3.15. In a rut to reduce the product life-cycle time, there is a danger in ignoring some essential CE steps. What do you think are important CE tactics and why?

3.16. What are the tenants for process improvements? Give some examples of metrics that are "productivity-related" and some that are "performance-related."

3.17. Why is Japan far ahead in product quality? Compare how Japanese master the "process management techniques" versus how Americans do.

3.18. How do you distinguish between continuous improvement versus quality control? Describe the requirements, obstacles, and benefits of CPI. How can one sustain continued adoption of improvements?

3.19. When layout design is complete, we are said to have locked in 70% of the product cost. How do you foresee changes in customer needs and technology advances after this point?

3.20. Identify and give examples of four types of tactics in the process reengineering activity? What are the critical tactics that can be employed during a design phase?

3.21. Why only 50% of the current activities are normally considered value-added in most enterprise. What can you do to fix that?

3.22. Discuss three ways of doing concurrent process reengineering.

3.23. What are the 6 Ms, 3Ps, 8Ws, and 7Cs?

3.24. Define the basic steps in a functional benchmarking. How do you establish checkpoints and milestones?

3.25. What are the pitfalls of benchmarking? Does this lead to being a leader or a follower?

3.26. How can organization increase the value of the products to the customer and to the organization? How would this type of activity be done?

3.27. How does efficiency change, if an original process is altered or improved? What are the major relative differences in efficiency when CPI, restructuring, and renovation are used?

3.28. How can we best manage the product development process? How can you accelerate the change process?

3.29. How could companies decrease their use of nonvalue-added activities without deteriorating the value-added activities?

3.30. What are the two common types of enterprise models? Why is enterprise modeling essential to advance a company into the twenty-first century? Identify some typical steps on which basis these models are constructed.

3.31. What roles do enterprise modeling or value-added analysis play in satisfying the technological and social aspects of the work environment?

3.32. What is work-flow mapping? What are the generic steps for creating a detailed flow-chart of a work-group process?

3.33. Discuss and compare the salient points of an "as-is" model versus a "to-be" model.

3.34. What are the two ways to achieve a "to-be" model? What are the major problems with a bottom-up approach to "to-be" modeling? When do you think a top-down approach can be beneficial to a company?

3.35. What do you achieve by combining top-down and bottom-up approaches to "to-be" modeling? What do you call this process?

3.36. The term "reengineering" is analogous to what type of information modeling? How can we identify areas where process improvement provides the greatest ROI?

3.37. Draw an enterprise modeling chart (both "as-is" and "to-be" models) for what you generally do every day.

3.38. What are the basic shape icons that can be used for capturing activities? Which are the basic flow icons for capturing flow types, and basic line icons for capturing data types?

3.39. Why is it important to define a number of icons to describe shape, flow, or line types? Are they all required in every process mapping situation?

3.40. Describe the process of consensus decision making as it relates to the "to-be" process. How can you gain consensus on what needs to be changed? How can you gain consensus on creating new processes and building a team to realize improvements?

3.41. What is TQM? Describe a process map of TQM that leads to a structured (or systematic) approach to problem solving.

3.42. How can you tell whether or not your project is on track? How one can apply project management, QFD, and CPI techniques to keep projects on track?

3.43. How can you maximize the use of management techniques and tools in the process (reengineering) flow-charts?

3.44. How can you use the specific measurement tools to analyze your old development process, measure your new process, and pinpoint what problems still exist? How do you get to the bottom of the root cause?

3.45. How can you identify quantitative data to measure timing and costs associated with product development activities?

3.46. How can you analyze available quantitative data to gain insight about your product development process? Does a ranking method provide this insight? Discuss one way to achieve ranking.

3.47. What are the different types of waste and rework that are commonly found in a manufacturing work-site? Give examples of each.

3.48. What is change management? Describe some key elements of a change management methodology. What, if any, is its relationship to configuration management?

3.49. Why use change control? Describe two aspects of change control methodology that are commonly used.

CHAPTER 4

CONCURRENT ENGINEERING DEFINITIONS

INTRODUCTION

The concept of concurrent engineering was initially proposed as a means to minimize product development time [Barkan, 1988; Evans, 1988; Stauffer 1988; Winner et. al., 1988]. Since then, many interpretations of "concurrent engineering" have emerged in literature. Today, CE is much more encompassing. Expectations range from modest productivity improvement to a complete push-button type automation, depending on the views expressed. CE is a paralleled approach—replacing the time-consuming linear process of serial engineering and expensive prove-outs. It is intended to elicit the product developers, from the outset, to consider the "total job" (including company's support functions). Some common definitions are as follows:

- Concurrent Engineering is "a systematic approach to the integrated, concurrent design of products and their related processes, including manufacture and support. This approach is intended to cause the developers, from the outset, to consider all elements of the product life-cycle from conception through disposal, including quality, cost, schedule, and user requirements" [Winner et. al., 1988].

$$\text{CE} \Leftrightarrow [\text{Paralleling of (life-cycle functions)}] \qquad (4.1)$$

- CE is "a systematic approach to integrated product development that emphasizes the response to customer expectations. It embodies team values of cooperation, trust, and sharing in such a manner that decision making proceeds with large intervals of parallel working by all life-cycle perspectives early in the process, synchronized by comparatively brief exchanges to produce consensus." [Cleetus (CERC), 1992; Ashley, 1992]

$$\text{CE} \Leftrightarrow [\text{Paralleling of (life-cycle functions)} + \text{Consensus} + \text{Cooperation}] \quad (4.2)$$

- The Computer-aided Acquisition and Logistics Support (CALS) office definition of CE [CALS, 1991] from Military Handbook-59 is "a systematic approach to creating a product design that considers all elements of the product life cycle from conception through disposal. . . . CE defines simultaneously the product, its manufacturing processes, and all other required life-cycle processes, such as logistic support. CE is not the arbitrary elimination of a phase of the existing, sequential, feed-forward engineering process, but rather the co-design of all downstream processes toward a more all-encompassing, cost effective optimum. . . . Concurrent Engineering is an integrated design approach that takes into account all desired downstream characteristics during upstream phases to produce a more robust design that is tolerant of manufacturing and use variation, at less cost than sequential design."

$$\begin{aligned}\text{CE} \Leftrightarrow [&\text{Cost-effective robust design (conception through disposal)} \\ &+ \text{Simultaneous design of (all downstream processes} \\ &\text{during upstream phases)}]\end{aligned} \quad (4.3)$$

- CE is a goal-directed effort, where "ownership" is assigned mutually among the entire group on the "total job" to be completed, not just "pieces" of it, with the understanding that the team is empowered to make major design decisions along the way [Turino, 1992]. This definition is more suited to a virtual CE enterprise, where each regional business unit is essentially its own separate business responsible for products sold in that region. Regional business units are empowered to make design decisions based on goals set with the parent organization (with constancy of purpose).

$$\text{CE} \Leftrightarrow \text{Minimization of (total product life-cycle time)} \quad (4.4a)$$

or

$$\text{CE} \equiv \text{Min.} \int_0^{\text{Product life-cycle}} [\text{Time}]\ d\ \text{life-cycle} \quad (4.4b)$$

- The U.S. National Research Council defines CE as "a design engineering environment in which computer-added design technology is used to assess and improve the quality of a product, not only during the active design phase but through its entire life-cycle." [Ellis, 1992]

$$\text{CE} \Leftrightarrow \text{Maximize Quality}$$

or

$$\text{CE} \equiv \text{Max.} \int_0^{\text{Product Life-cycle}} [\text{Quality}]\ d\ \text{life-cycle}$$

- Hartley [1992] broadened the CE definition by involving the concerns of design, production, purchasing, finance and marketing, and also by tying in lead times with improved quality and lower costs.

$$\text{CE} \Leftrightarrow \text{[Maximize Quality, Reduce Lead Time, and Lower Costs]}$$

- CE is a product development methodology where up-front "X-abilities" (such as manufacturability, serviceability, quality, . . .) are considered part of the product design and development process. X-abilities are not merely for meeting the basic functionality or a set of limited strategies, but for defining a product that meets all the customer requirements—see for example the work by Shina [1991] and Figure 1.9, Chapter 1.

$$\text{CE} \Leftrightarrow \text{[Up-front consideration of all X-abilities]} \tag{4.5}$$

- The narrow view of concurrent engineering is "integrating product and process design." The wide view is "integrating over the product life," and the wider view is "integrating over the enterprise" [Finger, 1993].

$$\text{CE} \Leftrightarrow \text{[Integrating Product and Process Design over the Enterprise]} \tag{4.6a}$$

or

$$\text{CE} \equiv \text{Minimize or Maximize} \int_{0}^{\text{Enterprise}} [F] \, \text{d life-cycle} \tag{4.6b}$$

Where, F could be time, cost, or quality.

The most commonly referred to definition is that of Winner [Winner et. al., 1988]. Some experts recognize influencing agents of CE as forces of change [Carter and Baker, 1992]. Prasad [1995] has chosen to divide forces that influence the domain of CE into seven agents (called here as 7Ts): talents, tasks, teams, techniques, technology, time, and tools (see Figure 4.1). One of the primary team issues is the decomposition of tasks. The people's issue is the composition of teams. *Teams* are often used to cooperatively solve the problem. *Technology* issues arise due to drive for competitiveness. Examples of popular technologies in CE are soft prototyping, visualization, product data management, design for X-ability, multimedia, electronic data interchange (EDI), etc. *Tools* include software, hardware, and networks that make CE practical in today's world of multinational corporations, multipartner projects, and virtual corporations. From the *time* point of view, CE is an initiative of the product development community that has the goal of reducing the length of the product design and manufacturing cycle time by allowing teams of engineers to develop design modules concurrently from their perspectives [Pennell and Slusarczuk, 1989]. *Training* also plays an important role in CE. A popular word in the business press is reengineering, meaning, in short, revamping the processes by which one satisfies customers needs. From a business angle, CE means reengineering the product development process so that tasks are organized concurrently. The Department of Defense

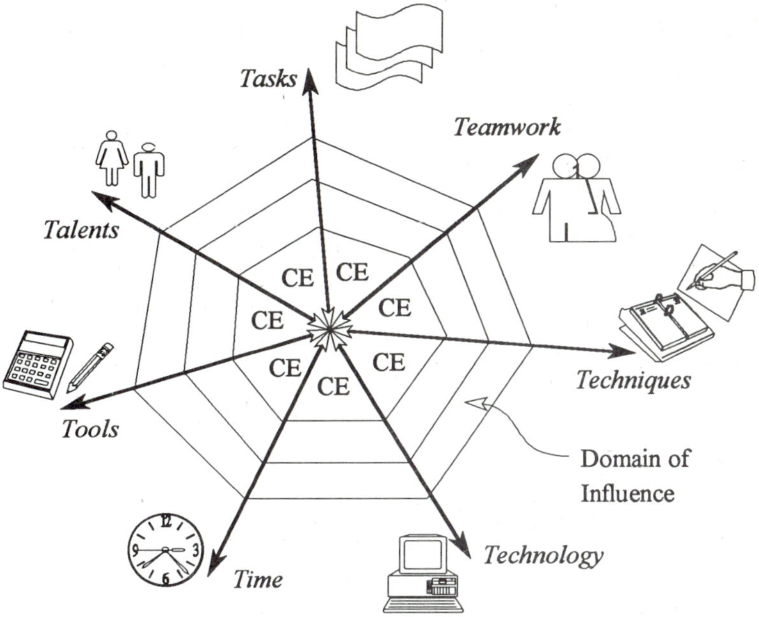

FIGURE 4.1 7Ts—Seven Influencing Agents of CE

(DOD) and some aerospace companies refer to CE as integrated product development (IPD) [Ahsley, 1992].

$$\text{IPD} \Leftrightarrow \text{Minimize (cycle time)} + \text{Paralleling of (life-cycle functions)} \qquad (4.7)$$

Concurrent engineering (CE) systems stem from computer integrated manufacturing (CIM) systems [Solberg, 1988; CIM, 1988] or engineering information systems (EIS), and are closely related to Computer-aided Acquisitions and Logistics Support (CALS) [Lewis, 1990]. Formal needs for these systems have been developed and published [Linn and Winner, Volume 1 and Volume 2, 1989]. The first tenet of CE is cooperation—from the human side as well as the data management side. Closely related are the ideas of "design for X-ability" and "concurrent part and process design" [Chao and Lu, 1989]. The latter can be met by teams cooperating early in the design process. Most traditional systems require complete or nearly complete part geometry before the start of any design iteration. CE systems are intended to work with approximate models of constraints and tradeoffs involving generic parameters common to multiple disciplines across the life-cycle.

CE has a major impact on the process setup and the way we conduct business. As shown earlier in Figure 2.26, concurrent engineering replaces the traditional sequential "over the wall" approach to a simultaneous design and manufacture spectrum with parallel, less interrelated processes. It aims at reducing the total effort in bringing the product from concept to delivery, while meeting the needs of both the consumers and industrial customers. The four major phases of the product development (as shown in Figure 2.26) have been detailed into nine tracks (shown in Figure 4.2) running in parallel. They show

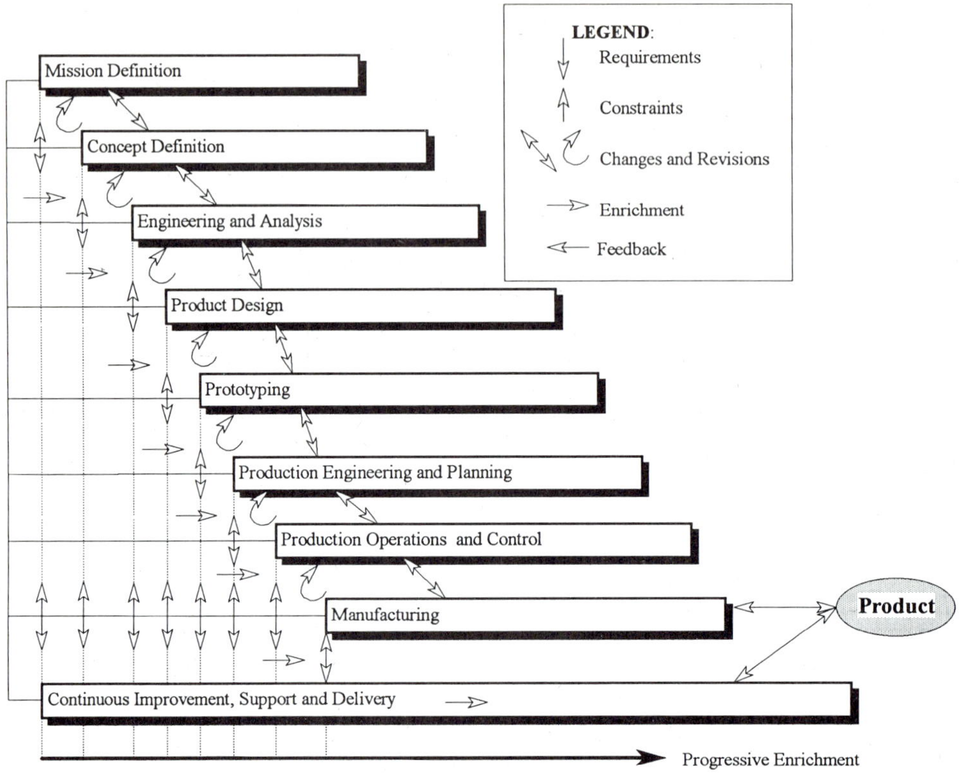

FIGURE 4.2 Concurrency During Phases of Product Design & Development Process

the different tracks of the development process. These tracks are mission definition, concept definition, engineering and analysis, product design, prototyping, production engineering and planning, production operations and control, manufacturing, and finally support and delivery. The continuous improvement—"support and delivery" is an ongoing coordination track, which runs for the full life-cycle. This track provides, besides normal project management functions, sequencing, cooperation, and continuous support to other tracks. These tracks are not unique to any particular product, and steps (overlaps) may differ from product to product. Once a product is decomposed into a set of tracks, and a track is decomposed into a set of tasks, they become one full set of steps leading to product realization. The staggering of their (steps) start points and overlaps is indicative of partial information sharing. Orders are indicative of their precedence. The amounts of overlap between two consecutive tasks are indicative of the degree of dependency that may exist between them [Krishnan, 1993]. If the tasks are completely independent, they all can be aligned along the left margin of the diagram keeping the precedence intact. The time-to-market in that case would be dominated by tasks that take the longest time to finish. This is a case of a true "simultaneity" or "simultaneous engineering" situation [Garrett, 1990]. The idea of "best concurrency," when the tasks are not completely independent, is to align each step to the farthest left of the diagram as much as possible while satisfying the following 3Ms:

1. Maintain the precedence of the tasks that were decomposed.
2. Minimize the horizontal overlap between the consecutive tasks.
3. Maximize the independence of the decomposed tasks.

In general, there is greater affinity and dependence between pairs of activities which are adjacent to each other. The farther away the activities are positioned from each other, the lesser would be the degree of affinity or the need for information transfer among them. For example, mission definition would be more closely related to concept definition but would have very little bearing on activities such as manufacturing. Similarly, manufacturing would be closely related to production operations and control but would be less sensitive to tracks such as engineering and analysis. The arrows in Figure 4.2 represent the interactions between the various tracks. The downward-slant arrows represent the net result of enrichment (a horizontal component) and net result of needs (a down-vertical component). The upward-slant arrows indicate the net results of feedback (reverse horizontal component) and net results of consequences and constraints (up-vertical component). A general description of the arrows is given in the Figure 4.2 legend. The concurrent approach is gaining worldwide attention at the moment. The paralleling of life-cycle activities is being deemed necessary by more and more industries to adapt quickly to changing market conditions and shrinking time-to-market targets. Additional savings may come from cost-avoidance—early recognition of possible faults, thereby preventing unnecessary issuance of so called "change orders" after the product is released.

4.1 BASIC PRINCIPLES OF CE

Timing is an important consideration in CE. A lot rides on the timing of making decisions and problem discovery. Approximately 80% of a product's life-cycle development cost is driven by decisions made in the first twenty percent of the program effort [DARPA, 1987; DARPA, 1988]. Concurrent engineering is founded on eight fundamental principles:

- ***Early Problem Discovery:*** Problems discovered early in the design process (particularly during the first 20% of the cycle time) are easier to solve than those discovered later.
- ***Early Decision Making:*** The "window of opportunity" to affect a design is much wider during an early design stage than at a later stage—when some of the decisions are frozen and when the design is matured. Teams often have the natural tendency of making quick and novel decisions, which is good, except those decisions should be lasting as well.
- ***Work Structuring:*** Human minds cannot practically work on multiple tasks simultaneously, parallel computers do. What a human mind is good at is systematically structuring the work or, more importantly, structuring the work-environment—so

that each task can be performed independently of each other either by a human being, a machine, or a computer.

- ***Teamwork Affinity:*** Teams of people working in separate groups are likely to create designs, which may be optimal in their individual domains but will seldom remain optimal in a combined domain which is a union-sum of those individual domains. Also, teams will have a better affinity if they trust each other. Trusting members, if they agree to accept responsibility for a task, prefer to work together rather than in isolation.
- ***Knowledge Leveraging:*** The domain of product design is often very large. It may be impossible to create a general purpose automated or knowledge-based system, which will use appropriate tools and knowledge-driven rules (mostly computerized) to guide decision making. Interlinking decision support tools with spurts of human knowledge-base will continue to be the most viable method for solving complex problems.
- ***Common Understanding:*** Teams will work better if they know what other members are doing. This includes operational understanding of all relevant interplay; for example, what constraints a team-member would encounter when certain parameters will be changed.
- ***Ownership:*** Teams will work enthusiastically to make a good product if they are empowered to make decisions in shaping the design and are given "ownership" of what they produce.
- ***Constancy of Purpose:*** Most departments have a natural tendency to make their departments look good to others—create false profits—even though it may be detrimental to the overall corporate goals. The whole corporation will do better if everyone works toward a common set of consistent goals irrespective of the departments they have allegiance to. This requires a change in thinking beyond the goals of one's individual department or teams to those of the SBU's or the company's goals. The obligation of any supporting unit is not to suboptimize its own goals (such as unit's profit potential or sales) without a clear and direct relationship to the company's overall goals. It must contribute its best towards the system goals. Aiming towards constancy of purpose results in everyone contributing his or her best—working toward a common set of consistent goals.

In summary, the benefits of CE stem from the eight basic principles—Early Problem Discovery, Early Decision Making, Work Structuring, Teamwork Affinity, Knowledge Leveraging, Common Understanding, Ownership, and Constancy of Purpose.

4.2 COMPONENTS OF CE

Although components of CE have not been formally documented, it is purported that it is the environment—spanned by its components—that yield a significant competitive advantage in product cost, quality, investments, and reduction in cycle time. Components distinguish whether a CE environment would be successful or not. The rewards and satisfaction in CE come from the success of its components. For instance, success of CE depends on the part of those who participate in it. In CE, every participant is expected to contribute to making the final product: from planners and industrial designers, analysts and detailers, to production engineers and support personnel [Garrett, 1990]. The following discussion describes the basic components of CE.

4.2.1 Multidisciplinary Setup

CE is structured around multifunctional teams that bring specialized knowledge necessary for the program [Owen, 1992]. The multidisciplinary setup—called product development team (PDT)—is composed of several distinct technical sub-units specializing in a variety of disciplines:

- Product planners (T_{pp})
- Product concept engineers (T_{ce})
- Engineers and analysts (T_{ea})
- Product designers (T_{pd})
- Prototype engineers (T_{pe})
- Production engineering planners (T_{ep})
- Management and control (T_{mc})
- Computer integrated manufacturing (CIM) and assemblers (T_{ma})
- Delivery and support (T_{ds}) teams

$$\text{PDTs} \equiv \cup\,[T_{pp}, T_{ce}, T_{ea}, T_{pd}, T_{pe}, T_{ep}, T_{mc}, T_{ma}, \text{ and } T_{ds}] \tag{4.8}$$

Where, $\cup$ indicates *u*nion-of and T stands for "*T*alents." In the above, nine concurrent teams are intentionally chosen to show the actual correspondence with each of the nine concurrent tracks of Figure 4.2. Each track is responsible for developing and integrating its own aspect to the product's life-cycle as the program requires. However, there could be as many tracks/tasks and teams as needs arise. For example, experts from the volume production area could be involved in prototype production to identify opportunities to improve process reliability as early as possible. By using this multifunctional team approach to merge design and manufacturing, GE Aircraft engine division reduced design and fabrication lead time for some GE engine components from twenty-two weeks to three weeks.

Figure 4.3 shows a bidirectional, sandwiched structure for an Integrated Product Development (IPD) System. In one direction, PDTs are supported by the customers on the top, and the company infrastructure (organization) at the bottom. In a perpendicular direction, PDTs are sandwiched between product and process on one side, and tools and technology on the other side.

$$\begin{aligned}\text{IPD System} \equiv \cup\,[&\text{Customers, Product, Process, Tools,}\\ &\text{Technology, Infrastructure, PDTs}]\end{aligned} \tag{4.9}$$

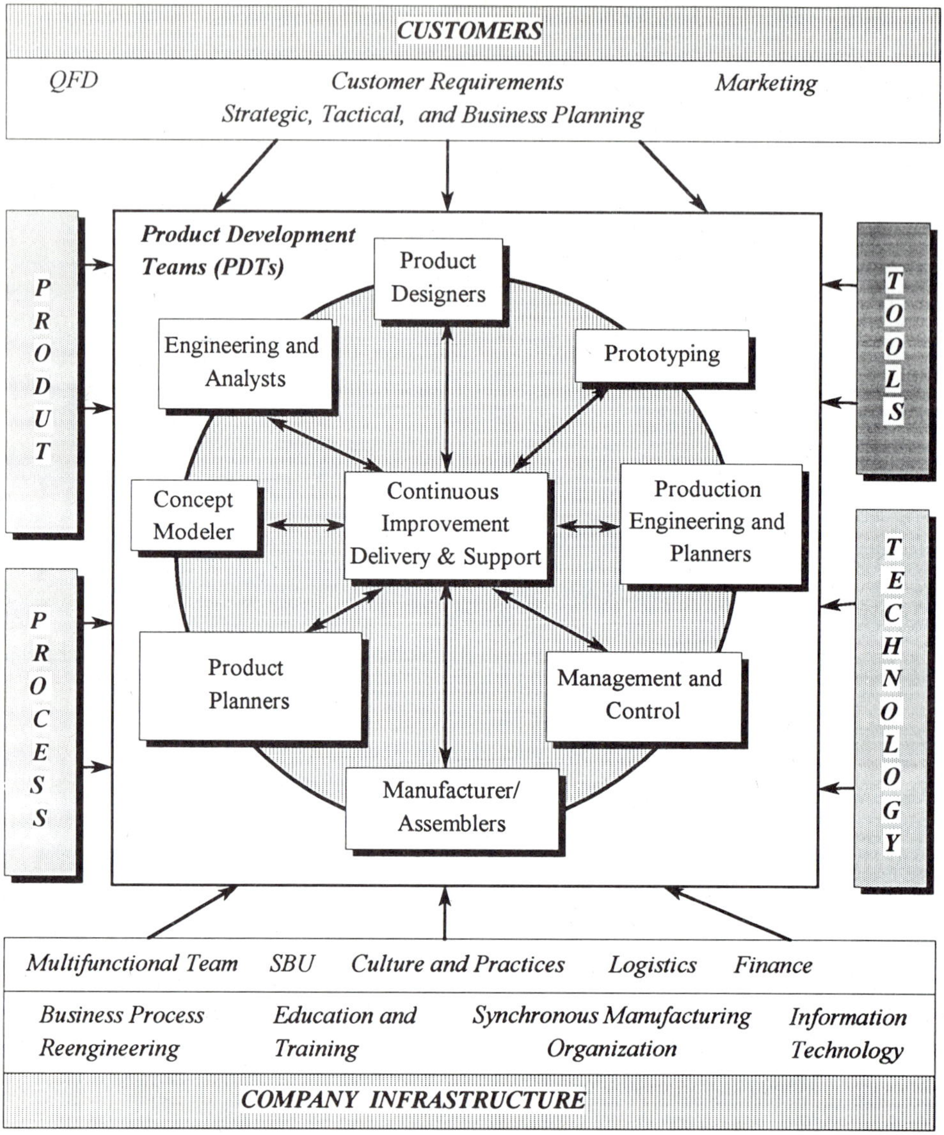

FIGURE 4.3 A Typical Product Development Team (PDT)—An Example

Where, ∪ indicates union-of and PDT stands for product development team. The infrastructure involves a wide range of disciplines including multifunctional teams, strategic business units (SBUs), culture and practices, business process reengineering, logistics, finance, information technology, education and training, and synchronous manufacturing organization. As shown in Figure 4.3, the customers help establish the requirements for IPD including QFD, marketing, and strategic/tactical business planning. In an IPD system, PDTs integrate the customer's inputs about their product and processes with their own experience in business solutions (tools), knowledge of future directions (technology), and the organizational infrastructure to provide worldwide competitive advantage. The PDTs, which replace the traditional functional department, are often organized along goal-oriented principles. Experts in the field of mechanical, electrical, industrial, chemical, and material engineering, as well as a variety of other fields, work together. Removals of barriers for cooperation and resolution of conflicts are responsibilities of the PDT manager [Deming, 1993].

The demands of today's ever-changing international marketplace are immense. Goals are moving targets, undergoing constant changes and shifting in response to market conditions. The diversity of disciplines in CE is essential to leverage core competency in order to address the growing complexity of today's product needs and global manufacturing trends. CE requires a new approach to project management. Each team must work closely with other teams to identify and develop techniques that are more cost effective, innovative, and simple to use.

4.2.1.1 Promoting the Spirit of Innovation

The multidisciplinary team must work in an environment in which innovative thinking is encouraged and teamwork is rewarded. The spirit of cooperation will encourage the team to use creativity in proposing solutions for difficult manufacturing situations. Suppliers must work with their clients and vice versa to design the manufacturing processes, tooling, equipment, and systems needed to fit the production and plant limitations. They must be selected and involved early enough to participate in the design and other critical decision making processes. By joining efforts with suppliers and partners, aimed at meeting the market needs or functions not yet served, everyone wins. Each unit gains the strengths provided by the resources of all the units combined. The customer gets better service or product at lower cost.

4.2.1.2 Early Supplier Involvement

Not too long ago, companies were reluctant to carve any work to an outside supplier before there was a valid contract establishing terms of their working relationship. This has since changed with the recent emergence of value managed relationships and long-term sourcing strategies. In the early 1980s, with increasing international competitiveness, profit margins declined sharply. Most of the manufactured products were very complex. It was beyond the scope of a single source (for instance an enterprise or a prime contractor) to design, build, and assemble the complete product. It became obvious that close working relationships were essential for the survival of both the product and the companies.

Today, most original equipment manufacturers (OEMs) recognize this need. They invariably involve the production source during the process of developing the design concept. Adequate expertise and involvement of key suppliers and specialists are vital to achieving the collaborative support required by CE. Good suppliers are capable and interested in participating in the development process. They should know and understand their process capabilities as well as their limitations very well and should be able to create an appropriate plan for their production. Early involvement is critical in seeking their vital inputs and in influencing the product's design as it evolves. All participants learn from the joint experience. Suppliers, through this process, get the current business or at least enhance their chances to become the most logical choice for the next model business. The company gets the early feedback, which might have some useful bearing in selecting other related components.

4.2.2 Synergy and Teamwork

Synergy—that is, combining team capabilities to produce results greater than any individual team-member effort—is the cornerstone of any CE organization. Teamwork emphasizes interpersonal relationships, cooperation, negotiation, and collaborative decision making. For synergy to take root and facilitate cooperation between teams, companies need tools that perform in concert. The human side of CE plays an important role in making the process work. It is the task of the team management to keep a balance between the "creativity forces" (such as empowerment, autonomy, etc.) and the "accountability forces." Creativity forces provide the teams the freedom to do what is right. The accountability forces want those to remain within the confines or goals of the organization. There are seven qualities to look for in a CE work-group:

- ***Empowerment:*** This means giving a design team freedom, authority, and the checkbook to get the job done right the first time. The management team must specify the major team goals, budget, and overall design task boundaries to limit their decisions within the organization and product goals.
- ***Staffing:*** This involves identifying what skills are required for what portion of the work and when to apply them. The success depends on keeping the team of experts lean to start with. Add members to the team when talents are needed to fit the task. Creating tasks to fit the talents of team-members is a backward process.
- ***Organization:*** This requires definition of the major objectives of the team including team-building procedures (such as decision making mechanisms), brainstorming, and other human-dynamics tools. The team interaction procedure ought to be set in place before taking up staffing or scheduling the tasks.
- ***Leadership:*** This involves tasks of motivating, guiding, and coaching both the technical experts as well as support personnel. Greater emphasis should be placed on "leading" the teams rather than "managing" them. The leadership role is rotating as opposed to appointing.
- ***Measurement:*** The team should focus on collectively defining the product development goals. Specific limits exemplified by "X-abilities," such as manufacturability, ought to be set as part of product development goals. The PDT should specify measurements for these goals including developing metrics for each goal.
- ***Autonomy:*** Autonomy is essential. Management should become involved in project reviews, but let the decision making power lie within the collective responsibility of the assigned teams.

- ***Technical Memory or Know-how:*** It is the responsibility of the technical core team to provide an environment fit to capture the "lessons learned" into some form of technical memory or history files to be used later. The management team must ensure that this information is archived and reused or communicated to others in the company, including those not directly involved on technical ("how to do") tasks.

In other words, the work-group outcome is a function of these seven quality indicators.

$$\text{Quality of Work-Group} \equiv f\,(\text{Empowerment, Staffing, Organization, Leadership, Measurement, Autonomy, Technical Memory or Know-how}) \tag{4.10}$$

Where, f indicates "*f*unction-of."

4.2.3 Global Participation

The concurrent engineering process for a product development is not a CE process unless it involves all parties that are responsible for its making, regardless of who they report to administratively. Subcontracting companies must be included as participants in the CE teams; at least until the interface design requirements have been determined, evaluated, and are somewhat firmed up. The distributors or retailers often tell the product manufacturers exactly what consumer wants are. The product manufacturers then are able to communicate upstream to the suppliers what parts or materials they would require to manufacture the product. For the organization to function as a unit and for the product to compete globally, all participants have to know what is expected from each other. Strong communication links should be in place. This will ensure a complete integration of the OEM's product needs with supply chain capabilities. With such integration, the subcontractors can influence requirements before it is too late. Requirements can be stated in joint terms which the subcontractor can effectively satisfy, and which are reasonably stable and unlikely to undergo any significant change.

4.2.3.1 Inclusion of Outside Trade Partners

An effective inclusion of outside (trade) partners in the cooperative development is frequently one of the underemphasized issues related to the implementation of a CE process. In today's environment, because of the growth in the complexity of consumer products and the increased reliance on specialized technologies and methods to manufacture them, partnership has become an increasingly important issue. Companies often rely on outside partners to supply expertise, services, and products in various specialized disciplines. Many examples exist [Womack, Jones, and Roos, 1990]:

- Ford and Mazda—Auto Alliance Plant, Flatrock, MI.
- General Motors and Toyota—NUMMI, Fremont, CA
- Chrysler and Mitsubishi—Diamond Star, Bloomington, IL
- General Motors and Suzuki—CAMI, Ingersoll, Ontario

- Subaru and Isuzu—SIA, Lafayette, IN
- General Motors and Isuzu—Luton, UK
- Toyota vehicles assembled by Volkswagen—Hannover, Germany

In conjunction with the United States Council for Automotive Research (USCAR), GM, Ford, and Chrysler are working to establish voluntary parts standards on items—such as light bulbs, car jacks, radiator caps, switches, handles and other noncompetitive parts—to reduce the influx of parts and boost global competitiveness. Taken to its extreme, this could mean the emergence of a concept called "virtual company," where the core company has only a limited staff [Kimura, 1994] mostly financiers, who are also the planners of the product ideas. The major workforce is comprised of individuals from various other companies having the appropriate skills to transform the ideas into a useful product. A list of typical participants of a virtual company is shown in Figure 4.4. They are mostly contract employees. However, they are responsible for delivering the services or products that they are contracted for, on time and at budget. There is a global partnership between the participants: the product manufacturers (stylists, designers, engineers, managers, white-collar workers), process planners (MRP, CAPP), robots/assemblers, machinists (blue collar workers), parts suppliers (vendors, contractors) or materials suppliers (operators, programmers), and product supporters (sales and service outlets, distributors, schedulers, deliverers, retailers).

$$\text{Virtual Company} \equiv \cup\ [T_{wc}, T_{pp}, T_{sd}, T_{ra}, T_{bc}, T_{cop}, T_{svc}] \tag{4.11}$$

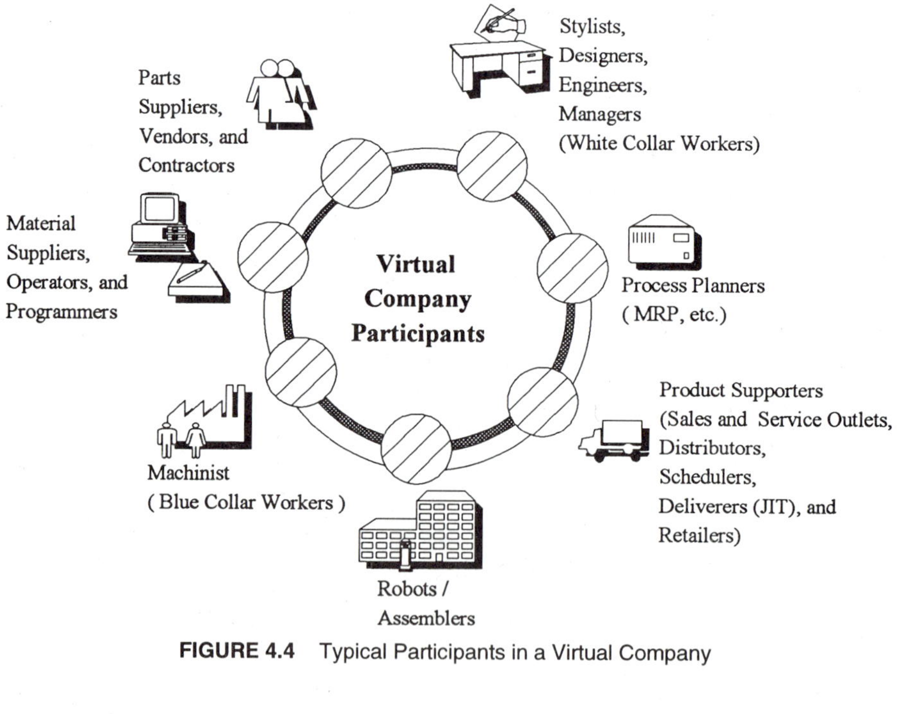

FIGURE 4.4 Typical Participants in a Virtual Company

Where

T_{wc} = white-collar workers
T_{pp} = process planners
T_{sd} = schedulers and deliverers
T_{ra} = robots assemblers
T_{bc} = blue collar workers
T_{sop} = material suppliers, operators and programmers
T_{svc} = parts' suppliers, vendors and contractors
$\cup$ = union of

It will do little good for a company to adopt a CE environment (or to control its product definition process) without including its trade partners, if a major or significant portion of its production is performed by outside suppliers [Owen, 1992]. Establishing a partnership can be strategically very important. It can eliminate or minimize the needs for in-house inspection. By establishing some type of partnership, where the certification program is a part of the deal, one can ensure the delivery of quality incoming materials. In that case, the cost-benefits of inspecting incoming materials and sorting out defective parts for return to vendors must be weighted against the supplier's cost of getting defect-free incoming parts. A successful partnership requires a harmonious communication environment characterized by rapid, accurate, and "paperless" business transactions. Other claimed benefits of partnership include greater satisfaction to the customer, simplified recycling, fewer computer entries, smaller inventories, and greater economy of scale. The increased use of electronic commerce technologies (such as electronic data interchange, or EDI, via wide area networks, value added networks, and electronic vendor bulletin boards) are paving the ways of making partnership seem quite painless. Partnerships are widely used in the auto industry to exchange purchase orders, shipping notices, and payments, particularly with the first-tier of supply-chain partners that deliver directly to OEMs. A first-tier supplier of instrument panels, for example, may be required to deliver product within a few hours of receiving an order, and deliver it in the assembled order needed on the assembly line. This close partnership could directly reduce inventory industry-wide.

4.2.3.2 Coordinating with Subcontractors and Suppliers

Subcontractors can share the product's life-cycle responsibility via three ways:

- *Specification and Formal Purchase Contract:* One way to share responsibilities with suppliers is via specification and formal purchase contract (FPC). In this traditional process, the customer of his/her own initially creates the requirements. Requirements are passed on to the suppliers (in the specification form) for a quotation (RFQ). The supplier sends back a bid and once the customer agrees with it, the customer issues a FPC or a purchase order. The suppliers design and develop the product as per their understanding or interpretation of the specifications and deliver to the customer the finished product in accordance with the contract. Such work order can proceed in a traditional fashion (see Figure 4.5a) as long as the suppliers are able to satisfy stated requirements and the requirements are not changed in the midterm. There are two flaws in this method. First, contractors are an outside entity looking in. Their positions are no better than two independent internal design groups (in a traditional process) operating at arms' length. Second, since specifications are developed without participation of the subcontracting companies, the chances that they will be able to satisfy the specifications fully and effectively are slim. The quality of the results would be very similar to that of two separate design groups operating at an arm's length. The other eminent danger is that by the time the work is done, it is very likely that the requirements may have to be altered. The two companies doing business in this way must, therefore, include legal and administrative provisions for possible requirement changes in the contractual agreement. The change-and-delivery process thus becomes lengthier and more expensive, since it will also involve more administrative-type personnel, and additional administrative chores.

- *Two-phase Contractual Agreement:* One of the major flaws of the traditional customer-supplier relationship is that no input is sought before specification is developed. To alleviate this, manufacturers sometime engage in a two-phase contractual agreement with their trading partners. First, they enter into a less formal agreement during an early definition phase that permits them to collaborate in product development while still preserving a truly competitive environment. This is called Phase I agreement. Later, after the requirements are firmed up, a more formal agreement, called Phase II, is signed between the partners. This is shown in Figure 4.5b. This ensures that the developed specifications can lead to a useful product. However, this type of agreement can be difficult under a government procurement.
- *Partnership and Mutual Learning:* A third alternative is for customers to include subcontractors and suppliers as partners from the beginning. This is shown in Figure 4.5c. There is collaboration at all levels. By participating as a member of a PDT, subcontractors and suppliers learn how to apply CE tactics in their portions of the product for which they are responsible for. This facilitates the adoption of CE philosophies by the subcontracting companies into the product definitions that both had vested interest in. This also reduces the overall cost and improves quality of the finished product. This type of contractor is commonly known as a "strategic partnership." While auto manufacturers have always set the standard for technology sharing and excellence, until recently long-term strategic partnership was not necessarily a part of the package. There is another type of partnership, which is called "tactical partnership." Tactical partners are generally not used for components or technologies that are of "core" type or are "strategic" in nature. The term "core" is meant here to include superior technologies for which the company is distinguished from its competitors and customers recognize. Operational partnership is a third kind of partnership, which is peripheral in nature—meaning least dependent or coupled with the main line of a company's business. Examples include transportation, shipments, raw materials, procurement, purchasing, legal, etc. To realize the maximum benefit, it is critical that targets for excellence (TFE) be implemented and uti-

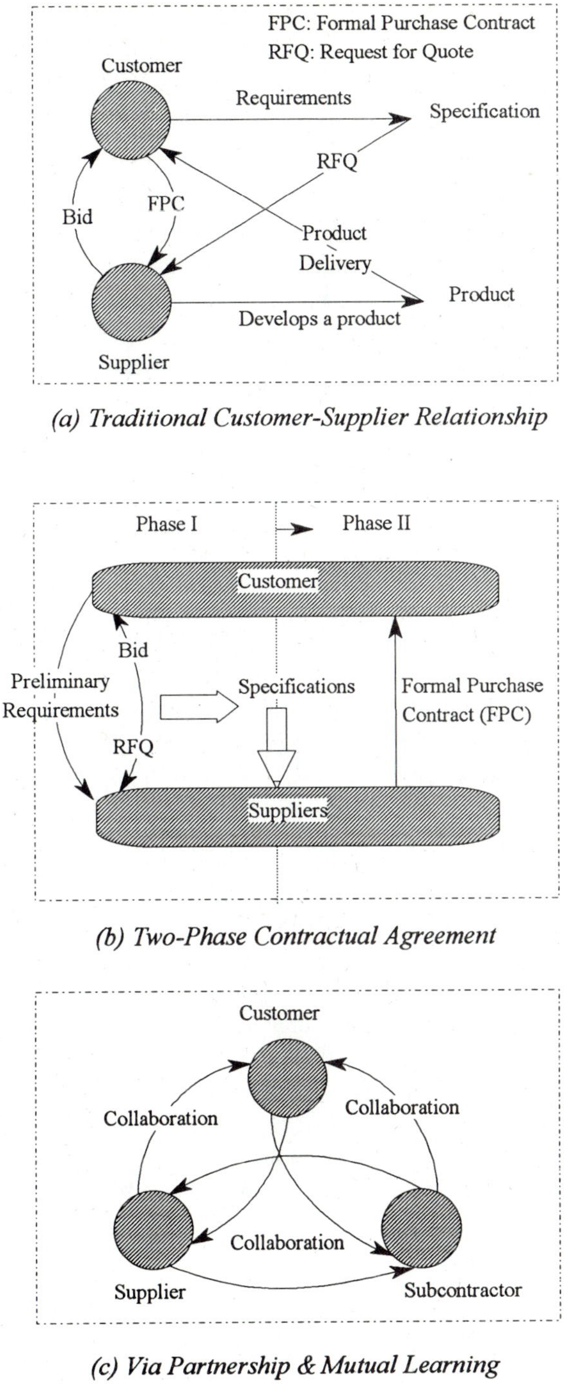

FIGURE 4.5 Sharing Responsibility with Suppliers

lized consistently by all parties of PDTs, the contractors, and the participating suppliers. The U.S. company that has moved most aggressively in this direction is Chrysler, in part because they rely more heavily on outsourced components than other auto manufacturers [Ohmae, 1989].

4.3 CONCURRENCY AND SIMULTANEITY

Concurrency is the major force of concurrent engineering. Parallel work-group [Barclay and Poolton, 1994] was one of the key elements of the concurrency described in Section 4.2.1. Paralleling describes a "time overlap" of one or more work-groups, activities, tasks, etc. Decomposing the product realization process into discrete activities is another essential element of CE. There are seven additional CE principles for which to aim:

- ***Parallel Product Decomposition:*** Product decomposition means viewing the product realization process as a part of the whole and then overlapping (aggregating) the decomposed sets to recreate or reconstruct the whole from its part. In other words:

$$\begin{aligned}\text{Product Realization} \Leftrightarrow [&\text{Decomposing (parts from the whole)} \\ &\oplus \text{Reconstructing (whole from the parts)}]\end{aligned} \tag{4.12}$$

  The term "whole" also includes multiple characteristics of life-cycle concerns (e.g., X-ability). Although not all life-cycle activities are independent or can be paralleled, many sets can be decomposed safely. For example, it is not necessary to delay the start of an activity if the information required for that activity is not dependent on the rest. Due to increased global pressure to bring a product into the marketplace early, parallel processing in CE is becoming a necessity. There are, however, many ways a product, process, or work information can be decomposed and overlaid in parallel. If a product, process, or work information activity does not affect other parameters or processes, it can be performed locally. If it does, it can be performed in a distributed fashion. Local or distributed processing, to a large extent, depends on how a product structure is originally broken up or decomposed. Do the decomposed parts exhibit independent or semi-independent characteristics? Decomposition allows the scheduling of activities to proceed in parallel.
- ***Concurrent Resource Scheduling:*** Facilitating the transfer of work information among work-groups is an essential organizational task of any company. Concurrent resource scheduling involves scheduling the distributed activities so that they can be performed in parallel. Paralleling is simple for activities exhibiting independent or semi-independent characteristics. However, it is not so simple for dependent activities. There are many cases when activities are dependent (not yet coupled) but need to be scheduled in parallel with other tasks. A simple case is that of an overlap. Even though a task is dependent on another, there is no need for one to wait until the other task ends. If a task precedes and generates the needed information for a later activity, the next task can start as soon as the needed information is made

available. There is no need to wait for the completion of the former task. If the two tasks are independent, they can be scheduled in any order necessary. The other options that address these issues more precisely are: optimal scheduling (minimizing time, resource, cost, etc.), backward scheduling (meeting target time), and team-based project management. Sanborn Manufacturing Company employed backward scheduling to set up major milestones consisting of hard and fast dates and worked back from those dates as a planning mechanism [Machine Design, 1993].

If product tasks, P_{tasks}, are independent, a pair of tasks (say task-*x*, task-*y*) can start immediately, meaning

$$\text{Start of } P_{task\text{-}x} \leftrightarrow \text{Start of } P_{task\text{-}y} \tag{4.13}$$

The symbol $\leftrightarrow$ means, the tasks coincide with in timing (starting time). *P* stands for *P*roduct, and

$$P_{tasks} \equiv \cup \, [P_{task\text{-}x}, P_{task\text{-}y}, P_{task\text{-}z}, P_{task\text{-}\ldots}, \ldots] \tag{4.14}$$

- ***Concurrent Processing:*** In concurrent processing, activities are staggered (performed simultaneously or overlapped) rather than carried out sequentially. Keeping track of those complex dependencies that vary with time is a critical task in concurrent processing. Appropriate synchronization efforts between different CE teams have to be made.

  If P_{tasks} is performed simultaneously (complete overlap), it implies that a pair of tasks can start together, meaning

$$\text{Start of } P_{task\text{-}x} \leftrightarrow \text{Start of } P_{task\text{-}y} \tag{4.15}$$

If P_{tasks} are overlapped (partial), this means

$$\begin{aligned} &\text{Start of } P_{task\text{-}x} < \text{Start of } P_{task\text{-}y} \\ &\text{End of } P_{task\text{-}x} > \text{Start of } P_{task\text{-}y} \end{aligned} \tag{4.16}$$

Where

The symbol $\leftrightarrow$ means that the two tasks "coincide with" in timing

The notation < means that the first one "starts before" the second one

The notation > means that the first one "starts after" the second one

- ***Minimize Interfaces:*** This entails reducing all sorts of interfaces (product, process and tools) required for the "product realization process" to a minimum.

$$\text{Number of } P_{interfaces} \Rightarrow \text{Minimum} \tag{4.17}$$

These include the interface relationship between management and design, supplier interface, design development interface, design manufacturing interface, production interface, etc. Such interface lists can be very long indeed and tend to depend on the size of the company and the product and process complexity.

- ***Transparent Communication:*** This provides virtual communication between the individual activities that are partitioned (decomposed).

$$\text{Tasks } P_{\text{communication}} \Rightarrow \text{Transparent} \tag{4.18}$$

- ***Quick Processing:*** means performing individual activities as fast as possible using productivity tools or design aids. It also amounts to speeding up the preparation time in building up the information content before and after an execution of a task. This emphasizes the mandate for shortening the pre- and postprocessing time and the time it takes for completing the decomposed tasks themselves.

$$\text{Quick Processing} \Leftrightarrow \text{Minimize (Lead time of } P_{\text{task-}i}) \text{ for } i = 1, n \tag{4.19}$$

Where, n is the number of tasks in the P-set, equation (4.14).

There is a difference between the complexity of the philosophies (such as product complexity, process complexity, enterprise complexity, or complexity of cognitive behavior), and the philosophies of their management. An organization committed to making such complex products in the shortest possible time need not require an equally complex management philosophy. An organization can still handle all that while following a simple management philosophy. This simple management philosophy is the philosophy of decomposition followed by concurrent processing. The concept is similar to what is called the European philosophy of "divide and conquer." To apply this to a complex product, a systematic decomposition of the product and process, including 7Ts (talent, tasks, teams, techniques, technologies, time, and tools) is required. In the following, the aforementioned seven CE principles are discussed in detail:

4.3.1 Product Decomposition

Smith and Browne [1993] describe decomposition as a fundamental approach to handling complexity in engineering design. The two (decomposition + concurrency) identify activities that can be overlapped or performed simultaneously. Decomposition and concurrency also allow one to formulate strategies leading to their separation, for example, indexing, alternate decomposition, teaming, or restructuring. Let us assume that an activity has been broken up into tasks: $P_{\text{task-}x}$, $P_{\text{task-}y}$ and $P_{\text{task-}z}$, etc. There are four possible ways such tasks can be related: dependent tasks (P_{dep}), semi-independent tasks (P_{sin}), independent tasks (P_{ind}), and interdependent tasks (P_{int}):

- ***Dependent Tasks:*** A pair of tasks (say $P_{\text{task-}x}$, $P_{\text{task-}y}$) is said to be dependent if a task requires information, which is an output from another task. The information required could be a complete output transfer or it may represent only a portion of the output. If the transfer of information is complete, they are usually run in a series. This is shown in Figure 4.6(a)
- ***Semi-independent Tasks:*** A pair of dependent tasks is said to be semi-independent if the transfer of output from one to the other is only a partial transfer (pseudoparallel). The pseudoparallel structure means that there exist weak interactions among

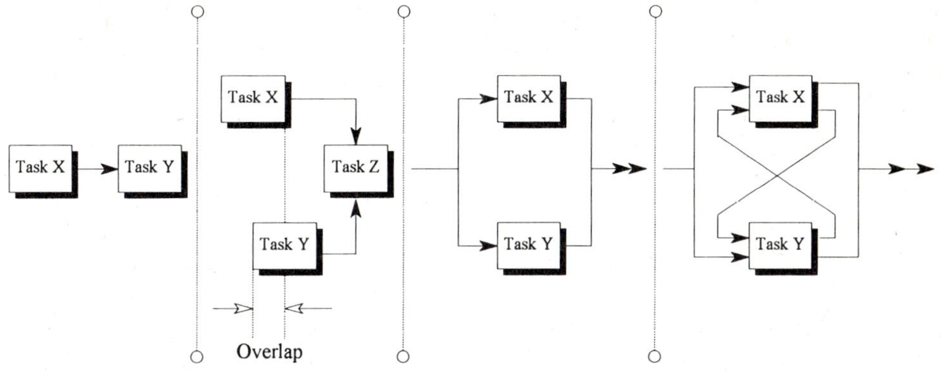

(a) Dependent Tasks *(b) Semi-Independent Tasks* *(c) Independent Tasks* *(d) Interdependent Tasks*

FIGURE 4.6 Possible Relationships Between Pairs of Tasks

groups of tasks. In Figure 4.6(b), a task $P_{\text{task-}z}$ is said to be dependent on the tasks $P_{\text{task-}x}$ and $P_{\text{task-}y}$ since partial outputs from both $P_{\text{task-}x}$ and $P_{\text{task-}y}$ are used to complete $P_{\text{task-}z}$.

- ***Independent Tasks:*** A pair of tasks is said to be independent if no portion of the output from one or the other is required for the completion of both tasks. Figure 4.6(c) shows a pair of tasks $P_{\text{task-}x}$ and $P_{\text{task-}y}$ that are independent.
- ***Interdependent Tasks:*** A pair of tasks is said to be interdependent, if a two-way information exchange is required for the completion of any task. Meaning, information from one task (say, $P_{\text{task-}x}$) is used to complete the second task (say, $P_{\text{task-}y}$) and the information from the second task ($P_{\text{task-}y}$) is used to complete the first task ($P_{\text{task-}x}$). This is shown in Figure 4.6(d).

In other words:

$$P_{\text{tasks}} \equiv \cup\, [P_{\text{dep}}, P_{\text{sin}}, P_{\text{ind}}, P_{\text{int}}] \tag{4.20}$$

The symbol $\cup$ implies "union-of".

Paralleling tasks and the amount of overlap depends on the types of relationship and the degree of dependency between them. The overlap between two intermediate tasks or specifications/outputs represents the time lapse to build the information required for the start of the subsequent tasks. Coordinating tasks that exhibit dependent or independent characteristics are quite straightforward. The dependent tasks are arranged in series and independent tasks are stacked in parallel. For the work-groups, however, the challenges of CE are extremely difficult when many tasks are interdependent; meaning they are coupled and cannot be separated explicitly either in a series or in a parallel mode. Interdependent (or coupled) tasks take more design time and many iterations (of information transfer back and forth) before they finally converge. CE strives for simultaneity and immediacy. In practice, however, mutually independent group of tasks seldom exist. Strategically, decomposing the interdependent tasks into a series of dependent, independent and semi-independent tasks can reduce the size of the working groups and the number of iterations that are required to obtain a reasonable solution.

$$P_{int} \equiv \cup \, [P_{dep}, P_{sin}, P_{inp}] \tag{4.21}$$

Where, symbol ∪ indicates "union of". Both of these steps can have a significant impact on reducing the time-to-market aspect. Furthermore, minimizing interactions between various tasks leads to more overlap and less preparation (build-up) time between the tasks. This could yield additional time reductions. Similarly, the composition of pseudo-parallel structures out of a large number of semi-independent tasks enhances concurrency. The semi-dependent and independent tasks are discussed in this Chapter. The strategies for dealing with coupled and dependent tasks are discussed in Chapter 8, Volume I.

4.3.2 Concurrent Resource Scheduling

Frequently, the "product and process" is radically redesigned to achieve parallelism [Costello, 1991]. Paralleling of activities provides management teams with opportunities to reorganize and control the resources applied during CE. These resources fall into three main categories: teams (for example, people, machines, facilities, outside firms); tasks (activities or projects they work on, knowledge of the projects, information they need to work with); and time. The trio provides a basis for defining a work breakdown structure (WBS). A WBS is really a series of interrelated work tasks initially set in motion by the planning track. New tasks are added or created by the subsequent tracks when put into motion. The latest series of tasks are mostly due to the support and delivery track. These tasks are over only when that product is finally disposed at the end of its useful life. A good WBS contains all three elements: paralleling of tasks, paralleling of teams (work-groups) and optimal time schedules. It uses as much knowledge as possible, which aggregates the existing evidence for concurrent work scheduling and tasks' decomposition that designers commonly use. Techniques such as optimal resource planning, cost accounting, level balancing, OPT, and other load management approaches are considered integral to WBS in achieving concurrent resource scheduling. The types of WBS required within an organization dictate how 7Ts should be developed and used. Figure 4.7 also shows how CE activities and work-groups should be organized into loops, linked (electronically connected) together by a product/process hierarchy (PtBS/PsBS). The product decomposition details have been integrated into such loops. Concurrent resource scheduling is shown in Figure 4.7 as a central block, where arrows to and from the nested loops or decision blocks either emanate or terminate. The outer loop starts with the multiple perspectives of design and the innermost loop ends with multiple analyses (or projects). There is a series of nested loops to prune the elements or the information envelope required to build a total product or process model.

One of the purposes of resource scheduling is to leverage the adequacy of a good supply system. Supply systems establish teams of organizations and processes, including manufacturing partners, vendors, distributors, transportation, delivery, and sales.

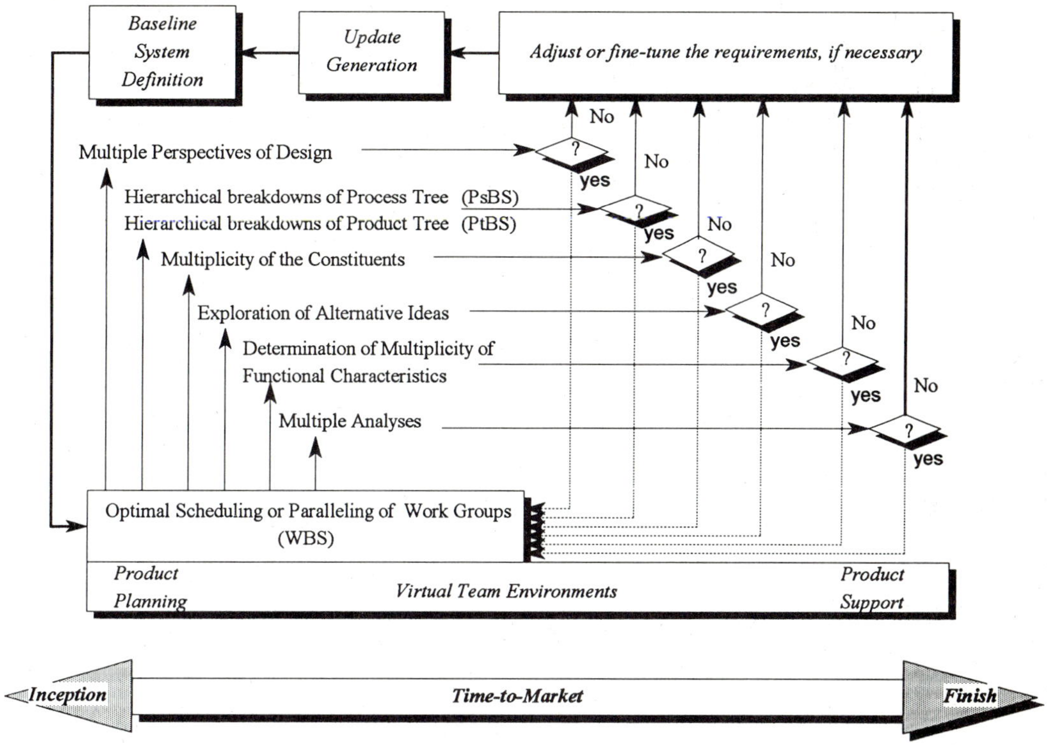

FIGURE 4.7 Schemata for Exploiting Concurrency

4.3.3 Concurrent Processing

Managing time is the fulcrum of concurrent engineering [Rosenblatt and Watson, 1991]. Some companies rely on milestones. Others use strategic routing and queuing as another way to manage time. Concurrent processing means optimal routing and queuing of activities both from the work-group distribution and the information build-up standpoint. This is essential to guide the design of the product and its processes toward a quality end. Concurrent processing is never easy, particularly in industrial settings where solvable technical problems are prevailed upon by cultural considerations. Resistance to change is predominant. This is seen, for example, in the automotive industry, and more generally, in companies where the age profile of the technical staff is high. Four important concepts are associated with concurrent processing: creation of "variable-driven" digital product/process models; route and queue management; developing bookshelf concepts; and integration with analysis/simulation tools. They are discussed next.

4.3.3.1 Create "Reuseable" Digital Product/Process Models

Creating digital product/process models is generally a first step in enabling concurrent processing. It encourages early data sharing between the teams that are involved. Most

CAD models provide a static, explicit representation of basic product geometry. Explicit-only modeling systems cannot quickly impact future design changes and productivity since its "reuseability" is very limited. Variable-driven modeling promises a "reusable" design system since it is based on parts written using variable constraints and features. Combining variable-driven modeling and "digital models" can go well beyond CAD to capture the "variability." Variability can include the functional intent of product and processes, encompassing product definitions, process definitions (process planning, execution, and production engineering), and production operation and control (schedules and build instructions). This is shown in Figure 4.8. Both routing and queuing enable and manage the execution of product/process models. There are many designers who believe that reuseable digital models will lead to more rapid design and concurrent processing. This is similar to the benefits experienced by computer programmers in using OOPS (object-oriented programming system) to write reuseable codes.

These common models give everyone access to current data and share the updated information instantaneously as the product evolves from one phase to the other. It encourages use of common models and database, eliminating potentials for wasted or duplicated efforts in traditional data recreation and reformatting.

4.3.3.2 Routing and Queuing

It is important to realize that many activities or processes are multifaceted, and that they cannot be introduced all at once. Routing means sequencing the activities of a variety of work-groups and CE teams. Routing management means paralleling the multifaceted ac-

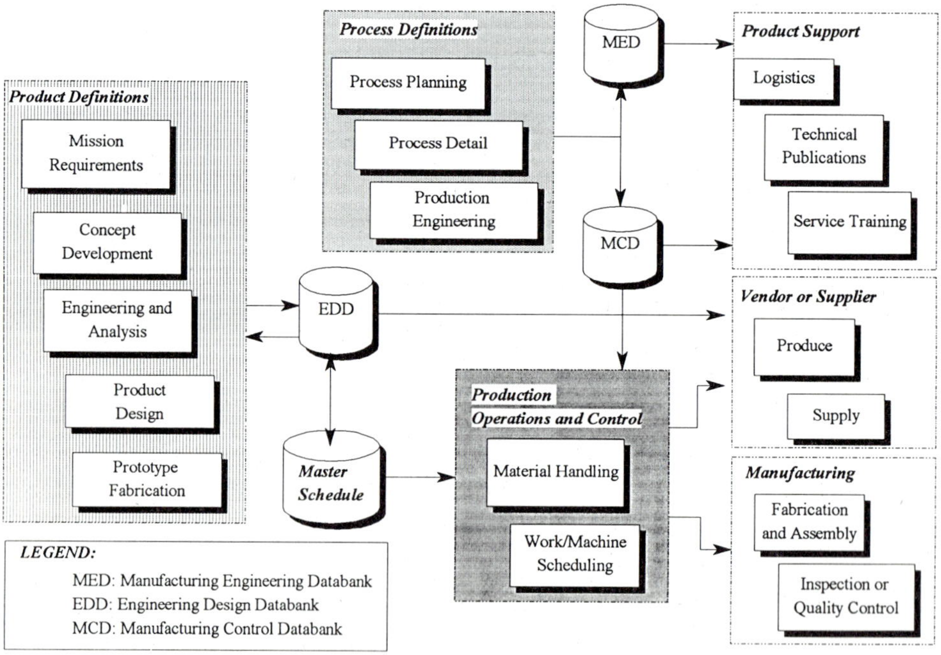

FIGURE 4.8 Common Digital Product and Process Definitions

tivities so that they overlap with each other. When a group of activities is routed, information starts building up. It does not make sense to build up separate packages of information which are not mutually supportive of each other or to a next activity level. It would be advantageous if the information generated from a prior activity were useful to build up the package of information for the definition of the product or process. Queuing management thus requires moving the pertinent set of information or the work tasks from one work-group to another. Concurrent processing means:

- Routing the activities to the work-groups and CE teams so that the package of information or knowledge-base is further enhanced beyond its present stage
- Queuing the next set of dependent activities that could utilize the information already built up at this stage of completion.

Whenever decisions are needed, the new set of activities is routed to build up the information until such decisions can be addressed. If a series of parallel activities takes different times for completion, priority can be assigned to the routing and queuing models.

Routing Models Routing models are chronological representations of planned, in progress, or completed activities. Routing models are typically shown as boxes interconnected with lines of relationship. These lines of relationship show which activity box precedes and which one follows, and show input, output dependency, requirements, and constraints (see Figure 4.7). These models can be built in a fashion similar to PERT (Program Evaluation and Review Technique) using any information modeling technique such as, semantic nets, schedule type, etc. Most schedule type routing models allow for definitions of priorities, sequences, and constraints. This helps built-in provisions to nest the activities and let the concurrent teams work on these nested units in parallel. This is further discussed in Chapter 5 (see Figure 5.15).

4.3.3.3 Developing Bookshelf Concepts or Technologies

Rather than creating new product concepts whenever they are needed during a concurrent product design and development (CPDD) phase, some new concepts could be developed off the board. Recognizing the future need for product enhancements, these concepts and features could be developed away from a particular product line, and tested up to a point of insertion into the actual product. Once it performs satisfactorily, it can be bookshelved until needed again. Since the concept is already proven in some experimental way, it is most likely that its actual implementation may just require fine tuning. This could eliminate a large amount of engineering and analysis, product design, and prototyping efforts. This step reduces the amount of interactions that would otherwise be required in a CPDD process.

4.3.3.4 Integrate Analysis/Simulation Tools with Digital Product Models

Analysis/simulation is often used as an evaluation aid or to verify a particular design situation. Analysis/simulation has seldom been an integral part of a design. It is not commonly used to support the product design and development. For example, initial methods like finite element analysis (FEA) and variation simulation modeling (VSM) are all CAE analysis tools used to verify the integrity of a structural design, fits, or clearances. VSM studies are used for analyzing problems with tolerances and dimensioning. They are not employed integrally to design a new part, to improve an existing design, or to eliminate these problems. Integrating CAE tools with digital models:

- Transfers the emphasis of analysis from an "evaluation" mode to a "prediction" mode
- Permits "analysis models" to be created from inputs that are closely tied to the "design models"
- Forces the analysts to think of the changes to be made in the designs as opposed to just viewing the results as possible modes of failures.

Integration encourages use of common digital product models and database, thus promoting cooperation and efficient communication. Its successful use will depend on the development of proper interfaces between programs and the creation of a close cooperative team of analysts, model builders, designers, and technicians.

4.3.4 Minimize Interfaces

Shared product design can be facilitated by introducing adequate interface management. The main focus is on identifying various sources of interfaces and determining whether they are actually needed. The goal is to reduce the number of design and manufacturing interfaces to a minimum. This saves time, reduces design costs, and gets the partners involved early in the process.

4.3.4.1 Minimize Product Interfaces

The product design problem is often decomposed into subdomains, each having its own design variables and constraints. These sub-domains can be quite independent of each other except in a limited number of common interfaces. The PtBS tree drives the product design to an interface-driven integration technique. The PtBS also serves as the model index structure and helps keep the digital equivalent organized and easy to cross-reference with other indexes. The problems of each subdomain can be solved in parallel and the results brought back to satisfy global needs at a later time. Such a decomposition of design, represented by the structured PtBS tree, can be achieved in a number of ways. For example, the design problem can be divided into a four-step process as shown in Figure 4.9. The first step is to develop a functional system. This yields system characteristics, which is fed to the next step to identify and develop subsystems. The subsystem characteristics are then input to the third step to identify and develop components. Finally, the components' characteristics are then fed into the fourth step to identify and develop parts (see Figure 4.9). There are four decision blocks corresponding to four loops: conceptual design, layout design, subassembly design, and an assembly design, which checks whether the corresponding design is satisfactory or not. The other aspect of the PtBS tree is the minimization of interfaces among these four steps: system, subsystems, components, and parts. If decom-

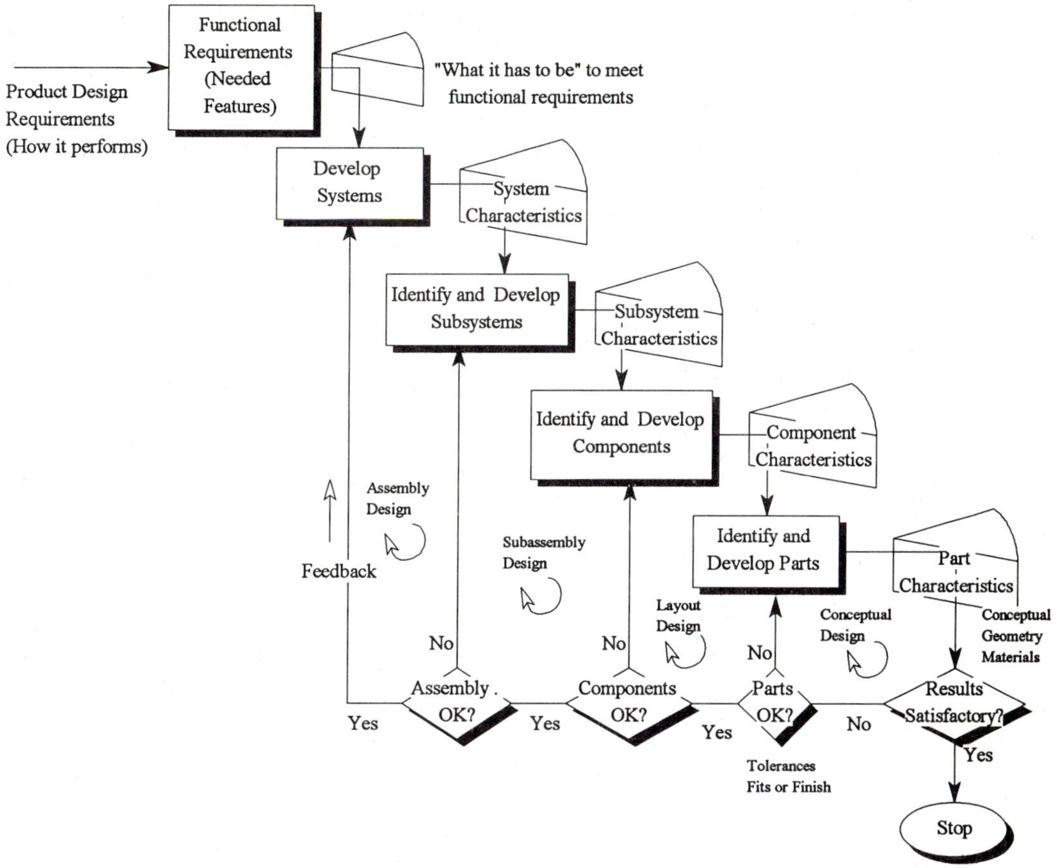

FIGURE 4.9 Minimize Product Interfaces

posed elements are decoupled (or loosely connected), the tasks of interface definitions are simple and straightforward. Tolerances, finish, and fit requirements have little or no impact on conceptual design, assembly, or components' functions. Material types, as represented by the structured bill-of-materials, can often be modified without jeopardizing the part, component, or assembly function. The convenience of processing design problems in parallel can lead to a converged design much faster than what is possible conventionally.

If decomposed elements of the PtBS tree interfere (cross areas' boundaries) or overlap, the interface definitions could be quite complex and intertwined. The major product development challenge in such cases is to integrate the many (decomposed) subproblem solutions into a well-connected system. Some organizations address this by assigning teams of analysts or conflict resolution engineers to handle the interactions between the decomposed subproblems. The trouble is that such interactions are rarely known in advance or their implications are not well understood. Interface management is the technique used to minimize interfaces. Interface Management implies preparing the PtBS tree or its content so as to preclude possible interfaces between the decoupled elements.

Through this approach, the design at the top most product level supports the next lower level of design, which supports the next lower level, and so on. The decomposition is consistent with interface requirements. This does not prematurely commit the product to a high cost. The following are three additional means to minimize product interfaces:

- ***Simplify Products:*** Companies that take the time to simplify processes—especially in the product design phase—before applying automated technology can maximize their return on investment [Mather, 1993]. Product simplification brought about by design for X-ability (DFX) and CE often leads to parts cost reduction that can significantly outweigh the reductions brought in by individual method, such as DFA (design for assembly) costs. By bringing in techniques such as DFX (an integral part of CE) the manufacturing team can bring a number of fundamental cost-cutting tips to the attention of the design team during the early part of product development. Unlike serial engineering approaches, where products are designed first and assembly considerations are determined later, design for assembly (in the CE context) looks at how a product will be put together from the outset of the design process. Let us name this process "Design for Concurrent Assembly." Designing for concurrent assembly cuts production costs by reducing the number of parts in the product and making them easier to put together. There are many spin-off benefits in jointly working as a team. Many companies have observed that these spin-off benefits are far greater than the primary objectives of simplifying product designs, such as easier manufacturability or assembly. Improvements can take place in quality and communication; inventories can be reduced; and employees can begin to think "smarter." Another way to improve assembly costs, in mass produced products, is through automation. Here, the benefits are realized by savings in assembly costs and time. It is not uncommon to drive the process of simplification to a point where assembly by hand is more practical than automation, which provides little or no return on investment. The IBM Proprinter is a live example of such an occurrence. There are many simplification techniques that are quite common. These are discussed in Chapter 2, Volume II. Examples include design for assembly and manufacturability, design for assembly, design for quality, etc.
- ***Simulate product performance values:*** Simulating product performance values provides means to decide, early on, the trade-off between the performance values and their features. It helps decide how the components will function and how the system will work as a unit. Digital mockups (another name for software prototyping), for example, replace the need for physical mockups, hardware prototype fabrication, or hardware assembly. Software prototyping can be used for checking interference and functionality before its actual fabrication, meaning before any actual part or a hardware prototype is produced.
- ***Commonalize product features that are invisible to the customer:*** In order to reduce the time to assemble all parts into a working product, some parts or some aspects of the product can be commonalized accounting for the interfaces that are required. This can allow for a great flexibility and a quick changeover to produce new products, if warranted. However, these commonalizations should be made so as to be invisible to the customer.

4.3.4.2 Minimize Process Interfaces

As in product design, the process design problem can be decomposed into subdomains. These subdomains can be quite independent of each other except for a limited number of common interfaces. Similar to the product design case, problems of each process plan domain can be solved in parallel and results brought back to satisfy global needs at a later time. Such a decomposition of a process plan—a Process breakdown structure (PsBS)—is shown in Figure 4.10. Here the process plan is divided into four stages: product planning, process planning, production planning, and production intent (manufacturable product or service). The first stage is to identify a functional set of product planning steps. It yields product specifications, which are input to the next stage to identify process planning steps. The resulting process specifications are then input to the third stage to identify production planning steps. Finally, its outputs are then fed into the fourth stage to obtain a production intent (see Figure 4.10). There are four decision blocks, corresponding to such

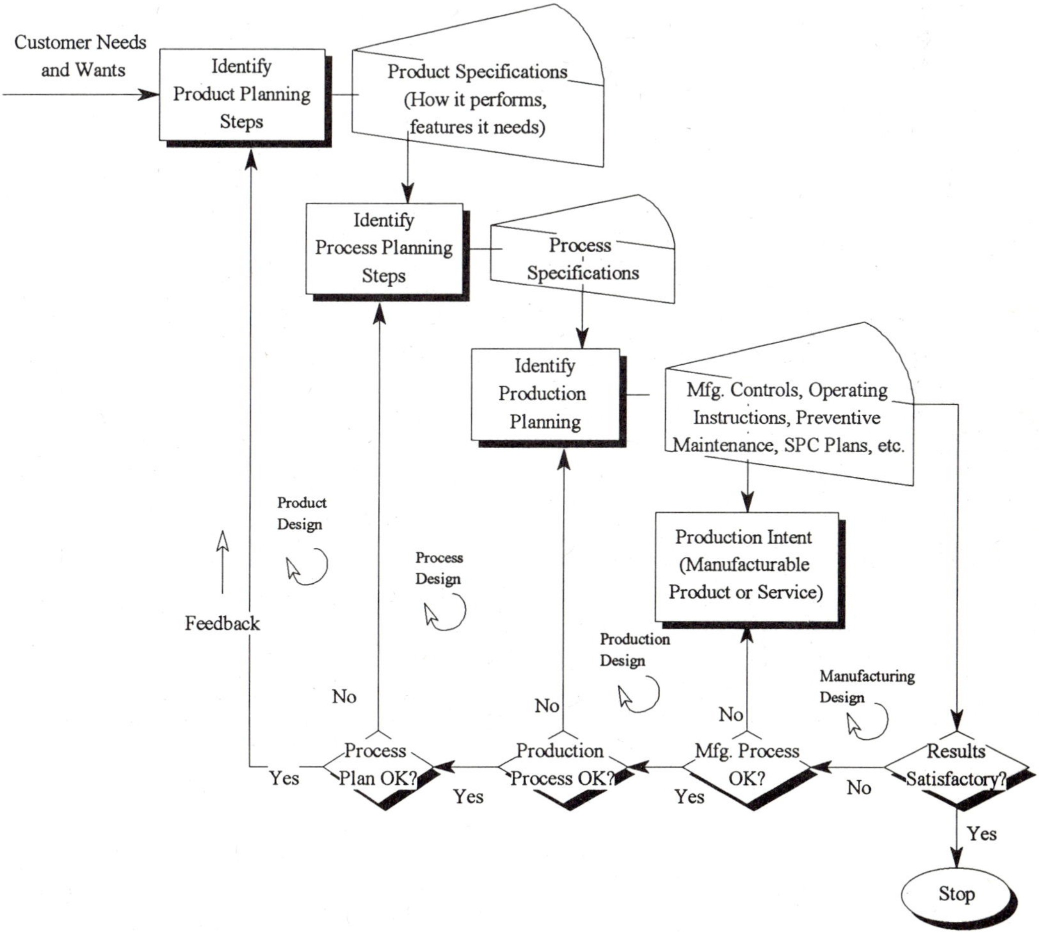

FIGURE 4.10 Minimize Process Interfaces

four loops: manufacturing design, production design, process design, and product design. The decision blocks check whether the corresponding plan is satisfactory or not. In essence, PsBS is the manner in which a company designs and manufactures its products, while PtBS is the means used to describe or capture the inherent complexity of a product. The following are eight additional means to minimize process interfaces:

- ***Characterize Process:*** Because product structure drives the process, there is an increasing demand for a well-characterized PsBS that can minimize the interface and manage changes at the source without requiring replication of product data. The process breakdown structure (PsBS) tree for the process, for instance, can be chosen similar to the product structure (PtBS) tree. This improves the interface needs by providing a structure similar to the eventual process (production-operation-manufacturing) structure required to produce the complex product. The two steps are process decomposition and process integration. The process decomposition establishes the PsBS tree, and the process integration brings in the detailed parts together to make up subassemblies. The subassembly parts, in turn, assemble and bring together a final assembly process, which assembles the end-product.
- ***Commonalize Process Features:*** Unlike the product features, where care has to be exercised since most are exposed to the customer, there is no such danger on the part of process commonalization. In order to reduce the amount of production engineering and planning, processes can be commonalized much more safely compared to product (features) commonalization. There is less danger to customer exposure, since the customer often does not care what methods or processes were used to manufacture the end-product. Similar to product, process can be commonalized accounting for the interfaces that are required. This can allow for a great flexibility and a quick changeover to an alternate process, if the market demands on the products are increased or decreased.
- ***Exploit Similarity of Tree Structures:*** The similarity of the product breakdown structure (PtBS) tree and the process breakdown structure (PsBS) tree has many advantages:
  - The methods for administering product design can be reused to interrogate the process designs.
  - Both product and process information can be electronically represented in a unified or generic tree structure.
  - In a digital environment, both the product and process definitions can be stored in a monodetail form (for example, a single part per digital model form, a single component, a single subsystem, etc.). The monodetail concept combines the product and process definitions into a single tree representation. However, it might be difficult to generalize this for all parts, components, subsystems, etc. By combining the two tree structures into a monodetail structure, the interfaces between product and process are satisfied automatically. The mono-detail model drives the product realization process to an interface-free integration schema. This approach carries a great promise, but the current capabilities of CAD, CAM,

CIM, and knowledge-based digital modeling need substantial revamping for this approach to be broadly successful.

- ***Simplify Process:*** This aims at reducing the manufacturing cycle and process steps within each cycle to their basic essentials. Reducing the process steps not only saves time and money but also improves workmanship quality. Reducing steps, by combining or eliminating the number of manufacturing processes, raises quality and lowers cost. For example, designing a component for net shape manufacturing, such as precision casting or powder metal, gives fewer possibilities for errors. There is a danger, however, that too much integration of parts may drive costs up.
- ***Avoid Variable Processes:*** Manufacturing processes with a high potential for variation are the most notorious in causing quality problems. Processes with a high degree of variability include the following:
  - The processes, which are difficult to monitor or control, such as those which are environmentally dependent, subject to impurity content, etc.—these processes are common with bonding, mixing, and curing
  - Processes that can change properties of materials
  - Skill-intensive processes such as manual soldering, welding, brazing, etc.
- ***Promote Just-in-time (JIT) Delivery of Defect-free Material:*** This requires that vendors and suppliers understand and support JIT concept both for parts and for information build-up and transfer. For example, they could receive orders generated through an electronic data interchange (EDI), produce small lots, and ship it just-in-time for assembly. This keeps inventory to a minimum at both vendor and production sites. This reduces waste and eliminates interface functions that do not add value to the product. However, this requires a new and closer relationship between manufacturers and suppliers. Some of these are being promoted today through strategic sourcing and supply chain.
- ***Use Proven Processes:*** Use processes that are proven in the field. If one is forced to introduce a new process, it is better not to combine it with any other process, thus limiting the degree of variability or unknowns. It is wise to adopt a "prove-as-you-go" type strategy—first apply any new process to the laboratory environment before finally implementing this to a production job on the factory floor. Also, it is wise not to implement several new processes for the same part or the product. The chances that anything can go wrong are proportional to sum of their individual probabilities.
- ***Use Process Simulation:*** If you have to use a new process, process simulation techniques can validate its outcome before it is put into place. Manufacturing simulation for prototype fabrication is equally important to understand how processes will work in actual hardware. This also helps to optimize the design of individual processes and production systems. It identifies potential dynamic process bottlenecks and can guide teams in determining the proper size and locations of process parts (such as accumulators) by maximizing total system up-time capability with least investment and floor space. Depending upon the quality of data and the process, it can provide a range of varying confidence. The good part is that if an

error is detected, it can be corrected immediately rather than waiting until the program is actually launched. Simulation also leads to early deployment of statistical process control and similar tools to understand process variables. It can detect process problems before churning defective parts, and can reduce dependency on inspection requirements.

There are numerous spin-off benefits, which are difficult to quantify, such as reduction in inventory, reductions in record keeping, "fast, flexible manufacturing," and improvements in material flow and production flow. Apple reduced its time to build digital product models by six to eight weeks through standardizing on a single CAD system. Overall development time was reduced 30% by using vendors that developed designs on the same CAD system.

4.3.4.3 Minimize Computer Interfaces

Too many computer interfaces can create problems with the smooth flow of information. Each program has its own data, input, and output format requirements. In order for these programs to run seamlessly, the inputs and outputs of these programs must work in concert with each other. Manual data entry is error prone. Moreover, there should be a single data source from where all inputs originate, so that if a parameter is changed, the correct value is passed on to all interface-programs using them.

4.3.5 Transparent Communication

This provides virtual communication between the activities that are partitioned, and between the team members.

4.3.5.1 Common Understanding

Transparent communication involves identification and definition of mission-critical data. All members of the CE teams need to have the same common understanding of frequently used terms and their meanings. It may require definition of "data dictionary and semantics" as a structured approach to resolving conflicts and for consensus building.

4.3.5.2 Global Access

Most organizations deal with interfaces on an ad hoc basis. No formal organization or link exists between work-groups like design and manufacturing to deal with design-manufacturing interface. Ad hoc personal relationships and communication equipment (such as print, newspaper, memos, voice-mails, meeting off-site, etc.) is used to deal with most common problems. A stronger interaction and a global access link between various interfaces, via multimedia or electronic systems, can contribute to a better cooperation. Getting a person on a team to share the knowledge with the rest makes that person an important team player. It makes the requirements of downstream operations available upstream without much loss of time. It can contribute to achieving a design based on requirements from assembly, service, and maintenance throughout the life of the product.

4.3.5.3 Universal Product Code

Universal Product code (UPC) is a bar code symbolism for identification, inventory tracking, and movement of merchandise. Most soft goods (like garments, accessories, and food items) and hard goods (furniture, tools, and auto accessories) carry UPC bar codes throughout their distribution cycle. UPC is also useful in bar code scanning at point-of-sale (checkout counter). Inventory can be tracked with a combination of bar code scanning at the point-of-sale, and stock keeping unit (SKU) data along with an electronic network.

4.3.5.4 Electronic Data Interchange (EDI)

EDI is the computer-to-computer transmission of information using standard formats. It sets up communication formats and protocols to interchange information between manufacturing and distribution outlets. Today EDI is used widely by the auto industry to exchange purchase orders, shipping notices, and payments, particularly with first-tier suppliers that deal directly with OEMs. New EDI technology is being developed by the auto makers to improve integration with the second- and third-tier of suppliers and to deliver customized parts at narrow time-frames.

4.3.5.5 Technical Memory

There are two ways a design team can approach a design problem. The first and the most common way is to design the part from the team's own memory, recollecting how the team designed the part earlier. In the case of a new part, this means designing the part from scratch with generous use of design axioms. The second approach is to begin with a set of existing parts that are stored electronically in some form, retrieve the existing parts from the technical memory, modify them if necessary, and create a new product. Technical memory enables team members to share design information more efficiently throughout a company. With technical memory, engineering teams can frequently find and reuse existing designs rather than redesign the same part. Even when the existing designs are not quite right, engineers can locate similar designs through technical memory and use them as a starting point. This way they will never have to design a new product from scratch.

There are four major components of a technical memory:

1. A classification and a coding system
2. A library of existing designs (history files)
3. A database linking these existing designs to the classification and coding system
4. A user interface, which allows the team to query and retrieve similar designs.

4.3.6 Quick Processing

Quick or fast processing is the process of devising means to perform individual activities as fast as possible. Fast processing can be accomplished through high-bandwidth technology or building flexibility into the process. Management techniques, which are the product of decades of corporate learning, can be captured as knowledge and rules. With high-bandwidth technology, such as object-oriented databases, technical memory, parallel computers, multimedia, X-window, a large amount of information exchange can take place at a very high speed. Using such means, product development rules can be coded into knowledge-based design and manufacturing software programs. Once these knowledge-based programs and technical memories are deployed as useful aids, they can provide considerable competitive advantage to the companies in terms of design speed, accuracy, and quality. The competitive advantage earned through process management techniques is not only retained in this case but the methods are also readily available for future use, when market conditions suddenly change or a competitor develops a superior product. The flexibility inherent in knowledge-based systems can be exploited to overcome any such short-term market fluctuations.

4.3.6.1 Efficient Experimentation

The essential idea is to provide a set of statistical techniques for studying the effect of uncontrolled variables on the product or process involving the smallest possible number of experimental trials. In the case of "design of experiment" approach, experimenting with design variables soon becomes an onerous task. For instance, a problem with fifteen two-valued variables would require 32,768 different experiments if we had to investigate all the possible permutations. However, using Taguchi's efficient design technique, only sixteen experiments would be needed to get the same effect. This experimental efficiency is achieved through the application of balanced array tables used to design the sixteen experiments. This is further discussed in Chapter 2, Volume II.

4.3.6.2 On-line Technical Reviews

The CE team approach may result in more project reviews and meetings. The need for reviews starts from product inception when requirements are defined with customers inputs and VOC. They continue through the entire process of product design and development cycles. It is quite intensive when product design reaches maturity. Previously, people wrote memos and sent them through the chains of command to keep everyone informed. Recent developments in collaborative techniques and multimedia equipments such as Groupware, have accelerated the penetration of electronic reviews and electronic forms of communication. Today, e-mail messages flow to executives and to co-workers without hesitation even across the organizational boundaries. They allow interactive reviews and real-time access of electronic data. Developments in network tools have made it possible to link customers and suppliers as a part of this review process.

4.3.6.3 Process Concurrence

Multiple processes or concurrent sessions are examples of process concurrence. Process-to-process communication, as the name implies, is an example of a one-to-one-relationship. In most distributed integrated systems, a single process must be in communication with, or be able to receive messages from, many other processes. This gives rise to two additional forms where concurrence is prevalent. The many-to-one relationship is where many processes talk to one process. The one-to-many relationship is where one process talks to many others. The one-to-many relationship is limited by system timing considerations. With newer workstations—SPARC stations, window-based DOS system, and Unix—several versions of the same design can be edited or viewed at the same time. Multiple sessions can be run: Each session starts from a command shell window and displays its own set of graphics and text windows. Different sets of designs (or the same design with different version levels) can be edited in each session or processor. Only one window is active for input/output, but selecting the active window is as easy as selecting the active view port in a DOS based CAD system. With the newer workstation, a session for the design of each component of a product-assembly can be started along with one for the total assembly system. The user can move from one component design to another and update the assembly design. Alternately, the user can simply display several designs and jump from one to another. With today's information technology, the same file can be viewed simultaneously by several teams, no matter where they are. New data types, like digital video and audio, allow a more personalized response. The simultaneity obtained using such visual tools reinforces the consensus reality, allowing it to soon reach a fair amount of maturity.

4.4 MODES OF CONCURRENCY

Concurrent systems tend to operate in one of two modes: an overlapped pull system or a linked system (see Figure 4.11)

The overlapped pull system uses an empty WBS, PtBS, or PsBS "tree node" as its "pull signal." This tree node can take the form of an empty work content, which requires more analyses (such as FEA, mechanism, linkage analysis, etc.) It can take the form of an empty product content (such as missing predefined subparts, features or materials content). It can also represent a designated process content on the factory floor (such as machine setup information or process identification, etc.). No downstream operation begins until an information content for a work (WBS), product (PtBS) or process (PsBS) breakdown structure for a tree is fulfilled. Likewise, no downstream operation begins until a signal is received to replace a previous content, an item, or a part (see Figure 4.11a).

Linked systems are most commonly used when work-groups are connected through workstations and computer networks (see Figure 4.11b). When a work-group communicates with another work-group, there are six steps involved. They are stated in the legend of Figure 4.11b and clearly marked as 1 through 6. Steps 1 through 3 take place due to the interactions from unit *Y*, and steps 4 through 5 occur due to unit *X*. The overlapped system is most effective if the design revision and control are managed over the networks having "electronic proximity."

4.4.1 Overlapped Pull System

The product development process is a process of gradually building up the right information and linking up the process activities with required skills so that the project can be finished in time. However, it takes time to gradually build up the information in a form that

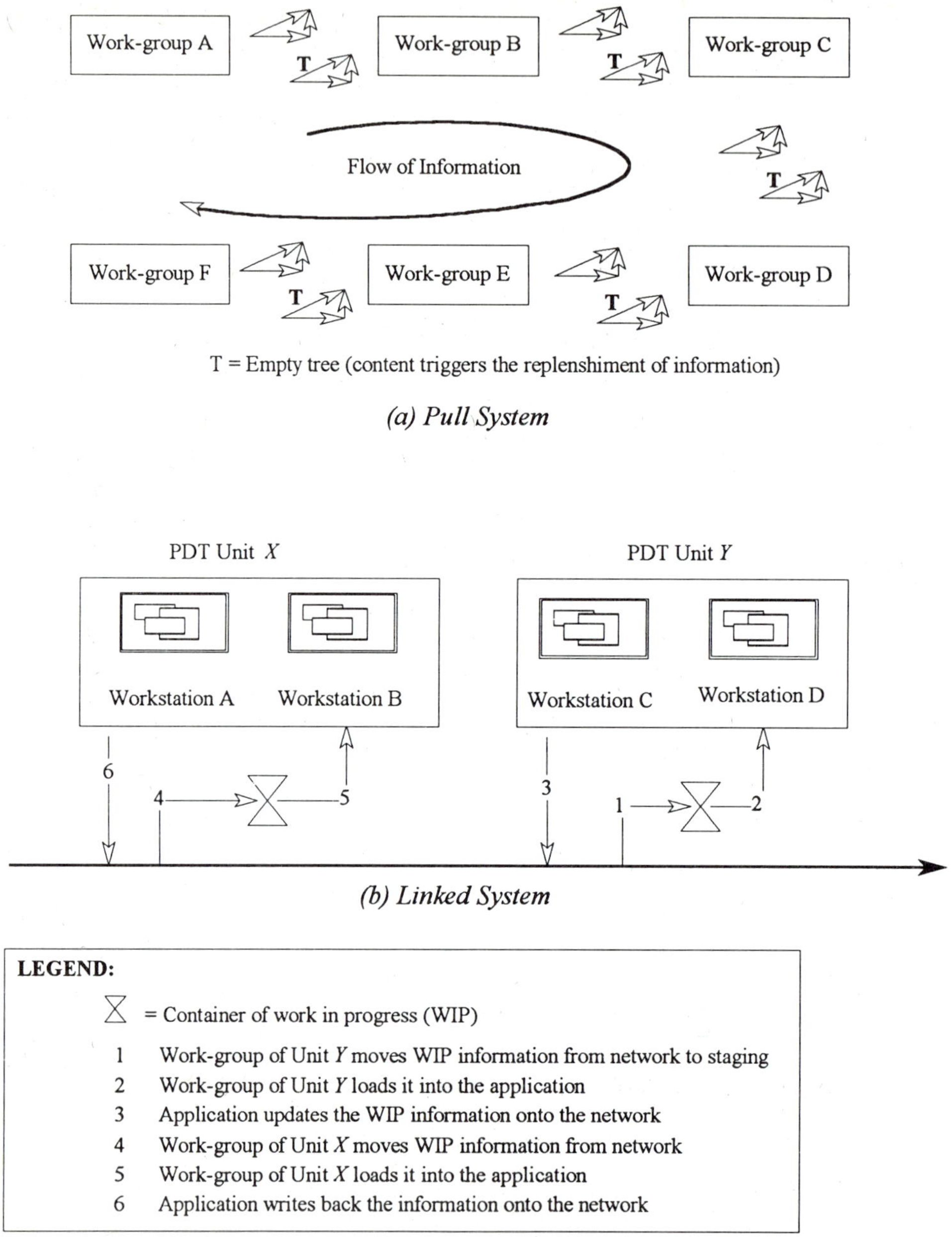

FIGURE 4.11 Modes of Concurrent Processing: (a) Pull System (b) Linked System

can be readily used by any skilled team member. Overlapping staggers these resources so that they are ready when information is complete and available in the form to be used. It would be a waste of resources if skills are available to work on a task but the corresponding information is incomplete. There are three distinct possibilities in which build-up and transfer of information can take place. Figure 4.12 shows these three forms of overlapping:

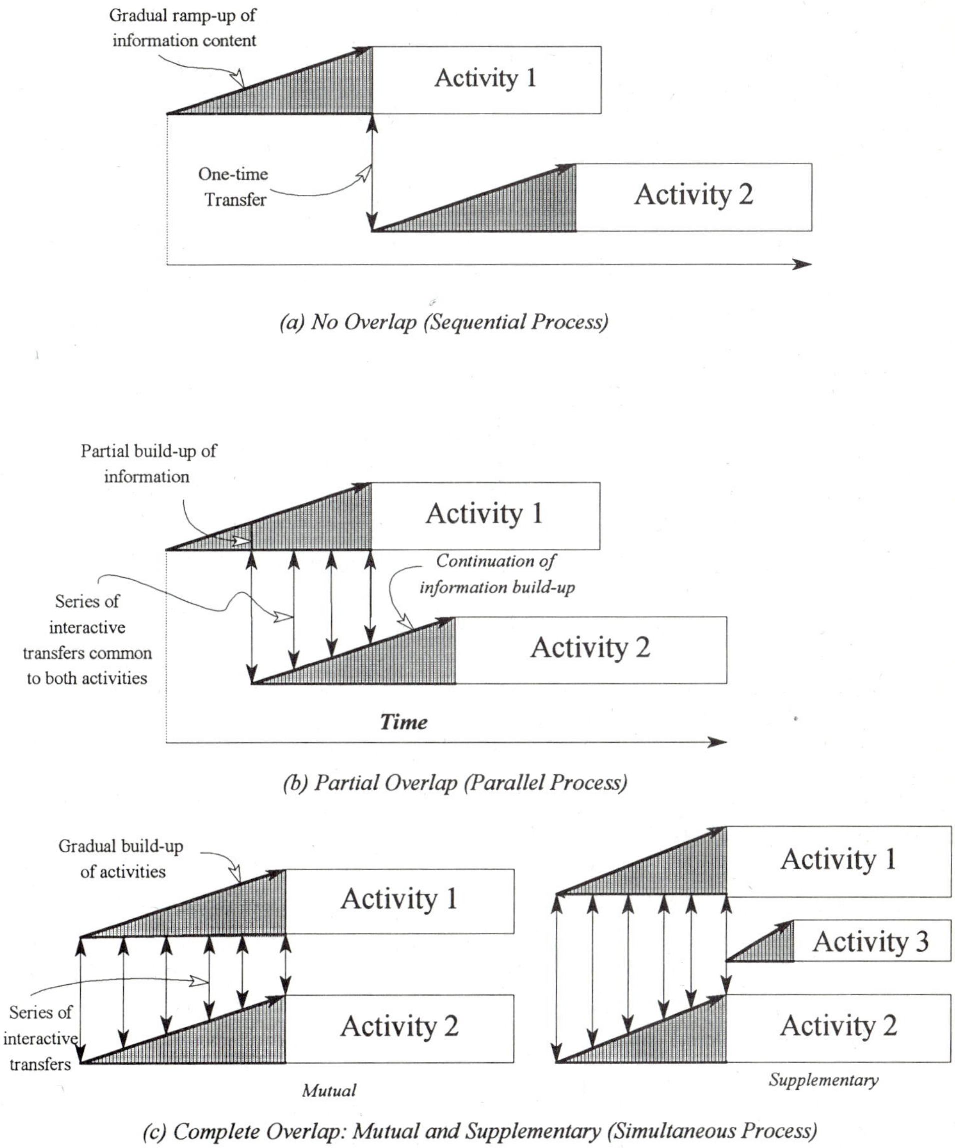

FIGURE 4.12 Overlapped Pull System: (a) No Overlap, (b) Partial Overlap, and (c) Complete Overlap

- ***No Overlap:*** There is no overlap; information is gradually ramped up for an activity until it is complete, then it is transferred to the next activity where it is used to build up a new grade of information to support subsequent activities. This is called series transfer and the resulting process is called sequential process. It is suitable when the two activities are completely dependent, that is, full information from one activity is required to support the second activity.
- ***Partial Overlap:*** The transfer of information takes place during a build-up of an activity, when information ramp-up is only partially complete. A series of interactive transfers continue to take place until the information ramp-up is fully complete. At that point the information transfer from one activity ceases but the information build-up for the second activity continues. This is suitable for activities that are semidependent. The degree of transfer depends on the degree of independence. If the two activities are completely independent, there is no transfer. If not, coupling is preserved and there may be a series of transfers. The resulting process is called "parallel process."
- ***Complete (100%) Overlap:*** If the coupling between upstream and downstream activities can be removed, the two activities can start at once and run simultaneously. A series of interactive transfers may take place while the information is being ramped up on each side. This is suitable when the two activities are independent. This type of parallel overlap is called "mutual" and the resulting process is called "simultaneous process". A supplementary overlap is when a series of transfers between two parallel activities takes place to supplement the start (an information ramp-up) of a third activity.

The suitability of one or the other forms of overlaps depends on the following instances:

- Whether or not there is a dependence between the activities
- How one activity provides information for a second activity to get started early
- Whether or not pertinent skills are freed-up or available
- Whether or not an activity is on a critical path.

Without the overlap, more time is needed because:

- Supplied information may not be in the right format.
- Information may not have the right content (as in the case of an information ramp-up).
- More time may be needed to digest the information supplied.

With some overlap, the time to completion can be significantly reduced.

4.4.2 Linked Form System

PDTs, at the various linked units, can follow a simple procedure to read and write information from one unit to the other. Teams within a unit can use linked workstations to pinpoint problems with any unit and can assist in resolving any problem over the network. A linked form of this CE system is shown in Figure 4.11b. When the procedure is well defined (such as WBS, PtBS, PsBS), there is no need to realign the work priorities, parts, or tree content.

4.5 MODES OF COOPERATION

One of the CE challenges is to find which combination of tools and technology (such as distributed computing, networking, information sharing, process concurrence, hardware, and software, etc.) will create an environment conducive to efficient communication, sharing, and cooperation. The good news is that in today's environment, there are many technological options available to choose from. The bad news is that most of these technical approaches to information sharing are primitives—that is, they are not based on CE principles. For example, the concept of cooperative problem solving has not been fully leveraged into newer C4 tools. There are many instances of cooperative problem solving in product design. Figure 4.13 depicts an interaction matrix between mode of cooperation and degree of involvement. The inner cells of the matrix show the various types of concurrency that are possible in a team environment. On the horizontal axis, in the top row, four possible sets of work-group configurations are chosen to represent different modes of cooperation. A work-group configuration may consist of the following make ups:

- ***Single User:*** A single user is responsible for design decisions.
- ***Co-operating Users:*** Often in a team environment, some processes are sequential. A co-operating user is a person who completes the work left unfinished by previous users. Such a person could be a customer, a member of another product team, or a member of any of the downstream organizations that may be required to work on the uncompleted task. Other responsibilities may include exploring or replaying the logic and analysis of the specified product form or understanding the processes that produced it.
- ***Simultaneous Users:*** This means multiple users or product developers are accessing the product information tools or applications. There are two situations in which this may occur. The users may access the same design, tool, or application concurrently, or different users may access or edit different versions of Product Information Tool or Application (PITA) at the same time.

On the vertical axis in the title column of Figure 4.13, the different possible degrees of involvement for these levels of users are shown. Each of the work-group configurations can have five levels of interaction, not all of which lead to concurrency. For example, a single user accessing his or her own PITA is sequential. Even if the user's partner (a co-operating user) accesses the information following the design, the process is still sequential. Similar situations occur when users try to run these functions against their own data. In other situations, concurrency is present in varying degrees, mild-to-strong. For example, concurrency occurs when a single user or co-operating user accesses data from other groups and runs them against the data from other groups in a computer environment tailored to their perspective on the design. Concurrency is further enhanced when users at

Modes of Cooperation → / Degree of Involvements ↓		Work-group Configuration: Single User	Co-operating Users	Simultaneous Users: Different Versions	Simultaneous Users: Same Version
		1	2	3	4
Access own product interaction tools or applications (PITA)	1	Sequential Engineering (SE)	(SE)	(SE)	(SE)
Run against their own DATA	2	(SE)	(SE)	(SE)	Sequential/ Concurrent Engineering (SE/CE)
Access PITA belonging to other work-groups	3	(SE/CE)	(CE)	(CE)	(CE)
Access DATA belonging to other work-groups	4	Concurrent Engineering (CE)	(CE)	(CE)	(CE)
Access both PITA and DATA from other work-groups	5	(CE)	(CE)	(CE)	(CE)

FIGURE 4.13 Interaction Matrix Between Modes of Cooperation and Degree of Involvement

geographically separated locations perform the aforementioned operations simultaneously.

The degree of concurrency increases as we move from top to bottom and from left to right as in Figure 4.13. The situation depicted by the bottom rightmost rectangle (location [5,4]) provides the largest degree of concurrency. It may be noted from the table that a style of concurrency, where simultaneous users run the same version of PITA against their own data (the location [2,4]), is characterized as both SE and CE. A similar situation occurs when a single user accesses PITA belonging to other work-groups (location [3,1]). Some computational environments for CE are such that they prevent their clients from editing a design module until another user is finished with that design. Even though the two users can work in parallel, the changes cannot be posted until the latter has had the chance to review the changes made by the first user. The latter continues only when the first user has finished with his or her version of the design module. For a critical module in the design, this particular slot may mean that CE is no faster than sequential engineering (see Figure 4.14). Auto-trol technology in Denver recently announced its Mozaic architecture that implements industry standards like STEP and the Object Management Group's Common Object Request Broker (CORBA) to provide vendors with a cross-platform, software-independent environment for collaborative cooperation. Some database systems, e.g., ROSE [Hardwick, et. al., 1990], thus support this type of concurrency by allowing multiple users and applications to edit different versions of the same design module (see Figure 4.15). Some mechanical design automation CAD/CAM vendors (e.g., SDRC, Milford, Ohio) are unveiling a new generation of software products to support concurrent associativity for a multiuser, multiapplication product development team. Concurrent associativity enables continuity among the master model and its linked application data sets. For example, a designer creates a master model for "version A." A draftsman can work on drawings, an analyst can do a stress analysis, a manufacturing engineer can do a process plan or a tool path all for "Version A," while the designer does refinements for an eventual "Version B." Yet when "Version B" comes around, each specified application will be automatically regenerated to reflect changes made to the master model. This reduces the length of the design cycle, provided that the benefit of concurrently editing the module is greater than the cost of having to merge multiple versions of that module at a later time.

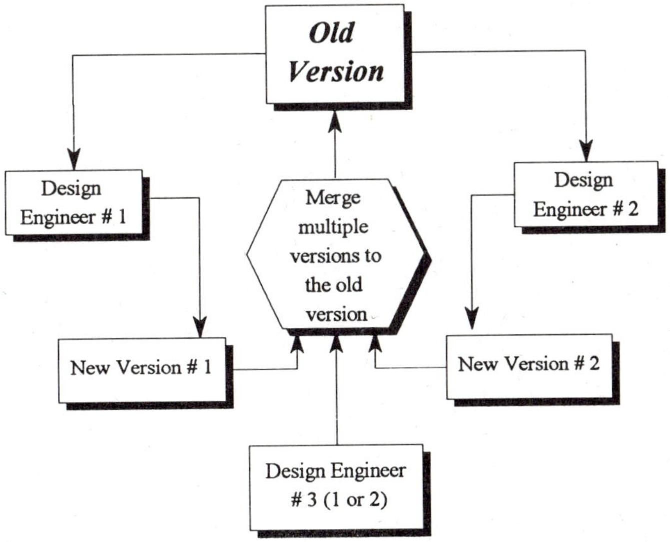

FIGURE 4.14 Simultaneous Users Working with Multiple Versions of Design (Sequential Engineering)

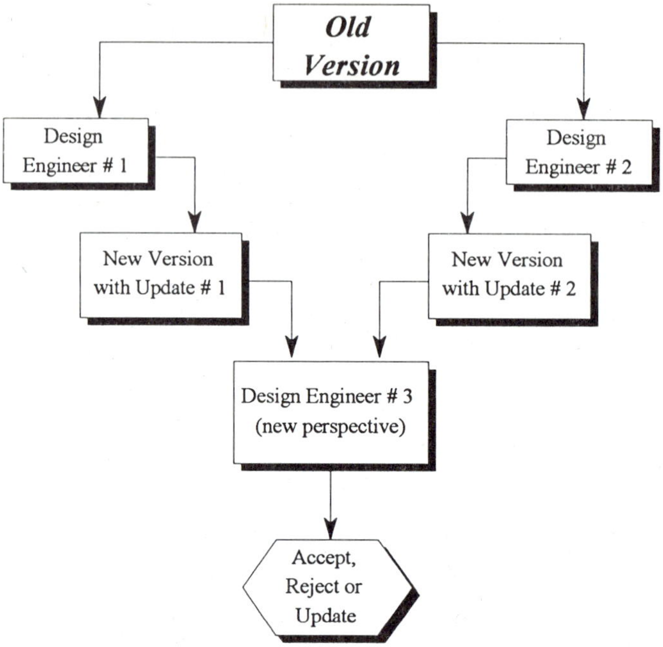

FIGURE 4.15 Simultaneous Users Working with Multiple Versions of Design (Concurrent Engineering)

4.5.1 Managing Interactions

Managing interactions between a user and an application such as PITA is not easy. Several methods are proposed to manage interactions. The two popular methods are (a) via linguistic description and (b) with icon displays and direct manipulation. A linguistic description provides an intermediary communication interface between a user and a PITA. Icon-based displays provide a model-world communication interface and enable the users to directly engage and manipulate PITA objects.

Interactions are further complicated by the multiplicity of media in which such interactions can possibly take place, such as voice, video, audio, and scripts. There are many techniques for coordinating the presentation of information in different media. Usually, types of information or knowledge determine what media would be ideally suited for expressing them.

Enabling tools play an important role in facilitating the cooperation between the teams that may not be colocated. With the lack of such cooperative tools, the performance of a team that works effectively face-to-face in a conference room deteriorates when they communicate via computers. We need all sorts of knowledge that provide or replicate the same feeling or environment that we get in an interactive conference room. We need both—the type of knowledge that is necessary to be an effective collaborative partner, and the knowledge that has proved useful in practice. In addition, we need an appropriate architecture for deploying these knowledge bases into tools that can act as an effective collaborative partner. Technology of collaboration, such as electronic mail (e-mail) systems, Lotus Notes, and work-flow based systems can assist in this purpose. The level of technological support for collaboration must correspond to the degree of cooperative effort required and the complexity of the CE environment.

4.6 BENEFITS OF CONCURRENT ENGINEERING

There are various advantages to concurrent engineering. Products get to the market faster at less cost, and quality is inherent in the product design rather than being an afterthought. Working with new tools, employees enjoy the camaraderie of teamwork, and the fact that their voices are heard when they make a positive impact during the development cycle. Suppliers, subcontractors, and partners benefit too, as projects run on time and become integral to a product's success. Many X-ability aspects of product development (such as manufacturability, quality, ease of maintenance, etc.) are built into the product. In a recent study the DOD researched concurrent engineering projects at Aerojet Ordinance, AT&T, Boeing, Deere & Company, ITT, McDonnell Douglas, Hewlett-Packard, and others. They found that just giving the decision making responsibility to CE teams helped resolve problems and improved efficiency and productivity. The payoff (upper and lower range) from CE in eight benefit areas is shown by shaded and unshaded blocks in Figure 4.16. The reported percentages are based on National Institute of Standards & Technology, Thomas Group Inc. (Institute of Defense Analysis) data, and published information in Business Week [1989, 1990]. Boeing reduced its product development cycle time by 40%–60%. AT&T reduced the process time for a new microprogrammed digital switch by 46%. Deere and Company reduced the product development time for construction equipment by 60%, its manufacturing cost by 30%–40%, and its scrap and rework costs by as much as 75% [Turino, 1992; Business Week, 1989; Ettlie and Stoll, 1990]. McDonnell Douglas reduced rework costs 29%; reduced scrap costs 57%; and reduced nonconformance by 38%. A typical distribution of savings in reducing the major life-cycle phases is shown in Figure 4.17.

One of the benefits of CE, shown earlier as a part of this group of eight, is reduction in engineering change orders. Two curves are shown in Figure 4.18 plotting the number of engineering changes against the number of months before job # 1 (this is a popular terminology used in the auto industry when the car is up for production). The solid curve shows a trend for a company using a CE approach and the dotted curve is for a company using traditional methods. The solid curve shows a temporary boost in the number of engineering changes in the beginning months. Later, it gradually decays to a small number range of 10% to 20% at job #1. The dotted curve initially had a late start but stayed fairly constant—showed little or no sign of slowing down. This is because traditionally, products, thrown over the wall from design, required numerous changes before they were ready for manufacturing. These alterations had to go back to design engineering, where they had to be squeezed in between works when activities on the next product were well under way. This led into a large number of needless revisions (frequent modifications of

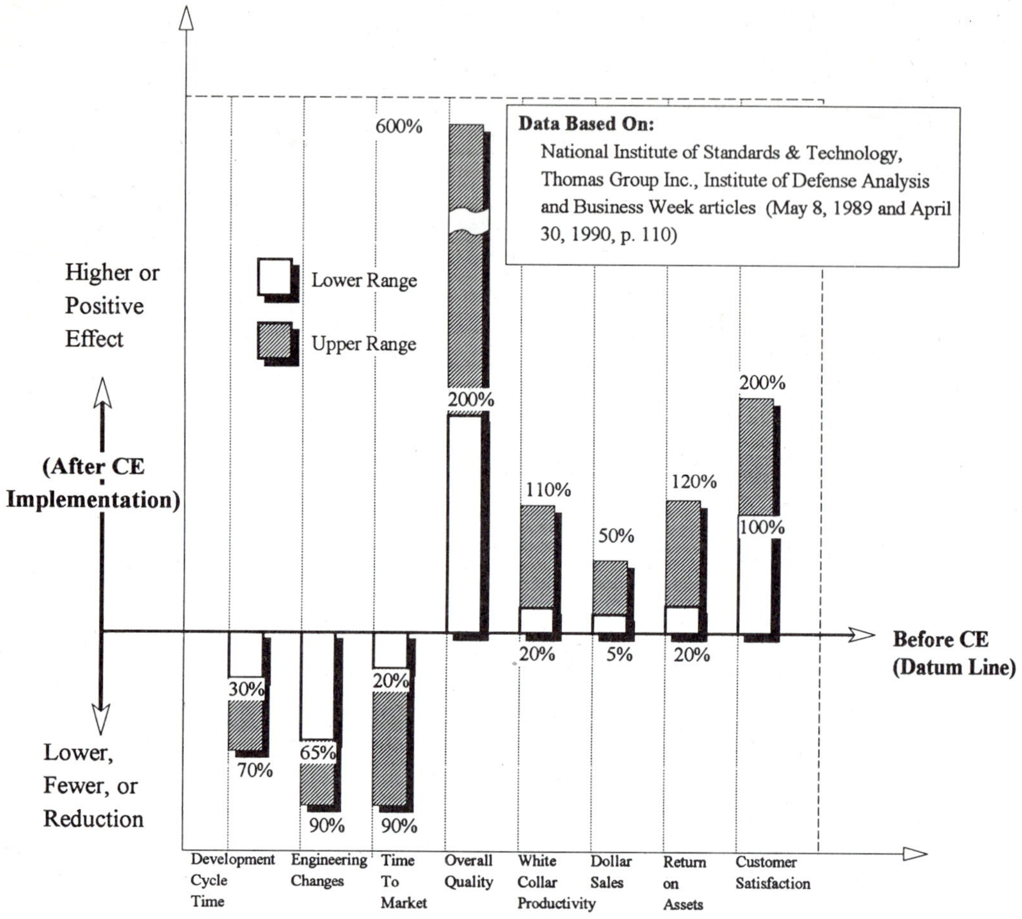

FIGURE 4.16 Range of Pay-off from Concurrent Engineering

prior orders) mustering the dotted curve to show no sign of slowing down. The temporary rise in the solid curve was due to the fact that during the initial starting months, a number of decisions in CE were purposely made based on incomplete and uncertain information. The slow-down of the solid curve, in the latter part, was from drastic reductions of the number of revisions due to CE providing a better information content and source. As the product realization progresses through the life cycle, maturity level of the design also changes. The process initially involves large topological changes in what is configured or not configured into the initial product model. However, the product data and information are at its infancy. This is shown by the small size of the balloon representing the maturity point. As design evolves, detection in constraint violation or errors in the original projections due to use of assumptions (such as incomplete data or inaccurate data) could lead to more changes. Later, in the process, as trade-off between the DFX values take over, this tends to shrink the size of the balloon showing better convergence and maturity. Figure 4.18 shows, through dynamics of changing balloon patterns, the progress in maturity level

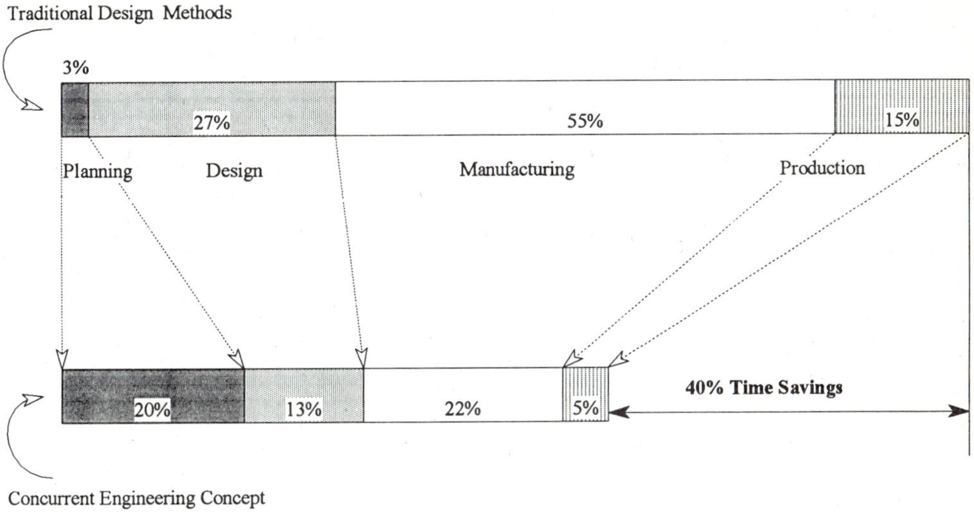

Source: US Air Force R&M 2000 Process Study Report (October 1987)

FIGURE 4.17 A Typical Life-cycle Distribution of Savings

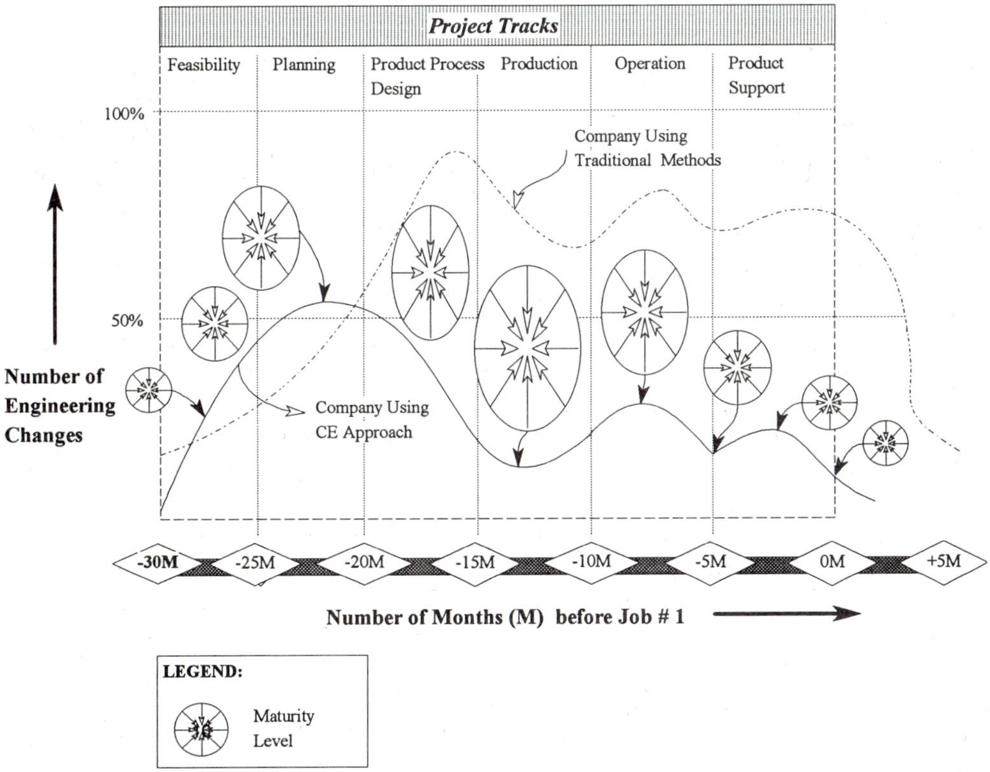

FIGURE 4.18 Effect of CE on Engineering Changes and Maturity

of the design with time. The degree of maturity is clearly proportional to the number of changes and granularity of changes present at a particular point in its life cycle. They are shown along the solid curve as product moves from a feasibility track to a product support and delivery track. The reduction in the number of engineering changes (made during production) was one of the key measurements to determine the success of Nissan's (Motor Co. Ltd.) CE program [Machine Design. 1993]. Other realized benefits from CE include the following:

- ***Rationalization:*** The CE approach ensures that design will be rationalized for the capabilities of the manufacturing system, which is going to convert the intent design to reality—a physical part that can be produced through a physical process. The team focuses on a common goal instead of conflicting departmental goals.
- ***Working parallel:*** The primary CE advantage is that many processes occur concurrently rather than sequentially—things are done simultaneously rather than in a linear fashion. For example, manufacturing engineers can begin to evaluate tooling while the design is still being finalized. Similarly, the documentation process can start virtually on the same day as conceptual design.
- ***Reducing lost time caused by communication breakdowns:*** Since the multifunctional teams have the same objectives, they strive to be more open and available to input each other. Communications tend to be more frequent and less formalized. In a linear process the team is divided. A lot of time is spent on needless conflicts.
- ***Preempting errors and spotting problems early:*** With concurrent engineering, there is a much greater chance of preempting errors and spotting problems while they are still easy to fix. When products are finally released for production there are fewer changes, while the ones that do crop up are usually minor.
- ***Flexibility to accommodate changes:*** When work is done in parallel, it provides the team with flexibility to accommodate changes and the ability to adapt quickly to changed situations or requirements. This ability to accommodate changes at any stage in the product cycle (beginning, middle, or end) provides a smooth flow of information among the involved organizations.
- ***Providing the best overall input:*** The benefits of concurrent engineering reach far beyond the usual collaboration between engineering and manufacturing. CE brings concentrated resources to bear on the task. For instance, various individual disciplines also benefit from working in parallel on product development. In a growing variety and complexity of the products that we manufacture today, close cooperation between mechanical and electronic design is critical to speedy development. Cooperation is fruitful to other disciplines as well—industrial design, conceptual design, structural analysis, technical documentation, etc. When teams and disciplines cooperate with each other, the mutual outcome is likely to be better and faster.
- ***Decreased occurrence of obsolescence:*** The large interval between system development and deployment during traditional product design increases the possibility of obsolescence before the product is ready to be marketed. In CE, since time-to-market is reduced, the effect of obsolescence on the life-cycle cost is greatly minimized. As a result, CE not only saves development cost due to shrinking time and schedules,

but it also reduces non-recurring costs (such as parts qualifications, tooling, inventory, stocks, manuals, support equipments, etc.) that would have incurred otherwise.

- ***Cross-training:*** A multifunctional group can help each other get up to speed quickly. Cross-training reduces the learning curve.
- ***Better use of scarce technical resources:*** If the product development cycle can be shortened, then project personnel can be made available for the next project sooner, thus improving the company's use of scarce technical resources.
- ***Other Benefits:*** In addition, several of the firms that have recently adopted the practice of concurrent engineering reported reduced lead time for creating bid proposals; shorter product development times (in design phases); reduced product development costs; parts reduction; lower inventories, fewer rework orders, smaller scrap piles, improved quality, and reduction in production costs [McKnight and Jackson, 1989] (during fabrication, manufacturing, and assembly).

Rover Group Ltd. claims the long-term success of CE by measuring the progress the company makes towards eliminating the "building and testing of physical prototypes" [*Machine Design.* 1993]. Moreover, these success stories are not typical for a particular industry. They come from automotive companies, aircraft manufacturers, heavy equipment producers, electronic firms, and computer manufacturers. The numbers are astonishing. For instance, using CE, General Electric (GE) has cut its delivery time of custom-made circuit breaker boxes from three weeks to three days. AT&T has reduced the product development time for telephones from two years to one. GE Aerospace, Government Electronic Systems Division used CE and a PDT for a ground-based radar system to compress configuration time for the proposal that once took between six to nine months into three months [*Machine Design.* 1992]. They used CE because they did not see any other way of getting a technically superior proposal finished before the drop dead time. For VLSI (very large-scale integration) circuits, AT&T has also reported part counts down to one ninth of their previous levels and quality improvements up over one hundred times in surface defects. Motorola used to make electronic pagers in three weeks from order, it now produces them in two hours. Rover was able to launch Land Rover Discovery in eighteen months as compared to forty-eight to sixty-three months for a similar product in Europe. At Navistar, product development for trucks has been reduced from five years to two and a half, and Hewlett Packard has brought Desk Jet printers to market in twenty-two months. Northern Telecom has reduced delivery time on its popular DMS switching systems by one half. A list of products and companies that have taken special efforts in this direction are shown in Figure 4.19 under two categories: "fast innovators," and "fast producers." The results were compiled from many sources [Trygg, 1992], [Fortune Magazine, 1989], [Carey, 1992]. A fast innovator is used here to indicate those companies, which require more lead time up front in terms of product definition, engineering, design, prototyping, etc., as opposed to a fast producer. Although each goes through the same or very similar tracks of product development, the beginning tracks appear to be more important to fast innovators. They seem to pay more attention to uniqueness or newness of each design. Likewise, fast producers seem to emphasize reaching to the tail-end (like production engineering and planning, production operations, and manufacturing,

(a) Fast Innovators					Major Contributing Factors
Company	Product	Best Development Time			
		Before (months)	After (months)	Reduction	
ABB	Switching Systems	48	10	79%	CMS
AT&T	Phones	24	12	50%	ACM
British Aerospace	Aeroplanes	36	18	50%	MS
Digital Equipment	Personal Computers	30	12	60%	ACMS
Ford	Cars	60	42	30%	---
General Motors (GM).	Engines	84	48	43%	MS
GM/Buick	Cars	60	41	32%	MS
Goldstar	Telephone Systems	18	9	50%	CM
Honeywell	Thermostats	48	12	75%	MS
Honda	Cars	60	36	40%	---
Hewlett-Packard	Printers	54	22	59%	ACMS
IBM	--	48-50	12-15	70/75%	ACM
Motorola	Mobile Phones	36	7	81%	ACM
Navistar	Trucks	60	30	50%	MS
Warner Electric	Clutch Brakes	36	9	75%	M
Xerox	Copiers	60	24	60%	ACM
Xerox	--	53	36	32%	ACM

(b) Fast Producers			
Company	Product	Order To Ship	
		Before (wks)	After (days)
Brunswick	Fishing Reels	3 wks	5 days
GE	Circuit Breakers	3 wks	3 days
HP	Test Equipment	4 wks	5 days
Motorola	Pagers	3 wks	1/4 day

Sources: Fortune Magazine, February 13, 1989; Automotive Engineering, November 1992; Trygg, L, "Simultaneous Engineering: A Movement or an Activity of the Few?," European Institute for Advanced Studies in Management, Brussels, May 1992

LEGEND: M: Multi-functional Teams C: Computer Integrations
A: Analytical Methods and Tools S: Suppliers in the Project Team

FIGURE 4.19 Time-to-Market Comparisons: (a) Fast Innovators (b) Fast Producers

etc.,) faster. The differentiation assumes that fast producers are not constantly designing and developing new products to the same extent as fast innovators are.

4.6.1 Some Common Pitfalls

There are some hard lessons to learn from concurrent engineering too. For example, sequential engineering is more suited for a matured design process. A design is considered to reach its maturity when each stage flows linearly—no apparent back tracking. When CE is applied to a matured design process, it can result in an "orbital process." The total time consumed, in such cases, could be larger than the sequential process—due to its inherent looping (a circular sequence of repetition). Other problems can occur as well:

- ***Degradation of turnaround time:*** If the organization is not equipped to process a large volume of information simultaneously, it may clog up the storage or degrade the turnaround time.
- ***Spurts not desirable:*** Release of large amounts of information in spurts is also inappropriate for concurrent mode of operation. Cross-functional teams require continuously updated information.
- ***Risk of wasted efforts:*** Furthermore, with CE approach there is a greater degree of risk of human resources doing something in parallel that is not required and thus creating intellectual inventory, which may clutter up the working environment. Intellectual inventory is also created in serial design processes. However, with concurrent engineering, where the goal is to cut the product development lead time, the consequences could be far more taxing.
- ***Possible increase of iterative costs:*** In certain circumstances, CE can probably increase some of the design costs because it is an iterative process and some portions of the work will have to be redone. For instance, one may have to repeat the process when better ways are found to configure the product or when manufacturing discovers a need for change in product design to enhance manufacturability. There is a slight possibility that with CE one can go overboard in iterations, especially during early phases, thereby taking more time than necessary. But its impact on reducing the total time can still be justified.
- ***Concurrent chaos:*** One of the benefits of CE is that incomplete or uncertain information can be used at an early stage. This means some processes, which were initially accomplished sequentially, can now be performed in parallel. However, sharing immature or imperfect information early may result in increasing the number of life-cycle areas. If such affected areas are not controlled, it can lead to what is normally described as a "concurrent chaos"
- ***Errors can build-up:*** There is a concern that engineers can go overboard and engage in more simultaneity than necessary. As more and more key activities, which lie on the critical path, are done in parallel, there may be no slack time left. This increases the risks of committing errors. It may take teams' valuable time away from key activities to correct errors, leading to the possibilities that corners will be cut. As a result, the overall quality of the end-product may suffer.

However, benefits for CE described here far outweigh any potential pitfalls. Possibilities are also very remote. Many organizations are thus embracing CE as the sure-shot way of winning the competitiveness war.

REFERENCES

ASHLEY, S. 1992. "DARPA initiative in Concurrent Engineering," *Mechanical Engineering,* Vol. 114, No. 4 (Apr.), pp. 54–57.

BARCLAY, I., and POOLTON, J. 1994. "Concurrent Engineering: Concepts and Practice," *International Journal of Vehicle Design,* Vol. 15, No. 3–5, pp. 529–544.

BARKAN, P. 1988. "Simultaneous Engineering," *Design News,* Vol. 44 (March 7), pp. A30.

Business Week. 1989. May 8. New York: McGraw-Hill, Inc.

Business Week. 1990. April 30. New York: McGraw-Hill, Inc.

CALS/Concurrent Engineering Task Group. 1991. "First Principals of Concurrent Engineering: A Competitive Strategy for Electronic System Development," *Review Draft,* Washington D.C., CALS Industry Steering Group, Feb. 1991.

CAREY, W.R. 1992. "Quality Tools for Today's Engineers," *Automotive Engineering,* Vol. 100, No. 11 (November), pp. 33–40.

CARTER, D. E., and B. S. BAKER. 1992. *Concurrent Engineering: The Product Development Environment for the 1990s.* Reading, MA: Addison-Wesley Publishing Company.

CHAO, N. H., and S. C. Y. LU (ed.). 1989. *Concurrent Product & Process Design.* New York: ASME.

Computer Integrated Manufacturing (CIM)/Manufacturing Technology Analysis Center (MTIAC). 1988. Chicago, IL, *MTIAC ST-88-02,* April.

COSTELLO, J. 1991. "Concurrent Engineering," *Computer Design,* Vol. 30, No. 1 (Jan.), pp. 21–23.

CLEETUS, K. J. 1992. Definition of Concurrent Engineering. *CERC Technical Report Series, Research Notes, CERC-TR-RN-92-003.* West Virginia University, Morgantown: Concurrent Engineering Research Center.

DARPA. 1987. *Workshop on Concurrent Design.* December 1–3. Key West, FL. Defense Advanced Research Projects Agency (DARPA), Washington, DC.

DARPA. 1988. *Workshop on Concurrent Design*, December 2–4. Key West, FL. Defense Advanced Research Projects Agency (DARPA), Washington, DC.

DEMING, W. E. 1993. *The New Economics.* Cambridge, MA: MIT Center for Advanced Engineering Study, November.

DIKA, R. J., and R. L. BEGLEY. 1991. "Concept Development Through Teamwork—Working for Quality, Cost, Weight and Investment." *SAE Paper # 910212,* Presented at Int'l Congress and Exposition, SAE, Feb. 25–March 1, Detroit, MI, pp. 1–12.

ELLIS, L. W. 1992. "Practice in Concurrent Engineering," Proceedings of the International Management Conference, (October), USA, pp. 15–118.

ETTLIE, J. E., and H. W. STOLL. 1990. *Managing the Design-Manufacturing Process.* New York: McGraw-Hill Inc.

EVANS, B. 1988. "Simultaneous Engineering," *Mechanical Engineering,* Vol. 110, No. 2 (Feb.), pp. 38–39.

FINGER, S., et. al. 1993. "Design Support Systems for Concurrent Engineering: A Case Study in

Large Power Transformation Design," Proceedings of the International Conference on Engineering Design, *ICED 93,* August 17–19, The Hague, WDK22, Roozenburg N.F.M. (ed.), Vol. 3, pp. 1433–1440.

Fortune Magazine. 1989. February 13, New York: Time Inc.

HARDWICK, M., et. al. 1990. "ROSE: A Database System for Concurrent Engineering Applications," Proceedings of the Second National Symposium on Concurrent Engineering. February 7–9, Morgantown, West Virginia, pp. 33–65.

HARTLEY, J. R. 1992. *Concurrent Engineering: Shortening lead times, raising quality and lowering costs,* Cambridge, MA: Productivity Press.

GARRETT, R. W. 1990. "Eight Steps to Simultaneous Engineering," *Manufacturing Engineering,* Vol. 105, No. 5 (Nov.), pp. 41–47.

KIMURA, F. 1994. "Virtual Manufacturing as a Basis for Concurrent Engineering," *IFIP Transactions of Applications in Technology,* Issue 17, pp. 103–108.

KRISHNAN, V. 1993. Design Process Improvement: Sequencing and Overlapping Activities in Product Development. *D. Sc. Thesis,* Massachusetts Institute of Technology, September, 1993.

LEWIS, J. W. 1990. "An Approach to Applications Integration for Concurrent Engineering," Proceedings of the Second National Symposium on Concurrent Engineering. February 7–9, Morgantown, West Virginia, pp. 141–154.

LINN, J. L., and R. I. WINNER, (eds.), and the EIS Requirements Team. 1989. The Department of Defense Requirements for Engineering Information Systems (EIS), Volume I: Operational Concepts. *IDA Paper P-1953,* Institute of Defense Analyses. Alexandria, VA.

LINN, J. L., and R. I. WINNER, (eds.), and the EIS Requirements Team. 1989. The Department of Defense Requirements for Engineering Information Systems (EIS), Volume II: Requirements. *IDA Paper P-1953,* Institute of Defense Analyses. Alexandria, VA.

Machine Design. 1992. "How to Make Concurrent Engineering Work—Part 1: Concurrent Engineering: The Need and Enabling Technologies," Vol. 64, No. 20, (August 6).

Machine Design. 1993. "How to Make Concurrent Engineering Work, Part 6—Measuring Success from CE," Vol. 65, No. 23 (November 26), pp. 77–80.

MATHER, H. 1993. "Product Design: Simplification before Automation," *Manufacturing Systems,* Vol. 11, No. 6 (June), pp. 58–62.

MCKNIGHT, S. and JACKSON, J. M. 1989. "Simulataneous Engineering Saves Manufacturing Lead Time Costs, and Frustration," *Industrial Engineering,* Vol. 21, No. 8 (August), pp. 25–27.

OHMAE, K. 1989. "The Global Logic of Strategic Alliances," *Harvard Business Review,* Vol. 67, No. 3 (May–June), pp. 64–75.

OWEN, J. V. 1992. "Concurrent Engineering," *Manufacturing Engineering,* Vol. 109, No. 5 (Nov.), pp. 69–73.

PENNELL, J., and M. SLUSARCZUK. 1989. An Annotated Reading List for Concurrent Engineering. Technical Report HQ 89–034130. Alexandria, VA: Institute for Defense Analyses.

PRASAD, B. 1995. "On Influencing Agents of CE," *Concurrent Engineering: Research & Applications—An Int. Journal,* Vol. 3, No. 2 (June), pp. 78–80.

ROSENBLATT, A., and G.F. WATSON, 1991. "Concurrent Engineering," *IEEE Spectrum,* Vol. 28, No. 7 (July), pp. 22–37.

SHINA, S. G. 1991. *Concurrent Engineering and Design for Manufacture of Electronic Products.* New York: Van Nostrand Reinhold.

SMITH, G. F., and G. J. BROWNE. 1993. "Conceptual Foundations of Design Problem Solving." *IEEE Transactions on Systems, Man, and Cybernetics.* Vol. 23, No. 5, September/October.

SOLDBERG, J. J. 1988. "Integrated Manufacturing Systems: An Overview." W. D. Compton (ed). *Design and Analysis of Integrated Manufacturing Systems.* Washington, D.C.: National Academy of Engineering, National Academy Press.

STAUFFER, R. N. 1988. "Simultaneous Engineering—What it is?" *Manufacturing Engineering,* Vol. 101, No. 3 (September).

TRYGG, L. 1992. "Simultaneous Engineering: A Movement or An Activity of the Few?" Proceedings of the International Product Development Management Conference on New Approaches to Development and Engineering, Brussels, May 18–19, European Institute for Advanced Studies in Management, Brussels, pp. 569–582.

TURINO, J. L. 1992. *Managing Concurrent Engineering: Buying Time to a Market: A Definitive Guide to Improved Competitiveness in Electronics Design and Manufacturing,* New York: Van Nostrand Reinhold.

WINNER, R. I., J. P. PENNELL, H. E. BERTREND, and M. M. G. SLUSARCZUK. 1988. The Role of Concurrent Engineering in Weapons System Acquisition. *IDA Report R-338.* Alexandria, VA: Institute for Defense Analyses.

WOMACK, J. P., D. T. JONES, and D. ROOS. 1990. *The Machine That Changed the World.* New York: Macmillan Publishing Co.

TEST PROBLEMS: CE DEFINITIONS

4.1. Why are there so many definitions of CE? Which one of these do you like the best? Explain reasons for your choice.

4.2. What CE tactics should be adopted to get the maximum benefits for a new product introduction (NPI)? Do any of these CE tactics deal with outside elements of the manufacturing organization? If so, what are those?

4.3. How does a CE process differ from a conventional SE (serial engineering) process? What is the major impact of CE? When is the "simultaneity" condition feasible?

4.4. Describe the typical structure of a CE system? What are its elements? How do you achieve "best" concurrency? What major benefits have industries realized using CE? Discuss instances when concurrent engineering degenerates into sequential engineering.

4.5. Why is it important to build relationships between suppliers and customers in implementing CE?

4.6. What are some different ways of partnering or sharing responsibility with outside suppliers? What benefits do you see in becoming a value-chain supplier? How does government use a two-phase contractual agreement?

4.7. Can PERT be used to describe a CE process? What task relationship in CE is not easily described in PERT?

4.8. Does CE promote the spirit of innovation? On what basic principles is CE founded? How are these principles used as a competitive weapon in capturing new market shares?

4.9. How can a company maximize their level of CE usage? Can this be done by making the steps along the stages of development fully (or nearly) independent?

4.10. What are the different ways a pair of tasks can be related? Why is it important to seek relationships between pairs of tasks?

4.11. As competitiveness grew during the 1980s and early 1990s, companies began to see the advantages of using outside partners to do business. List some of these advantages and give examples of successful partnerships.

4.12. Why is quick processing considered one of the eight governing factors facilitating CE?

4.13. How can a company maximize its level of concurrency? What are the governing factors that allow someone to execute a set of tasks concurrently?

4.14. What is a virtual company? Why, day by day, are more and more companies choosing to become more and more virtual? List some of the typical participants of a virtual company. Why does a virtual company prefer to create common digital product/process models?

4.15. What are the dangers of doing things concurrently without adequate independence? Does it increase the risks of hiding mistakes? Provide examples of dependent, semi-independent, independent, and coupled tasks.

4.16. In Figure 4.12, information build-up is shown to cease when a new activity begins. When is this valid? Provide examples of the overlapped pull system. Do you think, in R&D situations, that the build-up and collection of information occurs in the same manner? If not, describe what takes place.

4.17. By reducing the number of interfaces between sources, organization can save time and reduce costs. Compare the similarities of minimizing product interfaces with minimizing process interfaces.

4.18. What is the significance of product decomposition? Why does a team take a lot of pains in decomposing or breaking up tasks in the first place when the team has to regroup them again?

4.19. Product breakdown structure (PtBS) is one way of identifying the tasks associated with an integrated product development process. Create a PtBS tree for a lawnmower.

4.20. What is concurrent resource scheduling? Is this a part of a work break-down structure (WBS) or otherwise? How does it differ from PDM (precedence diagramming technique)?

4.21. How does a team plan and use PERT, Gantt, CPM charts, and choose the ones best suited to CE needs?

4.22. If you have more than one person working to modify the design on the "same version," which modification gets priority when making changes or decisions?

4.23. How would your company expect to benefit from CE? Relate these benefits to the benefits realized in the past—as shown in Figure 4.16.

4.24. How can you identify and avoid the three biggest pitfalls, which are likely to keep outcome of CE efforts living up to expectations?

4.25. What are the components on which success of CE depends?

4.26. Why does a product development team (PDT) have to be multidisciplinary? Describe a bidirectional sandwiched structure for a PDT.

CHAPTER 5

COOPERATIVE WORK TEAMS

INTRODUCTION

The basic intent of any manufacturing company is to employ a skilled (or trained and talented) workforce machinery and allocate resources to make quality products. The traditional hierarchical systems were designed to make it as easy as possible to keep track of people and the things they were doing, the machinery they were using, etc. The orientation was functional and vertical in nature. Experts made improvements within the confines of a department or a functional area. The result of that expertise gave organizations tremendous marketplace advantage for a short while. The use of certain titles such as manager, director, or supervisor, rather than leader, facilitator, or coach, reflects a bias. [Freeze and Aaron, 1990]

Today products are more complex than ever before. It is beyond the imagination of a single person, a single team, or even a single department to comprehend fully all the aspects of the product needs. The nature of the organization developing the product, however, has not changed as much. It has been a challenge for the design and manufacturing engineers in many traditional organizations to work together as teams to improve quality while reducing costs (capital, investment, etc.), weight, and lead time (time-to-market). In many organizations, teamwork, productivity, and efficiency gain have been very painstaking. There are many reasons cited for such difficult times. The most commonly cited reason is the "people" or the "human" component. The best illustration of this comes from the remarks of Roy Wheeler of Hewlett Packard, when he was asked:

> "*What tools does an engineer need to get started in CE?* Wheeler's answer was Pencil, paper, some intelligence, and a willingness to work with peers in other functional areas to get the job done" [Watson, 1991]

The "people component" involves many constantly changing variables that are more difficult to control than any other variable. This is because human behavior and corporate cultures are difficult to measure and quantify. There is a close association between the two. Changing the corporate culture by institutionalizing CE does not guarantee that human behavior will be changed or that the two will work in close (and mutual) synergy and vice versa. The interactions with such issues, in general, and the lack of synergy in particular, tend to be underestimated or even unaddressed as key organizational problems [Maliniak, 1991]. Other contributing factors that are commonly cited include the following:

- ***Lack of appropriate motivation:*** There is a lack of team-based motivation. Most reward systems, including incentives and sanctions, in traditional hierarchical organizations are based on individual creativity and contributions. Engineers are valued according to their ability to fix problems, not according to their ability to eliminate sources or causes of the problems. As such, there are not many incentives on the part of the individual employee to develop defect-free products or services or work as a team.
- ***Decision making style:*** The style of decision making is mostly top-down and most decisions follow a unidirectional path (such as a directive management style). By the time the design or process engineers get ready on a new project, many decisions are already made. The planner has chosen key product offerings, such as major specifications, gross weight, length, width, etc. The finance community has determined what the product must cost and how much the company can afford to invest. Marketing has decided how many of the product will be sold. The timing office has decided when it must be introduced. Under these circumstances, the rest of the organization's job ends up merely engineering the product under the aforementioned restrictions. Input or feedback from technical experts and the product feasibility team could not be taken into consideration in those early decisions. Neither the limitations of the manufacturing equipment at hand could be taken into consideration. Under such hard-pressed situations, often there is no room to maneuver for product realization. At times it becomes difficult to satisfy many of such demands and still meet product quality and timing standards.
- ***Lack of management commitment or action:*** The management commitment to empower participants to do what is right and to work as a team has been lacking. Often employees are given responsibility for a project but not the authority to introduce changes to it. For instance, employees are given a set of modeling, simulation, and analysis (for example, CAE, CAD/CAM) tools, with the hope that this will eliminate all their technical or organizational problems. Some management considers providing tools as empowerment. In management circles, there is a great lack of understanding of what cooperative team work is. There is also a lack of authority on the employees' part to take bold actions.
- ***Policies, practices, procedures (3Ps):*** Participants are required to follow an extremely rigid and complex procedure. The organization dictates most of these steps. Circumstances often change, people move, and the planned steps may not be valid any longer. Many of the participants may not know about the changes and confu-

sion prevails. Employees tackle what they think are the right things to do in such circumstances and trade the rest. Since these decisions are infrequently team-based, they are not in the best interest of the group, product, or the company.

- ***Lack of common understanding, commitment, or action:*** Making information available to team members does not guarantee that it will be used. It must be timely and meaningful. The information is not meaningful when the team does not understand how to use the information or how others will benefit from it. Often there is no prior common understanding or commitment between the recipients and the providers of information.
- ***Ineffective communication:*** Effective communication is the key to developing a knowledgeable and committed work-force and setting a common set of consistent goals. Clear and supporting goals provide constancy of purpose. They allow everyone to set aside frivolous issues and focus on what is really important to the "total system." Communication is a two-way street. Effective communication takes place both vertically (in spite of differences in responsibility or ranks) and horizontally (in spite of teams' functional differences). An ineffective communication environment (that is, giving partial information and holding the rest of it) discourages a free exchange of ideas up, down, and across organizational lines. Due to ineffective communication, there is a danger that deficiencies discovered in the downstream activities may not be communicated to the upstream activities. This inhibits innovation, retracts teamwork, and strangulates opportunities for continuous improvement.

The aforementioned six points are typical reasons cited for numerous project delays, inciting horror stories, general chaos, or ultimate productivity loss in many industries. But things do not have to come to this. It would be preferable to use the collective experience/knowledge of project teams to develop design and manufacturing concepts so that, if circumstances change, decisions can be altered quickly. It is also desirable to collectively develop a reasonable set of specifications and objectives that are feasible and fully understood by all parties before they are finally committed and deployed by management. The success of rapid product realization will depend on the team's ability to handle change. Change occurs at all levels: during the people management, product management, process management, or enterprise management. "Knowledge for managing change" is becoming a strategic tool for competitiveness. CE teams must manage change carefully, whether it occurs upstream (for instance, during strategic planning), or downstream (for instance, during deployment) levels. One of the greatest challenges in managing change is to figure out how to get people to work with each other and as a team. Human beings, by nature, tend to be territorial and look out for their own (personal) interests.

We must recognize that not only individuals, but work-groups and departments all operate out of their own self-interest. Some groups even fight to protect their own turf instead of working toward a common set of consistent goals. This generates a number of controversies (See Figure 5.1). Early indicators or signs of controversies are avoidance, nonaccommodation, conflicts, personal goals, egos, etc. A number of these are listed in Figure 5.1. The defensiveness, foot dragging, and "so what" attitudes are all potential hin-

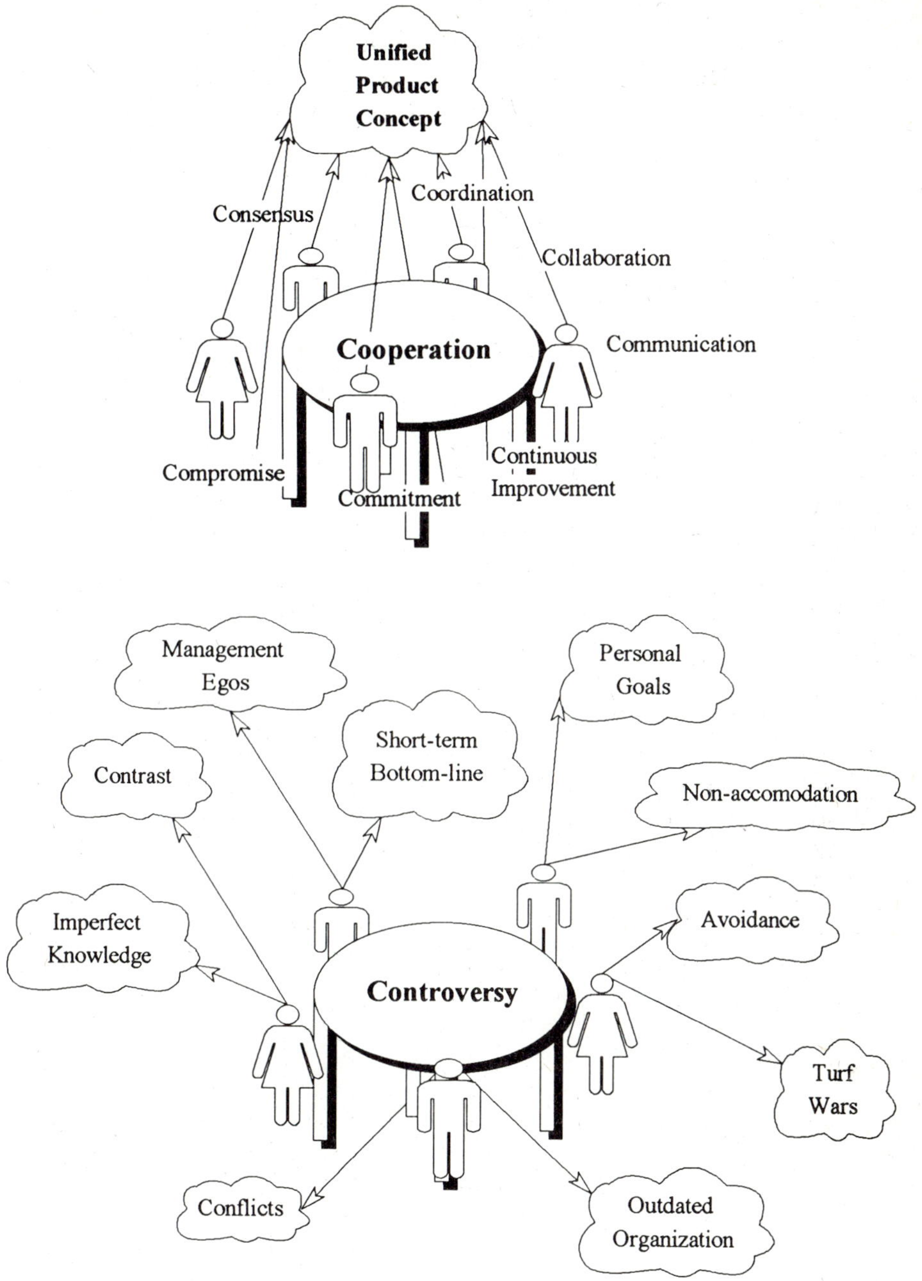

FIGURE 5.1 Controversy versus Cooperation

drances to implementing change. For example, personal attitudes may become more important than the computer productivity tools for effective communication. When controversies occur and something goes wrong, finger pointing begins along some familiar refrains: "If they only built it the way we designed it, we would not have these problems!"; "If they had listened to what I said regarding . . . !"; "I told you so . . . what is in it for me," etc. These counterproductive arguments are merely a reflection of who we were, our heritage, and our cultural history, and there is not much one can do to significantly change that. It might be easier to exploit ones' own sense of self-interest or pride and apply it toward the team's or the group's interests. It would be nice if management could create an environment where teams could feel that it would be in their best "self-interest" or "self-esteem" to cooperate with one another rather than compete. This would not be an unusual change. Cooperation has been and usually is a part of the daily work environment. In Japan, for instance, cooperation has been a way of life for many years. All parties are able to share information right from the conceptual stage. There is no hidden agenda or designers' secrets in the work. For many years (as late as the 1980s), teamwork was not encouraged in American academic institutions. The culture and the curriculum discouraged people from cooperating. In the past, when students worked together, it was referred to as "cheating." But in recent years this has changed. Many schools have remodeled their curriculum to emphasize teamwork, open-ended problem solving, design lessons, and communication skills. The change we are seeing is not surprising. Professionals all over the world have communicated for years. They openly share their views in journal articles on new ideas, new theory, and new applications. Team skills are very much encouraged and rewarded in the industry. Competition, on the contrary, leads to loss. People pulling in opposite directions on a rope only exhaust themselves—they go nowhere [Deming 1993]. Cooperation is the key lynch-pin of achieving teamwork. There are seven elements (called 7Cs) to this team cooperation philosophy:

- ***Collaboration:*** This describes a process of value creation that a traditional structure of communication and teamwork cannot achieve. Instead of focusing on methods of communication (such as teams with definite roles and sets of operating procedures), collaboration seeks out the unplanned and unpredictable.
- ***Commitment:*** Empowered teams define the tasks and prioritize areas to make breakthrough opportunities. Goals and objectives, duration, utility, complexity, expected results, and key success factors are outlined as much as possible. Management is fully committed to meeting the goals.
- ***Communications:*** Effective communication is the precursor to meaningful collaboration. Communication is the exchange of information. Whereas the collaboration is a commitment to create a shared understanding [Sullivan, 1988].
- ***Compromise:*** There is compromise and input from every discipline so that simultaneous development of the product, process, and associated tooling can be achieved.
- ***Consensus:*** Project team and management members may disagree, but team support of the requirements and a commitment to objectives at the very outset of a project is essential. These common objectives are reinforced throughout the life of the project.

- ***Continuous Improvement:*** Product or process design teams work toward the total elimination of waste. The concept focuses on enhancing productivity and profitability through the improvement of product quality and reduction in product development cycle time.
- ***Coordination:*** The most cited definition of coordination is by MIT Sloan School of Management—"Coordination is the act of managing interdependencies between activities" [Malone, 1991]. Coordination involves actors performing interdependent activities that achieve goals, and its analysis includes goal decomposition, resource allocation, synchronization, group decision making, communication, and the preparation of common objectives. Partnerships are formed among all disciplines involved in the project and communication links are formally established and utilized. Suppliers are involved in the early stages of the project.

Figure 5.2 lists these 7Cs characteristics for achieving cooperation. CE teams must examine the extent to which the organizational culture or "self interest" supports or detracts from achieving a unified product concept (or deviates from a common set of company goals).

Let us assume that the symbols a, b, c, d, . . . designate the contributions of the teams A, B, C, D . . . in a company. There are two types of contributions the teams can make to the company's overall throughput or its productivity:

- Individual contributions
- Team contributions

Individual contributions will be the algebraic sum of a, b, c, d . . . :

$$\text{Individual contributions} = a + b + c + d + e + \ldots\ldots \tag{5.1}$$

The team contributions will be from interactions—teams working in pairs (ab, ac, . . .), trios (abc, abd, . . .), etc.:

The total team interactions will be the sum of these team contributions.

$$\text{Team Interactions} = \langle ab \rangle + \langle ac \rangle + \langle ad \rangle + \langle\ \rangle + \ldots \tag{5.2}$$

$$\begin{aligned}
&\langle bc \rangle + \langle bd \rangle + \langle\ \rangle + \ldots \\
&\langle cd \rangle + \langle\ \rangle + \ldots \\
&\langle abc \rangle + \langle abd \rangle + \langle abe \rangle + \langle\ \rangle + \ldots \\
&\langle acd \rangle + \langle ace \rangle + \langle\ \rangle + \ldots \\
&\langle ade \rangle + \langle\ \rangle + \ldots \\
&\langle bcd \rangle + \langle bce \rangle + \langle\ \rangle + \ldots \\
&\quad \langle abcd \rangle + \langle abce \rangle + \langle abcf \rangle + \langle\ \rangle + \ldots \\
&\quad\quad \langle\ \rangle +
\end{aligned} \tag{5.3}$$

$$\begin{aligned}\text{Company's productivity} \equiv\ & \text{Individual contributions} \\ &+ \text{Productivity gains due to team Interactions}\end{aligned} \tag{5.4}$$

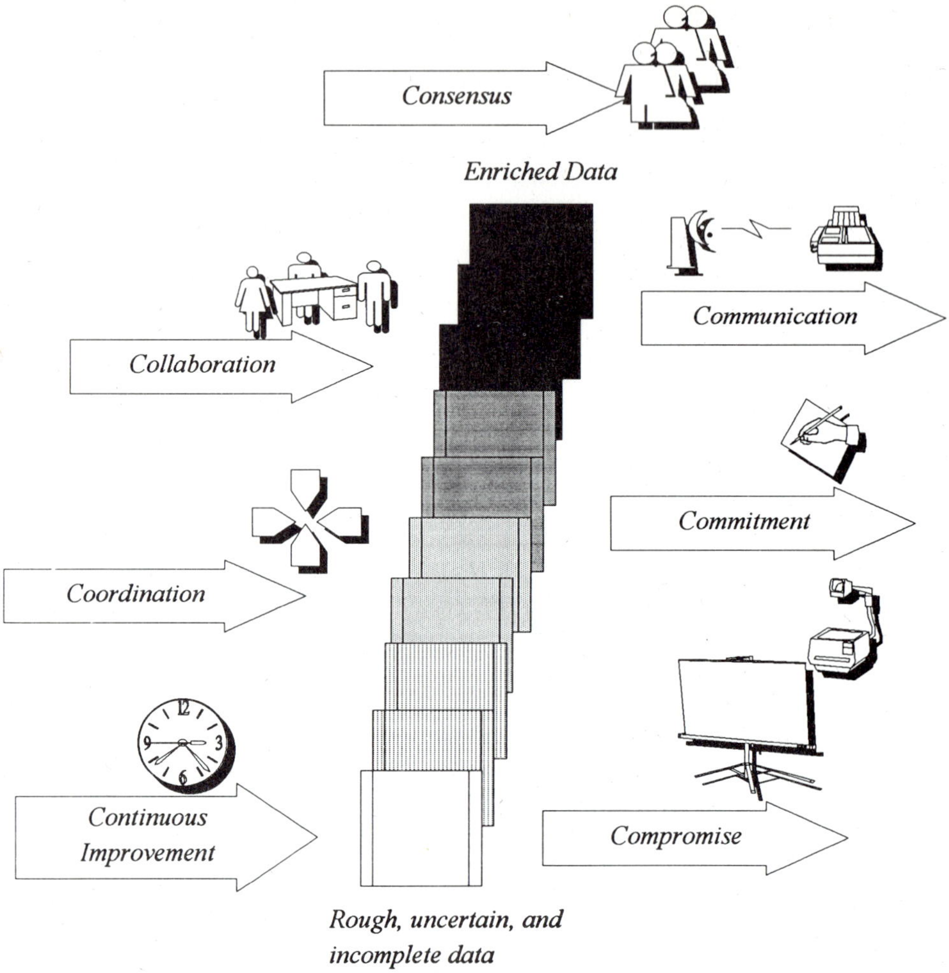

FIGURE 5.2 Optimization of 7 C's—Cooperation

The angle-parentheses < > denote interaction between teams of two, three, or more. Such sets of interactions can be shown schematically through a Venn diagram. Interaction may reinforce or nullify efforts of the interacting teams. If a magnitude of an interaction between the teams is positive, it is called cooperation. If it is negative, it could be due to controversies or unconstructive competitions between the parties involved. Unconstructive competition tends to pull the teams apart. If an individual team is self-motivated, its contribution will be positive. Thus, the net contribution of all the working teams will depend on the signs of the interactions—whether the effort is in cooperation or in controversy. Clearly when the interactions are all positive, or at least some are zero (nonparticipative), the contributions will all be additive, giving, perhaps, the maximum net result.

Negative contributions tend to cancel the positive contributions and the net results are smaller. In any organization, including CE, the efforts of all work-groups will be mixed—some additive and some subtractive.

How to entice positive interactions (i.e., cooperation) from teams is what the CE management needs to explore. Professor Göran Ekvall [1993] has made creativity and innovativeness in organizations his research focus for over thirty years. He has developed models which describe interplay among the organizational climate, the organizational and psychological processes, incoming resources, and the expected outputs. Depending upon the climate or the degree of cooperation in an organization, the results could be supportive, detrimental, or even destructive [Ekvall, 1993]. One of the elements that is essential to cooperation is the identification and nurturing of an effective CE organization. This means one must develop methods to work toward the "self-interests" of the teammates in such a way that the net outcome is in the best interest of both the teams and the company. Methods that promote self-esteem are when teams find joy, dignity, and are curious in whatever they do. One method could be to align organizational functions with "intrinsic or extrinsic motivations"—items of self-interests to promote teaming and discourage turf wars [Deming, 1993]. Another useful technique is to involve members in mapping the major steps (for instance, decomposing the tasks) of the CE project/work and in identifying the time, concurrency, and resources required for each step. Most people are born with a natural inclination to learn. Researchers have found that a team goes through four major stages as it organizes to get under way. Bruce Tuckman in his 1965 article identified these as forming, storming, norming, and performing [Tuckman, 1965]. Professor Kim Clark of the Harvard Business School has said that highly productive organizations are characterized by the existence of "early conflict." The concept development phase, for example, should raise the major issues of potential conflicts between organizational areas relating to each of the design alternatives studied. Each team member should not only express his or her concerns, but should have the opportunity to influence the design, when it is still early [Clark and Fujimoto, 1991]. Teams should be reasonably confident (on each other) that the supporting goals can be met with hard work.

It is not unnatural, therefore, for self-directed teams to enjoy jobs, new challenges, find joy in learning, problem-solving, and discover some of his or her innate abilities. One role the management can play is to facilitate the interactions between the CE work-groups. There are a variety of conceptual tools that can ease this planning process; they are discussed in Section 2.8 (see Figures 2.21 through 2.25). Examples include Project Evaluation and Review Technique (PERT), Gantt charts, etc. QFD is a helpful technique to sort out the various types of inputs. There are also motivational techniques (for example, group-dynamics) and extrinsic sources (for example, team rewards, bonus, etc.) and intrinsic sources of motivation that can assist in implementing change. Several books, including Dalziel and Schoonover's *Changing Ways* [1988], provide practical and useful suggestions for diminishing the natural resistance to change and implanting self-esteem. General Electric focused team efforts on team goals by establishing work-group (team) incentives. Edward Deming says that good management knows how to nurture and preserve these cooperative behaviors and build self-esteem attributes of cooperating teams [Deming, 1993].

5.1 COOPERATIVE CONCURRENT TEAMS

Cooperation is a structured process for an honest three-way cooperation between employees, the organization (such as management), and the customers. The three-way cooperation involves exchange of information up, down, and across the organization. This creates a deep, common understanding of personal, organizational, and customer goals. The process combines personal values with organizational values (profitability, cost, quality, return-on-investment, time-to-market, customer satisfaction, etc.) to build a balanced team-force commitment towards achieving mutual success. As once said by Dr. Deming, personal values are of two types: intrinsic (jobs, salary, bonus, incentives, reward, personal security, etc.) and extrinsic (self-esteem, dignity and pride in quality work, innovation, etc.). A team cooperation is shown in Figure 5.3 through a four-step structured process:

- The ***first step*** is the gathering of data, materials, and behavior information. During this step a three-way cooperation takes place between employee, management (or organization) and the customers.
- In the ***second step,*** three-way cooperation leads to the understanding of CE goals.

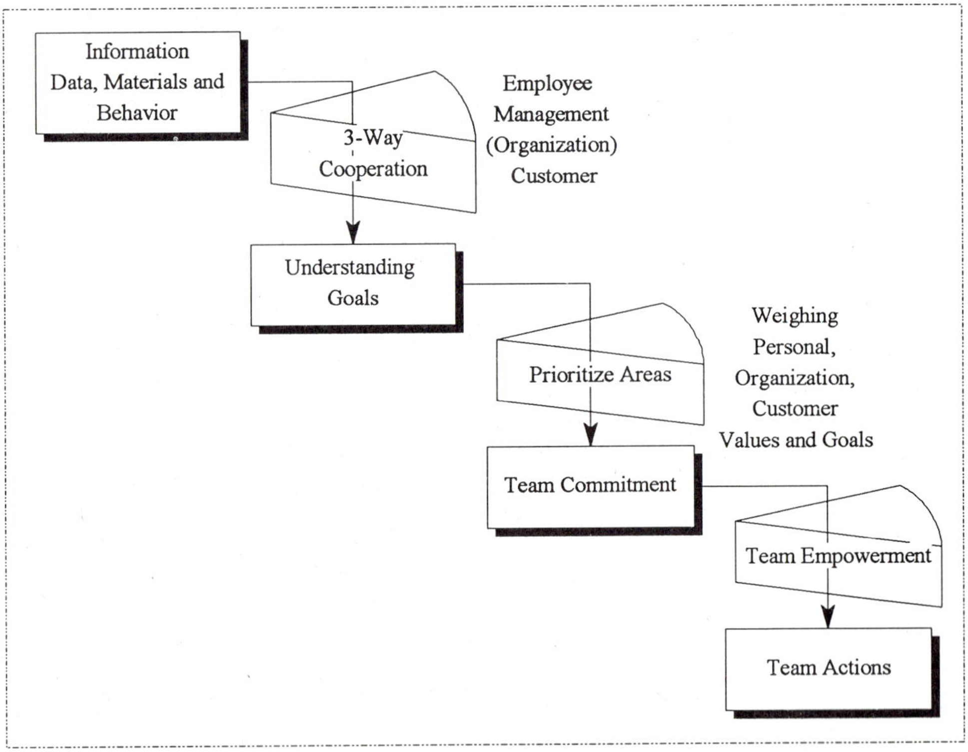

FIGURE 5.3 Structured Process for Team Cooperation in an Effective Organization

- In the ***third step,*** weighting factors—personal, organization, and customer values and CE goals—are added to obtain priority for team commitment.
- The ***last step*** is the team's actions. In this case, individual teams are empowered through supportive action, personal responsibility, and freedom to make decisions and act on them.

An integrated product development can be viewed as a cooperative work-group environment spanned by four multi-dimensional concurrent teams (see Figure 5.4). The work-group concept has grown out of the recognition that CE is a multiperspective phenomenon involving complex interplay among 7Ts (talents, tasks, team, technique, technology, time, and tools). More than one work-group is necessary for product realization, and within each group, a number of concurrent teams are required to support the various perspectives. For example, a personnel team only provides the required talents—examples being customers, product designers, process planners, processors, assemblers, leadership and management staff (managers), and others. They were shown earlier in Figure 5.1 as a part of a cooperation team to achieve unified product concept. *P*ersonnel team must be supported by other processes (mostly unmanned). They are named in Figure 5.4 as *L*ogical teams, *V*irtual teams, and *T*echnological teams. Figure 5.5 shows these four cooperating

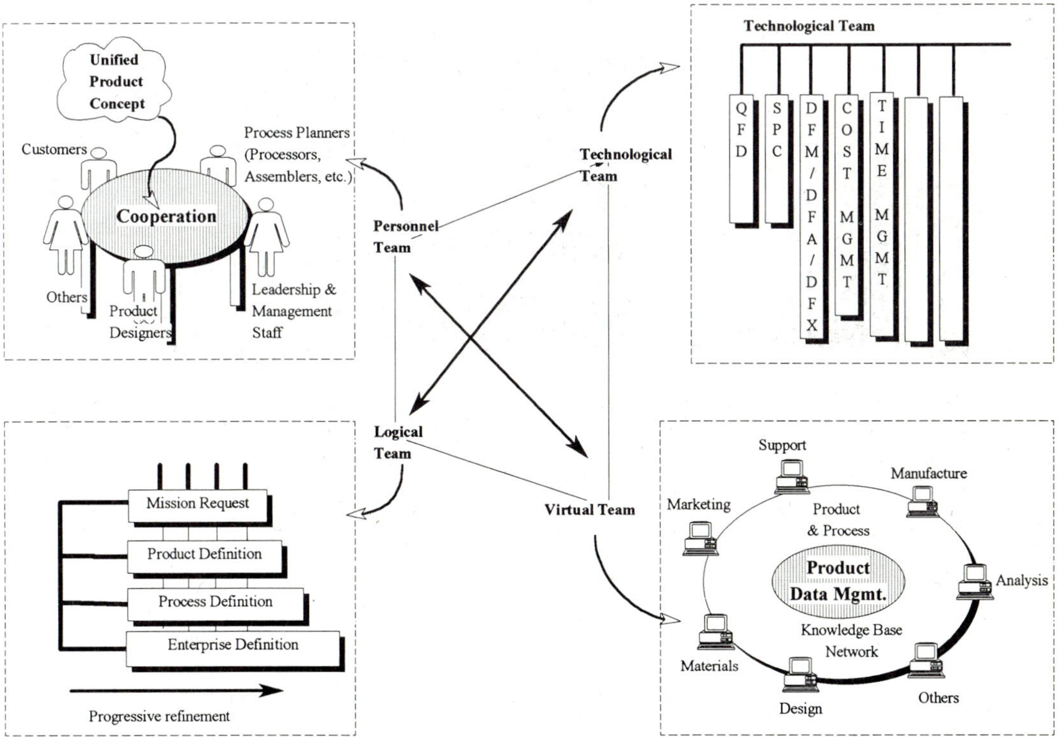

FIGURE 5.4 Concurrent Teams in Product Development

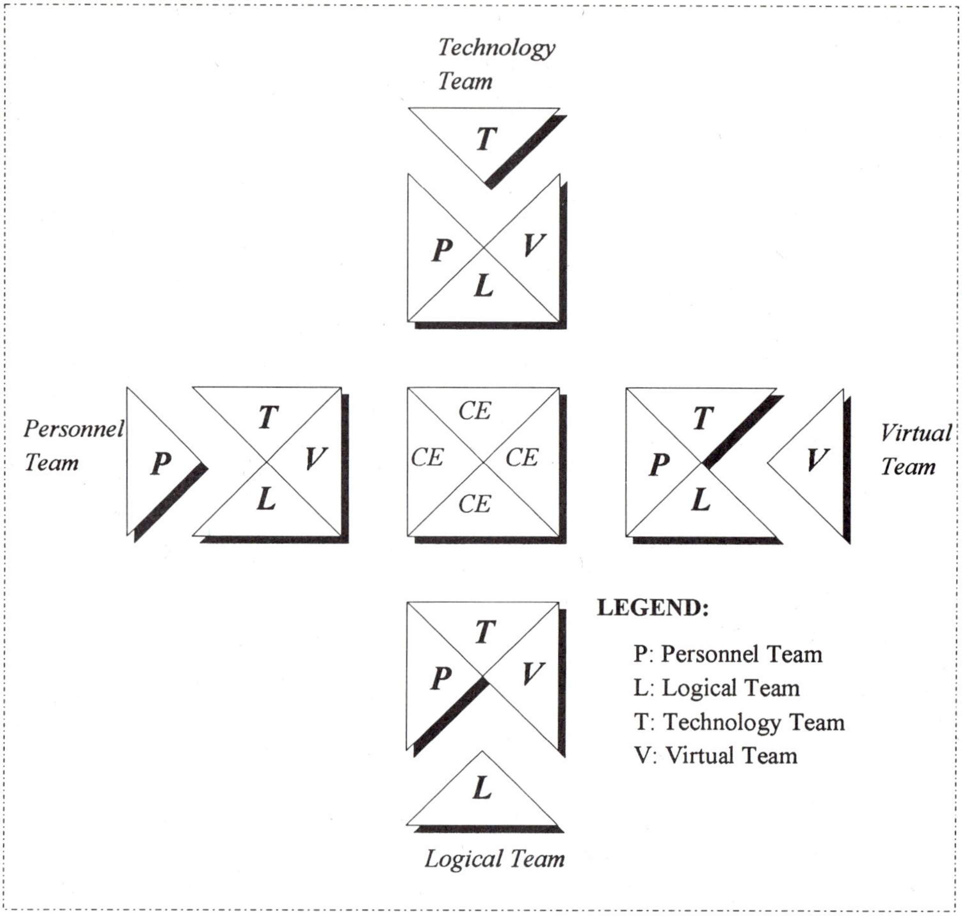

FIGURE 5.5 Cooperative CE Teams

teams as a set of four triangles, which when put together form a single square. Conversion from four triangles (with twelve sides and nine nodes) into a square (with four sides and four nodes) is indicative of fusion, which takes place when four contributing teams coalesce into one CE team. The letters: L, P, T, V on the triangles represent the first letter of the four concurrent teams. The dependency is mutual for example, other processes must support the personnel team as well. Virtual team employs right tools and computer techniques to speed up each of the tasks and connects them together (makes them available). Technology team defines the rights tasks to do in each category. Logical team defines the tasks and the order (task's timing) in which these tasks should be executed. Personnel team executes the tasks effectively. Personnel team may call upon the technology team to provide them with right techniques to use. Personnel team does this concurrently by calling upon the logical team to arrange the tasks in parallel. Personnel team executes the tasks efficiently. It does this by calling upon the virtual team to speed up the process (information access and execution). A new technology would not be successful without considering its implications on human (such as work force, expertise, culture), organizational, and integration (computer-based) aspects. In developing a product concurrently, the aim, therefore, is to infuse cooperation among the four contributing CE teams:

- ***Logical Teams:*** Logical teams break up the process into logical units (such as operations, activities, or tasks) so that they can be dealt with adequately. Logical teams not only partition requirements between subunits and define their interfaces, but must also follow up to ensure that the interface boundaries are not crossed (meaning, tasks are not coupled). This team checks whether or not customer expectations are properly analyzed and factored in the product to be designed, and that functional specification is partitioned out to the right work-groups to work on concurrently. Figure 5.4 shows a simple example of breaking up a product development process into four logical process units: mission requests, product definitions, process definitions, and enterprise definitions. The logical teams must ensure that the interface change-requests in the break-up process are properly evaluated and mediated between parties (internal or external). This change request assignment is customarily given to system engineering teams with responsibility for such things as mission analysis, system definition, functional specifications, requirements partitioning, and hierarchical breakdown of the product tree structure. The product tree structure (PtBS) may be logically partitioned into functional subunits (such as subsystems, components, parts, materials, and features, etc.). Each constituent may further be broken down into subunits, a portion of which may be designed internally while the rest may be subcontracted.
- ***Personnel Team:*** The personnel team identifies the roles of contributing PDT members, experts, and advisers to aid the product development process and to provide expertise for the logical team process breakups. Included here are the means of organizing, hiring, training, motivating, measuring, and interfacing to ensure teamwork.
- ***Virtual Team:*** The virtual team aids the personnel team in understanding the interactions and tradeoffs among the different, and even conflicting, mission requirements such as cost, quality, and performance. Examples in Figure 5.4 show a set of seven computer icons as virtual teams linked to a computer network. Teams are called virtual since they are in the form of computer modules and can reside anywhere (each is interconnected with one another via a computer network).
- ***Technological Team:*** Technological teams generate strategies and concepts on design perspectives (for example, strategies relating to its manufacturability, ease of assembly, serviceability, reliability, etc.). The technology team is responsible for ensuring the integration and consistency of the total definition of the product. Early focus is often on analytically assuring the highest product quality at the lowest inherent cost. The technology team is responsible for performing (or overseeing) the high-level analyses (for example, Total Quality Management, QFD, Statistical Process Control (SPC), X-ability, time management (responsiveness), cost manage-

ment, infrastructure management, etc.) associated with the overall product. A partial but important mix of CE teams' potential capabilities are shown in Figure 5.4.

$$\text{CE Teams} \equiv \cup\ [\text{Logical Team, Personnel Team, Technological Team, Virtual Team}] \qquad (5.5)$$

The next section provides insight into the mechanisms, which enable effective team cooperation and participation.

5.1.1 Team Design

The key to the success of the CE process is team design—selecting, training, motivating, and forming team membership and structure [Huthwaite, 1991]. A work-group is as effective as the participating groups that are working together as a team. The effectiveness of a team contribution could be much greater than contributions from all the individual groups combined. Individually, each group's actions reflect a specific contribution to a technical or business practice, process, or technique that can improve customer satisfaction or the bottom line. However, because of strong interactions (relationships among the actions), the total contribution will be much larger. Let us also assume that "team *i*" is working with "team *j*" in pairs. The *n*th way interaction ($n > 2$) among the teams are not considered in the following calculations.

Define a_i as the individual *a*ctions (or contributions) of team *i* with no outside interactions.

Define α_{ij} as the normalized factor of improvement in team *i*'s output as a result of team *j* interaction,

If α_{ij} # 0, the team *j* does produce an appreciable effect on team *i*.

If $-1 < \alpha_{ij} < 0$, team *j* has a disruptive effect of team *i;* the team's contribution is reduced but the net could still be substantial.

If $\alpha_{ij} < -1$; it means team *j* causes team *i* to be counterproductive; or it may imply that team *j* is working against the current understanding.

α_{ii}—represents the effect of team *i* onto itself (self-learning or self-improvement).

If $\alpha_{ii} = 1$; it means team *i* is present and is fully active. Team *i* is contributing to its full extent.

$\alpha_{ii} = 0$; means that team *i* is either absent or is not contributing to its full potential.

The total effect of mutual interactions (two-way and in pairs) on the team outputs can be expressed as:

$$\sum_{i=1}^{n}\left[a_i + \sum_{j=1}^{n}\{\alpha_{ij}a_i\}\right] \qquad (5.6)$$

which can be simplified as

$$[\sum_{i=1}^{n}(a_i) + \sum_{i=1}^{n}\sum_{j=1}^{n}\{\alpha_{ij}\, a_i\}] \tag{5.7}$$

In order for the whole to be far greater than the sum of the individual parts, the following must be true:

$$[\sum_{i=1}^{n}(a_i) + \sum_{i=1}^{n}\sum_{j=1}^{n}\{\alpha_{ij}\, a_i\}] \gg \sum_{i=1}^{n}(a_i) \tag{5.8}$$

or

$$\sum_{i=1}^{n} a_i \sum_{j=1}^{n}\{\alpha_{ij}\} \gg 0 \tag{5.9}$$

Comparing this to the sum of parts shows that the "whole will be far greater than the sum of the individual parts" as long as most of the coefficients (interactive effects) are positive. This can be stated in a mathematical form as follows:

$$[\sum_{j=1}^{n}\{\alpha_{ij}\}] \gg 0 \tag{5.10}$$

The above is not a necessary condition but sufficient for the "whole to be far greater than the sum of its parts" condition to be true. Where α_{ii} is the improvement made by an *i*th team to itself (self-learning), and α_{ij} is the improvement in *i*th team due to an action initiated by the *j*th team.

An attempt is made to represent this graphically in Figure 5.6. The initial actions, a_i, are represented by the sides of a polygon. To show the effect of interactions due to itself, the sides of the polygon are shown to be extended in the same direction by an amount equal to a product of α_{ii} and a_i. Both α_{ii} and α_{ij} are called interactions. Interaction may reinforce efforts, or it may nullify efforts.

The left hand side of Equation (5.8) can be written as

$$\sum_{i=1}^{n}(a_i) + \sum_{\substack{i=1\\}}^{n}\sum_{\substack{j=1\\ i=j}}^{n}\{a_i\, \alpha_{ij}\} + \sum_{i=1}^{n}\sum_{\substack{j=1\\ i\neq j}}^{n}\{a_i\, \alpha_{ij}\} \tag{5.11}$$

The first part of Equation (5.11) is the individual contributions. The second part represents the contribution due to self-learning, and the third part is due to team interactions.

In Figure 5.6, the first two terms are plotted. All self-interactions are assumed self-

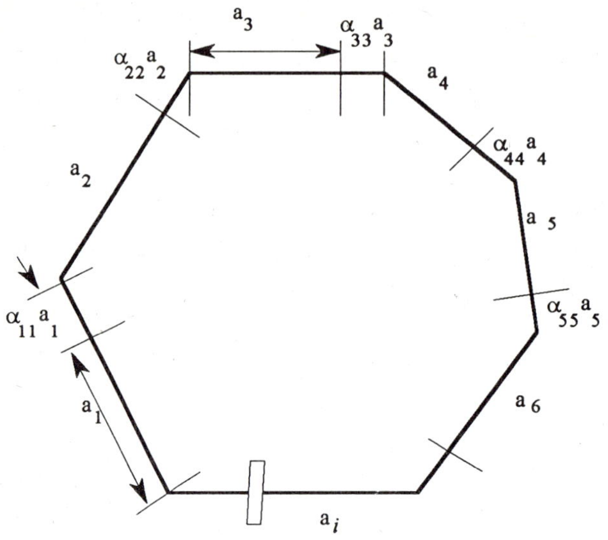

LEGEND:

a_i : Individual Action of Team i

α_{ij} : Nornalized Effect of Team i on Team j

$-1.0 < \alpha_{ij} < 1.0$

FIGURE 5.6 The Whole is Far Greater than the Sum of Its Parts

assertive, that is, positive. Negative interactions will, of course, reduce the length of the corresponding side by an equivalent amount. The net effect of the third term, even though it may be positive for cooperative team, is assumed negligible. For an action a_i, if an interaction (say α_{ij}) is positive, it is often due to cooperation between the *i*th and *j*th team. If an interaction (say α_{ij}) is negative, it is due to the teams having competitive, controversial or adversarial relationships. Examples of such controversial overtones were discussed in Section 5.0.

An effective team is like a peak-performing symphony orchestra: A diverse group of specialists can create an inspirational performance through mutual harnessing and cooperating process. Like an orchestra, a work-group is a collection of specialists brought together to achieve a common set of consistent goals. Teamwork cross-pollinates teams' ideas and gives members of a work-group a better understanding of their approach and methods of common problem solving on the whole project. In each work-group, there are four groups (L, P, T, V) of teams cooperating:

1. Logical Team (for example, distribution of functions)
2. Personnel Team (for example, people experts)

3. Technological Team (for example, QFD)
4. Virtual Team (for example, computer-supported functions)

This is schematically shown in Figure 5.7. The personnel team consists of experts in subject areas relevant to a work-group such as product designers, process planners, customers, suppliers, and leadership staff. In an organizational setting, they are surrounded by a multitude of support personnel who take care of other logistics, such as financial control, organizational management, education and training, marketing, work practices, and industrial relations. Management of the CE organization must commit to and support the necessary changes in responsibilities and execution, and empower the personnel teams with major decision making responsibility. The management must begin to balance the teams' autonomy and their self-interest with a reward system based on teamwork—so that as a whole it builds a more effective organization. It is hoped that each work-group (team of teams) in such an organization will enthusiastically accept the new responsibility and enjoy the changed environment as if it was their own creation.

Technological teams (advances in tools and technology) and virtual teams (ad-

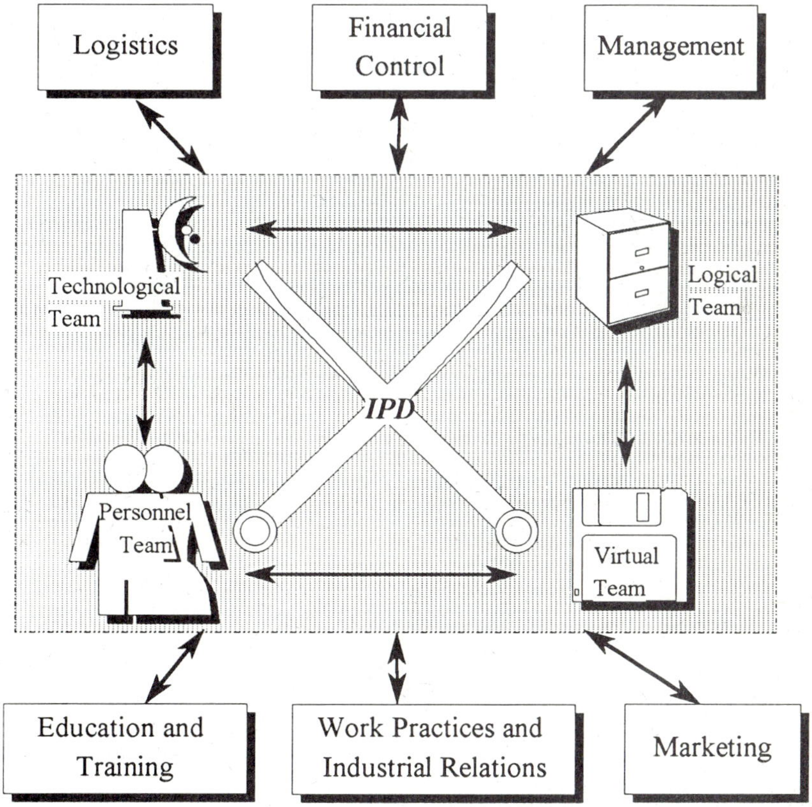

FIGURE 5.7 Team Approach to Integrated Product Development

vances in computer-aids and methods) mostly help during decision making for integrated product development (IPD). IPD is shown in Figure 5.7 as a fulcrum of a scissor type mechanism. Mechanical advantage of the scissor type mechanism is not complete without the "push" from the personnel team or the virtual team. This does not mean that any two (say, the technology and the logical team) will not do the job, but the two by themselves may not be able to meet the required challenge. Another analogy of the cooperative IPD is a stool. The four CE teams are like a four-legged stool (see Figure 5.8). Without any one of the legs, the stool would not be stable. Likewise, if any of the CE teams is missing, the CE organization would not be effective. As shown in Figure 5.8, the product development team (PDT) manager is at the center of the stool. If one leg lacks in capabilities (not able to share the expected team load), the stool can still stand but it would be unstable (will not be as effective). So if there is any instability in the PDT environment, it is due to the controversy from one of the teams. The logical and personnel teams provide the environment (process and methods), while the technological and virtual teams provide the automation tools and technology. Once the right environment is in place, automation (technological or virtual teams) can help speed the practice, but it cannot compensate for the wrong environment. Therefore, the personnel team must work together closely and with the help of the logical team to produce the highest quality product definition possible. The logical team receives the highest attention of all topics addressed in CE implementation. As shown in Figure 5.8, typical logical team topics include decomposition, systematization, interface development, unification, common systems, and standards. A typical personnel team configuration includes teams of engineers, analysts, and planners from the various disciplines and departments—marketing, engineering, design, manufacturing, etc. Examples of the virtual team include a variety of networks (inference, service, etc.), agents (receiving, requesting, affecting, etc.), concepts (agent-based) and tools (computer-based).

5.1.2 Team Composition

Not only are the teams important, but their composition (make-up) must also be considered. The composition of each team depends on the talents (expertise) required to support the total production of that portion of the product for which the team assumed responsibility. Education and training is a part of talent development. The personnel selected must not only possess the knowledge of the discipline they represent, but must also be willing to work effectively, exercising good judgment, in a team atmosphere [Margolias and O'Connell, 1990]. The role the participants play is more important than their sheer numbers. The numbers vary depending upon the nature of the product, the industry, and a slew of other factors, such as level of experience, capability, emphasis, etc. The level and depth of knowledge differ depending upon the CE work-groups they represent. Each team member, however, must have basic product knowledge so that he or she is able to analyze key issues involved in product development. He or she must have the basic organizational experience needed to recognize opportunities to improve the process, as well as a basic knowledge of popular analytical or decision support tools. Shrinking cycle time for product realization partly means shortening the ramp-up time for staffing the project. This

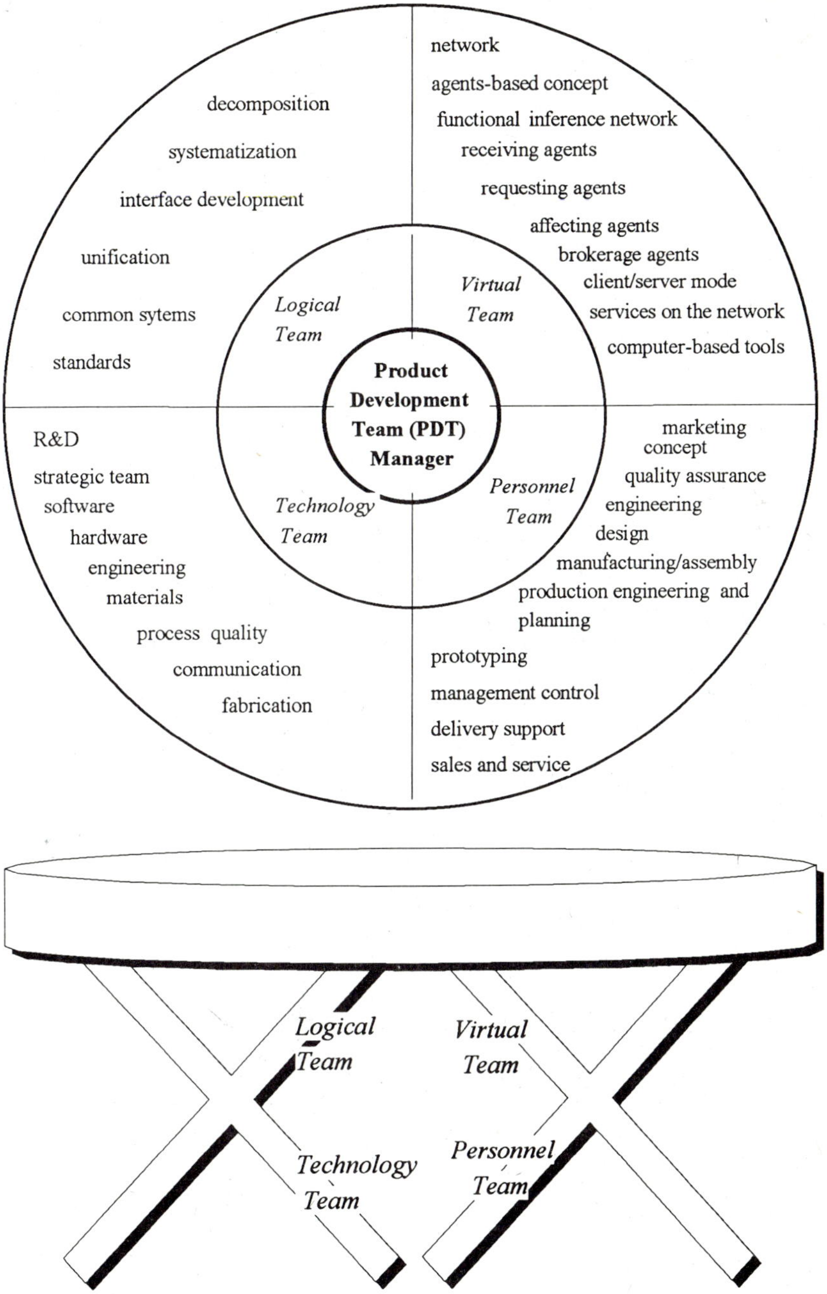

FIGURE 5.8 Typical Composition of Each Team

changes the basic structure (composition) of teams in terms of who is selected, educated, and trained. A four-dimensional layout of a CE team composition is shown in Figure 5.9. The four dimensions are shown corresponding to four independent coordinate axes. The virtual team is shown along the vertical axis and the technological team is along the horizontal axis. The personnel team and logical team are shown along the longitudinal axis. The four-dimensional composition space provides a unique framework for the following functions:

- Defining an array of computer tools (virtual teams) for the information modeling domain
- Relating their functions and features from the value characteristics (technology) perspectives
- Identifying definitions during various stages of product development (logical)
- Requiring types of the personnel teams to meet the product realization goals.

The matrix spanned by first two axes describes the relationship between the computer tools and their technological content (functions and features).

5.1.2.1 Logical Team Composition

The members of the logical team may have worked both at design and manufacturing, and are capable of developing an overall production strategy, including tooling policy, structure partitioning, and purchasing. Major assignments include mission statements, product definitions, process definitions, enterprise definitions, etc. (see Figure 5.9). A member of this team is capable of assisting "make versus buy" decisions. He or she can suggest "off-the-shelf" components (such as a computer store software, raw materials, and catalog-ordered parts), and know if they can be procured to reduce the time and cost of developing them internally. This team has some exposure to product support, marketing, or commercial manufacturing, and can develop an overall supportability strategy, including maintenance, repair, and customer training in use of the product or the tools.

5.1.2.2 Technology Team Composition

The members of this team include persons with overall system level knowledge and are possibly experts in various design technology:

- Life-cycle capture
- DFM/DFA/DFX
- Constraint management
- QFD/SPC/TQM
- Stereolithography
- Taguchi robust design
- Product data management (PDM), etc. (See Figure 5.9)

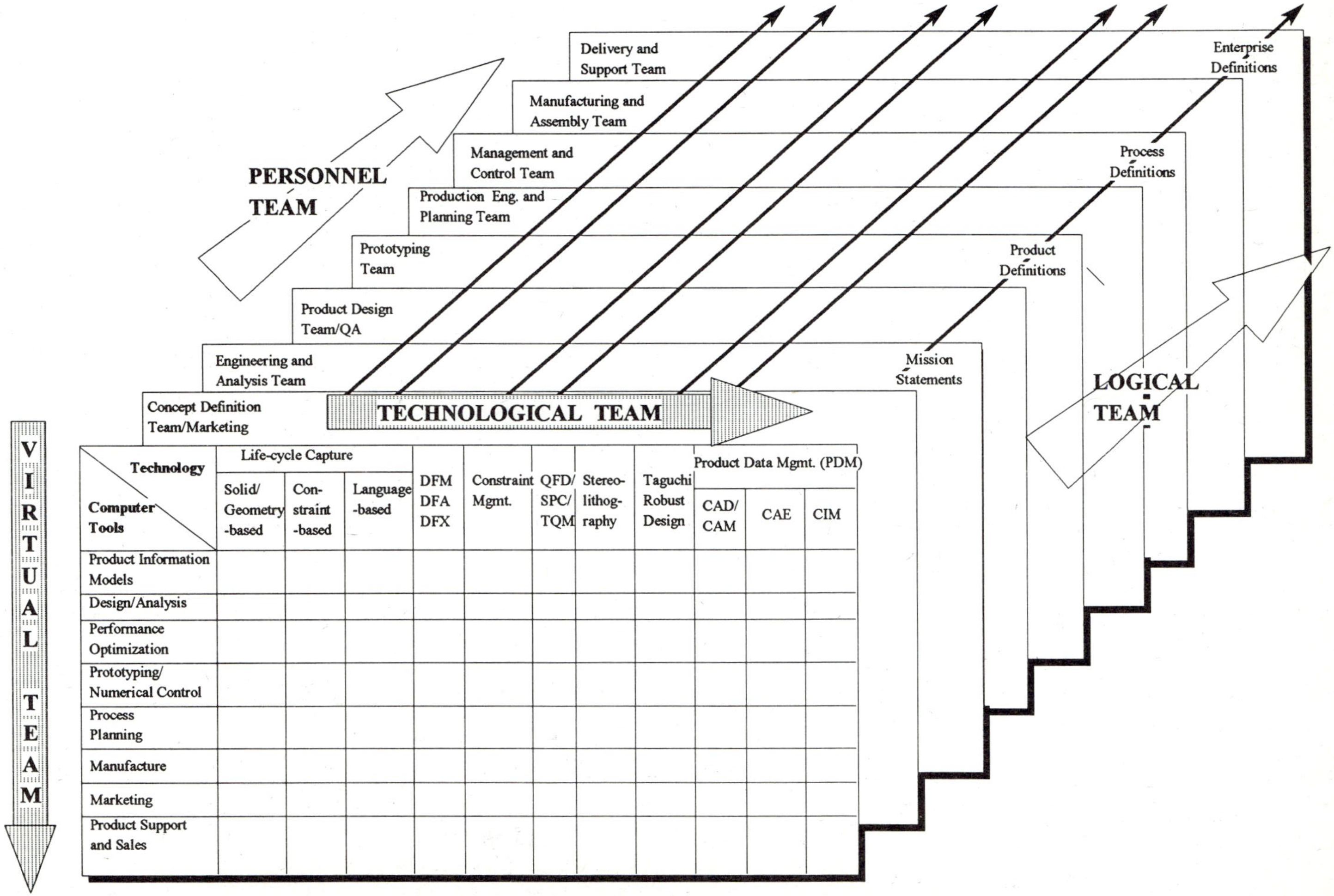

FIGURE 5.9 A Four-Dimensional Breakdown Structure of CE Teams.

This team, by its very nature, is composed of senior employees possessing the authority to formulate strategies for the disciplines they represent, and the overall knowledge to either develop a new strategy or know who to contact in case help is needed [Margolias and O'Connell, 1990].

5.1.2.3 Virtual Team Composition

A set of utilities composes a part of the virtual team which enhances collaboration, for example, a cooperating, multiple-agent inference network. The agents base contains the functional knowledge of the product and the process. Such functional knowledge may contain the following:

- Product information models
- Design/analysis
- Performance optimization
- Prototyping/numerical control
- Process planning
- Manufacture
- Marketing
- Product support/sales, etc. (as shown in figure 5.9)

An example of an intelligent network for virtual processing is described in Volume II. These agents interact on a demand basis requesting and providing services, with no predetermined path of information flow and with no dominant players. Each agent performs the requested functions and returns the result to the requesting agent. The receiving agent checks the results for any constraint violations and informs the affected agents of any such occurrences. These agents cooperate with each other in a client-server mode switching their roles as many times as necessary. This virtual team concept can include negotiation by "brokerage agents" that have knowledge of specific services on the networks. These brokerage agents compare the deficiencies of designs with specifications imposed by personnel teams and issue requests for services needed to complete a given task.

The need for free-flowing communication is not limited to personnel teams. It applies to virtual teams (computer-based tools) as well. In modern CE, virtual teams form the foundation upon which all important elements of the Concurrent Engineering Intelligent Information System can be built. This is discussed in detail in Chapter 8, Volume II.

5.1.2.4 Personnel Team Composition

The personnel team forms the human side of concurrent engineering. The participants' role is more important than their actual number. The best personnel team is composed of representatives whose input is crucial yet small enough in number. The team members must have the following characteristics:

- Product knowledge required to consider all of the key issues involved in product development
- Organizational experience needed to recognize opportunities to meet the change
- Authority to make decisions against departmental lines and make them stick.

The appropriate subgroups of the personnel team generally include (see Figure 5.9):

- *Concept definition subteam:* This subteam brings in new product ideas, new design alternatives, and features that customers may like to see in the product.
- *Marketing subteam:* This subteam provides insight into product design features desired by the prospective customers.
- *Engineering and analysis subteam:* This subteam evaluates systems, circuits, structures, design for X-ability evaluations, load determination, mass properties, and the like.
- *Product design subteam:* This subteam defines product concepts and develops design specifications and standards.
- *Quality assurance (QA)*: This subteam ensures design and production conformance to quality.
- *Prototyping subteam:* This subteam establishes a testing plan and strategy that supports the product life-cycle verification requirements from art-to-part, including part testing, system testing, structure verification testing, circuit-type verification testing, and supportive type design developmental testing.
- *Production engineering and planning subteam:* This subteam establishes procurement policies relevant to the development, including high level subcontracting strategies, materials management, etc.
- *Management and control subteam:* If any portion of the product is subcontracted out, this subteam should also include a subcontract administrator and key selected suppliers. Suppliers can assist manufacturing efforts by offering recommendations about materials, machining, tooling, and other fabrication issues.
- *Manufacturing and assembly subteam:* This subteam suggests design modifications that can streamline and expedite the piece part manufacturing processes, such as tooling, producibility, manufacturability, etc.
- *Delivery and support subteam:* This subteam offers input regarding product design issues that cause problems during installation, service, and field support. The subteam can also provide insight into possible product features and functionality that could be easier to maintain and support. One function of the support subteam is to produce technical publications. This includes creating different levels of product design specifications, users' manuals, service or maintenance documents, and to develop customer training materials.

The contributions of the various subteams towards the product's key X-ability aspects are given in Table 5.1.

TABLE 5.1 Contributions of Various Subteams on X-ability

Key Subteams	Key X-ability Aspects
Concept definition subteams	Marketability, design functionality, risk, features, feasibility and quality assessment, . . .
Marketing subteams	Usability, cost, features, functions competitiveness, customer satisfaction, . . .
Engineering and analysis subteams	Integratibility, compatibility, testability, FMEA, materials/features substitution, performance, . . .
Product design subteams	Reliability, maintainability and supportability, interchangeability, simplicity, . . .
Quality assurance subteams	Cost-to-quality, robustness, quality, variability, verification, . . .
Prototyping subteams	Feasibility and formability, producibility, simulation, rapid prototype, machinability, automation, . . .
Production engineering and planning subteams	Supportability, process planning, serviceability, flexibility, repairability, . . .
Manufacturing and assembly subteams	Manufacturability, assemblability, disassembly, . . .
Management and control subteams	Investment, tooling, maintainability, . . .
Purchasing subteams	Affordability, materials usability, . . .
Delivery (or distribution) and support subteams	Deliverability, recyclability, environmentability, . . .

5.2 PROGRAM ORGANIZATION

Figure 5.10 shows a program organization headed by a strategic business unit (SBU) vice-president, structured around the key resources required to be a self-supporting business organization. Six supporting directors such as operation engineering, business management, product development teams (PDTs), technology management, certification testing and mission management, and resource management, are shown reporting to the vice-president. Central to the SBU is the establishment of product development team (PDT) organization. There are five elements to this PDT organization structure: multifunctionality, matrix responsibility, core competencies, teamwork, and cooperation. The vertical axis represents units of core competency for a specific system/unit deliverable (shown as X, Y, and Z in Figure 5.10). The horizontal axis spans the matrix responsibility structure. The work groups within each system/unit must be multifunctional and be given clear responsibility. Each work-group must be empowered to make design decisions. To cite a real life example, Timex claimed to have reduced its watch development time from a maximum of two years to six months by organizing a program along such product lines. Vertical breakdowns of the remaining key organizational functions can be done similar to what is shown in Figure 5.10.

5.2.1 Product Development Team (PDT)

All PDT units are led by the equivalent of a unit/system manager and may report directly to a PDT director. A PDT unit is made up of both horizontal and vertical groups. Horizontal groups are those which typically make up standard product decomposition trees, such as

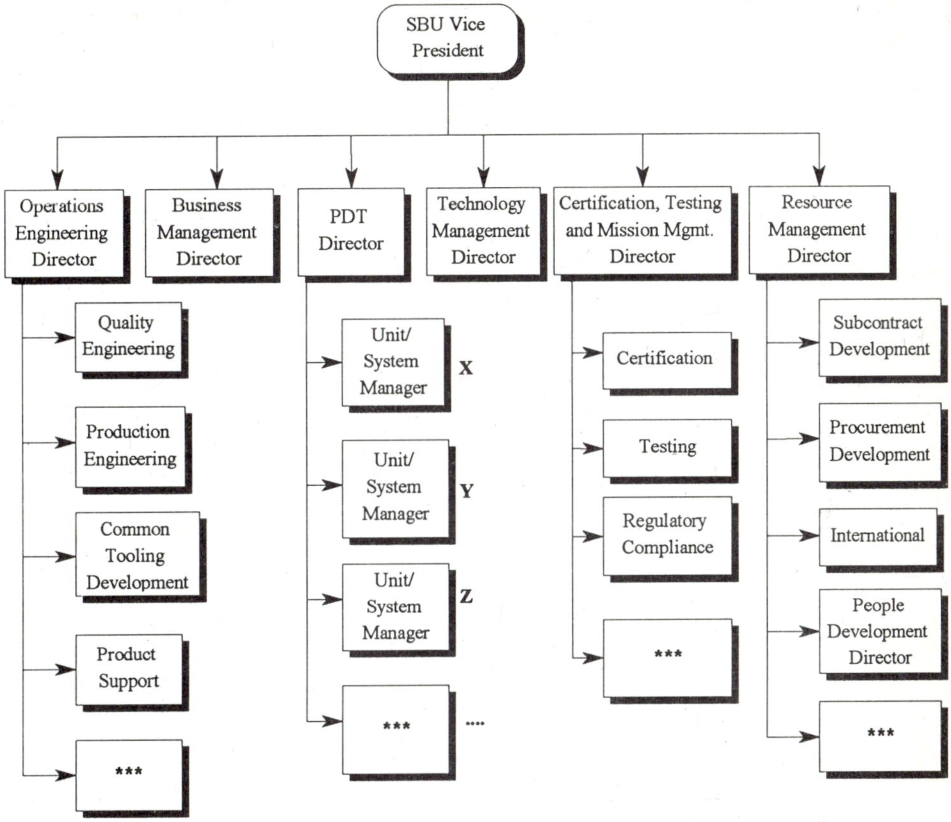

FIGURE 5.10 A Typical Organization Chart for a Strategic Business Unit (SBU)

system, subsystem, components, parts, materials, etc. Vertical work-groups are those groups that look at the problems from different perspectives or disciplines. They bring the needed level of experience and proficiency to the job. Work moves between groups based on skill needs, impact, priority, and disciplinary requirements. The groups may consist of one or more of the following: product, process, assembly, design review expert, cost reduction, product improvement, warranty reduction, SPC, reliability, design for X-ability, quality improvement, value engineering expertise. When an individual is a member of the PDT, he or she is much more likely to participate in defining the work objectives. Thus, the objectives are likely to be more useful, relevant and better understood by the participants. A sample of a typical project development team reporting structure is shown in Figure 5.11. The responsible managers from the horizontal teams (lead product manager, life-cycle manager, and work-group manager) work together under the leadership of a PDT unit/system manager. They act as information conduits between the PDT unit/system manager and work-group leaders. These managers make sure that communication among work-group members happens smoothly and continuously. The PDT unit/system manager is designated as the product development process owner. The schedule of various activities in a PDT team

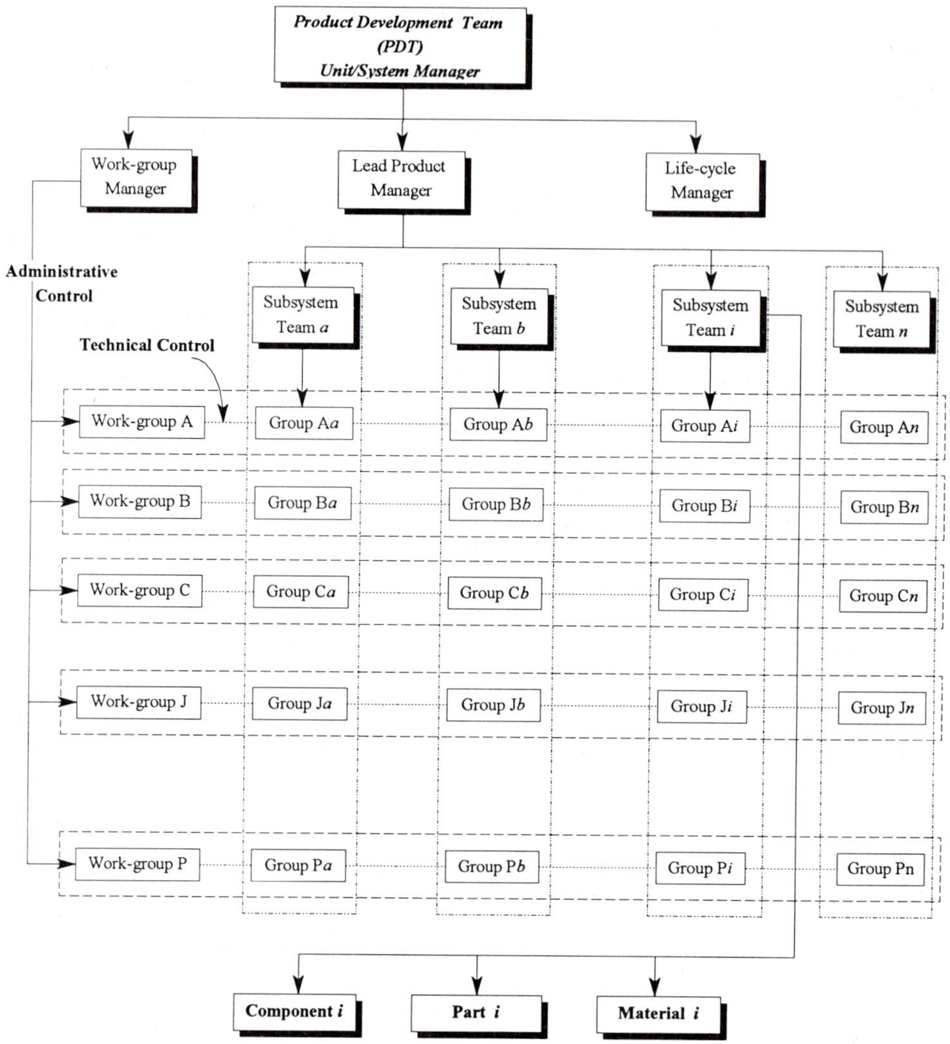

FIGURE 5.11 A Typical Product Development Team (PDT) Matrix Structure

is based on priorities and the lead product manager is responsible for the product development schedule. The PDT unit/system manager ensures that the number of iterations among its lead product units (such as subsystems, components, etc.), work-groups (say A, B, C, etc.), and life-cycle units are minimized.

5.2.1.1 Lead Product Team

The lead product manager performs the role of a system manager for the various PtBS (product breakdown structure) tree nodes he or she represents. Each of the product or subassembly nodes to be designed are treated as a decomposed object. The lead product manager is responsible for overseeing the integration of the decomposed objects back into the PtBS tree structure. Of course, if the product unit is sufficiently large and complex, it is likely to be decomposed into finer hierarchical subunits (subsystems, components, parts, etc.) and work-groups (e.g., A, B, C, D. . . . J), and the groups can work on these subunits and work-groups in parallel (see Figure 5.11). All groups involved in the product design process are responsible for inter-group cooperation and the design of any new methods to ease communication. The life-cycle manager will integrate and coordinate the results of the various disciplinary groups.

5.2.1.2 Life-cycle Team

The life-cycle manager is usually a senior person possessing the authority to formulate strategy for the process (life-cycle) breakdown structure (PsBS) tree nodes he or she represents. This manager possesses the overall knowledge either to develop such strategy or know whom to contact if help is needed. Most teams under the life-cycle manager (see Figure 5.12) include feasibility study analysts, preliminary designers, prototype engineers, production engineers and planners, manufacturing engineers (including quality control engineers), product support, and field (or logistics) support engineers. The numbers and type of personnel will depend on the size and complexity of the unit or system being developed. Figure 5.12 shows a typical distribution of seven life-cycle tracks—from feasibility study, preliminary design, prototype, production, manufacturing, product support to field supports. In addition, work-groups are made out of groups that comprise the life-cycle track activities and the product's decomposed unit responsibilities. Work-groups are created (see Figures 5.11 and 5.12) to provide a cross-fertilization of ideas from the product structure tree—PtBS—to the life-cycle decomposition tree—PsBS. PtBS members are specific to a particular product line, such as body assembly, powertrain assembly, chassis assembly, interior, etc., whereas PsBS members are applicable across all product lines. This *"matrix within a matrix"* provides the necessary checks and balances within the program.

5.2.1.3 Work-group Team

An important step in setting up a successful PDT organization is determining work-group membership. Work-groups are organized as a pool of experts from product decomposition—PtBS hierarchy, and the life-cycle decomposition—process (PsBS) hierarchy. The participating teams may be the same or different. The members of a work-group thus carry expertise in at least one of the product units or life-cycle areas. It is very possible, therefore, that a member from a work-group (say A, B, C, . . . , J, . . . , P) may belong to a product team (say team a, team b, team c, . . . team *i*, . . . , team *n*) and/or a life-cycle team (say 1, 2, 3, . . . , *k*, . . . , *m*). The members are thus distributed across two organizations and may carry multiple responsibilities. The aggregate number of workers in each work-group, when the groups are added at that level, would be the same.

That is,

$$\text{Work-group A} \cup [\text{Group A}_i] \text{ for } \forall \text{ A's, where } i \text{ ranges from } i = \text{a, b, c, \ldots through } n \tag{5.12}$$

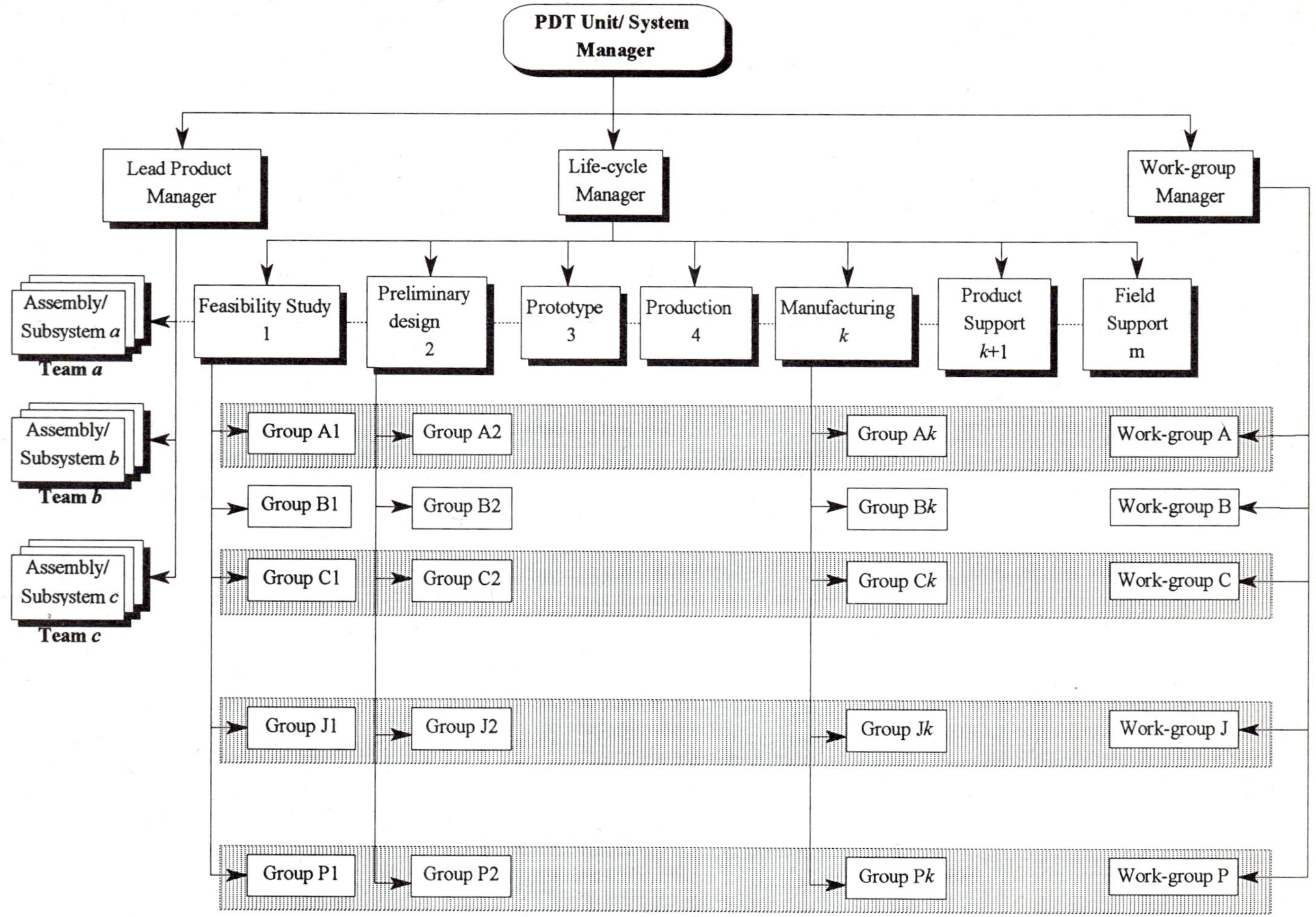

FIGURE 5.12 Organization Details Associated with PDT's Life-cycle Management

Where Group A_i belongs to lead product teams (see Figure 5.11).

also
$$\text{Work-group A} \cup [\text{Group } A_k] \text{ for } \forall \text{ A's, where } k \text{ ranges from } k = 1, 2, 3, \ldots \text{ through } m \tag{5.13}$$

Where Group A_k belongs to life-cycle teams (see Figure 5.12).

Similarly, for any work-group, J; the above can be rewritten as,

$$\text{Work-group J} \cup [\text{Group } J_i] \text{ for } \forall \text{ J's, where } i \text{ ranges from } i = a, b, c, \ldots \text{ through } n \tag{5.14}$$

also
$$\text{Work-group J} \cup [\text{Group } J_k] \text{ for } \forall \text{ J's, where } k \text{ ranges from } k = 1, 2, 3, \ldots \text{ through } m \tag{5.15}$$

Where J takes a value from A to P.

Toyota limits the participation of engineers into as few groups as possible to keep a clear team focus. Each functional work-group is responsible for monitoring only the support of that particular function to the various groups. Each work-group team may appoint an area team manager.

5.2.2 CE Staffing

The levels of staffing needs are different for concurrent engineering than for a conventional organization based on serial engineering. In a multi-department traditional organization, when a product is initiated, staffing needs are usually low. New members of the groups from design, engineering, manufacturing, and central support departments are added, depending on the phase of product completion. When the design is finished, members of engineering groups are added. When engineering is done, members of manufacturing groups become the dominant players. Such conventional staffing practices force development efforts into phases that tend to promote sequencing of work loads. The staffing level is dominated by specific functions—a conducive feature of serial engineering (SE). The major attraction of this (SE) practice is that staffing needs are typically small at the beginning. In reality, staffing grows manifold in SE at the later stages as shown in Figure 5.13(a). Since CE is built on overlapping, mutual cooperation and sharing of responsibility, the staffing needs for CE are very different from SE. This is shown in Figure 5.13(b). Apart from an initial ramp-up period, the level of staffing stays fairly constant. Though the total time (for example, days) spent in each case is not expected to be very different from each other, the lead time requirement in CE is expected to be smaller and the quality higher. In CE, most of the teams are forced to be involved from the beginning and stay involved until production is done. The advantage of involvement from the start is that teams learn from each other; they are forced to find opportunities in each other's area, and may develop a sense of cooperation. For example, if manufacturing is involved from the start, they will have to work with other teams such as design, engineering, marketing, etc. or else, they will have to find opportunities in design and engineering areas to overlap with so that they can begin work on areas of their own expertise—manufacturing. Some would

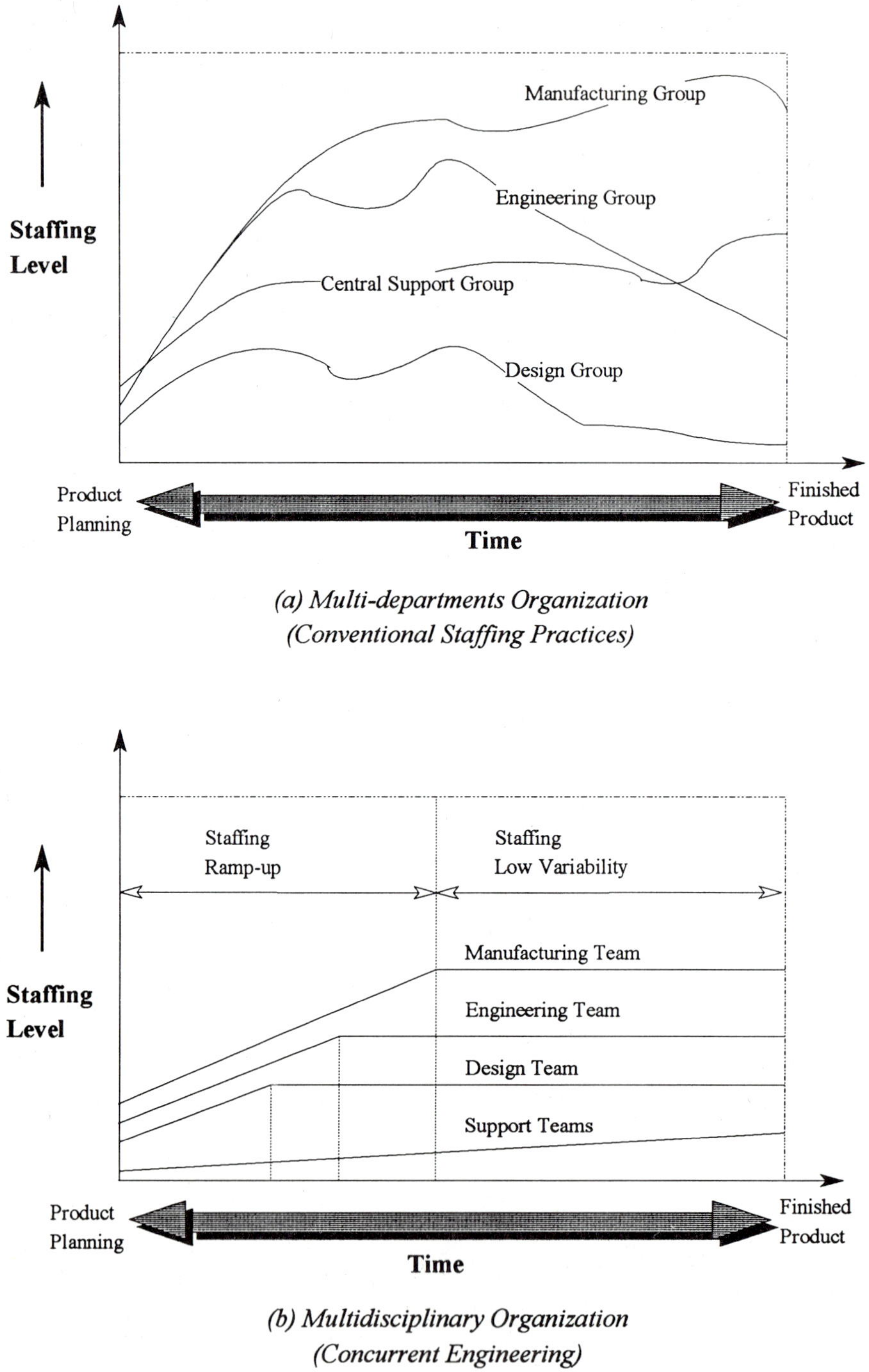

FIGURE 5.13 Change in Staffing Levels with Time: (a) Conventional Staffing Practices (b) Concurrent Engineering

identify opportunities to collaborate on even if the tasks have to be completed with only partial information. The search for ways to get started earlier encourages new ideas, independent thinking, and tends to keep the focus on company and customer interests. For instance, some teams may be too close to a problem to see any different ways of doing things whereas people from other disciplines may find it easier.

5.2.3 Routing and Queuing Models

The routing and queuing models provide a mechanism to match the loads evenly with available resources. Since the activities are interdependent, some activities may start earlier than others. As the information envelope containing the design is moved around, it is possible that some groups may end up having the envelope "passed on" to them a number of times than they have time or resources to complete. If such a resource is critical, it may end up consuming more real-time than is available. A dynamic queue management can sense the possibility of such occurrences and take appropriate actions to ease overcrowding.

5.2.3.1 Queue Management

The need for queue management arises when one of three events (see Figure 5.14) occurs:

- ***Peak:*** The level of workload can temporarily peak while the available teampower is fairly level; see Figure 5.14a. This occurs generally in the following circumstances: unexpected routing, refocus on the technical approach, needs for more analysis (simulation, and testing proof of the concept), etc. Queue management is needed to redistribute the workload so as to minimize its impact on time-to-market.
- ***Valley:*** This occurs when the level of the work-in-progress queue is actually in accordance with a previously committed workload (fairly constant—see Figure 5.14b), but the individual or group is unavailable at the moment. Work shifting is needed since the individual or group may be on vacation, ill, in training, etc. A valley may also occur (see Figure 5.14b) when the support individual or work-group cannot accept queued work due to impending loss of personnel, unanticipated increase in work-load, or possible new business opportunities.
- ***Oscillation:*** In a CE environment, a "balanced" workforce or workload, where everyone in the team and across the teams is equally busy, does not happen. The workload oscillates while the available team power is constant—see Figure 5.14c. This is often due to the variation in teams' experience or expertise, uncertainty of finding a reasonable solution (especially in invention or new product introduction cases), and due to inaccuracy in time estimates. A balancing of work across multiple work-groups and among teammates within a work-group is required. A priority setting factor is applied in queue management from time to time to achieve a balanced schedule. A routing and queuing model is a mechanism used to balance the workload (an example is shown in Figure 5.15). This is further explained in the following section.

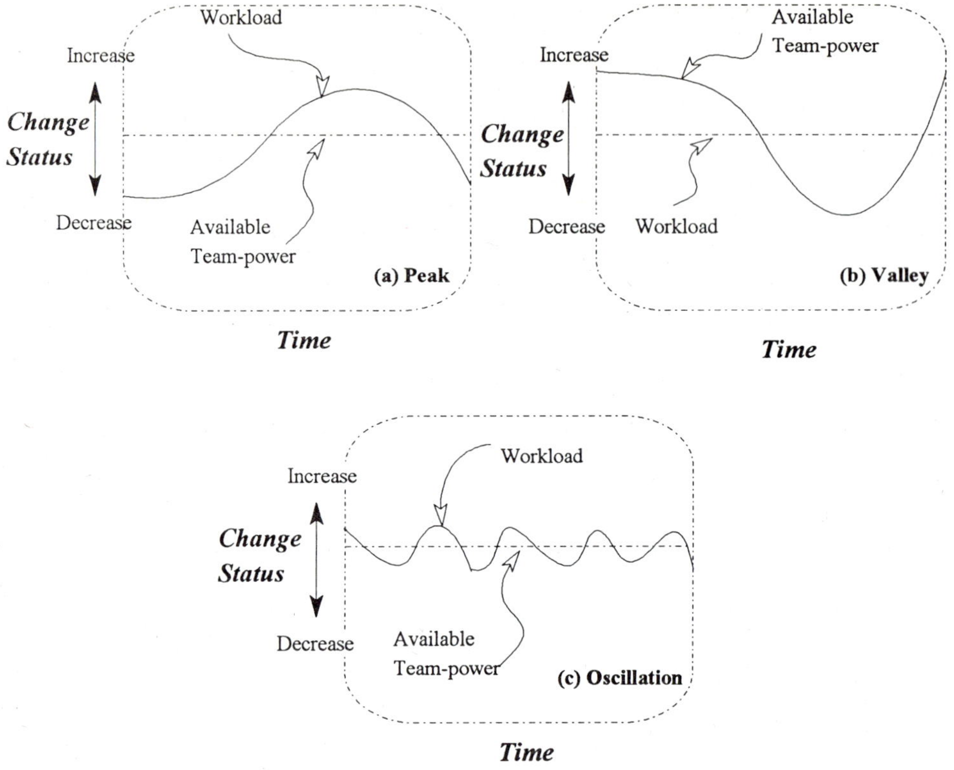

FIGURE 5.14 Possible Variations in Workload and Available Team Power

5.2.3.2 Responsibility Management

The members of the work-groups report to the (work-group) area team manager/leader who, in turn, reports to the work-group team manager (see Figure 5.15). The area team manager/leader shares a defined product management responsibility for a set of partitioned work-groups A to I. He or she is not relieved of his or her responsibility until his or her portion of the product is successfully built, verified, and integrated into the overall product. The area team manager/leader for a work-group assumes a mentor role and is responsible for maintaining a work-group standard of expertise or skill:

- The work-group team manager must involve his teams' area managers or leaders in a number of business management processes:
  - Multiproject planning
  - Defining work breakdown structure (WBS)
  - Task scheduling
  - Cost estimating
  - Project management responsibilities
- The work-group team manager is responsible for formulating the overall strategy for the interaction between various teams or groups of teams.

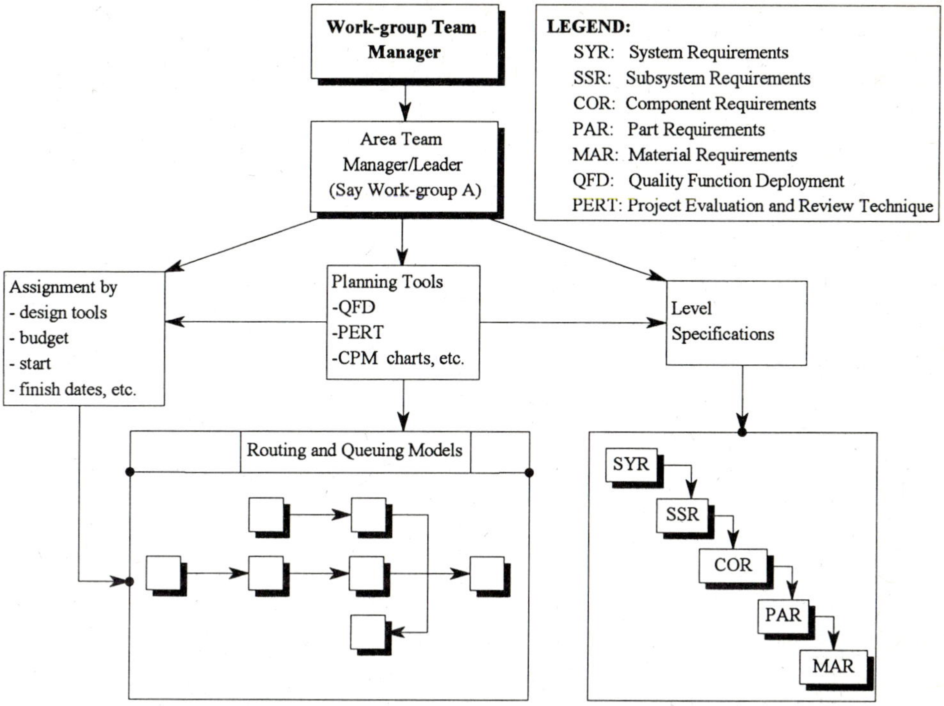

FIGURE 5.15 Routing and Queuing Models

- The work-group team manager sets criteria for performance measurement, progress reporting, plan of corrective actions, and resource reallocations as needed.

The area team manager/leader must not only mediate and lead the team-of-teams (personnel, logical, technological and virtual), but must also take responsibility to make the team successful. The leader has to be both technically superior and personable (a get-along individual). Usually it is difficult to find someone with both of these qualities since one is generally built at the expense of the other. The team manager/leader must have a broad skill-base to lead the whole project, not just a part of it.

5.2.4 Team Manager/Leader Characteristics

A team manager/leader must be capable of switching roles if needed and have the following four key skills.

5.2.4.1 People Leadership Skills

The team manager must have strong leadership skills. The team leader should be able to sort out people's problems and correct them on a timely basis. The leader must be able to capture the worker's imagination and inspire others to achieve what they are individually

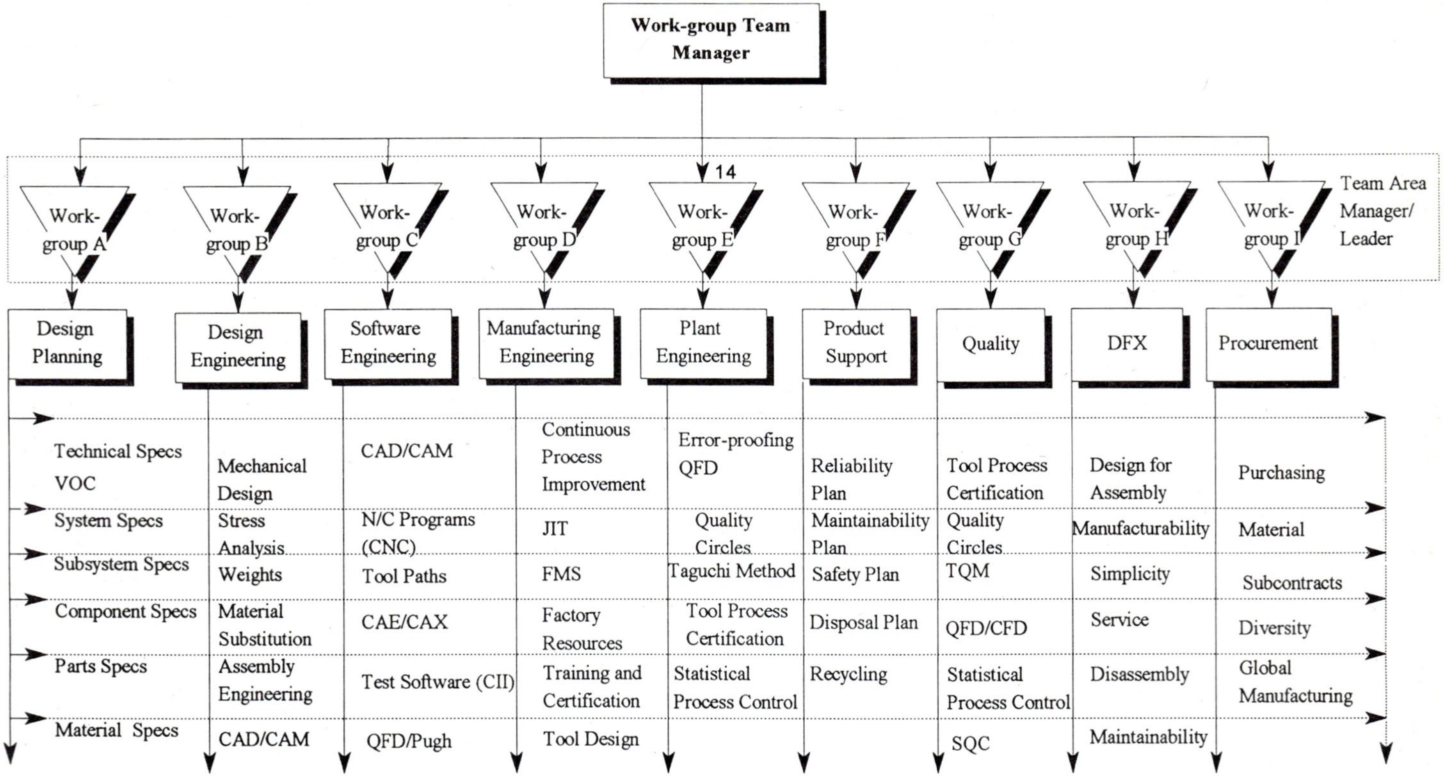

FIGURE 5.16 Skill Set Required to Manage Work-group Responsibilities

capable of doing. The team manager or leader should be able to identify the expertise required for a job, the specific personnel to be added to the team, negotiate such availability with the parent organizations, and decide when someone must be replaced.

5.2.4.2 Vision Skills

The team manager or a leader must possess a clear vision of how the product will be designed and developed, its characteristics, key success factors, and what it offers compared to the competition. Other vision skills include competitive assessment or analysis, and keeping the team focused on both its business and quality objectives. Leaders with this quality are able to see the "big picture," recognizing the extent and limitations of a specified knowledge in the context of an overall problem.

5.2.4.3 Problem Solving Skills

The most obvious role that a team manager or a leader plays is in problem solving. The leader must possess many of the lines of expertise needed by the team and fill in when employees are ill or on vacation.

5.2.4.4 Project Management Skills

The team member or a leader must follow a constancy of purpose type project management style (managing the project in the context of its overall purpose and not just based on its short-term gains). As W. Edward Deming said in his book *The New Economics,* setting a particular numerical goal accomplishes nothing—only the method to achieve a common set of consistent goals is important [Deming, 1993]. Clear and consistent goals provide "constancy of purpose." Without a common set of consistent goals, there is no system. Each team and its members must contribute to the success of the company mission or its purpose by participating in the methods—setting layout schedules with all of its tasks and due dates, including resource requirements. The team leader is also responsible for administering the budget and maintaining the schedule.

5.2.5 Skill Management Matrix

An orderly product realization process requires an orderly work-force. People in life-cycle and work-group management (say work-group A, B, C, D, . . . I, P) should be organized in a project by a "skill management matrix." Figure 5.16 illustrates examples of skill sets for major work-groups to manage the life-cycle aspects of PDT. The composition of each of these work-groups should be consistent with the disciplinary expertise required to satisfy their assigned responsibilities. Many organizations follow a matrix reporting structure so that the team members can work with as many groups as possible so that their expertise is fully deployed. This also provides for the carry-over of knowledge and requirements from one portion of the product or process to another.

CE team formation and work-group selection thus requires identifying many interdisciplinary skill sets in team members as opposed to multidisciplinary skill sets. Group selections in CE projects may cross functional boundaries. Figure 5.17 illustrates the difference between the multidisciplinary, interdisciplinary, and cross-disciplinary skill sets. Most product development needs are multidisciplinary—meaning it requires a number of teams or team members who specialize in different aspects of product development. Interdisciplinary means the team possesses some portion of a skill set (partial knowledge) belonging to two or more multidisciplinary areas that are needed for the product development. Shorter development cycle time in CE forces PDT members to learn many new

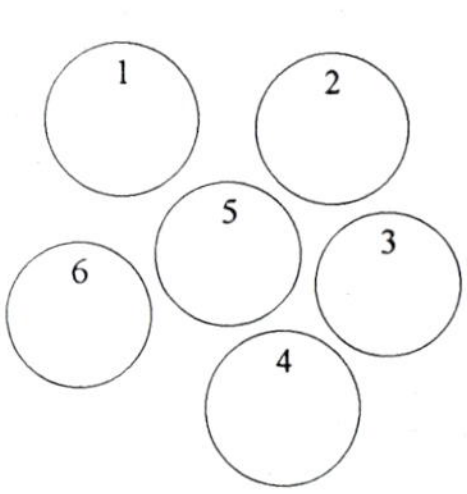

Expert/Talent	Nos.
Multidisciplinary	6
Cross-disciplinary	0
Interdisciplinary	0
Total Team	6

(a) A team consisting of multidisciplinary experts.

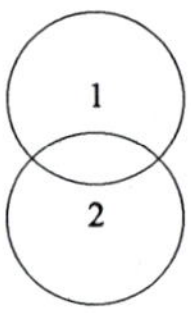

Expert/Talent	Nos.
Multidisciplinary	0
Cross-disciplinary	0
Interdisciplinary	2
Total Team	2

(b) A team of two, each has knowledge of what the other does.

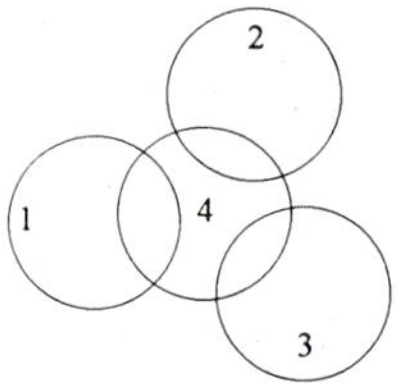

Expert/Talent	Nos.
Multidisciplinary	3
Cross-disciplinary	1
Interdisciplinary	0
Total Team	4

(c) A team of four, only one has profound knowledge of what the other three do.

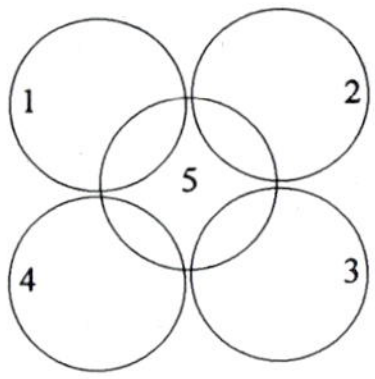

Expert/Talent	Nos.
Multidisciplinary	4
Cross-disciplinary	1
Interdisciplinary	0
Total Team	5

(d) A team of five, only one has profound knowledge of what other four do.

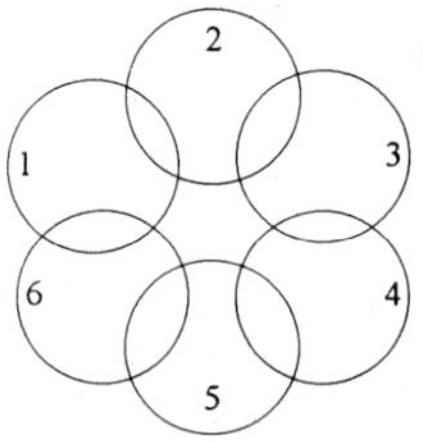

Expert/Talent	Nos.
Multidisciplinary	0
Cross-disciplinary	0
Interdisciplinary	6
Total Team	6

(e) A team of six, each has knowledge of what other two do.

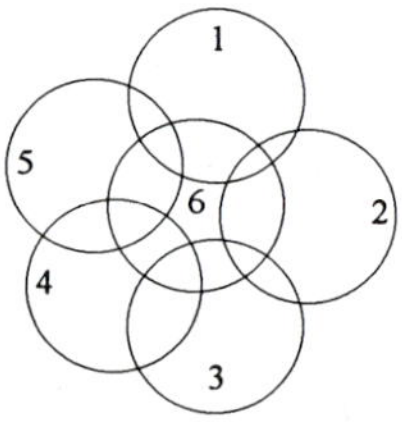

Expert/Talent	Nos.
Multidisciplinary	0
Cross-disciplinary	1
Interdisciplinary	5
Total Team	6

(f) A team of six, each of the five has knowledge of what the other two do and the sixth knows what the other five do.

FIGURE 5.17 Team Composition and Disciplinary Talents

skills, technologies, and procedures. Beyond these there may be a few important skills that are not so readily available in the PDT group and may require specialists:

- ***Specialists:*** Most specialists are multidisciplinary. A team may not have a need for a full-time specialist. Moreover, the specialist's narrow focus may not fit well with the environment. The expertise might be too narrow to be fully occupied. However, it might still be very important for the project. The specialist's contribution might make a crucial difference and provide the edge that is needed to gain a competitive advantage.
- ***Generalists/Consultants:*** These people are visionaries who serve as a window to outside sources such as technological newsbreaks, technological tidbits, etc. From time to time they participate in a "walk-through" and give advice on what path or recourse to take. They also serve as a gatekeeper to watch the development and suggest changes before things get out of control. The ideal situation would be that each member of the team plays a consultant role in their own field while doing their work. However, depending on their exposure and background, they may not be able to keep abreast of outside developments. Most consultants are interdisciplinary. A interdisciplinary outlook comes through working (several years of experience) on diverse projects in two or more disciplines.

Though in other places, multidisciplinary, interdisciplinary, and cross-disciplinary terms are used interchangeably, distinction is made here to emphasize the importance of collaborative thinking, deep common understanding, and convergence. The latter topics are discussed in Section 5.5.3. In order to maximize the chances of overall project success, one additional form of expertise must be sought from inside or outside the project-team:

- ***Product champion or sponsor:*** A project must have a project sponsor or a champion. This person is typically an individual external to the team who has significant clout in the organization. One of the principal functions of this individual is to make sure the needs of the team are met in a timely fashion. The champion, therefore, must genuinely believe in the cause and be its supporter. This individual should have the clout to intervene when unknown problems arise. A person high enough in the organization to have some influence but not so high as to lack time to spend with the team when needed, will be a good candidate.

Difficulties can arise if a project is large and the activities in the matrix of teams are closely interdependent. There will be a need for skills at each vertical level corresponding to each horizontal function. This situation leads to a skill matrix requiring a communication network that may be fully populated causing cross-links in all directions. This is shown in Figure 5.18. In a real situation, this may translate into a large number of time-consuming communication links that are required for CE. The following provides a basis for computing communication links in a matrix of teams.

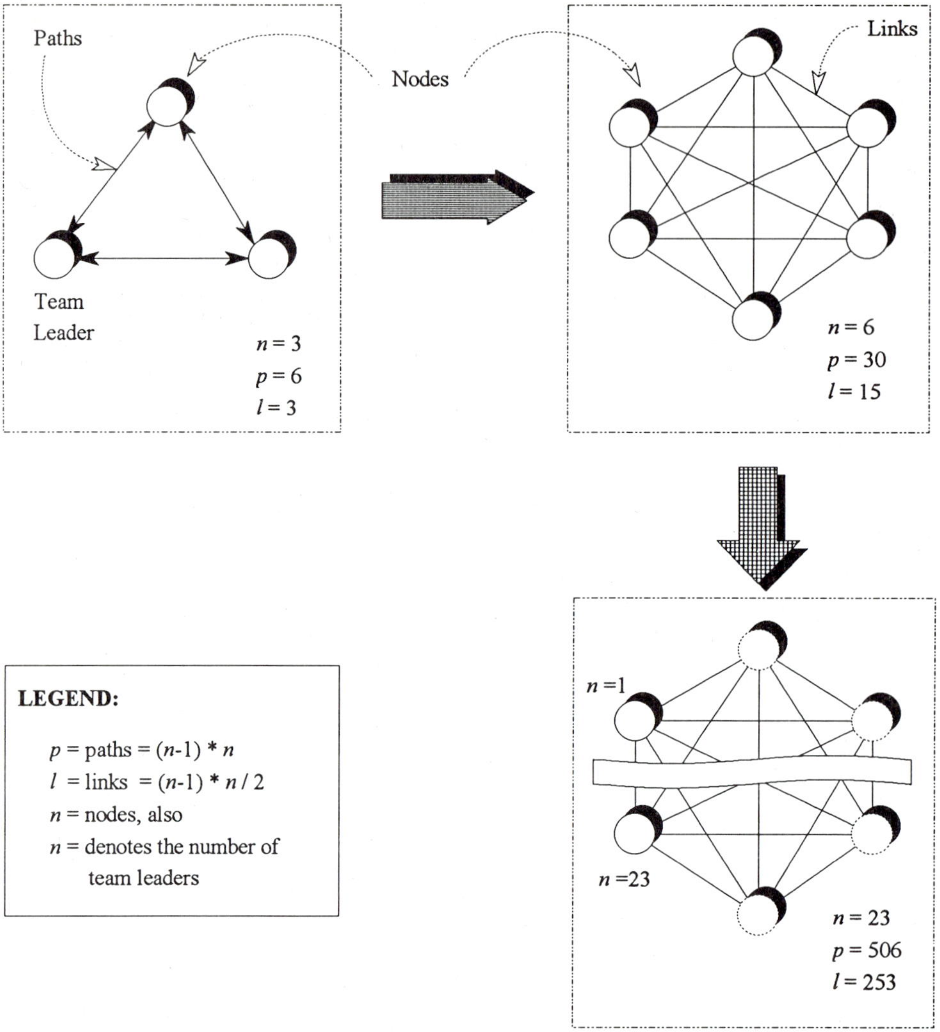

FIGURE 5.18 Growth in Communication Paths in a Matrix Organization

5.2.5.1 Matrix of Teams

Let us assume that a set of vertical functions (v columns) is interacting with a set of horizontal skilled groups (h rows). The matrix organization forms a matrix of teams. Each cell of the matrix represents a communication node, that reports back on two different paths: one horizontal and one vertical. For example, team (1, 1) must communicate with teams (1, 2), (1, 3), . . . , (1, h) along the horizontal row, and with teams (2, 1), (3, 1), . . . , (v, 1) along the vertical column. Each team requires [$(h - 1) + (v - 1)$] communication paths. Since the total number of teams is $h*v$, the total number of paths p:

$$p = \text{Total number of communication paths} \equiv [\ (h + v - 2) * h * v\ ]$$

or

$$p = [h^2 * v + h * v^2 - 2 * h * v] \tag{5.16}$$

$$l = \text{Total number of links} = p/2 \tag{5.17}$$

Link is the union of the two paths, one going in a forward direction, and the other coming back in a reverse direction.

$$\begin{gathered}\text{If } h = 12; \text{ and } v = 11, \text{ such that } (h + v) = 23; \\ \text{Total number of paths } (p) = 21 * 12 * 11 = 2772 \text{ and } l = 1386\end{gathered} \tag{5.18}$$

5.2.5.2 Cooperating Matrix of Teams

In the above discussion it is assumed that each group, represented by a cell or node is required to communicate along the horizontal or vertical lines. However, if we elect a team leader in each respective horizontal or vertical functional groups in such a way that each group cooperates and communicates only with their respective horizontal or vertical team leaders, the number of communication links can be reduced as follows:

There will be h team leaders along the horizontal row, and v team leaders along the vertical column. The communication can occur in three ways:

1. Between "v vertical functional groups" and a representative team leader
2. Between "h horizontal groups" and a representative team leader
3. Between "the n team leaders"

Where n is the total number of team leaders in this **matrix of teams**, for an (h, v) matrix,

$$n = (h + v) \tag{5.19}$$

Number of communication paths between v vertical functional groups and the team leader $\equiv v$

Number of communication paths between h horizontal functional groups and the team leader $\equiv h$

$$\text{Number of communication paths between the team leaders} \equiv (n - 1) * n \tag{5.20}$$

$$\text{Total number of possible interactions} = (h + v) + (n - 1) * n = (n * n) \tag{5.21}$$

$$\text{Number of Links} = (n * n)/2 \tag{5.22}$$

Thus, for $n = 23$, the number of paths could be as large as 529. As illustrated by the equation (5.22), the number of possible interactions is proportional to the square of the sum of the rows and the columns of a matrix. This number (for example, 529) is however, five times smaller compared to communication path numbers (for example, 2772) for a **matrix of teams.** The number of such interactions can be used as an indicator to determine whether or not the size of a matrix organization is practical. More thought is necessary in organizing such interactive projects. One solution is to divide the projects into smaller subunits of several matrix organizations that can be tackled relatively independently. This situation is shown in Figure 5.19. By dividing the initial matrix organization of node-size 23 into five submatrix organizations (say node-sizes 3, 4, 5, 6, and 5), the number of interactions can be

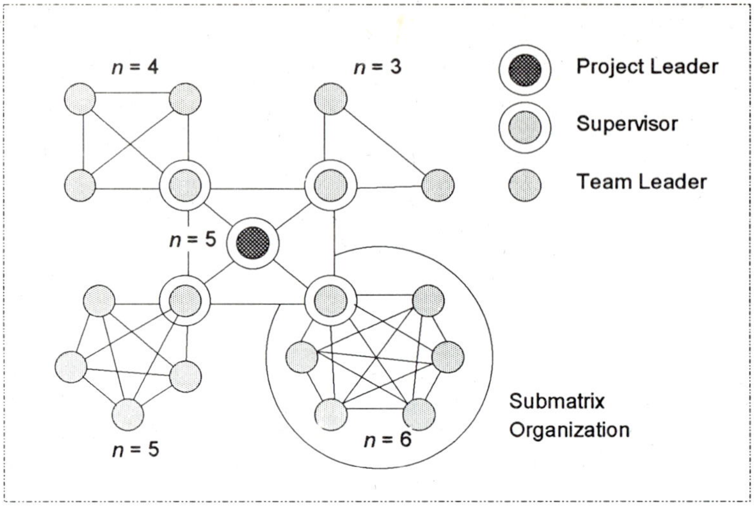

Relation	Communication Path Counts					Submatrix Total	Flat Total
n =	3	4	5	6	5	23	23
p = n * n	9	16	25	36	25	111	529

FIGURE 5.19 Concept of Cooperating Matrix Organization

significantly reduced. The number of communication paths are reduced from 529 to 111. This leads to a concept of cooperative matrix organization where the number of communication paths is manageable. With this cooperative setup, the matrix organization no longer represents a bottleneck in getting the project done on time.

5.2.6 Team's Roles

Many roles must be fulfilled to make a personnel team successful [Cox, 1991]. Some roles must be carried out by team members, some by the team leaders, and the rest by outsiders, such as suppliers or vendors. Concurrent teams play a supporting role to help the personnel teams in decision making.

5.2.6.1 Role of a Teammate

Each member of a team functions as a contributor in maintaining the constancy of purpose in goal setting. Team members must have a true peer-level relationship, with authoritative powers exercised only as an exception by the counsel when no compromise or consensus can be achieved.

5.2.6.2 Team Commitment

All members of the PDT should have an equally strong commitment to the "constancy of purpose" goals. Project subteams must be multidisciplinary, self-directed, highly focused, and fully committed to the project from start to finish. Qualified vendors and suppliers must be included as PDT members early in the development cycle. Product developers may form partnerships with select suppliers that show the same commitment to quality and continuous improvement as embodied in the CE concept. Organizations that are in a highly competitive situation could develop strategic advantages from such partnerships.

5.2.7 Team Collocation

Communication is the cornerstone of success in CE. The ability to interact constantly with PDT teammates, while the design is taking shape, lays the foundation for consensus building and rapid decision making. There are many ways to communicate. One approach is to move the teammates together; that is, physically locate them (PDTs) on the same premises. This creates physical proximity. An equivalent measure would be "electronic proximity." Electronic proximity is a class of virtual links that could make the travel involved quite painless. Proximity could be achieved through electronic interchange of messages, CAD models, drawings, sketches, and other communication modes (data, voice, video, etc.). Virtual team approach, networking, and groupware tools are available to aid PDT members in communication and decision making. However, tools may not be matured enough to replace a natural interactive environment, which comes from physical proximity. Meetings and physical proximity maximize various modes of communication (such as body language, group dynamics, personal touch, etc.). Geographical collocation develops a natural appreciation of each team's capabilities, problems, and opportunities. When people work side by side, they begin appreciating each other's opinions. Cultural or language barriers that traditionally prevent them from cooperating and optimizing their overall output begin breaking down. However, as soon as the groups get larger, the advantage diminishes because there may be too many people to deal with at one time. Teammates from diverse disciplines (such as electronics, mechanical, civil, industrial, etc.) also face difficulties in prioritizing the importance of their aspects to the problem. The impact of physical proximity on turf wars and territorial issues has produced mixed results. It is not so obvious when members stop seeking credit for their own contributions and when they begin working toward a common set of consistent goals. Breaking down the barriers, getting people outside their organizational boxes, and working together as coherent teams are some primary challenges most CE management is facing today.

5.2.8 Management Commitments

Senior managers often show little interest in a project until it is in the prototype phase. Figure 5.20 shows three profiles. Profiles (a) and (c) show "effort versus duration" trends. Profile b shows the "potential to change outcome versus duration" trend:

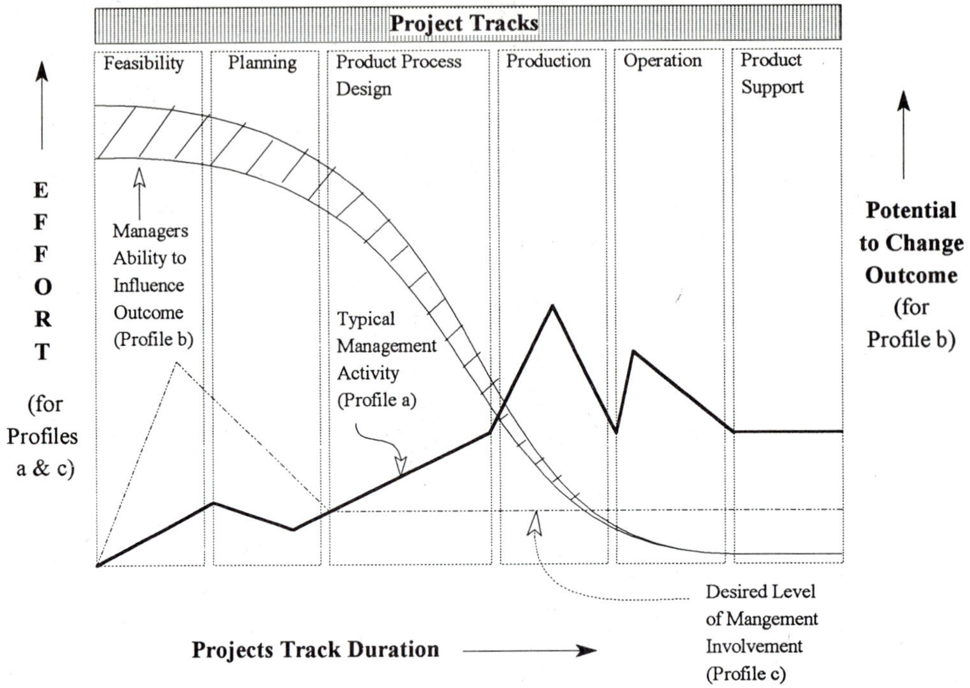

FIGURE 5.20 Level of Management Commitment

- Profile (a) shows a typical management activity during a project life-cycle. Looking at profile (a), the following points can be inferred:
  - Very little interaction occurs due to the passive interest of management during the product planning and feasibility stages.
  - Management interventions occur, in most part, late during prototype production and testing, or during manufacturing ramp-up.
- Profile (b) shows a band displaying management's ability to influence the outcome.
- Profile (c) shows a desired (an ideal) level of management involvement during the product's life-cycle.

Low management involvement during the early phases is both good and bad. Profile (a) is good in the early phases because it allows the empowered employees freedom to make decisions without being routinely hampered. Profile (a) is bad because, if anything would have gone wrong, managers would not have a major impact at this late point without upsetting the project momentum. Low management involvement is not accidental either. The management involvement in early phases is low because the overall product cost commitment is low. It is higher in later stages since cost commitments are higher. Managers also get directly involved towards the end, since in traditional systems (SE) they need to mediate on products that are late, have design revisions, cost overruns, or are afflicted by similar problems. High management involvement in the later stages does not pay high dividends. There are many peaks representing spurts of unnecessary management time spent. It also slows down the pace of the product development teams (PDTs), since managers are not familiar with the previous processes. Managers take extra time to fully comprehend the progress made so far to make sound and rational decisions. Since the cost of waiting this late is much larger than anytime before, there could be a large cost penalty associated with it. A desirable thing to do would be to keep a higher level of management involvement in the early phases, at least during the feasibility and planning phases. This is shown by profile (c) in Figure 5.20. In the latter phases, a constant but reduced level of involvement is maintained. Higher manager involvement in early phases empowers the teams while keeping management synchronized with decision making. During the latter phases in profile (c), management primarily assumes vigilant roles in matters that are related to overall strategy and policy decisions. This complements the goal-based management style that is discussed in Section 5.4.2.

5.3 SUPPLIER RATIONALIZATION

There could be an excessive growth of communication, transportation, or shipping requirements if there are a large number of suppliers, shipping locations, and manufacturing plants. Like what we did with skill management matrix, there is a similar need to reduce the number of suppliers, the suppliers' shipping locations, and the transportation locations to the manufacturing plant. Supplier rationalization is a process of rationalizing (evaluating-qualifying) and certifying (partnering with) those who demonstrate their commitment to quality and customer satisfaction, and share the strategic values such as commonality of standards, purpose, etc.

5.4 TYPES OF CE ORGANIZATION

One of the top-brass aims of an organization is to satisfy the customer with the highest quality, highest value products, brought to market in the shortest possible time. The CE challenge is how to blend all the skill-sets of the functional experts with good but static processes of product realization into a dynamic, highly integrated team so that the leadership aims are fulfilled. This requires a change in the current organizational structure—which is often vertical with "control" residing with the top management staff. Flexibility to react to unknown situations comes if the "control" is placed in the work area of the organization. A horizontal or cross-functional orientation provides a most common form of organization setup. Here, the control is a pull-type—meaning decisions are need-based on individuals or departments at the worker level. Ideally, besides the pull-type feature, the control should be adjustable to the task of meeting the customer requirements. In other words, the organization structure must include process optimization and continuous improvement skill-sets that are focused on the customer needs. These needs should be common knowledge throughout the organization, not just within a department or a work-group.

The assertion that global competitiveness begins and ends with an organizational structure and employee skill-sets is not news to anyone. During the craft era, the skill power of the craftsworker was the deciding factor in product marketability. Today, setting up an organizational structure is still a crucial step in implementing a successful CE program. Bart Huthwaite [1994] suggests that the design of organizational structure to manage product development should be treated as a strategic issue.

5.4.1 Functional Setup

Hierarchical organization is a good example of a functional setup. Most traditional or serial organizations are aligned along functional lines of control. Such a structure is not conducive to CE. In such a setting, employees are recognized on the basis of their individual contributions and are rewarded accordingly. This does not provide the needed incentives for employees to work as a team. A cooperative work-group and teamwork environment are needed to make the CE process a success. Similar to the multifunctional organizational structure discussed earlier, cooperative teamwork also involve cultural changes. The following sections discuss the various ways by which a traditional organization can be transformed into a CE setting. Each brings with it some advantages and some disadvantages. In functional organizations, the objectives are set to provide the best functional support and services to the supporting customers, internal or external to the organization. In cross-functional organizations, objectives are set to a common set of consistent goals.

5.4.2 Goal-oriented Project Management

Classic project management techniques such as Program (or Project) Evaluation and Review Technique (PERT) are not appropriate for CE. This is because, in PERT, the only way to parallel an operation is to first divide the problem into program phases, subsystems, and disciplines, and then divide each into smaller and smaller tasks. We can visualize this as a multidimensional structure with n_3 cells, where n is the number of tasks (assumed equal in number for each). Furthermore, in PERT, three time estimates are given to each task (optimistic, pessimistic, and the best guess) [McGrath, 1984]. In a complex system, this can easily create several thousands of cells. This method quickly becomes inoperable and, therefore, counter-productive. PERT is also not suitable for providing adequate intelligent support to the routing and queuing model construction, overlap determination, and sensitivity and trade-off analysis.

On the flip-side, PERT is useful in identifying the "tasks" on the critical path. The critical path method (CPM) performs a linear time or cost tradeoff for activities on the critical (longest lead time) path. An example of a PERT chart is shown in Figure 5.21. The critical path is highlighted by a chain of bold circles with a letter C etched in it. In CE, many tasks can be done simultaneously. Five alternate paths are shown in Figure 5.21. However, only those tasks, which lie along the critical path (denoted here by bold circles) will determine how long it will take to complete the total project. The only way to

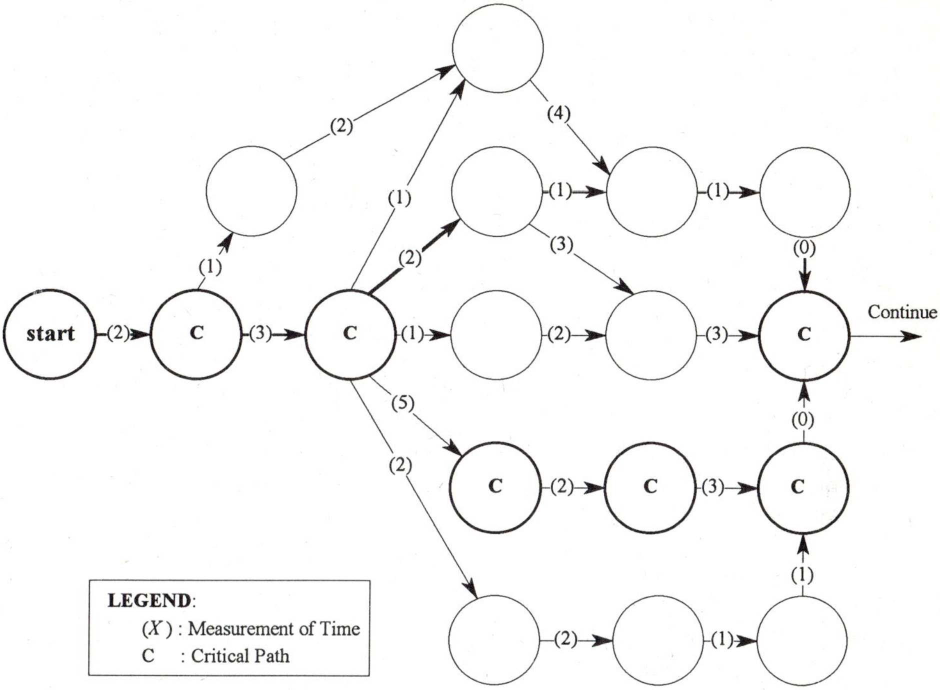

FIGURE 5.21 Determination of Critical Path (PERT Diagram)

reduce total project completion time is to reduce the time needed for the activities on the critical path. Speeding up processes that are not on the critical path will not make any difference in total project completion time.

Cross-functional teams must be driven using a common set of management styles, goals, or philosophies. A system level goal for a company, for example, could be to stay in business for the long haul or to provide maximum benefit to themselves, to their customers, suppliers, and to society [Deming, 1993]. The goals can be further broken down into lower rank goals (each supporting the former from the "constancy of purpose" viewpoint).

5.5 MANAGEMENT STYLES OR PHILOSOPHIES

Developing a coherent management style or philosophy seems a daunting task, given the wide range of possibilities and practices that must be addressed. The job of managing is becoming very difficult due to a rising complex web of changes and competitive pressures. Many managers would like to swim through the currents of the change, but they do not know what policy they need to follow to be successful. Others have realized that the currents are simply too fast for any old style of management to withstand. Changes are forcing companies to adapt to a more flexible management style and structure. Empowerment and responsibilities are shifting from the usual vertical setting to the horizontal setting. The chain of command is shifting from the tall silo structure (or pyramid) to peer networks and cooperative teams [Hamel and Prahalad, 1989]. There are three primary management styles: (a) directive management; (b) supportive management; and (c) constancy of purpose oriented Management. These are shown in Figure 5.22.

5.5.1 Directive Management Style

Directive management style has no theory. It is a grab bag of techniques, rituals, customs, and superstitions. Some examples of such techniques are setting a pyramid for reporting—as in management by objectives (MBO) or in management by results (MBR) [McGrath, 1984]. The concept is not based on any sound theory of cooperation or any theory of system optimization thus leading to the fragmentation of the organization. Deming noted that in the course of fragmentation, each component becomes the individual profit center, destroying any hope of how they would contribute to make the system work better. Other techniques that belong to this style are buying materials and services at lowest bid,

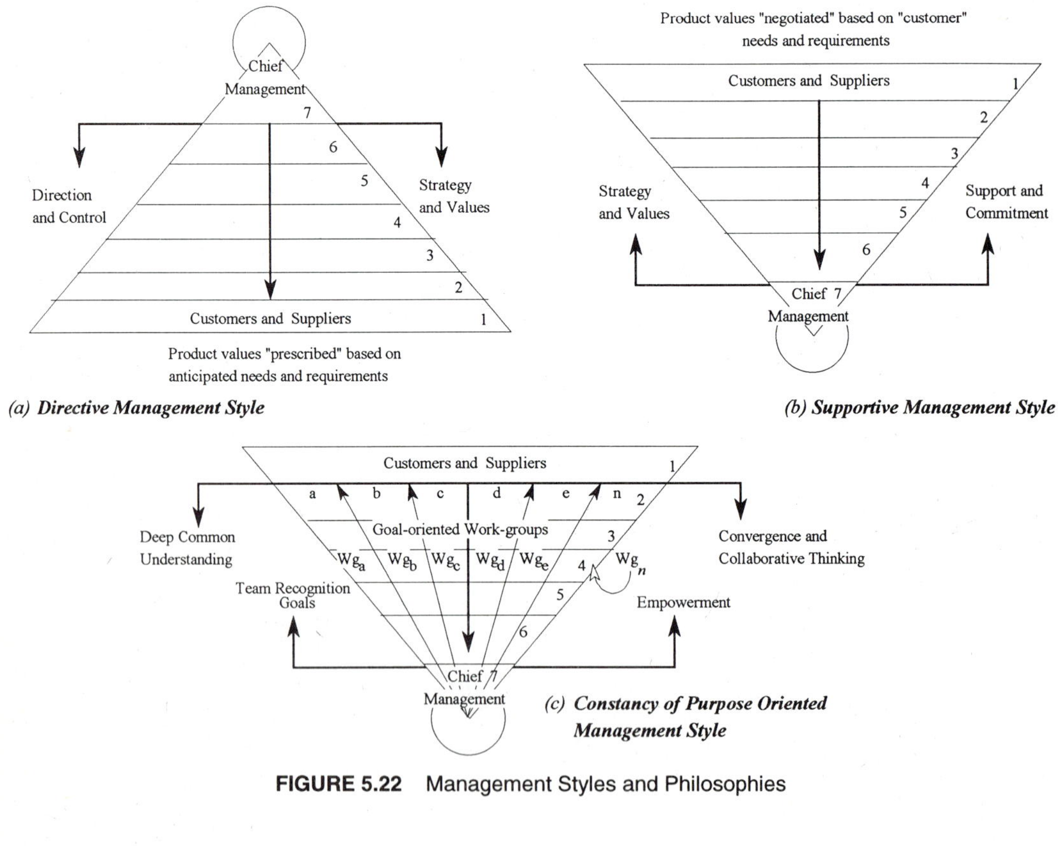

FIGURE 5.22 Management Styles and Philosophies

setting numerical quota for sales, measuring performance, ranking, etc. In this style, the chief management sets the direction and control. Management defines the strategy and values from the top down, for the customers and suppliers. The differences among the three management styles are contained in a matrix in Figure 5.23. The three columns list the distinguishing qualities of the management styles against eleven measures of merits. In directive management style, product values are "prescribed" based on anticipated needs and requirements.

5.5.2 Supportive Management Style

Supportive management is based on the principles of total employee cooperation and involvement. Product values are "negotiated" based on customers' needs and requirements. The effects are reflected in Figure 5.22b by an inverted triangle—depicting the customers and suppliers at the top and management at the bottom. A strong multidisciplinary team effort, with continuous interaction between the customers and teammembers, and a clear focus on goals is the most important. Simultaneous processing and problem solving, along with effective project management, are the keys to continually improving cycle time. Instead of setting numerical quotas, management works with the teams and defines

Qualities	***Directive Management Style***	***Supportive Management Style***	***Goal-Oriented Management Style***
Power	Authoritative	Teams	Goal-oriented teams
Information	Withheld/Screened	Support and Commitment	Open/Available
Representation	Political	Selective/Mutual	Open/Need-based
Competencies	Position-based	Technical	Technical/Interpersonal
Partnership	Personal	Mutual	Goal-oriented
Commitment	Management - high Team - low	Management - medium Team - medium	Management - high Team - high
Product Values	"Prescribed" based on anticipated needs and requirements	"Negotiated" based on customer needs and requirements	"Defined" based on goals
Customer Priority	Management on the top	Customers and suppliers first	Customers and partners first
Boundaryless	Low	Medium	Maximum
Colocation	Needed	Needed	May not be neccessary
Participation	Empowerment	Convergence and collaborative thinking	Deep common understanding

FIGURE 5.23 Key Features of the Three Management Styles

a method to improve the process. The long-term vision of supportive management style involves 7Cs: *collaboration*, *commitment*, *communication*, *compromise*, *consensus*, *continuous improvement*, and *coordination*. These are referred to as cooperative team characteristics and are shown in Figure 5.2.

5.5.2.1 Management by Fact

The supportive management style is frequently based on "management by fact" (MOF). This means giving information to the supported teams so that decisions are based on facts (Dr. Deming calls this a theory of profound knowledge in his book *The New Economics*) rather than a "gut feeling." The information provides the first principle of supporting management style, but all by itself MOF is not enough. We need to know where we are starting from (facts) and the current levels of our products and services that are in our customers' hand (profound knowledge). Having the facts or profound knowledge necessary to manage the business at all levels is the second principle of supportive management style. Having both principles, a company is in a better position to manage and discharge its responsibilities based on customers needs.

In the past, some companies have institutionalized supportive management styles by planting "tiger teams." A tiger team consists of pulling in the best and the brightest talent from the different areas or disciplines that are essential for the project. NEC followed this approach for developing a new laptop computer. They set a ninety-day time limit to prevent apparent loss of market share to their competitors. They instituted a "tiger team" consisting of experienced management and personnel from various computer development projects. They gave this tiger team full authority over all aspects of product development. A "backward scheduling" technique was used to assure that the product would meet the ninety-day target. The success of the NEC tiger team is a significant example of what a highly motivated group with a strong experience base from related disciplines, a highly cooperative team with decision making authorities, and a high urgency and strict enforcement of target, can do in a short while. It was certainly a challenge since the infrastructure was not in place. Does this mean that planting a tiger team is the answer? Many argue that the tiger team worked well because they were the best and brightest people, there was ample peer respect and management visibility. People knew it was a one-time deal. One cannot expect the same result by applying this scenario to everyday work life because not everybody works well together, or possesses the same level of competency and respect among their peer groups. The question is what could we accomplish in a tiger team, how could we achieve the same through an empowered team system? The answer is the establishment of an infrastructure that facilitates C7 (*collaboration*, *commitment*, *communications*, *compromise*, *consensus*, *continuous improvement*, and *coordination*) on a regular basis. There is no need to strictly enforce all targets. As a part of the CE infrastructure, among other things, the team can receive the training on how to work together and achieve an understanding of product, processes, tools, teamwork, capabilities, and limitations. Once the team members have a common understanding, they can work together better. The infrastructure will be equipped to minimize the impact of variation on any of the above elements. It dampens the outcome so that effects will not be felt as much.

5.5.3 Constancy of Purpose Oriented Management Style

Constancy of purpose management style is a variation of supportive management style in which most of its structure, including an inverted triangle style are shared (see Figure 5.22). Here, the individual goals are targeted toward providing a constancy of purpose—goals are supportive of other goals. This style requires the most significant change and it is quite a departure from the traditional system. Here, all personnel may report technically to the same manager and work toward a common set of consistent goals. For instance, one goal is to manage a large and diverse organization to operate as if they were one intimate and cohesive workforce. This may require a shift in allegiance of personnel from their parent functional organization to a strategic business unit or a PDT. If the personnel are not members of the same PDT, the probability is higher that some of the eight CE principles will be violated (see Section 3.1 for the listing of these principles). All members of various teams are expected to owe their allegiance to the company's or SBU's goals (a constancy of purpose oriented management). This requires a change in thinking beyond the goals of one individual department or work-group to those of the SBU or the company's goals. The arrangement is very much the same as the inverted triangle style of Figure 5.22b. The customers and suppliers remain on top and the chief management on bottom. The roles of management are, however, changed. The obligation of any supporting unit management is to empower the unit so that it can contribute its best toward the system's goals. The aims of the units are not to suboptimize their own performance (such as units' profit potential or sales) without a clear and direct relationship to the company's overall goals. The project's goals must be supportive of the team's goals. Team goals must be supportive of the PDT unit goals. PDT unit goals must be supportive of company's goals, and so forth—ensuring a constancy of purpose. This way everyone contributes his or her best toward a common set of consistent goals. Within a team, for example, everyone must understand the team's objectives, which could be to produce a high quality product on time and within the budget. The management role is to continuously improve the processes that work toward ensuring a better set of constancy of purpose objectives. Management should reject compromise when decisions are detrimental to the company's goals, even though it may be supportive of team or project goals. Participants should accept compromise when it is permissible and is a better strategy overall (when everybody wins—stockholders, employees, suppliers, customers, community, environment—over the long-term). Engineers work closely as teams, orchestrating their special expertise, talent, and experience, while retaining their individuality. In the beginning, this may not be easy for some, while for others it may provide opportunities to have equal say in decision making and allow them to become contributing partners in the growth of the company. In the constancy of purpose oriented management style, among other things, team training is directed toward agreeing on a mission. This consists of goals, and the role of each individual, work-group, team, department, and management, the processes of getting things done, communication media, and the CE relationship with 7Ts (talents, tasks, teams, time, . . .).

The best laid constancy of purpose oriented plans and the most prodigious efforts, however, will not prove effective without four key elements: convergence and collaborative thinking, empowerment, team recognition, and deep common understanding. As

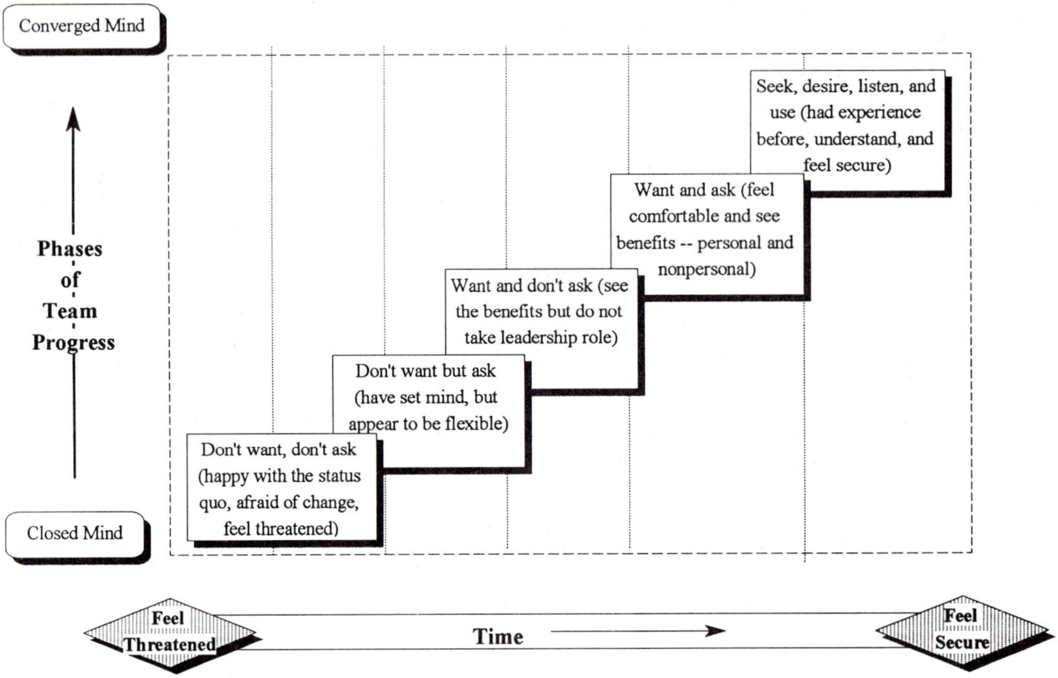

FIGURE 5.24 Phases of Team Progress Leading to Convergence in Thinking

shown in Figure 5.22c, these are critical elements that are considered part and parcel of a successful constancy of purpose oriented workforce.

5.5.3.1 Convergence and Collaborative Thinking

This is an important feature of a constancy of purpose oriented work-group. Figure 5.24 describes the stages of team progress through which convergence of collaborative thinking takes root. Five stages are chosen to describe a progress from a closed mind to a converged mind set. These are merely pointers or intermediate steps and do not necessarily represent distinct phases.

- *In the beginning stages,* most teams possess a closed mind, have a "don't want, don't ask" attitude, are afraid of unknowns and often feel threatened. With time, members of each team develop an understanding of each others' point of view. They begin to appreciate the importance of their disciplinary contributions at various points along the way and their impact to the product goal's realization. The attitude sails through a series of changes.
- *In the second stage,* the team moves from the "don't want, don't ask" attitude to a "don't want but ask" attitude.
- *In the third stage,* the attitude goes from "don't want but ask" to a "want and don't ask" attitude.
- *The fourth stage* goes from "want and don't ask" to "want and ask" attitude. Collabo-

rative thinking extends teamwork concepts and communication capabilities. It applies these concepts to formulate how work will actually be done. Considering this, the teams form an expanded group knowledge of the whole thing in such a way that it allows each teammember to completely understand the needs and goals of his fellow members. This collective mindset allows members to grasp each others' differing points of view, which results in confidence building and increased levels of cooperation.

- *The fifth stage* moves to "seek, desire, listen, and use" (see Figure 5.24). At this stage, teams' minds converge—teams feel more secure than ever before. With convergence and collaborative thinking, the impact of cross-functional teaming on the product realization is maximized.

One of the places where such stages of convergence and collaborative thinking take place, is during a design review. Design review (made out of select cross-functional review teams) is an efficient method to

- Monitor the progress of a project
- Facilitate reporting and appraisal of results to management
- Keep team interest in line with the common set of consistent project goals.

Design reviews promote a team-oriented review strategy, which optimizes the team's collective talents in problem solving. A carefully timed and organized design review is not an engineering inspection, but rather a value-added process of team building and a step toward convergence and collaborative thinking. To obtain meaninful results, it is important that teams stick with a standard review format and timing.

5.5.3.2 Empowerment and Involvement

Empowerment is often referred to as power of involvement. No matter whether an involvement is planned or is inadvertent, the result has always been positive—truly involved people can do anything. Empowerment does not mean that the line or functional managers have to give up control. Rather, it refers to the sharing of responsibility in matters of process definitions, such as policy making and overall strategy formulation. It also means transfer of needed authority and assigned responsibilities so that the team, that is close to the problems can make decisions. If the management has rightly empowered the work-groups by establishing a self-governing "check and balance" process (in terms of assigned responsibilities, established purpose, capabilities, targets, benchmarks, and review process), the goal-directed process can move smoothly. The work-groups will understand the boundaries and parameters within which to operate. The project will get the benefit of collective inputs in formulating decisions—technical or nontechnical—with less management barrier. In the beginning, this may appear restrictive for some experienced employees, since they may have to

- Learn to speak a technical language they never spoke before
- Learn to work side-by-side with people they may not get along with
- Work closely and cooperatively with people, who at one time might have worked

for them, now for a good part of their own success, they must rely on their teammates' performance (team reward)

- Acquire more interpersonal and group dynamic skills.

5.5.3.3 Team Recognition

Recognition and reward are mechanisms used frequently by employers in one form or the other. However, these are often awarded to an individual, even though he or she may be part of a team. In CE, team recognition carries more weight to entice the entire team to work together. Teams freely collaborate in setting the project's goals with the "constancy of purpose" in mind, and meeting them on time and within budget. The reward for this change on the part of the team is quite promising for the company. Constancy of purpose oriented management is expected to provide the maximum productivity gain and cost advantages.

5.5.3.4 Deep Common Understanding

Empowerment and rewards are useful for team motivation. Deep common understanding is useful for creating highly charged employees and PDTs. Such a group of PDT members has a high level of confidence in each other and is able to quickly create an informal atmosphere of human networking. They communicate with each other and can come up with a highly reliable plan in a short time. They form a pyramid of cooperating teams—the so called "learning organization." It has three sides to it. This is shown in Figure 5.25. The first side contains a common set of consistent goals—from its corporate vision to the

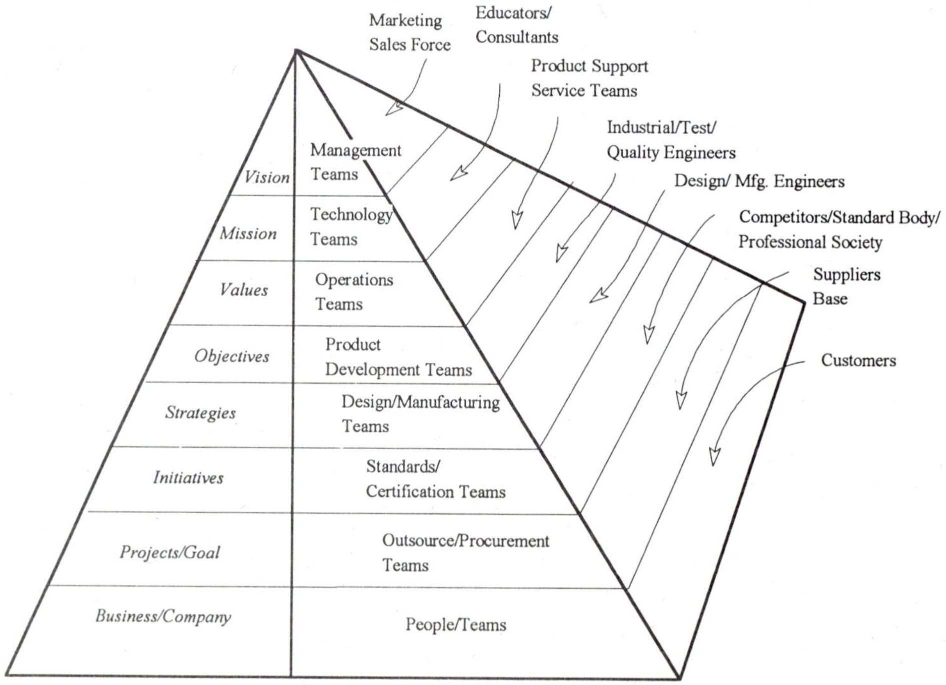

FIGURE 5.25 Pyramid of Cooperating Teams

project goals, following the SBU's constancy of purpose management plan. On the remaining two sides, the PDT skill sets corresponding to life-cycle management and work-group management are listed (see Figure 5.25). The levels are insignificant.

Effective collaboration requires that the various PDT members of the pyramid do the following:

- Create a state of cognitive mind (mindset or a deep common understanding); this must account for the company interests as opposed to members own interests while PDTs participate in their company efforts to improve or design new products
- Follow an occupational culture so that PDTs can develop shared views of problems that need to be resolved, collectively utilizing the knowledge of their members "native-views"
- Harness the education and background strengths of each contributing member in shaping the product from all different angles so that all useful views are represented
- Follow a value system that ties them together, such as sensitivity to other members' capabilities, respecting member "native-view" opinions, setting priorities that cross functional boundaries, and institutionalizing means to resolve conflicts.

5.6 WORKPLACE ORGANIZATION AND VISUAL CONTROL

5.6.1 Plant, Machine, and Office Layout

In CE, teams are often collocated. The layout of an office, in such a case, plays a significant role in short-circuiting the objectivity with subjectivity. The face-to-face arrangements force the teams to interact freely on a regular basis, thus taking away the hesitation or formalities. Four typical layout designs are shown in Figure 5.26. Depending on the layout of the plant, machine, or the office, each is designed to minimize the time loss in travel. The sharing of information between teammembers is also facilitated if there is a common terminal/workstation they can use to share resources, software, manuals, procedures, standards, etc. The teams can sit together and work on shared terminals or workstations. These can be arranged in a number of patterns [Kusiak and Belhe]:

- Cluster arrangements
- Angular or semiangular shapes
- Geometrical patterns
  - Triangular
  - Rectangular or square
  - Hexagonal
- Swivel (turntable) patterns

The layout design of a CE office and the sitting arrangements are shown in Figure 5.26. Layout design is intended to maximize the use of equipment and time, and to facilitate collaborative decision making.

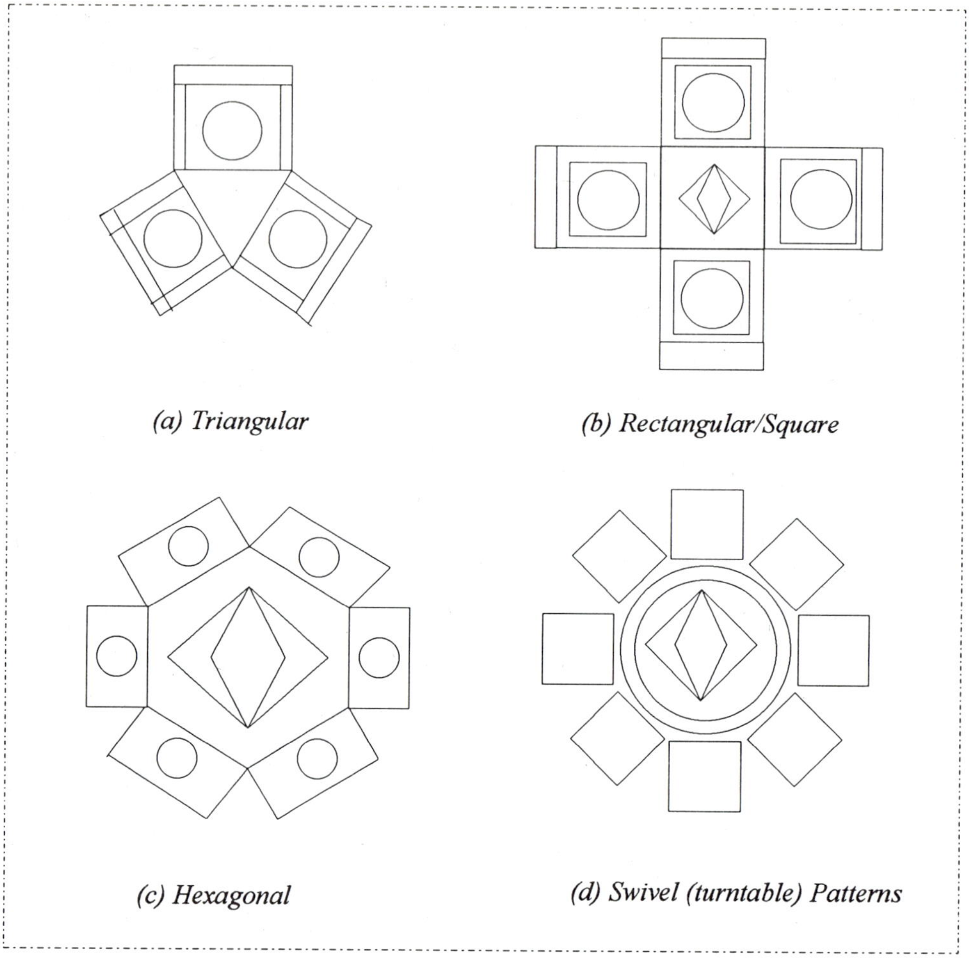

FIGURE 5.26 Layout Design of a CE Office (Physical Proximity)

5.6.2 Visual Control

A visual control (VC) system is a mechanism which makes the status or operating conditions known to the management and workers. One form of VC is called ANDON, which is of two types. One resembles a scoreboard employed during production scheduling. The other is similar to a traffic signal with green (running), yellow (needing attention), and red (stop) lights. The colors signify the state of quality, or problems, at each processing station. Action board is another form of visual control. It is mostly used to depict current operating conditions. Like the status display, as in traffic signals, an action board incorporates both warning lights and audible signals to draw attention. VC is quite useful in CE, since it provides a visual communication channel to promote teamwork and 7Cs.

5.7 EMPLOYEE EXCELLENCE DEVELOPMENT (NEW TECHNOLOGIES AND TEAM CAPABILITIES)

Most of the participants in CE teams are employees who come from an atmosphere where they operated in isolation from other disciplines. For instance, they may have coordinated with others only through the exchange of information supplied on official company letterheads, forms, or documents. The CE environment, with its highly cooperative team base and many new technologies that are used to enhance its functioning, will be a completely foreign experience to many of the CE personnel, especially during its initial implementation in the company. The following discussion outlines some developmental programs (such as employee rotation programs, suggestion plan, strategic sourcing education, and training, etc.) to impart excellence in teamwork and cooperation.

5.7.1 Employee Rotation Program

During the launch of CE, the company commonly embarks on an extensive and aggressive rotation program in which engineers from different units are sent to work on projects with other disciplinary flavors. This way, PDT members can get experience and training on the job, and can acquire a much broader background in product development. At the same time, the rest of the company can find out what information will be needed to support a CE product development environment. The PDT members begin to appreciate the problems and difficulties others face in executing their portion of the responsibility. Both technical persons and management functions should form a part of this rotational cycle. This tends to create an environment conducive to a "learning organization." Senge [1990] cites "mental models" as one of the four core disciplines required for building a learning organization. "Mental models" are deeply ingrained assumptions, generalizations, or even pictures or images that influence how we understand the world and how we take actions.

5.7.2 Suggestion Plan

The suggestion plan underscores the need to invite the PDT members, including suppliers, to be full partners in the business, recognizing them as the greatest resource. Since working teams are closer to the problems, PDT members will be able to make a better suggestion plan based on their mental models. Recognizing, unearthing, and scrutinizing PDT members' mental models is an important step in CE.

5.7.3 Strategic Sourcing

The list of participants in CE should not be limited to the members of PDTs. In CE, suppliers and customers are considered partners to achieve what is called "strategic sourcing." Manufacturers in CE teams include representatives from such suppliers forming a "network of strategic partners" called a supply chain (see Figure 5.27). Three tier suppliers are shown in Figure 5.27. First-tier suppliers are generally subsystems or finished

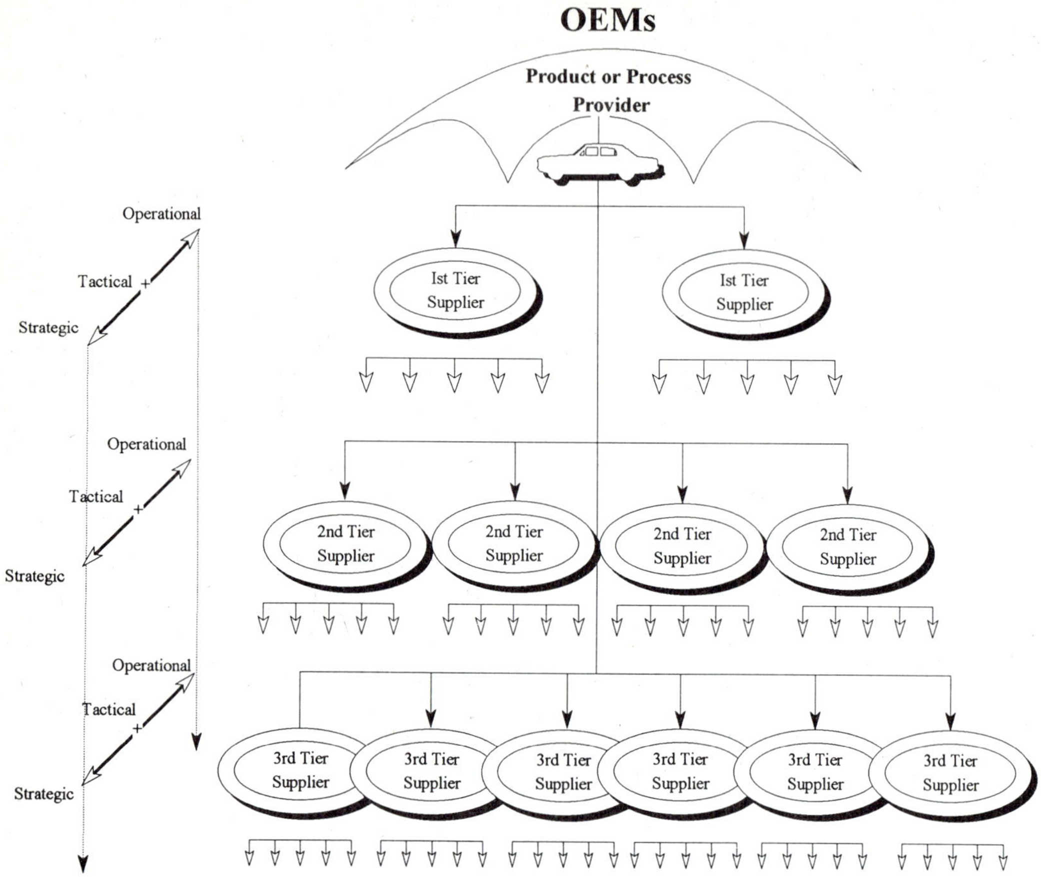

FIGURE 5.27 A Typical Supply Chain in an Automotive Industry

components' suppliers—they deliver directly to OEMs. Second-tier suppliers deliver parts and raw materials to the first-tier suppliers, and third-tier suppliers deliver to the second-tier suppliers. There is a collaborative relationship at each level, which may be of strategic, tactical, or operational nature. The continuum of strategic to operational ties indicates the level of commitments in sharing both the risks and rewards. The sharing of risks and rewards is more for strategic partners and less for operational partners.

$$\text{Risks and rewards}_{\text{first-tier}} > \text{Risks and rewards}_{\text{2nd-tier}} > \text{Risks and rewards}_{\text{3rd-tier}} \tag{5.23}$$

When supply chain partners understand the CE process and adopt the process improvement techniques into the designed parts, the company gets the benefits. Each supplier is able to pass through the delivered parts the quality improvement that each can muster. Collective supply chain advantage is the joining of individual participant strengths into a network of partners, each contributing to meeting the (constancy of purpose) enterprise goals. In such a partnership, the measure of success is not from "how a company is competing against another" or from what they are doing. Measures of success are from "how a supply chain is competing against another supply chain." A supply chain succeeds together, or fails together, based on its ability to provide better value jointly to the ultimate customer. Similarly, inviting selected customers as a part of the supply chain can strengthen the product and the company, which in turn can strengthen the customer. This could amount to possible repeat business. For most companies, supply chain initiatives go hand in hand with elaborate internal education and training in effective communication with customers and suppliers. Honda of America has an extensive training program for suppliers and has given monetary incentives (such as grants) to establish smooth interface and communication links with the supply chain.

5.7.4 Education and Training

Education and training are essential for building knowledgable talented employees. They are an important part of integrated product development [Dziuban and Knepell, 1992]. Preparation for the future includes lifelong learning for employees [Deming, 1993]. An educated employee in CE is more likely to be open to change and have an appreciation for real-world benefits. Through education in CE , employees get the required knowledge, enthusiasm, and the empowerment to affect change. Among other things, the training in CE must include the following major topics:

- Concepts for concurrent engineering
- The deep common understanding in cooperation characteristics—7Cs
- The seven teamwork characteristics—7Ts
- Group problem solving techniques and pointers to develop helpful team behavior
- Defining a common set of consistent goals and objectives
- Improvement in social, economic, and technical environment
- Need for innovation: creation of new product, new process, or new service
- CE tools and how to use them
- Productivity concepts, for example TQM
- Process improvement technique such as benchmarking, CPI, restructuring, renovation, reengineering, etc.

Smart businesses realize that they must take responsibility for developing a talented workforce and imparting training to their workers. Motorola, whose sales were a record 13.3 billion in 1992, spent $120 million on education. Motorola calculates that every $1 spent on training delivers $30 in productivity gains within three years [Henkoff, 1993]. Federal Express has used computers to revamp the way it educates its employees. At least once a year, every one of the company's 40,000 couriers and customer service agents plug into an interactive, PC-based program that tests their job knowledge. Many companies today, including Ameritech, use interactive videos to train its employees [Henkoff, 1993]. The key to successful training is to describe a step-by-step sequence in process improvements, using the best practices of product development based on sound software tools. Many tools are available to solve the problems; the important thing is to teach how to use a particular tool to solve a particular problem. In-class training must be accompanied by plenty of hands-on training—the more the better. Many courses teach either just specific techniques or just specific software tools. Courses must combine all three elements: the quantitative techniques, the software tools, and a step-by-step approach for achieving process improvements. A CE team requires all three and it is advantageous to reinforce them by including all three in the same course. By employing realistic beginning-to-end process scenarios, participants can gain a lot more confidence and skills than otherwise possible, and learn how to achieve revolutionary process improvements. This can lead to faster returns since participants will be able to improve processes with a minimum of training time and costs.

The teacher of the course is important. Training by independent consultants is usually general purpose and restricted to one or more "of-the-self" software packages. Often, participants do not get any insight as to how to apply these principles to solve actual organizational problems. Training by software vendors is usually designed to bring experienced engineers up to speed on their particular package, rather than to teach the techniques. To be effective, training must be oriented to use the techniques and tools for process improvements, rather than limited to the techniques or tools as ends in themselves [Dziuban and Knepell, 1992].

REFERENCES

CLARK, K. B., and T. FUJIMOTO. 1991. *Product Development Performance.* Boston: Harvard Business School Press.

COX, A. 1991. "The Homework Behind Teamwork," *Industry Week,* January 7, pp. 21–23.

DALZIEL, M. M., S. C. SCHOONOVER. 1988. *Changing Ways,* New York: American Management Association.

DEMING, W. E. 1993. *The New Economics.* Cambridge, MA: MIT Center for Advanced Engineering Study.

DZIUBAN, S. T., and P. L. KNEPELL. 1992. "Manufacturing Improvements Through Computer-based Training on TQM/SA Tools," In the proceedings of the SME Autofact '92 Conference. Nov. 9–12. Detroit, MI, pp. 1–10.

EKVALL, G. 1993. "Creativity in Project Work: A longitudinal study of a product development project," *Creativity and Innovation Management,* Vol. 2, No. 1, pp. 17–26.

FREEZE, D. E., and H. B. AAARON. 1990. "Customer Requirements Planning Process CRPII (beyond QFD)," SME, Paper No. MS90–03, Presented at the Mid-America 90 Manufacturing Conference, April 30–May 3, 1990, Society of Manufacturing Engineers, Detroit, MI.

HAMEL, G., and C. K. PRAHALAD. 1989. "Strategic Intent," *Harvard Business Review,* Vol. 67, No. 3 (May–June).

HENKOFF, R. 1993. "Companies that Train Best." *Fortune.* Vol. 127, No. 6, March 22, pp. 62–75.

HUTHWAITE, B. 1991. "How to Design Team Excellence," *Design News,* Vol. 48, No. 6 (June), p. 162.

HUTHWAITE, B. 1994. *Strategic Design: A Guide to Managing Concurrent Engineering.* Rochester, MI: The Institute of Competitive Design.

KUSIAK, A., and U. BELHE. 1992. "Concurrent Engineering: A Design Process Perspective." *Concurrent Engineering.* ASME. *PED.* Vol. 59, pp. 387–401.

MALINIAK, L. 1991. "Teamwork is the Key to Concurrent Design," *Electronic Design,* Vol. 39, No. 1 (January), pp. 41–43.

MALONE, T. W. 1991. "Towards an Interdisciplinary Theory of Coordination." Proceedings of the Ist International Conference of Enterprise Integration Modeling Technologies, ICEMIT. USA, June 8–12, South Carolina.

MARGOLIAS, D. S., and M. H. O'CONNELL. 1990. "Part 3—Implementation of a Concurrent Engineering Architecture." Paper No. MS90–205, Presented at the WESTEC'90 Conference. Los Angeles, CA, (March), SME.

MCGRATH, J. E. 1984. *Groups: Interaction and Performance.* Engelwood Cliffs, NJ: Prentice Hall Inc.

SENGE, P. M. 1990. *The Fifth Displine: The Art and Practice of the Learning Organization,* New York: Doubleday Currency.

SULLIVAN, L. P. 1988. "Policy Management Through Quality Function Development," *Quality Progress,* (June), pp. 18–20.

TUCKMAN, B. W. 1965. "Development Sequence in Small Groups." *Psychological Bulletin.* No. 6, pp. 384–399.

WATSON, G. F. 1991. *Concurrent Engineering.* Special feature of *IEEE Spectrum*, Vol. 28, No. 7, (July).

TEST PROBLEMS: COOPERATIVE WORK TEAMS

5.1. What are the four staffing rules for building an effective CE team? Describe the involvement of the personnel team in CE. What kind of qualities must the personnel team possess?

5.2. How can we recruit, select, or build a winning PDT team? What are the benefits associated with PDTs? What are the duties of PDT? Give some real-life examples of PDT.

5.3. Why is communication one of the most important factors? What is the best way to communicate relevant data to the CE teams?

5.4. How can teamwork and synergy be effectively used in solving manufacturing and quality problems? What are the common characteristics of an effective team?

5.5. Develop a concurrent work plan for company wide-development of a product. How do you deal with different work-groups? How do you ensure teamwork and entice cooperation?

5.6. Teamwork is an essential component of CE in providing a set of consistent goals and objectives. Discuss the seven qualities necessary for a concurrent work-group.

5.7. Describe the composition of the four CE teams. Where should we involve suppliers?

5.8. List four basic components that depict the philosophy of the constancy of purpose oriented management style. Of these components, which possesses the largest obstacle when the company is first trying to implement CE?

5.9. Illustrate the four types of CE teams involved in the making of a theater production. What is the difference between working in a team versus working in a group? Is it synergy?

5.10. Although it seems logical that teamwork is necessary for CE, list four skills that a team leader/manager must have. Do you think large companies will have difficulty keeping these people as team leaders with all these skills? If so, why?

5.11. There are those who argue that no matter what industry you are in, the employees are the ones that make or break an organization. Do you agree? Explain why. How can a company keep people motivated?

5.12. If CE does not produce the right results, who is to blame? Choose one: the employees, management, customer, foreign competitor, or none.

5.13. Referring to Figure 5.24, describe pros and cons of what this figure represents. Does team progress always go through these stages? Give a set of examples showing stages of team progress leading to convergence in thinking.

5.14. How does your work load and your experiences at work or home differ or resemble that which is shown in Figure 5.14.

5.15. Describe four events of queue management. How can peaks and valleys in available workforce and oscillations in workload be balanced effectively?

5.16. Graphically show the possible differences in staffing levels and trends of a CE organization versus serial engineering. What can you do to maintain a fairly constant staffing level? What are its advantages?

5.17. What is the right size of a product team? What are the causes of peak, valley, or oscillations to occur?

5.18. Figure 5.14 shows three possible variations in workload and available workforce. Have you encountered any one of these in your work or social life?

5.19. What are the number of communication links and paths required for establishing communications among forty groups of single individuals. Can you suggest an alternative organizational setup, which requires less numbers of interactions? Discuss how to ensure transparent communication.

5.20. How many communication paths and links are needed for a project with h horizontal functions and v vertical functions? Assume that when intersecting functions form a matrix of teams, each cell of a matrix denotes a communication node that reports back on two different paths. One is horizontal and one is vertical.

5.21. Are there logical reasons and possible explanations for late management involvement? What can you do to rectify these situations?

5.22. What problems are caused by management when they become involved too late in the product development stage? Can CPM help identify such problems? Does this have any effect in elongating activities on the critical path?

5.23. What is the best way to reward a team? How can you instill a high degree of motivation in CE teams?

5.24. A labor strike is an indication of failures to overcome conflicts or controversies. What factors do you think lead to such a nonproductive conflict?

5.25. Why isn't a matrix organization always better? What can we do to minimize communication paths?

5.26. Why is functional organization generally not conducive to CE philosophies? Discuss some of the key reasons why traditional organizations may find the task for teamwork, productivity improvements, and efficiency gain to be very painstaking?

5.27. Discuss potential difficulties expected in transitioning a functional organization to a goal-oriented based organization.

5.28. What are some common management styles? What are some characteristics of each style? Describe some pros and cons.

5.29. What could be the major pitfalls of a brute-force conversion of a traditional organization to a goal-oriented organization? How should we do it right?

5.30. Assuming that self-interests of individuals cannot not be relinquished, what is the best way to harness this quality?

5.31. What are the controlling mechanisms for dealing with human behavior and overcoming corporate culture? Does motivation play a role in the CE environment? If so, how strong?

5.32. What are the major concurrent teams required in product development and what key role does each team play in the CE process? What is the purpose of having a logical team in product development?

5.33. Why is team design considered a key to the success of the CE process? Present an argument for and against setting team management in organizations.

5.34. What are the seven principles of supportive management style? How do they differ from 7Cs? Explain the importance of each.

5.35. What are the three key managers that make up a PDT? Do the product and process breakdown structures described in Chapter 4 relate to PDT? If so, in what way?

5.36. Is there a need to establish, for each PDT team, a common vocabulary, priority, and purpose?

5.37. Are there any similarities between Figure 5.20 and Figure 2.5? Explain, if any.

5.38. Create a composition matrix (5*5) for the four CE teams The column header lists the four CE teams. The row header lists the types of cooperating teams, membership/expertise fields, functions they perform, and the expected output/assignments.

5.39. How can visual control help improve quality and productivity in the workplace?

5.40. Why is working as a team considered better than working as individuals? List seven factors that contribute to team development and effectiveness.

5.41. What are some characteristics of an effective work-group or a team? Who should lead a product team? What are the team leader characteristics? What are the advantages of the employee rotation program?

5.42. What are some obstacles that prevent effective communication? Can you describe three strategies to overcome conflicts between teammates?

5.43. When and how does the technology team play its role? What can we do to get our technology teams to share their knowledge early on?

5.44. A company has recently institutionalized CE work-groups, but after months of work there seems to be a great deal of conflicts within the group. What suggestions do you have for reducing the conflict and improving the effectiveness of the team?

5.45. You have been asked to recruit a person for a PDT position. What qualities do you look for in the person? What would be an ideal and what would be acceptable?

5.46. Must a design team be collocated? Does the physical layout of a CE office affect the success of the team or the work-group? How does collocation help team performance?

5.47. Is the whole always far greater than the sum? If not, when does that occur? At what points would you see no benefits from cooperation or teamwork?

5.48. Describe the critical elements to develop a successful workforce. Does team size have any effect on the team performance? What things can you do to cause behavioral change so that your company can reach the standards of the best-in-class?

5.49. Why is management involvement an important part of the design process of a product? What are the approaches to obtain management buy-in?

5.50. What factors do you see in your workplace that need changing in order to improve teamwork? What is a typical composition of a virtual team?

5.51. What are the techniques to use for avoiding conflicts? How are these conflicts resolved when they occur? What should we do when team members do not see "eye-to-eye"? List seven most common sources of conflicts.

5.52. What are some of the reasons for tense relations between automotive companies and their unions? What can be done to improve this situation and to incite enthusiasm and collaboration?

5.53. How can someone set up a company to strike a good balance between empowering the employees and yet maintaining some control over the process?

5.54. What are some of the things management should be concerned with other than how the team is doing? List three contingencies in the project plan. How can you use a safety net to make sure nothing falls through the cracks?

5.55. Answer the following questions on empowerment.

Questions	**Select the most appropriate answer(s)**				
In CE, who makes the design decisions?	(a) supervisors and managers	(b) single-disciplinary team	(c) mixed-discipline team	(d) every member of the enterprise team	(e) all of the above
Who participates in the design decisions?	(a) customer	(b) third-party vendors	(c) employees	(d) managers or supervisors	(e) all of the above
Who is responsible for (a), (b), (c) and (d)?	(a) scheduling and completing their tasks on time	(b) outcome of their tasks	(c) outcome of their group's tasks	(d) outcome of the entire project	
Who is responsible for the development of engineering specifications and their correlation to interdependent specifications?	(a) PDT	(b) employees	(c) customers	(d) managers and supervisors	(e) mixed-discipline team
Who is responsible for system design specifications?	(a) enterprise team	(b) PDT	(c) customer	(d) each employee	(e) managers and supervisors
Who should the management reward for good contributions and why?	(a) individuals	(b) team	(c) managers and supervisors	(d) suppliers	(e) all (f) none of the above

CHAPTER 6

SYSTEM ENGINEERING

INTRODUCTION

Over the years, unfortunately, many core industrial functions have grown to be component-centered. Various engineering functions in a product are conducted by autonomous component groups. The complexity of a product such as an automobile, helicopter, or airplane, has been one of the underlying reasons for compartmentalizing. System engineering does not disagree with the notion of compartments or divisions of works, but it emphasizes that the interface requirements between the divisions (interdivisional) and across the levels should be adequately covered [Prasad, 1991]. That way, when the time comes to modernize other components of the system, one has the assurance that previously introduced technologies and processes will work logically in a fully integrated fashion, thereby increasing the net efficiency and profitability. This amounts to both horizontal integration and vertical integration. This is, however, not the way products are traditionally designed today. Making tradeoffs, using disjoint goals, has been the way of life for engineering designers for decades. Such tradeoffs are often made based on limited goals of a particular unit, or a subset of the decomposed criteria. The reason for this is a phenomenon called psychological inertia. It is a tunnel thinking—a human tendency to "keep the blinders on" and not wander outside the tunnels of previous experience, or technologies beyond one's own "comfort zones." One antidote to psychological inertia is applying system engineering principles [Prasad, 1992]. It consists of framing a system's goal hierarchy (for a trade-free system) based on the constancy of purpose and applying the tradeoff criteria at appropriate levels in this hierarchy. Often the culture of a mature company, such as is common in the automotive industry, gets in the way while attempting to apply such system engineering methodology. Fortunately, many aggressive companies are waking up to this and are changing their structure to facilitate system engineering (SE) principles [Chelsom, 1994; Sawyer, 1994]. A glimpse into the production and manufacturing of an automobile reveals why.

6.1 AN AUTOMOBILE MANUFACTURING PROCESS

Most auto companies introduce a new car model every two to three years at a cost of a billion dollars per vehicle. The new car development program in the United States now ranges between three to four years, whereas in Japan it takes less than three years. Development is generally the responsibility of the operating platform groups, with new product sold by one or more of their marketing units. The major elements of an automobile include the following:

- ***Outside Body:*** It includes major designs for outside body parts, structures such as roof, moveable roof, body glass, quarter panel, fender, A-pillar, B-pillar, C-pillar, decklid, trunk, apron, shot-gun, vehicle tools, paints, etc. They are often designed by a staff group:
  - *Styling* is done by a central design staff with support from components' divisions and outside suppliers. They come up with the design of the outside contour, look, and feel, mostly from aesthetic considerations.
  - *Detail design* of the parts and body panels are done by the CAD/CAM shop contractors.
  - *Analysis* is proposed by the engineers but often performed by the analysts on contracts.
  - *Tooling and dies* are handled by process engineers internally and by outside prototype shops.

The following elements of an automobile are done by a prototype shop, one or more of the components' groups, or by first and second tier suppliers:

- ***Interior Systems:*** Instrument panels, air bag, steering wheel, door trims, door modules and related hardware, latching, window regulators, power closures, power sliding door, seat systems—seat trim, adjusters, recliners, frames, head rest, arm support, etc.
- ***Vehicle Wiring Systems:*** Ignition wiring, fiber optic data transmission, fiber optic lighting distribution, electrical/electronics connection, multiplex, wire harness—integration of electrical electronics into modular structures, temperature sensors, electronic modules and switches.
- ***Brake Systems:*** anti-lock brake, traction control, intelligent brake control, power brake assemblies, electric brake, disc and drum, corner assemblies, wheel spindle bearings, knuckles, calipers and rotors, etc.

- ***Suspension Systems:*** Suspension assemblies, controlled suspension, structural composites, integrated chassis, module suspension, powertrain mounts, etc.
- ***Climate Control Systems:*** Heating, ventilation and air conditioning, condensers, compressors, accumulator dehydrators, evaporators, heater cores, etc.
- ***Engine/Transmission Cooling Systems:*** Radiators, oil coolers, engine cooling module, etc.
- ***Engine Management Systems:*** Air fuel, ignition, fuel handling and evaporative emissions, electronic control modules and algorithms, exhaust system, valve train, etc.
- ***Energy Management System:*** Power generation and storage, batteries, generators, sensors and solenoids, electric vehicle, etc.
- ***Lighting Systems:*** Forward lighting, signal lighting, center high mounted stop lamps, distributed lighting, high intensity discharge lamps, etc.
- ***Vehicle Control Systems:*** Advanced steering, power steering, pumps, gears and hoses, variable effort steering, standard and adjustable steering columns, intermediate steering shafts, etc.
- ***Driveline Systems:*** Axles—front and rear, propshaft, halfshaft assemblies, constant velocity joints, intermediate drive shafts, boot seals, etc.
- ***Engine:*** Structural, crankdrain, valve train, cam drive, accessory drive, lubrication system, cooling system, air intake, PCV, combustion, exhaust, sealing and fastening assembly, etc.
- ***Transmission Systems:*** Transmission (auto and manual), torque converter, case and cast components, gears and shafts, mechanical components (clutches, free wheelers, chain drive), cooling and lubrication, sealing and fastening, dress components, transfer case, etc.
- ***Powertrain controls and diagnostics:*** Diagnostics, electrical, electronics, software, driver display, driver controls, sensors, actuators, other miscellaneous systems.
- ***Others:*** Entire program is supported by thousands of second- and third-tier suppliers that provide interior parts, bolt-in parts, and hundreds of other components and materials.

An operating group or platform normally is responsible for a particular line of automobiles—small or sporty cars for example. A corporation generally has engineering facilities at multiple cities and has assembly plants in multiple countries, (such as South America, Canada, and Mexico). Many of its plants are spread throughout the United States (e.g., the Midwest and South). Operations within an operating group are supported by an extensive vendor network or a supply chain.

During this three-year cycle of a new vehicle or product development process, an operating group or a platform must also build other car lines. This means, in the current year $199X$ (1995, if $X = 5$), manufacturing engineers would be building $199X + 1$ (1996, if $X = 5$) models' car. While process engineers will work on $199X + 2$ (1997, if $X = 5$) models, and product engineers will concern itself with $199X + 3$ (1998, if $X = 5$) product lines.

Other groups within the company must support these four groups: design group, process group, manufacturing group, and the operating group. For instance, design support groups may seek a balance among piece cost, manufacturing, assembly, fuel consumption (mileage), emission and safety regulations. The planning group may balance investments with budgets. Marketing groups may seek competitive concerns, such as styling, vehicle content, quality, and numerous other issues. These groups are often matrixed to each other to address common concerns. Since many of these groups are independent of each other, no one manager is likely to own the right or control of the total program. Funding and control of resources are usually decided through committees. Each group, thus, ends up doing (suboptimizing) their own things with lack of overall coordination between the groups. The problem is typical of a company where groups have too much independence and not enough coordination. System Engineering is found to be quite useful in such cases. General Motors, for instance, has recently announced a reorganization of its automotive components group worldwide (ACGW), now called DELPHI into six supporting units, seventeen technical centers, spanned over 190 locations in thirty-one countries around the globe.

SE techniques are really a concatenation of many good practices and concepts that have existed before in some form or other [Sawyer, 1994]. More and more of these are now being computerized. In previous chapters, we discussed techniques for cooperative work-groups, process reengineering, 3Ps, and life-cycle management. In fact, these four are the key components of systems engineering. Concurrent Engineering embraces SE, as a means to tie in the loose components.

$$\begin{aligned}\text{SE} \equiv \text{Concatenation-of [Cooperative Work-groups, Process Reengineering,}\\ \text{3Ps, and Life-cycle Management]}\end{aligned} \tag{6.1}$$

In this chapter, we describe this methodology.

6.2 SYSTEM ENGINEERING

System engineering is a structured methodology for designing and developing a product. It does not require changing the existing process setup (like what paralleling of tracks/tasks in concurrent engineering does to the conventional process) as much as it emphasizes the need to appropriately decompose and classify distributions of tasks. The methodology requires that the product realization process be viewed as a *"systems-centered"* problem as opposed to a *"components-centered"* problem [McPherson, 1986]. SE enforces interface compatibility and apriori deals with interface situations, before solutions are tackled (at any level). It requires allocation of team responsibilities to all logical levels to assure that the levels do interact when needs arise. Such up-front allocation of responsibility and authority in system engineering is critical to CE success. It challenges the traditional way of doing things, which has been mostly components-centered. In a components-centered philosophy little attention is paid to the interfaces or the interfacing work-groups developing these components. Subsystems, components, or parts of a system are more or less considered as a series of independent units. Most groups diligently work to optimize their subsystems, but due to lack of incentives they tend to work independently of each other. This results in a product which is often suboptimized at each decomposed level. If the components are independently optimized, it is very rare that optimizing at each decomposed level in its hierarchy will insure an optimized system at the end. Most components in reality do interact. A component is far more than a decomposed element of a product. A product is far more than a collection of components. And a component is far more than a collection of parts. Ideally, after a components' suboptimization, subsystem optimization is required with variables that are common to the components making up the subsystem. The product, in that case, is bound to perform well if the components are also orchestrated to perform best—first at the subsystem level and then at the system level.

In Japan, extensive attention is paid to the design of the processes. The Japanese tend to study and improve the process through environmental knowledge, team cooperation, and consensus [Whitney, 1992]. This practice follows Galbraith's theory of contingency that there is no one best way to organize [Galbraith, 1974]. The best organization depends upon the tasks' uncertainty and their mutual dependency. Such a difficult and unpredictable nature of the tasks makes the "product and process design jobs" everybody's job. In European countries and the United States, more attention is being directed toward development of new concepts and tools for integrated design and manufacturing. With emphasis now shifting from "product development practices" to "integrated system engineering," CE is getting more and more aligned as a concept of total "product realization." CE, as it stands today, is not just limited to a paralleling of activities or simultaneous actions in product development. CE is shaping up as an integrated, highly cooperative, team-oriented, computer-supported tool communicating on the network for total product realization.

$$\begin{aligned}\text{CE} \Leftrightarrow\ &[\text{Paralleling of activities or simultaneous actions in product development}\\ &+ \text{Integrated, highly cooperative, team-oriented work-groups}\\ &+ \text{Computer-supported tool communicating on the network}]\end{aligned} \tag{6.2}$$

6.3 SYSTEMS THINKING

System engineering is a systematic approach to addressing a product problem or a process situation. It is aimed at looking at the problem as a whole rather than looking at the individual subpieces. It provides a consistent frame of reference and a way of fitting the constituents together as they are approached one by one. Most people view the product realization problem from one particular angle—the angle with which they are most familiar. They seem to pay attention to those constituents of the problem that happen to fit their views depending on their own specialized understanding or knowledge of the problem. However, the wisdom of one's choice depends on what is happening in the rest of the whole. If there is no understanding of which part influences which constituents of the system, (that is, of the big picture) then it is difficult to come up with a balanced decision.

This concept is best illustrated in an article written by Draper L. Kauffman, Jr. [1980]. He defines a system as "a collection of parts, which interact with each other to function as a whole." The air conditioning system in a car consists of a heater coil, freon, a compressor, a thermostat, a heating element, fan, and several pipes and connectors. Together they keep the inside area of the car fully conditioned. Though individually these parts serve a purpose in their own right, they cannot provide a system function until they are all connected. To do the job, "all" the parts must be present and they must be arranged in a proper way. Thus, when parts are taken apart, this assembly knowledge is lost. Deming also refers to "the system as a network of interdependent constituents that work together to accomplish the aim of the system" [Deming, 1993].

6.3.1 Elements of a Functional System

In order for the system to be functional or work repeatedly, the following must be true:

- ***Parts of a Whole Function:*** Each part must provide or contain a set of partial but distinct functions by itself—functions that can be distinctly identified as associated with each part of the whole.
- ***Whole of the Parts:*** A part or a combination of parts should work together to accomplish the function (aim or purpose) of the whole system. Each is capable of contributing a functional build-up leading to the system's functionality when every part is in place.
- ***Structure or Constancy of Purpose:*** There is a definite order (in some instance priority) in which these parts need to be arranged to work as a unit. The system requires a structure or a "constancy-of-purpose" to yield the system's functionality.

$$\text{Structure} \Leftrightarrow [\text{Enterprise, Product, Assembly, Work, and Process Hierarchy}] \tag{6.3}$$

Without a structure or a "constancy-of-purpose" there is no system. Take for example, an exploded view of a product's subassembly. The view maintains the assembly sequence by displaying a principal axis of a subassembly and by placing the mating parts adjacent to each other with respect to a base part (datum). The precedence information is explicit in the exploded views. Without proper assembly plans or sequences, there would not be any functional assembly. Structures, like a cellular manufacturing organization or robots in a flexible manufacturing system, are what makes factories work.

$$\text{System} \Leftrightarrow [\text{Parts of a Whole Function} + \text{Whole of the Parts} + \text{Structure or Constancy of Purpose}] \tag{6.4}$$

If something is made of parts and it does not matter how these parts are arranged, then we are dealing with a collection (a "heap") and not a system. A pile of sand, a collection of stones, a mob, parts in bins, candies in a candy dispenser, or a pail of water are examples used to describe heaps. There are three differences between a collection and a system, as shown in Figure 6.1: A collection or heap has no structure; it exhibits randomness; the size and location of the particles have no influence on the outcome. A heap is uncorre-

What is a System?	
Collection	**System**
Heap: A pile of stones A glass of water A mob Parts in bins, boxes, piles Garbage Randomness	Order: A row of stones A measuring glass A parade Parts in pallets, trays Carrier, tapes, rows Structure
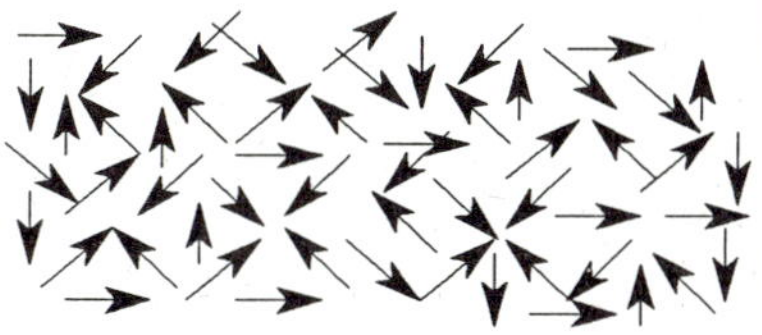	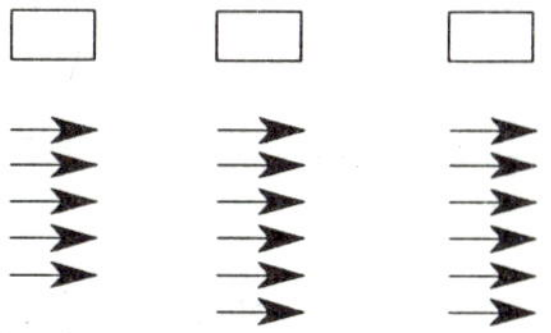
No Correlation	Constancy-of-Purpose
Collection has no structure, exhibits randomness, disorder, jumbles, improvisation, etc.	System has a unique structure, exhibits predictability, order alignment, prearrangement planning, and goals governed by "constancy-of-purpose."

FIGURE 6.1 Difference Between a Collection and a System

lated. If we stir a heap, it will remain a heap, since each part does not have a unique enabling or supporting functional role. However, when the external environment is controlled, for example, when water is poured into a measuring glass or a container, it becomes a system. The glass, along with its markings, now serves to define the volume the substance occupies. The measuring glass, markings, and the weight-summary chart together provide a purpose. This leads to a very interesting postulate.

Postulate: What is important or best for the whole system is nearly identical to what would be the best in the long run for each of its constituents. However, what is best for each individual constituent may not be the best for the whole system.

This is best illustrated in the example of a throughput maximization for a production system. Figure 6.2 (a) shows a production system in which raw materials—Mat1, Mat2, and Mat3 enter through the bottom and are processed through the system to become "product 1" and "product 2" (shown at the top of Figure 6.2). The production process first starts with

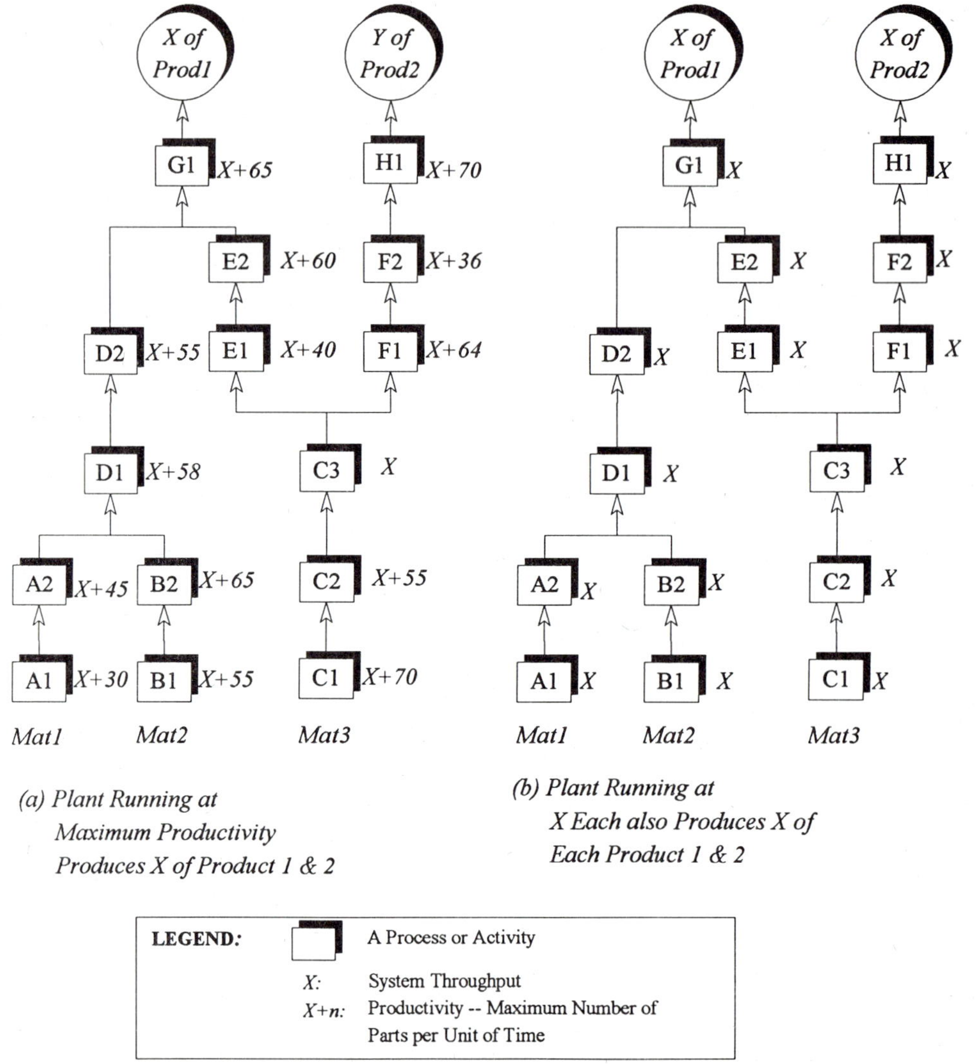

FIGURE 6.2 System Throughput Limits Individual Productivity

three parallel independent units: A, B, and C. Mat1 is processed by operation series A; Mat2 is processed by operation series B, and Mat3 is processed by operation C concurrently. An operation is shown in Figure 6.2 by a rectangular block. Later operations A and B are assembled to form a new operation series D. For example, the process D1 is an assembly operation, which requires one unit from A2 and one unit of B2 to be assembled together. The series Cs are decomposed into two operation series Es and Fs. Later, series Ds and Es are combined to create a new operation G1. The process G1 is an assembly operation requiring one unit from E2 and one unit from D2. The output of the production system is two separate products. The operation G1 results in a product 1 and operation H1 results in a product 2. The number shown next to each process or operation is the capacity—the maximum number of parts per unit of time (for example, per hour or day) that an operation can produce. As an example, operation A1 can produce up to $(x + 30)$ parts in an hour. The question is what is the throughput of product 1 and 2? Examining the system, clearly the maximum throughput of this production system cannot exceed X units. It is based on the realization that system throughput is limited by the operation with the least capacity. In this example, operation C3 is a bottleneck. Even though all of the other operations in the system are capable of producing more than X units, the system itself cannot produce more than X. This means that there is no difference between throughput of a production system, shown in Figure 6.2(b), where capacity of each plant is kept strictly at X units. Though, in practice, there may be other reasons to keep this number above X. Inevitably, any call for increasing output from each individual unit, team, or operation will have no apparent effect on the throughput of the system. In a normal situation, it will contribute to an excess WIP (work-in-progress) inventory that will keep on growing.

Each system has a particular function, and only with all parts present and arranged in the proper manner can the system function as a whole. The proper manner could take the form of a common set of consistent goals defined for the parts with constancy of purpose. Instead of pouring water into a glass, if we put the glass into water, it will merely sink and will not serve any purpose.

Just as measuring can be a part of measurement systems, there can be a unit, which consists of a family of larger units. Consequently, the very nature of a system makes it a part or subsystem of another larger system and in turn an umbrella system of its own subsystems. If one is called a subsystem, then the larger system, of which it is a part of, can itself be a subsystem of a still larger system. In fact, this pattern of systems being part of a larger system is very common in the concept of object-oriented programming, where one object encapsulates many other objects. One layer of this hierarchical pattern of a system's structure is also referred to as a resolution level. Whatever is defined to be "the system" at a particular time is a choice of resolution level or level of detail at which the activities are described at that particular time [Hales, 1993]. At any particular level of resolution, there is a number of alternate viewpoints to consider. If a resolution is considered as the vertical structure of the hierarchy, the viewpoint can be considered as a horizontal extension to that structure. By mapping a problem or a project into resolution levels and by mapping a particular resolution level into viewpoints as shown in Table 6.1, it

TABLE 6.1 Mapping of a Problem into Vertical and Horizontal Tasks

Product Hierarchy	Vertical Structure	Horizontal Extension to Vertical Structure
System	Resolution Level 1	Viewpoint A_1, B_2, C_3, . . .
Subsystem	Resolution Level 2	Viewpoint A_2, B_2, C_2, . . .
Components	Resolution Level 3	Viewpoint A_3, B_3, C_3, . . .
Parts	Resolution Level 4	Viewpoint A_4, B_4, C_4, . . .
Materials/Features/Parameters	Resolution Level 5	Viewpoint A_5, B_5, C_5, . . .

decomposes the original problem into a set of manageable vertical and horizontal structures. This in turn provides a systematic way to predict the likely influences of 7Ts characteristics (explained in Chapter 4, Figure 4.1) on the overall project goals.

6.3.2 Advantages

There are many advantages of the system engineering approach to concurrent problem solving:

- ***Easy to grasp or learn new things:*** By relating constituents of a system (say a product or a service) and studying their dependencies on other subsystems or their constituents, a basic knowledge is built as to how things behave in reference to each other. Such accumulation of knowledge and its analysis builds the product theory. It is then easy to predict (or understand) differences in behavior if a constituent is modified, replaced, or eliminated.
- ***Establish a global view or perspective:*** Applying a system engineering approach to a product helps create a pattern of behaviors when rules of association between parts are changed or varied. In other words, it establishes a product theory. From the product theory, the order selected will determine whether such a pattern will function as a unit or not.
- ***Basis to analyze new phenomena:*** System thinking provides a CE theory which in turn provides a basis for concurrent prediction. Without prediction or theory there is no learning—experiences and examples teach nothing [Deming, 1993]. Learning is the systematic revision or extension of the theory based on the comparison of the old predictions with the new measurements or observations. As children grow up they first pick up things in pieces, they learn to socialize, to laugh, respond, obey, etc. Once they understand the basic rules of behavior (theory), they can tackle (predict) each new subject or activity by building (comparison) on what they have already learned.
- ***Understand complex behavior or phenomenon:*** By dividing the product or the problem into manageable pieces, the cause and effect of each part on the related parts becomes clear. This is true even if a problem or a product is dependent on the behavior of these pieces. The chances of an accurate prediction, even about an unknown subsystem, also increases even if these pieces are used as a basis for deriving the results of this unknown subsystem. This is because, in this physical world, most phenomena are interdependent.
- ***Identify effective paths:*** A system engineering approach helps to determine the relative effects of following different design or process alternatives. Since we can predict the outcome by changing the constituents of a system, it is possible to identify differences. On this basis we can come up with a suitable path (design or process) to reach the destination by considering various possible routes that may become visible.
- ***Help in decision making:*** System thinking can help reach a consensus on what might work and what might not, when applying X-ability rules. Many X-ability

agents or subteams have partial knowledge of how bits and pieces work, but they cannot comprehend how a combined system would behave when all these bits and pieces (X-ability rules) are assembled. The system engineering approach provides a consistent frame of reference for reaching an amicable decision.

Constancy of purpose, or structure, is a property that gives a product or a process robustness in functional realization. The job of CE teams is to figure out how to create a structure (enterprise, product, system, work, and process hierarchy). The hierarchical breakdown structure is further explained in Section 8.1 (Chapter 8). The systems approach provides a reference point irrespective of whether a product has been broken into constituents using a top-down approach or a bottom-up approach. The results of these approaches are usually the same:

- In the **bottom-up approach,** the in-line functional models and the processes it drives can be used to assure system-level performance when subsystems are being designed. Similarly, the functional models and processes can be used to assure the subsystem level performance, when components are being designed. The constancy of functional models and processes assure the component's level performance, when parts are being designed. The assurance of performance for a higher level constituent continues when a lower level constituent is being designed.
- In the **top-down approach,** the constancy of the in-line functional models and processes also ensures that the appropriate requirements are addressed at each level. Following this, the corresponding constituents' functionality is ensured in succession. The in-line models for subsystems ensure components-level functionality, when subsystems are being designed. The in-line models of the components ensure the parts' level functionality, when components are being designed and so on.

Chapter 9 describes the techniques for capturing the product and process specifications into a set of taxonomies.

6.4 APPROACHES TO SYSTEM COMPLEXITY

There are a number of approaches to dealing with the system complexity that have evolved in the last couple of decades. The initial attempt goes as early as in 1964, when Alexander proposed partitioning the design process into minimally coupled groups [Alexander, 1964]. Simon [1981] later suggested that complexity can be handled by dividing the original problem as a set of "nearly decomposable groups." These groups could be organized as hierarchical structures such that the strongest interactions occur within groups and weaker interaction occurs across groups [Simon, 1981]. One most widely used approach to deal with complexity is to decompose the system into a set of hierarchical structures. The system may span into sets from subsystems to components, to parts, to materials/attributes/features/parameters, and then to common representation and stan-

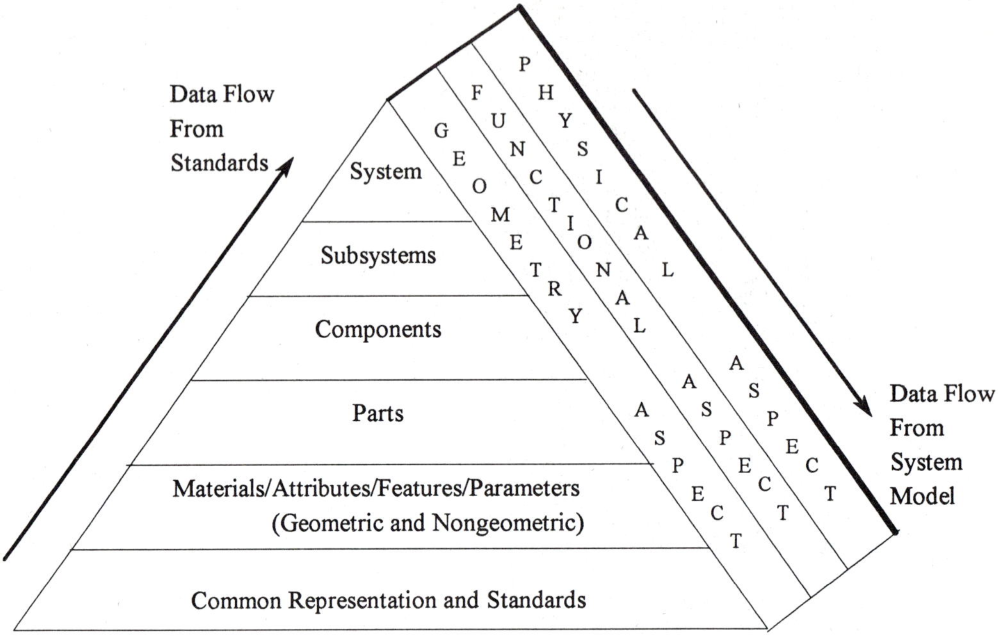

FIGURE 6.3 Hierarchical Decomposition of a Product

dards (see Figure 6.3). The product breakdown structure (PtBS) forms the basis for hierarchical descriptions of the CE design process. This can be done whether the CE design process is tightly coupled or highly integrated.

$$\begin{aligned}\text{PtBS tree} &\Leftrightarrow \text{[Class descriptions of the product tree breakdown structure]}\\ &\Leftrightarrow \text{[(Subsystems, Components, Parts, Materials/features/}\\ &\quad \text{parameters, Common representation and standards)]}\end{aligned} \tag{6.5}$$

Each parent class in the above set can be further decomposed into their children. Breakdown of the part definition tree, for example, has two main elements: the primary features class and the materials specification class (see Figure 6.4).

$$\begin{aligned}\text{Parts tree} &\Leftrightarrow \text{[Class Description of the part]}\\ &\Leftrightarrow \text{[(Primary Features, Materials, etc.)]}\end{aligned} \tag{6.6}$$

Primary features define the basic shape of the part by progressively chaining the lower level characteristics (a set of form features and assembly information). Nongeometric information, such as tolerance and surface finish quality, can be assigned as attributes to the form features class. Five types of form features are readily discernible in mechanical products: atomic, simple, compound, pattern, and nary [Taylor and Henderson, 1994].

$$\text{Form Features} \Leftrightarrow \text{[Atomic, Simple, Compound, Pattern, and Nary]} \tag{6.7}$$

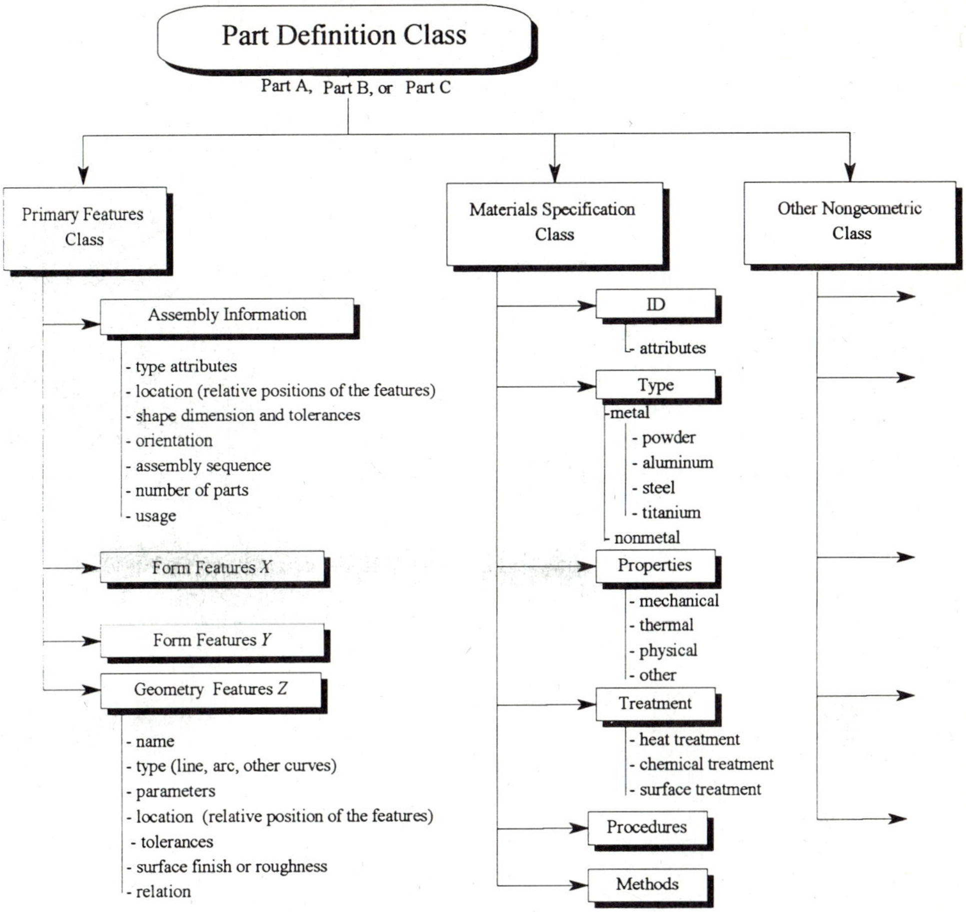

FIGURE 6.4 Elements of Part Definition Class

The material specification class defines the identifying characteristics such as material ID, type, compositions, material properties (E, ν, ρ, σ, . . . , etc.), and material treatments (heat, chemical, and surface treatments). Process information (such as equipment, procedure, and method used to manufacture the parts), is assigned as an attribute to the material class.

Interactions among the various product constituents lead to a set of desirable behaviors, but they also produce a set of undesirable characteristics or behaviors that need to be controlled. This leads to imposition of a set of necessary constraints on each design. Such constraint impositions are essential to control any adverse design outcome and to ensure a sound economic solution corresponding to some good CE manufacturing practices. The finer the decomposed tasks, the less complex it is to solve each, because the corresponding sets of specifications for solving each of these smaller sets are smaller. However, the amount of efforts required for managing the information resulting from decomposed parts increases proportionally. A large number of tasks require an equally large number of teams or work-groups to solve them. Depending on the teams and the teams' interdependencies, there will be an equal number of interfaces. The greater the interdependence between tasks are, the greater will be the need for communication and cooperation between them [Deming, 1993]. It is shown in Figure 5.18, that if there are "*n*" decomposed tasks, the communication effort for the decomposed set is proportional to:

$$\text{Communication Effort} \propto [n * (n - 1)] \tag{6.8}$$

Where n is the number of communication nodes.

Figure 6.5 shows the variation of the product or process complexity and the corresponding level of communication efforts required. As shown, there is a particular level of granularity in the decomposed parts that provide an optimal balance. When the level of decomposition reaches this optimal level, the effort of communication is manageable and at the same time, the level of complexity is also reduced. Additional means, such as an interval for time dependencies, impact analysis, risk assessment, knowledge processing, and neural techniques, can predict unknown domains and ease the communication burden. If the optimal grain falls within a zone (50% granularity point with 10% deviation from this point), the relative increase in the complexity and the amount of efforts are considered

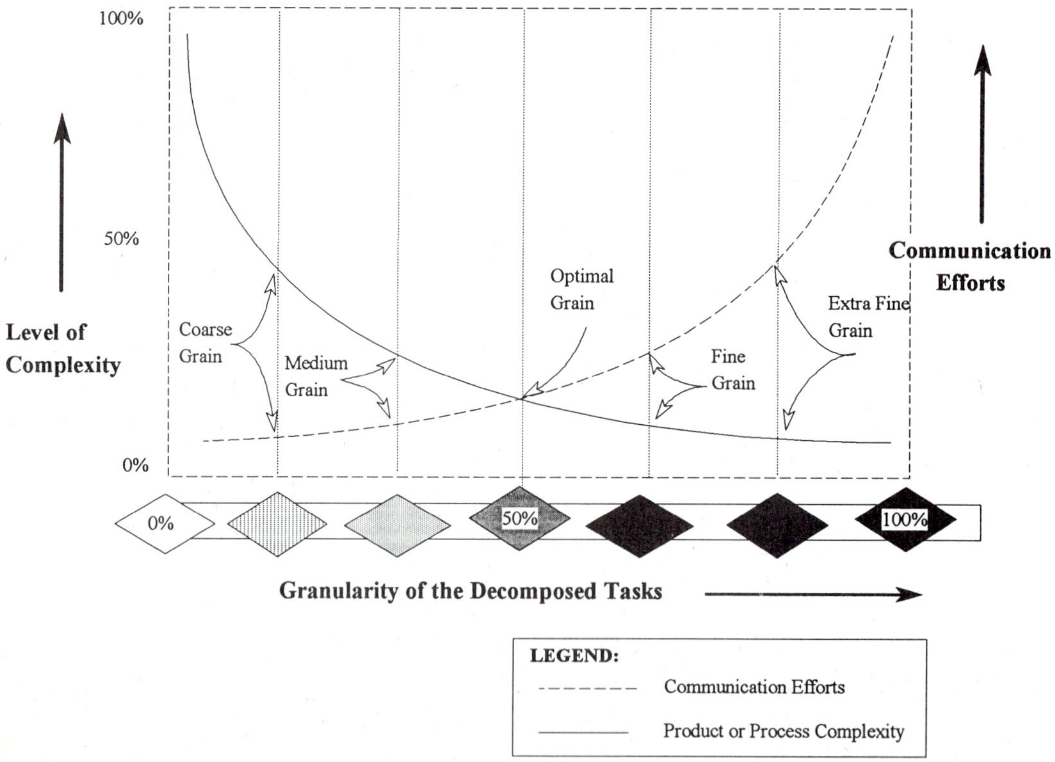

FIGURE 6.5 Right Amount of Decomposition to Manage Both the Complexity and the Communication Efforts

under control. As shown in Figure 6.5, the corressponding increase tends to stay within a 10% range too.

6.4.1 Product Decomposition

Decomposition is widely used as a basis for exploiting concurrency in IPPD [for example, Arai, and Iwata, 1990; Kusiak and Wang, 1993]. A product can be decomposed in a number of ways:

- Physical-based decomposition (PhD)
- Function-based decomposition (FuD)
  Engineering Functions
  Manufacturing Functions
- Activity-based decomposition (AcD)

$$\text{Decomposition} \equiv \cup\ [\text{Physical-based, Function-based, Activity-based decomposition}] \tag{6.9}$$

6.4.1.1 Physical-based Decomposition (PhD)

It is the complexity of the products and of the processes, present in the system, which compels a product manufacturer to look for the product and process breakdown structures. This breakdown is necessary to exploit any inherent concurrency so that the individual tasks can be overlapped (run in parallel). Physical-based decomposition is one way to achieve this parallelism as shown in Figure 6.6. Perspectives represent the first level of physical-based description (PhD). PhD is also commonly referred to as product holistic decomposition. Other possible levels of PhD (into which product can be decomposed to exploit concurrency) are hierarchy, multiplicity, alternatives, characteristics, and projects (see Figure 6.6). Please note that "decomposition" is not intended here to mean clustering the problem parameters in different ways. A typical case of this type occurs when a problem is decomposed simultaneously into a number of ways (such as program phase, subsystem and discipline). The term decomposition is used here to mean "product holistic decomposition (PhD)." Parameters are not fragmented into separate, decomposed sets. All parameters belonging to a particular class or family stay together (after decomposition) and collectively influence the decision making process.

$$\text{PhD} \equiv \cup\ [\{\text{Perspectives}\}, \{\text{Hierarchy}\}, \{\text{Multiplicity}\}, \{\text{Alternatives}\}, \{\text{Characteristics}\}, \{\text{Projects}\}] \tag{6.10}$$

Where the curly bracket—$\{x\}$ denotes the set of quantities of type x. The following's a list of some typical decomposition scenarios of a PhD:

A. ***Perspective:*** A design problem usually involves multiple perspectives. Each may have its own set of constraints and could interact with each other. At the highest level, different work-groups can work in parallel on separate competing perspectives of product life-cycle concerns. Such concerns are often required for product evaluation or assessment. These perspectives include the intellectual process of commonality or class hierarchy between different families of products:

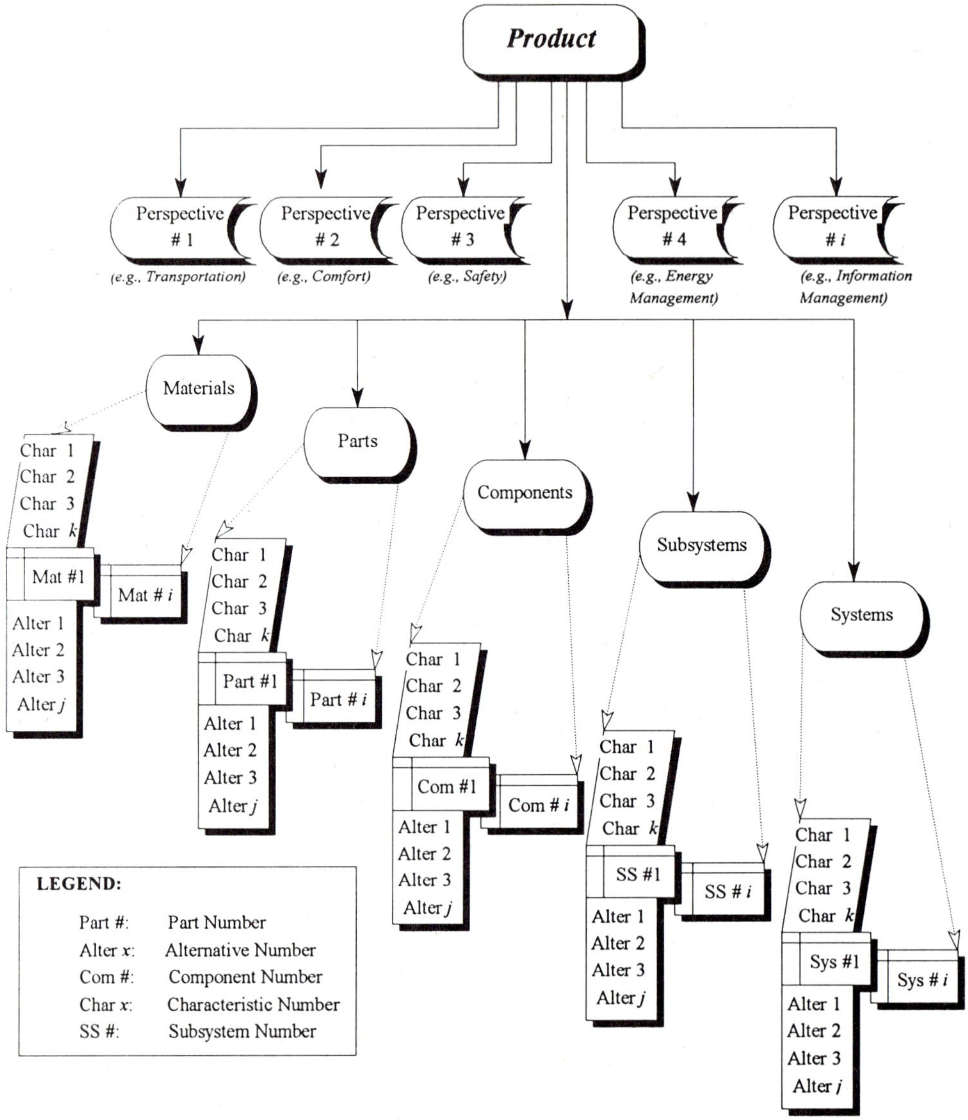

FIGURE 6.6 Areas of Concurrency During Product Synthesis (Bottom-up Representation)

- Size-wise (large, medium, or small)
- Model-wise (sporty, coupe, or luxury)
- Engine-wise (4-cylinders, 6-cylinders, 8-cylinders, etc.)

$$
\begin{aligned}
P_{\text{size-wise}} &\equiv \{\text{extra-large, large, medium, small, } \ldots\} \\
P_{\text{model-wise}} &\equiv \{\text{luxury, sporty, coupe, } \ldots\} \\
P_{\text{engine-wise}} &\equiv \{\text{4-cylinder, 6-cylinder, 8-cylinder, } \ldots\}
\end{aligned}
\tag{6.11}
$$

Where, the letter *P* denotes the perspective. A class-hierarchy can be based on usage (passenger car, commercial vehicle (jeep/truck), and bus). One common perspective used during organization and management of information is a combination of size, usage, and marketing perspectives. This, for the automotive industry has transpired into a triad spanned by three axes [Davis, 1994]. This is schematically shown in Figure 6.7:

- The vertical axis shows the division by platform types. The common platform types for automobiles are small car, mini-van, large car, jeep/truck. The subclasses of jeep/truck can be jeep, pick-up, vans, tractor-trailer, multipurpose vehicles, etc. Powertrain (engines and transmissions) is also categorized as a type of a platform, since it cuts across all major car lines.
- The horizontal axis lists the division by program types, such as P90, W, or C car programs. There is usually a product development team (PDT), similar to what is described in Chapter 5, that is responsible for each car program. They are generically named here as PDT A, PDT B, PDT C, PDT D, . . . , *X*, etc.
- Concurrent to each program there usually is a third dimension now commonly called centers. These centers perform tasks such as marketing, design office, engineering, finance, procurement and supply, manufacturing, service and parts, and product support. They are either dedicated services to a program or are matrixed across several programs: such as PDT A, PDT B, PDT C, etc.

Defining product breakdown structure (PtBS) tree to perspectives attaches additional meaning and order to the complex product design process. A passenger automobile's basic product structure, for example, (four wheels, 4–8 cylinders, reciprocating gasoline engine in front, round steering wheel, two to four doors, one to two

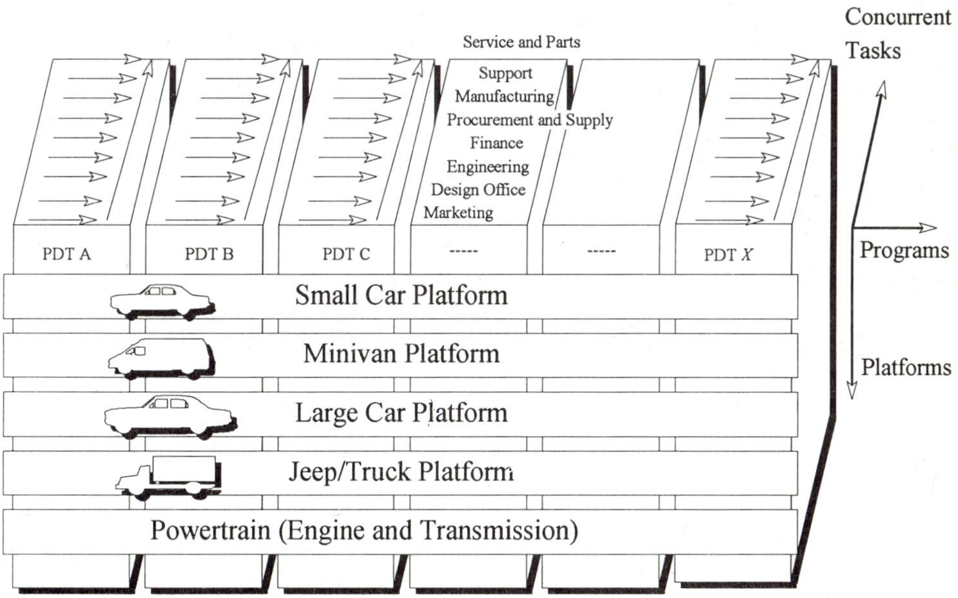

FIGURE 6.7 A Three-dimensional Concurrent Triad (Automobile Manufacturing Example)

rows of seats, interior instrument panel, trims, etc.) has not changed much in three decades. Many new models have been introduced but these inherited the basic concept of the automobile.

B. ***Hierarchy:*** The physical product or the "product system" may be divided into several logical hierarchical blocks or classes depending on its complexity. The advantage of this logical division is that different people can work in parallel in these different hierarchical blocks. The associated teaming between groups of people in a large manufacturing organization is discussed in Chapter 5. If separate teams are assigned to each class and subclass, they can work concurrently. A product breakdown structure (PtBS) example for an automobile class is shown in Figure 6.8, an example for an aircraft class is in Figure 6.9, and an example of a helicopter class is shown in Figure 6.10.

$$\text{PtBS tree} \equiv \cup\,[\{\text{system}\}, \{\text{subsystems}\}, \{\text{components}\}, \{\text{parts}\}, \{\text{materials}\}, \{\text{characteristics}\}] \tag{6.12}$$

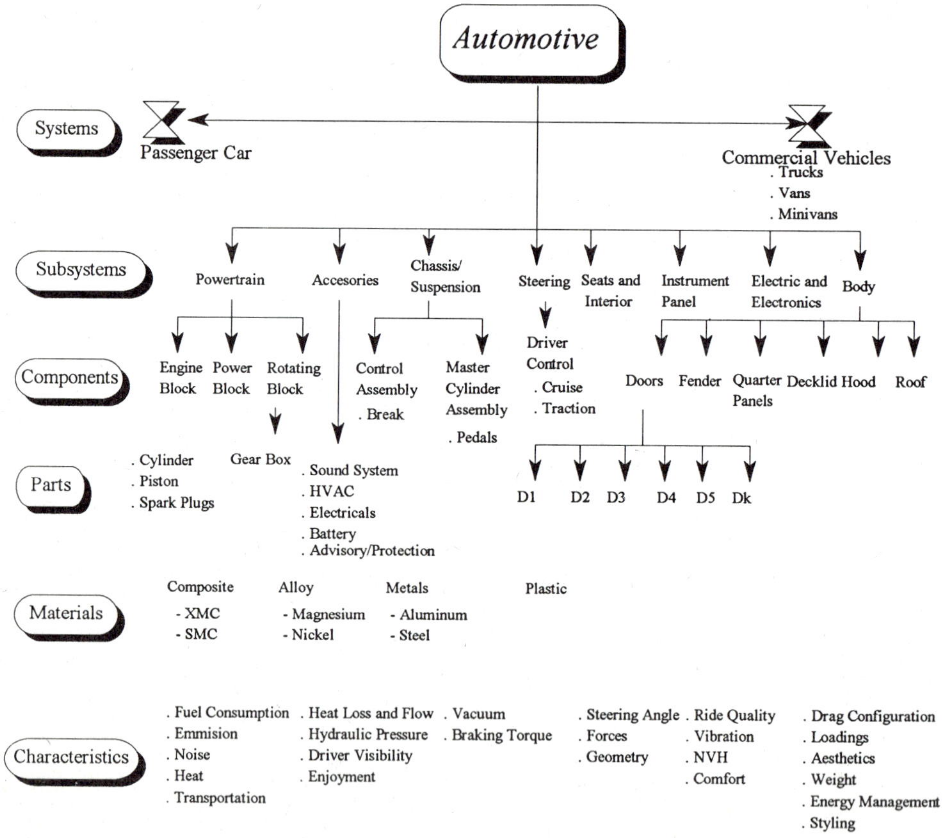

FIGURE 6.8 Areas of Concurrency in an Automotive Top-down Decomposition—An Example

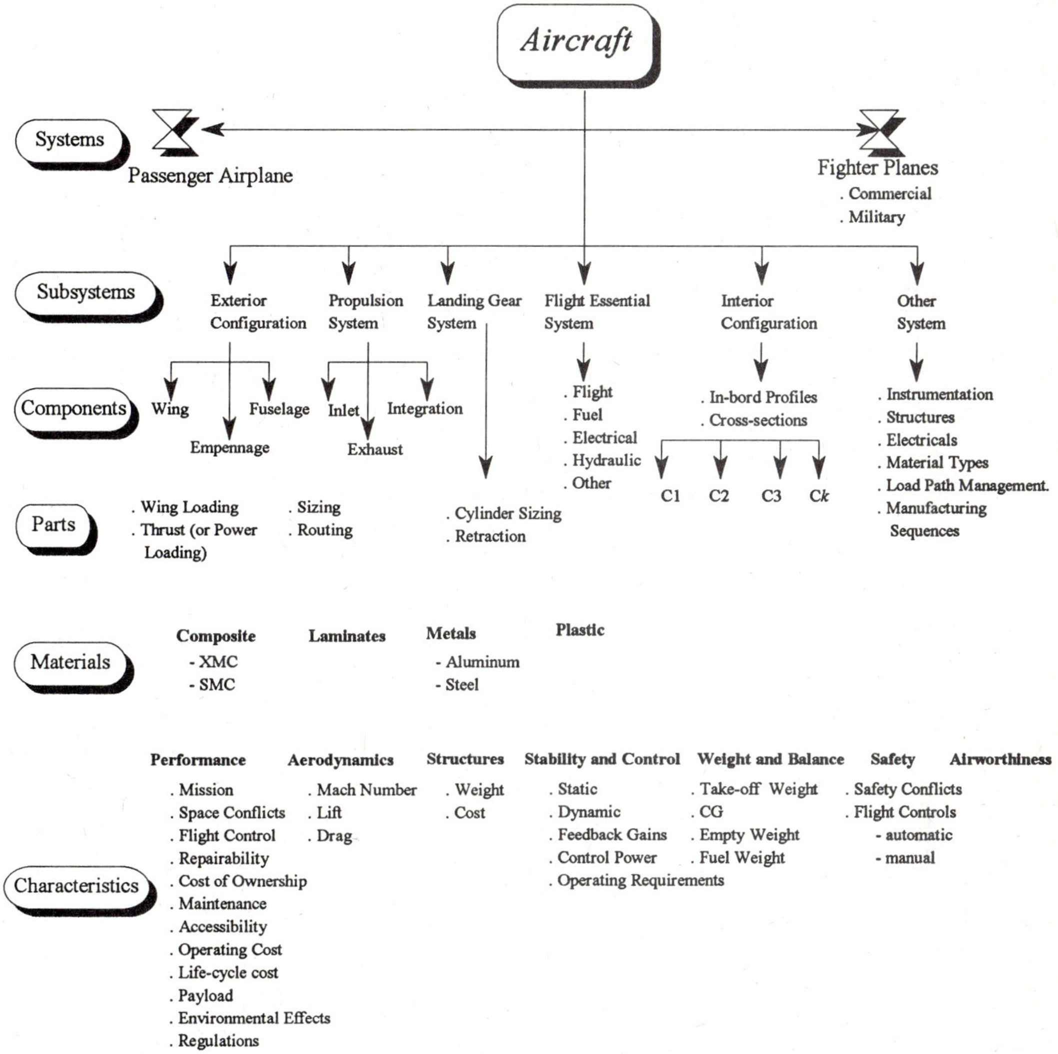

FIGURE 6.9 Areas of Concurrency in an Aircraft Top-down Decomposition—An Example

The PtBS can be superimposed on work-groups involved in the system design, with supporting subteams dealing with subsystem design, and another set of subteams handling the remainders, such as components, parts' design, materials, form features, etc. A nested routing work-flow model can be drawn starting from the bottom and showing the activities of each of the PtBS's trees leading up to the system model as information builds up. Some dependencies can exist between the branches. An important job of the CE work-groups is to recognize and manage the interdependency between the PtBS nodes. Establishing common interface standards

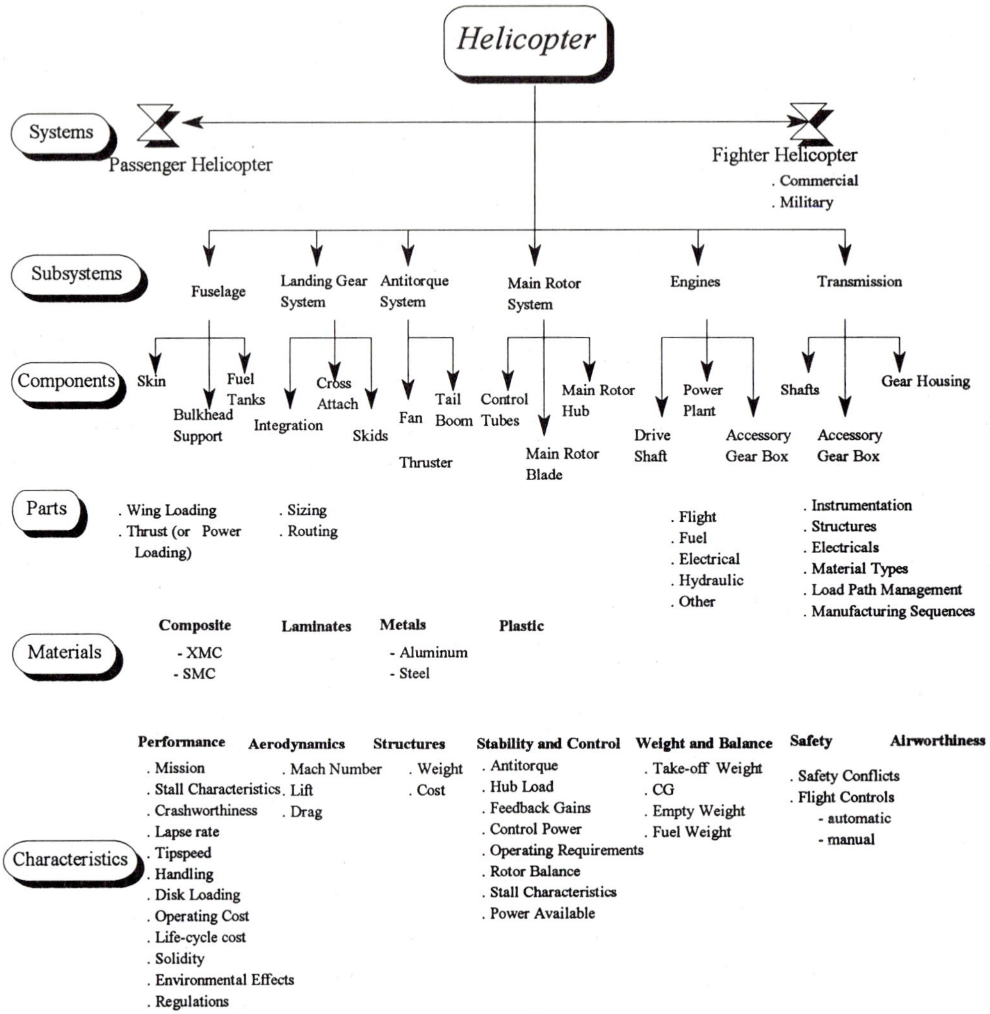

FIGURE 6.10 Areas of Concurrency in a Helicopter Top-down Decomposition—An Example

for communications and common dictionary definitions of problem parameters and checkpoints can allow parallel groups to work concurrently. Checkpoints are essential to ensure the smooth coupling of completed tasks. This is accomplished by staggering the product breakdown structure (PtBS) tree as shown in Figure 6.6. For example, the system level tasks can only begin when tasks for subsystem track are already well underway. The subsystem level tasks can begin only when tasks for the component's track are well underway and so on. PtBS organizes a product hierarchy by using a step-wise refinement and differentiation technique. Step-wise refinement adds hierarchy to the structure and differentiation adds distinguishing details at a particular level. Product or process features, materials, attributes, and parameters provide the lowest level of hierarchical abstraction. The amount of granularity present at each level is usually a function of the product and process complexity and their knowledge, such as knowledge of objects, functions, design cases, and needs. Object knowledge for product (i.e., topology, geometry, etc.) provides attributes, structures, assembly, and their relationships. Functional knowledge produces evidence for hierarchical decomposition (systems, subsystems, components, parts, . . . , etc.). Design cases or case histories provide additional evidence of breaking the hierarchy into alternatives, characteristics, etc. During the differentiation technique, different characteristics and alternatives can be assigned to a PtBS tree as shown in Figures 6.8, 6.9, and 6.10. The PtBS tree drives the product or the process design to a manageable set of units and nodes that can be independently worked on by the work-groups or the concurrent subteams.

The illustrations in Figures 6.8 through Figure 6.10 show how a hierarchy of system decomposition would look if we started with system assembly of an automobile, aircraft, or a helicopter and worked our way down from this top level. Each decomposed element combines with other decomposed elements of about the same level to make up the next larger level. Strategy (what services to render and to whom) and processes (how to convert inputs to outputs and how to deliver outputs to the customer) practically determine expected quality level, productivity, costs, and profitability.

C. ***Multiplicity:*** Within each different hierarchical group, for example, a part or a component group, multiple parts or components going into the final product may be worked upon simultaneously (see Figure 6.11).

$$\text{Parts of a PtBS tree} \equiv \cup\,\{\text{part\_1, part\_2, part\_3}, \ldots, \text{part\_}n\} \tag{6.13}$$

As such, a decomposed element of a PtBS tree can be a quantified set.

D. ***Alternatives:*** Within one hierarchy level, a group of designers guided by its hierarchy leader may work on several alternative ideas in parallel.

E. ***Characteristics or Aspects:*** Each alternative idea may involve integrating some life-cycle aspects, that is, validating output through compliance from multiple characteristic views. Where each characteristic view may represent a different life-cycle aspect such as aerodynamics, noise, ride quality, NVH (noise, vibration, and harness), strength, stiffness, energy management, packaging, etc. Subteams from different disciplines and background may be needed to support these aspects. These subteams can work in parallel on each characteristic view.

F. ***Projects:*** Multiple projects, such as predictive analyses, fault tree analysis (FTA), DFMA/FMEA analyses, may be required to evaluate product compliance to functional specifications. Many analysis subteams may be working in parallel to determine the integrity of the design with respect to these specifications. Additional details, increased accuracy, and other aspects of alternative designs may be considered as examples for projects.

6.4.1.2 Function-based Decomposition (FuD)

The function-based decomposition (FuD) is similar to physical-based decomposition (PhD). In FuD, the breakdown distribution is based on decomposing the products into a functional hierarchy. In PhD, distribution is largely based on existence of physical parts.

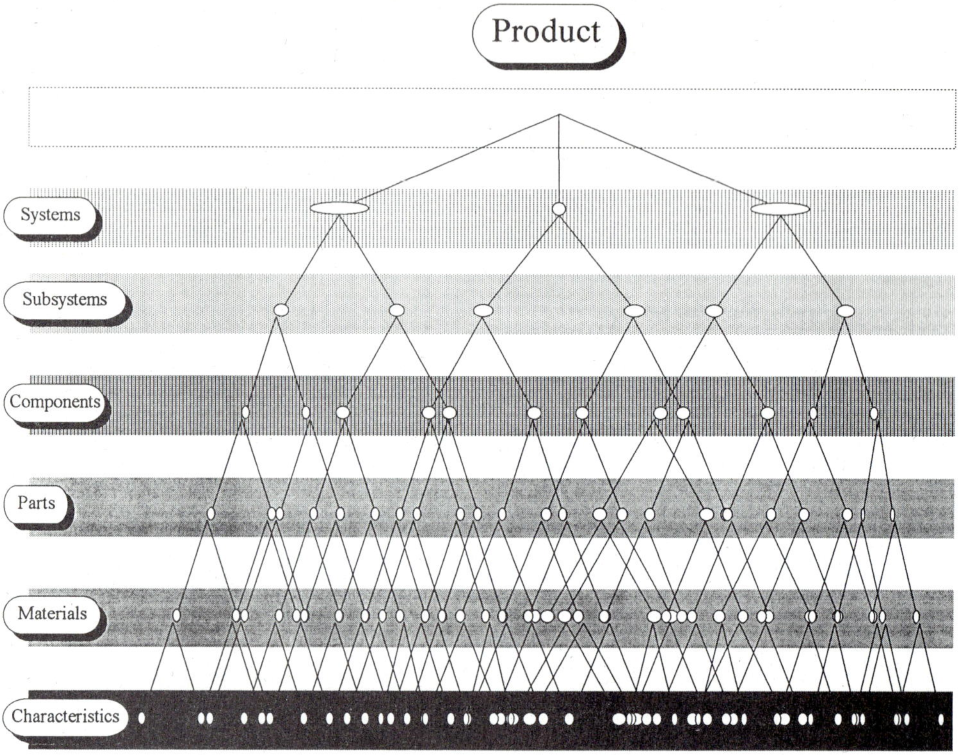

FIGURE 6.11 Tree Structure Branching Showing Multiplicity

Similar to physical decomposition, in FuD product comprises a number of functional subsystems, each of which further consists of a number of functional components. The difference is instead of identifying the decomposed sets by their physical names (in PhD), functional names are assigned to these sets in FuD, as shown in Figure 6.12. The situation is more pronounced in the case of a software product, where compositions of physical parts are rare or nonexistent. A functional decomposition for a software product is shown in Figure 6.12. The degree of overlap complexity goes up when the number of sequential links (say for instance) in the horizontal chain increases. An example of an horizontal chain can be:

$$\begin{aligned}\text{Horizontal Chain} \Leftrightarrow [&\text{User interface} \rightarrow \text{Graphics window} \\ &\rightarrow \text{Graphics icon} \rightarrow \text{User-pick}]\end{aligned} \tag{6.14}$$

The degree of overlap complexity also goes up when the number of dependent functions within a particular link (say for instance, command line, menu-typed windows, graphics window, etc.) increases. The degree of overlap advantage on the other hand is governed by the complexity of the tasks performed by the link in question. The degree of overlap advantage decreases as complexity of the tasks increase.

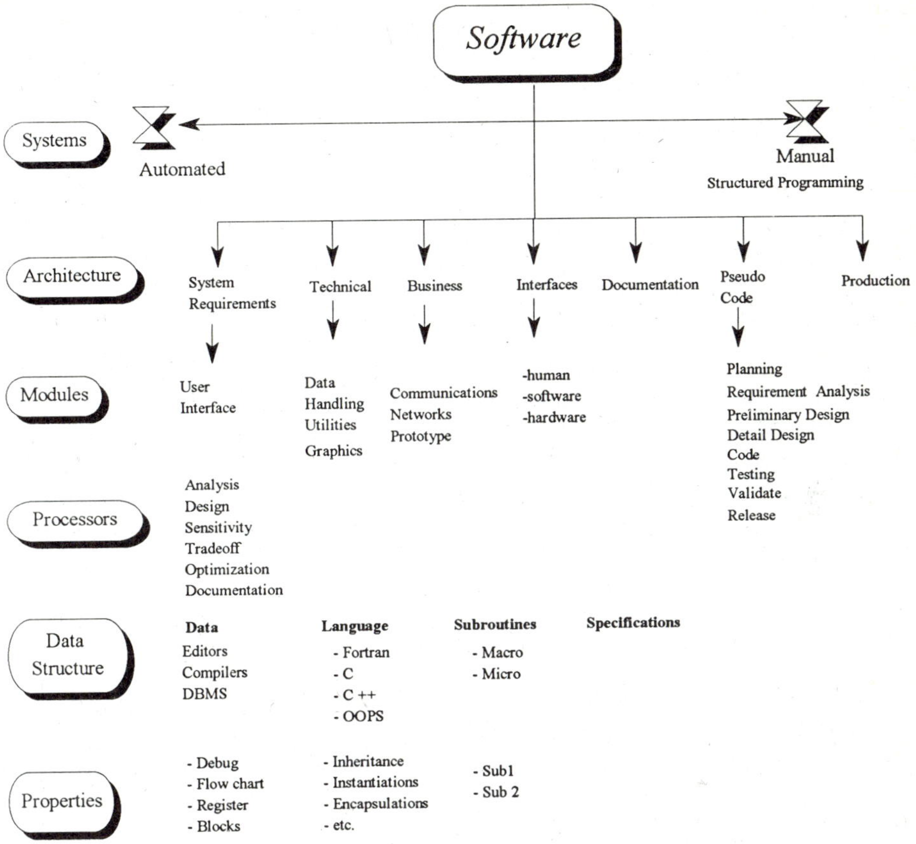

FIGURE 6.12 Functional Decomposition of a Software Product

6.4.1.3 Activity-based Decomposition (AcD)

Here the product is decomposed into several activities instead of supporting functions or a number of physical parts. ABC (Activity-based Costing) is an example of an activity-based decomposition. Specification decomposition can be considered as another example of activity-based decomposition. Highly effective CE teams convert important customer requirements into specifications (that is, what does the "system" or product have to do?). A set of customer requirements is first converted into high level specifications. These specifications are then decomposed into a set of low level specifications first in the functional space, then into another set of low level specifications in the physical space.

6.4.1.4 Degree of Overlap

The degree of "overlap possibilities" of the decomposed subsystem is governed by a set of its components and their dependencies—functional, physical, or otherwise. In other words, if the degree of overlap advantage of a decomposition scenario is given, we could construct a set of decomposed components to make up the desired subsystem having a given or desired degree of overlap. One can further argue that the development, production, or assembly effort of such a subsystem could be a function of this degree of overlap. In fact, for the sake of computational simplicity, it is quite safe to assume that the developmental, production, or assembly effort of one such subassembly would be proportional to its degree of overlap advantage. When the system specification hierarchies are directly traceable to customer requirements' hierarchy, teams will be designing around system functions and features that customers want, crave for, or are excited about.

6.5 SHARING AND COLLABORATION IN CE

Concurrent engineering implies the sharing of information, process, or knowledge. Division of work into classes and then to the work-groups or to the subteams was one form of sharing discussed earlier. Other forms of sharing are governed by the state of computer communications, workstation, and database technology. In the information age that we live in today, civilization depends on information and the dominant weapon for success is timely sharing.

6.5.1 Types of Knowledge

Today, electronically it is easy to connect to anyone in any part of the world. We are bombarded with all sorts of information. However, we have very little knowledge of how to use it effectively. As Deming said [Deming, 1993], "information is not knowledge." Knowledge comes from theory. Without theory there is no rational basis to apply the information. The types of knowledge that are useful in meeting product realization goals are

- ***System knowledge:*** Knowledge of how the system works in terms of concept, fact, principle, methodology, or technology.
- ***Interface knowledge:*** Knowledge of how the underlying components and their interfaces work in terms of concept, fact, principle, methodology, or technology.
- ***Transformation knowledge:*** Knowledge of how to transform a set of specifications into a physical artifact or service.
- ***Behavior knowledge:*** The basic behaviors of the computer models or concepts as some of the model attributes change.
- ***Environment knowledge:*** The attitudes, values, visions, and beliefs of the participating teams, (including virtual team, logical team, and technology team), and those of the organizations.

In other words:

$$\text{Knowledge} \equiv \cup\,[\text{System-k, Interface-k, Transformation-k, Behavior-k, Environment-k}] \tag{6.15}$$

In this age of competitive advantage, what differentiates one CE work-group from the others is how one does the sharing between teams, machines (or computers), and processes. These forms of sharing and collaboration are discussed next.

6.5.2 Forms of Sharing and Collaboration

Figure 6.13 shows various forms of knowledge sharing and collaboration between concurrent teams, machines (or computers), and processes:

- **Team-to-team sharing** involves intrateam and interteam communication (between personnel, technological, logical, and virtual teams). For instance, there is a need to establish an effective method for communicating knowledge and information among

Modes	**Team**	**Machine/ Computer**	**Process**
Team	Team-to-Team Sharing (e.g:, Help Desk, Customer Support, Intra-teams, Work-group Applications, and Inter-team Cooperations)	Team-to-Computer Sharing (e.g., User Interface, Information Retrieval, Data Inputs, etc.)	Team-to-Process Sharing (e.g., Interactive Modeling, Analysis, CAD/CAM, CAE, CAD Conferencing, etc.)
Machine/ Computer	Computer-to-Team Sharing (e.g., Display, On-line Help, Desktop Applications, Computer Graphics, Plots, etc.)	Computer-to-Computer Sharing (e.g., Inter-computer, LAN, Internet, e-mail/Messaging, Network, etc.)	Computer-to-Process Sharing (e.g., Concurrent Multiple Processes, DNC/Postprocessing, Concurrent Sessions)
Process	Process-to-Team Sharing (e.g., Computational Agents, System Adm., Network Adm., Groupware, Virtual Teams, Blackboard, Expert Systems, Advisors, etc.)	Process -to-Computer Sharing (e.g., Feedback, Trouble Shooting, Remote Diagnostics, Fault Mgt., Process Control, Date, Time, Warning Flags, Storage, Retrieval, File Management, etc.)	Process-to-Process Sharing Inter-process (e.g., Groupware, Remote Applications, X-window Applications, Multimedia, Automation, etc.)

FIGURE 6.13 Forms of Sharing and Collaboration

all personnel teams (such as operators, team leaders, and application support personnel). The larger the organization, the more important it is to establish a formal method of sharing within (interteam) and between (intrateam) disciplines. A proven old method is to hold group meetings at frequent intervals that include all the personnel team members. This can be time-consuming and may get out of hand if not effectively managed.

- **Team-to-computer sharing** involves human-machine interactions, such as user interface, information retrieval, data inputs, natural language, programming, etc.
- **Team-to-process sharing** entails interactive modeling, capture of geometry, such as CAD/CAM, or analysis (for example FEA/FEM/CAE) techniques including conferencing.
- **Computer-to-team sharing** involves display, on-line help, desktop applications, computer graphics, output plots, etc.
- **Process-to-team sharing** includes communication capabilities between virtual teams, such as computational agents, system administration, network administration, groupware, blackboard architectures, advisors of all types, expert systems, knowledge-based engineering, optimization, etc. Ability of the personnel team-members to have virtually immediate access to in-process or released data, as required, is an example of process-to-team communication.
- **Computer-to-computer sharing** involves intercomputer communications or intermachine interactions such as LAN, penetration of e-mail, Internet, networks of all types, messaging, etc.
- **Process-to-computer sharing** involves trouble shooting, remote diagnostic, fault management, feedback, process control, date, time, warning flags, etc.
- **Process-to-process sharing** entails interprocess communications, such as groupware, remote applications, X-window applications, multimedia, Lotus Notes and workflow-based systems, process automation, etc.
- **Computer-to-process sharing** means machine communications such as DNC/post-processing, multiple processes, and concurrent sessions. The effectiveness of collaboration is discussed next.

6.5.3 Effectiveness of Collaboration

One of the premises on which CE is based is the "paralleling" of activities so that multiple groups of people (work-groups) can work in teams. A work-group is a collection of subteams or people having common interests or expertise. Work-groups usually have a short life in a CE organization. Such work-groups or subteams must collaborate with each other in a timely fashion so that they all are working toward a common goal. The need for collaboration and communication grows as more and more subteams are involved. It is usually not advantageous to have multiple process and concurrent sessions without a coherent communication pipeline. More processes can slow down a collaborative decision making process due to increased "loads" for timely communication and coordination.

Thus there is a limit to how far paralleling can go. It would seem logical to adopt a modular communication infrastructure, which can grow in an evolutionary manner. As the work-group grows and needs arise for more power and timely communication, new infrastructure can be added. The efficiency of communication is one of the barriers to "paralleling" in CE. There exist a number of other similar barriers to CE. They are discussed at length in Chapter 8, Intelligent Information Systems of Volume 2.

6.5.4 Paralleling of Responsibility

Distribution of responsibility between the teams, machines (or computer), products, and processes should be such that they maximize the effectiveness of the "joint cognitive system." A joint cognitive system refers to a complex product manufacturing environment:

- Business partners and other CE teams (personnel, virtual, logical, and technology) share design responsibility.
- The product is differentiated using the systematization approach.
- Modular manufacturing and production concepts are utilized.

Work breakdown structures (WBS) establish a framework through which work-group resources can be allocated. The following factors influence how such allocations can be made:

- ***Communication and coordination:*** Types of communication and coordination necessary to collaborate on a task determine the actions and basis of how to distribute responsibility between the agents so that they can work in parallel. Examples of coordination include delegating responsibility, reporting, or evaluating results.
- ***Knowing human behavior:*** Studies have shown that humans interweave problem-solving tasks with problem-setting tasks. They create personalized workspaces to organize their thoughts and tasks.
- ***Level of skills:*** The shift of responsibility from human to the intelligent machines (computers) depends on the types of work, such as routine or repetitious tasks. Repetitive tasks can be better handled by a computer system. However, if it is a skilled job or one of a kind, it can be handled better by a human being with high intelligence.
- ***Efficiency of communications between the agents:*** The best way to figure out inefficiencies of communications is, first, to ask the individual departments how they are doing their business. A "flow charting" methodology, as discussed in Chapter 3, may be used to capture the flow of information through various departments. One way to gain some insight is for everyone to cooperate and communicate in real-time what they are doing and who benefits from it. A full commitment and determination on the part of management may be required for its success. It is vital to agree on

common symbols and methods of depicting the information derived from the team members. There is also a need to establish an integration methodology for doing CE from system perspectives. This is discussed next.

6.6 SEAMLESS INTEGRATION

The essence of concurrency in CE comes from both decomposition and integration. Integration is a focal point for companies to position themselves for the twenty-first century. Indeed, integration has long been recognized as a potential contributor to improving productivity and efficiency [Pawar and Riedel, 1994]. To many, integration has implied fast, customer-driven responses with open communication and ready access to data, information, and knowledge. Effective integration requires people (or subteams) working together within a well-thought process or a system of management. Integration should permeate the entire organization practically from the "ground up." It has less to do with computer systems but more to do with logistics. That's why it is commonly referred as "logistics integration."

6.6.1 Logistics Integration

Logistics are an important part of the integration equation. Logistics integration deals with the logistic aspect; policy, procedures, and practices that have to be in place for the smooth functioning of the organization as a whole. The capability to manage one's operations (internal or external) is crucial for achieving time-to-market. There are two types of logistics integration: horizontal and vertical integration. Figure 6.14 shows an integration matrix, where six levels of vertical integration and eleven levels of horizontal integration are taking place concurrently. The interactions are shown by a "cross-hair" arrow icon. Examples of the vertical integration are: providing customer satisfaction to the consumers, and sales and marketing support to the distributors. It also means giving empowerment to the workers, establishing business relationships/partnerships with suppliers/academia/vendors, developing strategies for CE management, and securing finances from the investors.

- ***Vertical integration*** deals with processes that involve departmental management. It involves long and close associations or relationships with investors, management, suppliers/academia/vendors, workers, distributors, and the consumers. It means establishing a strategic business relationship with selected suppliers/vendors and their products. Vertical integration is like putting the house in order. It puts the emphasis on management—policy, procedures, and practices (3Ps) to align companies' hierarchies—from investors to the consumers (see Figure 6.14). It shifts the emphasis from operational to strategic partnership. The purpose is to streamline the investor-to-consumer process pipeline so that product finance can lead to customer satisfaction.

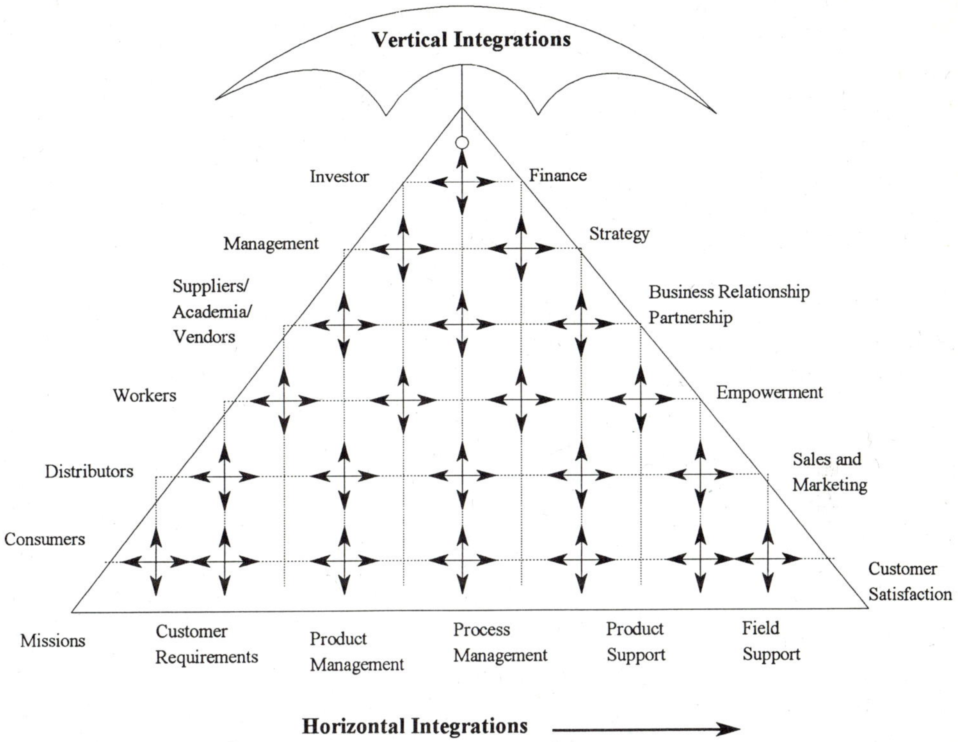

FIGURE 6.14 Elements of Logistics Integration

- ***Horizontal integration*** deals with connecting standard business functional activities such as mission, customer requirements, product management, process management, product support, field support, etc. (see Figure 6.14). The target is to integrate the units so that the raw materials are transformed into finished goods or products with the highest enriched value and are meeting the customer needs (see Figure 6.15). Horizontal integration is like building on the mutual strength of its constituents. It shifts the emphasis from individuals to collaborative cooperation. The purpose is to streamline the product development or realization process from customer requirements to field supports (see Figure 6.15).

Work moves between team's co-workers, PDT leaders, and among executive levels of an organization based on the level of talent, empowerment, authority, responsibility, or content. Customer-driven manufacturing requires seamless integration irrespective of information flow direction. A total integration from consumers (customer or user of a product), in house product processing groups (work-groups, subteams, designers, plants, processor, manufacturer), to distributors (dealers, sellers, brokers, suppliers, or retailers of the product). Seamless integration implies a total integration of all the PDTs' members or the work-groups who can take the customer requirements and the company mission (in

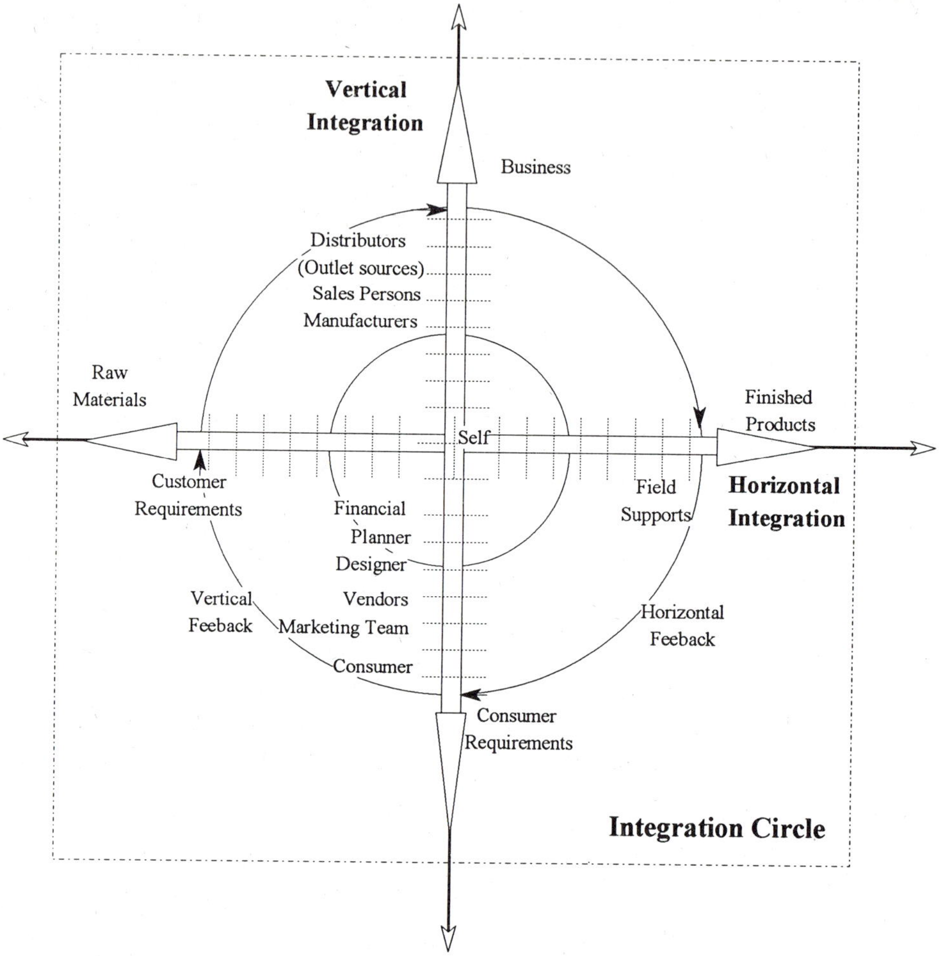

FIGURE 6.15 Horizontal and Vertical Integrations

terms of VOC) and produce a real-life artifact (see Figure 6.16). The two management cycles involved in producing the artifact are product management and process management cycles. They are integrated with the product support and delivery. Integration also includes providing field support for the duration the product is in use.

Effective communication takes place both vertically—in spite of differences in responsibility or ranks—and horizontally—in spite of functional differences. The new approach to collaboration not only includes the industry players (employees, suppliers, workers, etc.) but the combined and focused efforts of industry, community, government, research and development (R&D) consortia, academia, and research institutions. Government collaboration may come from tax credit or other incentives that it provides for research and development consortia to focus on advanced technology development.

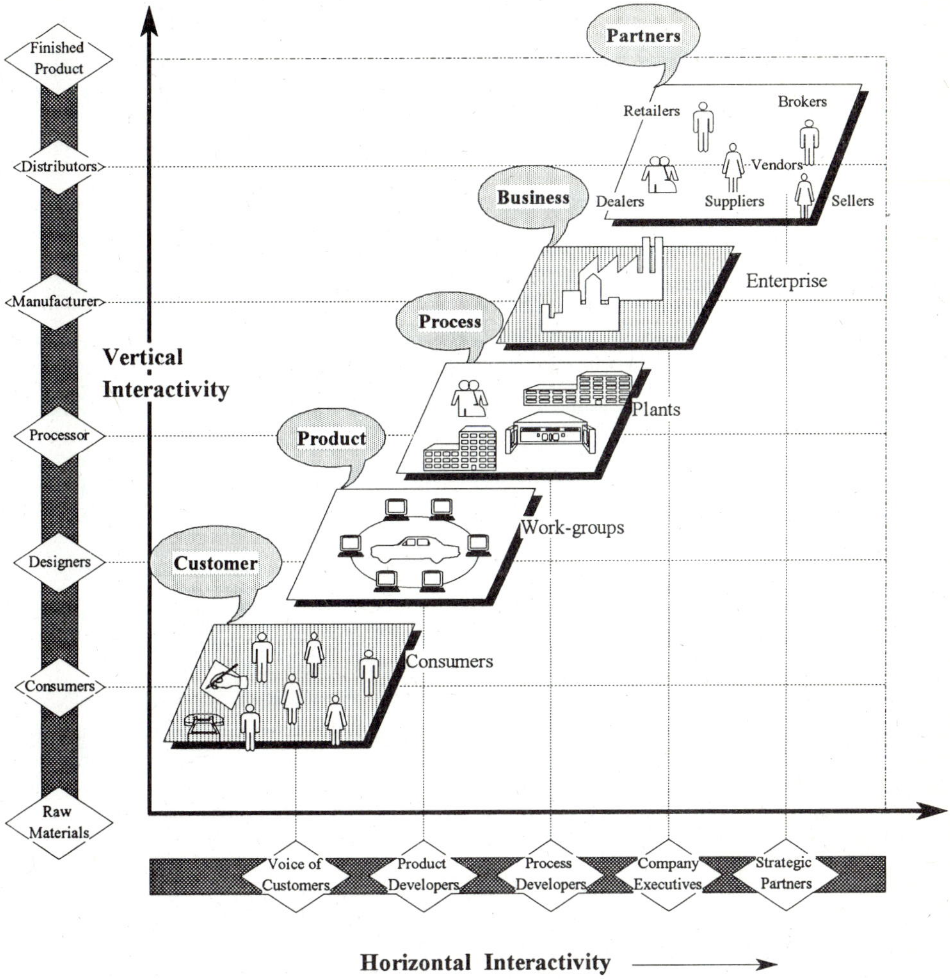

FIGURE 6.16 Total Seamless Integration

Today we are entering an era of flexibility. Organizations need agility and flexibility to exploit market anomalies and to overcome the very short technology obsolescence cycle. Today, organizations need capabilities to produce a multiplicity of interrelated products at accelerated speed of development from a concept idea to a finished product. This move toward enterprise-wide information integration is breeding flatter organizations. Most organizations are moving away from silo, or tall, multilayered structures. Figure 6.17 shows a typical six-level vertical pyramid of a manufacturing operation, which was flattened to just three levels: manufacturing management, manufacturing execution, and infrastructure management. This flatter organization will have a loose set up vertically across various teams and an agile, virtual coupling through a world of collaborative and electronic networks.

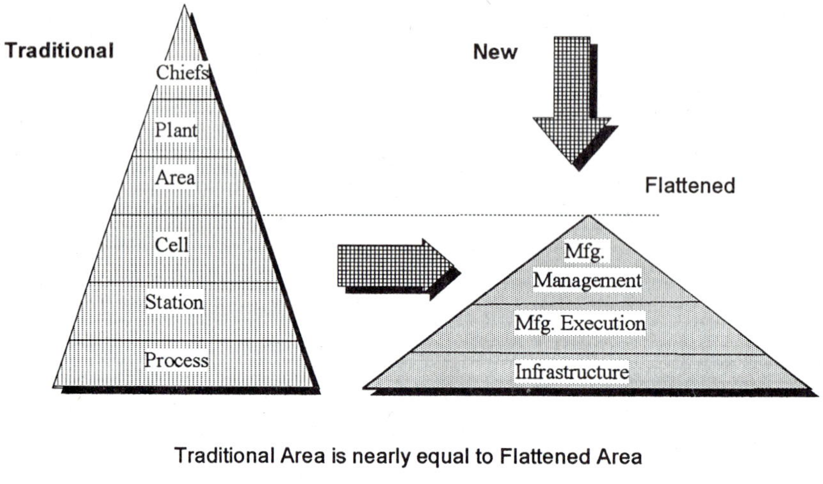

FIGURE 6.17 Flattening of a Manufacturing Operation

6.7 MANAGEMENT AND REPORTING STRUCTURE

Companies have been vertically oriented for so many years that it has become extremely difficult to work across departmental or functional lines.

6.7.1 Management Structure

The CE process is usually organized on a cross-functional basis. A traditional (functional) organizational setup can be reorganized into a CE setup in two ways: (a) goal realignment with product focused restructuring; and (b) goal realignment with life-cycle focused restructuring. Realignment refers to objectives or project goal setting. In a cross-functional organization, administrative functions are a thin functional layer around the periphery. Management structure helps bring in administrative functions (and an order) into the enterprise and sustain a common set of consistent goals and key processes.

6.7.1.1 Goal Realignment with Product Focused Restructuring

This is perhaps the easiest of all. The functional organization is not changed but its objectives are redefined along product lines. This is also called product focused organization. Each of the departments has product focused organizational structure. There may be one or more groups of people assigned to that product, depending on the types of analysis and design expertise required during each product development. In such restructuring, there is a danger in promoting and rewarding performance in one area, even when it is accomplished at the expense of another, since there is no focal point of control for managing people.

6.7.1.2 Goal Realignment with Life-cycle Focused Restructuring

The old department is not dissolved. Some of the employees are moved out and some are added. It is not a brand new organization, but some of the traditional lines or functional organization objectives are realigned with CE. This will result in the traditional engineering organization absorbing the other participating disciplines of the CE teams such as the manufacturing engineers and planners, the quality control engineers, and the like. The organization's name can be changed from functional (for example, engineering, manufacturing) to a product life-cycle definition (such as feasibility study, preliminary design, technical design, business design, prototype, testing, etc.). Many pitfalls remain:

- Though the name is changed, the organization would still be dominated by the remaining people and past disciplinary practices.
- It could alienate incoming participants. This could also be true for what is left of the old organizations (in manufacturing, quality assurance, material, test, and product support).
- The danger is that the manufacturing engineers and planners, for example, would be viewed as members of an "outside" organization by their former brethren. The latter would begin distrusting them in the same way they formally distrusted the design engineers, and would replace them (thereby duplicating their numbers and expertise) with new manufacturing engineers and planners in the manufacturing organization that could critique and, at least partially, redo what was done in the "product definition" organization.

6.7.2 Reporting Structure

Any one of the above organizational structures may have an associated reporting structure as well. The following reporting structure can be employed in any of the discussed CE organizational settings.

6.7.2.1 No Restructuring but Temporary Duty Assignment (TDA)

Temporary duty is often established in many CE settings when the resources needed are project-specific and for a limited duration. Personnel with right CE expertise (for example, product definition, system engineers, etc.) are drawn from functional organizations and loaned to the project for the duration required. Their allegiance, including their reporting command structure, is transferred to the new organization. Ties with their parent functional organizations are temporarily severed but, when their assignments are complete, they are returned to their specific functional organizations. This way they remain active players in sharing the advances of technologies and ideas with known colleagues who speak the same language and have the same basic interests. The performance of the project may be boosted temporarily with such TDA, but there are usually inadequate benefits to the overall development program; for instance, knowledge is not transferred unless it is planned that way.

6.7.2.2 No Restructuring but Matrix (or Dotted Line Reporting)

Another form of temporary duty is called matrix setup when there is a dual or trio-reporting structure. In temporary duty, persons are loaned to the matrix organization, but the person now reports to two or more managers. In most cases, a dotted line relationship is maintained with the parent organization and a solid line is maintained with the loaned organization. However, the reverse is also true. In case of the solid line, allegiance is transferred to the new organization, but ties with the worker's functional organization is not totally severed. The worker may schedule visits to their parent organizations and act as a focal point of contact at both ends. This way, experiences and advances in technology are shared, creating satisfaction for both involved parties.

Problems that crop up in both types of arrangements might include the following:

- How to fairly evaluate the performance of the temporarily displaced employee
- How to judge members' contribution and compensate them based on that
- How to treat the employees fairly compared with those who did not transfer or matrix, and have continued to work in the functional organization; often, employees who have been away from their parent organizations receive different treatments than those who remained behind.

6.7.2.3 Control Shift Management: No Matrix, No Restructuring, but Agreeing to Collaborate

On the other end of the spectrum, the current functional organizations could essentially remain intact, only agreeing to better collaborate in the product definition function. Unless there are exceptionally talented and strong-willed personnel to lead every one of the CE teams, this solution stands a real chance of slipping back into the traditional development mode. It does not have the needed facility or the structure to keep the expertise together so that the sharing of ideas and the advances of technology, indigenous to that discipline, can be promoted and shared. It has the advantage of not displacing large numbers of personnel. However, it is easy to slip back into the traditional product development mode and may return control to the functional organizations.

No matter what organizational and reporting structure is chosen, the change will always be difficult to manage. There are bound to be resentful people, either because they feel personally betrayed or because a significant amount of influence is lost (in the case of some management positions). It will take a great deal of time and management training to resolve these problems in such a way that everyone feels as a winner and the success of the CE process is enhanced. Until a change in employee attitude takes place to match the realigned objectives, it will be a bumpy ride for all concerned.

6.8 AGILE VIRTUAL COMPANY

The term global emphasizes *"economy of labor."* The term "mass" emphasizes *"economy of machine power."* In the mass production era, competition was one-dimensional and centered mainly on price. The success of an organization was measured by the ability to produce the product at the lowest cost. Hence, the investment in the production resources was the prime focus. Product was designed to suit the organizational mass needs—to utilize the existing production system (capability or limitations of the plants), the marketing resources. The products were later pushed to the distributors or retailers, who sought the interested customers.

PRODUCTION CAPABILITY
>> WHAT PRODUCT TO PRODUCE
>> PRICE-CENTERED PRODUCT (6.16)
>> MARKETING
>>> INTERESTED CUSTOMER SOUGHT

The term agile emphasizes *"economy of flexibility or ease"* and virtual applies to *"economy of teamwork or cooperation"* and *"economy of seamless flow of information."* The trio contributes to the *"economy of change."* With agile virtual coupling, the degree of competitiveness, unlike mass production, no longer remains one-dimensional. It is judged by a value tree, which is important to both the customer and the company. Price remains only one element of this value tree.

$$\text{Agile and Virtual Competition} \propto (\text{Values to the Company} + \text{Values to the Customers}) \tag{6.17}$$

Where $\propto$ means proportional to. Other elements of the value tree which are relevant to a company include flexibility, time-to-market, productivity, inventory, management, etc. Values which might be important to the customers include cost, quality, customer excitement, functionality, ease of use, customer satisfaction, etc. The common sense for agility is based on distributed information and services the company provides, less control, flattened organization structure, knowledge-based systems, and continuous improvement. Together they translate the intellectual power of the company into new and constantly evolving products, processes, and services. An agile virtual company is thus built on the intellectual strength and cooperation of the concurrent teams to constantly produce new products quickly to meet evolving market demands.

INTERESTING PRODUCTS
<< PULL IN RIGHT RESOURCES
<< INTELLECTUAL STRENGTH + TEAM COOPERATION
<< CUSTOMERS WANTS SOUGHT (6.18)

In agile virtual competition, a concurrent team pulls in the customer wants. Other teams seek the right resources from a virtual organization to create an interesting product that the customer desires [Goldman, Nagel, and Preiss, 1995].

They compete on core competencies and on values derived from combining "thinking and learning" environments with virtual decision support tools. The agile and virtual products are customized to the needs of its customers as opposed to mass manufacturing where product is customized to the needs of the corporation and/or of the factories. The role of people to carry out tasks, the need and ability of the people to make decisions, and to undertake acceptable risks varies from mass manufacturing to the agile virtual company. With agile virtual companies, authority to take actions, to make decisions, or to take risks are diffused (and concurrent) throughout the enterprise. Such decisions benefit from a seamless flow of information among diverse life-cycle work-groups, for example, engineering, manufacturing, marketing, etc.

To succeed, an agile virtual competitor requires a shift into a dynamic thinking organization, however, major technical breakthroughs may not have to be shifted into an "agile-virtual" gear. This is because in the past none (technical breakthroughs) had to be shifted for lean or synchronous manufacturing trails. It may be enough to create an environment that encourages or aids concurrent teams to take speedy actions, to make informed decisions and to manage calculated risks [Goldman, Nagel and Preiss, 1995]. A common understanding of the critical factors that influence the quality, time-to-market, customer satisfaction, cost, efficiency, and productivity, is essential. A technological push can help the concurrent teams within a "virtual enterprise" to make the best decisions on how to respond quickly to changes. In the "agile virtual" world, partnership might dominate the marketplace because no single company is likely to have all the intellectual strengths to be able to compete on its own.

On the manufacturing front, the definition of "flexible" resides in the material handling and production equipment, systems, and the workforce. Flexible machines means multi-purpose machines that can be set to do a variety of jobs. Flexible machines are not the same as agile machines. Agile machines means more new product with less new process. The key attributes that make a machine agile are reusability, reconfigurability and scalability. Figure 6.18 compares five manufacturing traits: craft, mass, lean, agile, and virtual. They are compared against a number of organizational factors that are essential to manage change. Figure 6.18 reveals that there are several commonalties among the first three in terms of: team cooperation, policies, size and structure, and organizational focus. The difference lies in the means of communication and the level of cooperation sought from the work-groups and the subteams. One of the communication differences is the change in emphasis from "getting an individual access to a computer" to "putting teams or work-groups on the network." However, these traits seem to build on each another. For example, once a company makes the transition from mass to lean manufactur-

Manufacturing Traits	Craft Manufacturing	Mass Manufacturing	Lean Manufacturing	Agile Manufacturing	Virtual Manufacturing
Economy-basis	Economy of skill	Economy of scale	Economy of Kaizen	Economy of flexibility	Econony of cooperation
Organiz.-Focus	Short-term	Short-term	Medium-term	Long-term	Long-term
Reconfigurability	Performs within a fixed production cycle	Performs within a fixed production cycle	Performs both within and across production cycle that exists	Reconfigures a factory for unplanned production	Partners with VAR to exploit an unexpected market opportunity
Scalable Size and Structure	Takes account of specific aspects of components manufacturing	Automates the specific aspects of components manufacturing	Takes account of entire production chain from customer to disposal	In addition to all elements of "lean" scope -- includes process reengineering and corporate strategies	In addition to all of "agile" scope, includes partnership, teamwork,cooperation, collaboration and information highway links
Decision Making Style	Top-down (1-D) directive management style	Top-down (1-D) directive management style	Mostly top-down (1-D) directive management style	Top-down, bottom-up (2-D) supportive managment style	Bottom-up, top-down (2-D) goal-oriented decision making
Mgt. Participation and Commitment	Bottom-up driven, no definite actions	Little management commitments and actions	Bottom-up driven, some mgts. commitments and actions	Top-down driven, medium level mgt. commitment and actions	Both bottom-up and top-down driven large mgt. commitment and actions
Partnering: Strategic Approach	Personal strength and skill	Machine's capabilities, technology utilization and insertions	Teamwork and collaboration within groups	Temporal partnership of workteams, cooperation across groups, suppliers and customers	Teamwork and cooperation across groups, intraorganization, business relationships and teamwork
Interface-comp.	None	None	Small	Medium	Large
Plug-in Compatibility	None	Little	Small	Distributed computing	Cooperative computing
Communication Network	Through voice and paper	Through machine language and instructions	Through information network or pipeline	Through seamless network across and within organization	Meeting-on-the network: Free exchange of ideas up, down, across
Empowerment	No empowerment	No empowerment	Transfer of decision making to top level managers	Transfer of decision making to mid-level managers	Transfer of decision making as low as possible in employee ranks
Policy, Practices an Procedures	Extremely rigid, complex procedures, organization dictated most of the steps	Less rigid and more flexible than craft manufacturing	Less rigid and more flexible than mass manufacturing	Less rigid and flexible than lean manufacturing	Extremely flexible, simple procedures, teammates dictate most of the steps
Team Cooperations	No team cooperation, high personal contribution	Low team cooperation, medium personal	High team cooperation, low individual contributions	High team cooperation and individual contributions	Very high team cooperation and fewer "hidden" goals
Motivation and Reward System	Based on individual creativity, contributions	May be little better than craft manufacturing	May be a little better than mass manufacturing	More team-based and less individual-based	A reward system that weighs both team and individual contributions

FIGURE 6.18 Differences in Manufacturing Traits

ing, it would be quicker, less disruptive, and less costly to make a further transition, say from an agile to a virtual company.

6.8.1 Organizational Factors

In previous sections, two factors: Economy-basis and Organizational focus for agile virtual company were discussed. Other contributing factors for agile virtual company to be successful are (see Figure 6.18) as follows:

- ***Reconfigurability:*** This means the ability to upgrade a product or process easily, and to react quickly to changes in the market conditions. Reconfigurability is a term also described to contain the following sub-features:
  - Self-organizing: Examples include dynamic alliances, teaming, self adapting, dynamic scheduling and bidding.
  - Non-Hierarchical: This means capabilities for direct negotiation, communication and interaction among units that are on the same level.
  - Dynamic Relationships: The relationship that can be reversed in real-time and is late-binding.
- ***Scalable Size and Structure:*** These show the extent over which change can be applied. They define the unit populations of the product life-cycle or production chain over which the change can be managed. It has two subfeatures:
  - Unit redundancy—it provides capabilities for capacity fluctuation
  - Extensible unit framework—that accomodates any type of unit-legacy, common or completely new.
- ***Decision Making Style:*** This ranges from top-down to bottom-up decision making style (e.g., from a firm directive management style to a goal-oriented style), where constancy of purpose is maintained.
- ***Management Participation and Commitments:*** A steady level of participation, activities, and commitment from the beginning to the end of the project are required.
- ***Partnering:*** This is a strategic alliance approach, where closer relationships are established with the supply chain (suppliers) and customers.
- ***Interface Compatibility:*** This represents the ability to interface with a work-group (or a partner) through standardized policy, procedures, and manufacturing practices.
- ***Reusability:*** This is a term described to contain 2 sub-features:
  - Modularity: This implies that the system is composed of distinct but seperable units
  - Facilitated templates: This implies that the system is equipped with standardized unit description templates and modification tools.
- ***Plug-in Compatibility:*** This shows capabilities to plug into any hardware and software environment, irrespective of the brand name and applications, such as open system with standardized units.

- ***Communication Networks:*** This shows the ability to provide interoperability (robust communication network) capabilities up, down, within, and across organizational lines in a multivendor environment.
- ***Empowerment:*** This is the freedom of decision making as low as possible in employee ranks, at the point of maximum information.
- ***Policies, Practices, and Procedures (3Ps):*** These define the ability to electronically use 3Ps to prevent the possibility of undesirable conditions that could possibly contribute to a problem.
- ***Team Cooperation:*** This includes cooperation from everyone, including supplier-base, so that simultaneous development of the product, process, and business operations can take place.
- ***Motivation and Rewards:*** A system of rewards, which is based on both team success and individual creativity.
- ***Responsiveness:*** This includes ability to reduce the total life-cycle time and the individual lead time for each operation. Lead time is the time an operation takes to finish.

In the future many manufacturing companies will yield to the concept of fluid or agile manufacturing. The technology, communication, transportation, and social structure are likely to keep on changing, creating a sense as if one is living in a "dynamic paradigm shift" environment. Those companies, that had prior experience mass producing single or specific products, will acquire a newer array of tools and technologies to give them added flexibility. Most firms and enterprises have to keep on shifting from one transitional paradigm state to another. Those who become comfortable with certain aspects of one transitional paradigm and dare to stick with it may find themselves losing the market share and profit potentials.

REFERENCES

ALEXANDER, C. 1964. *Notes on the Synthesis of Form.* Cambridge, MA: Harvard University Press.

ARAI, E., K. IWATA. 1990. "CAD with Design Specification Decomposition and Its Application." *Annals of the CIRP.* Vol. 39 (Jan.), pp. 121–124.

CHELSOM, J. V. 1994. "Concurrent Engineering Case Studies: Lessons from Ford Motor Company Experience." Chapter 2 in *Concurrent Engineering: Concepts, Implementation and Practice.* ed. by Syan and Menon. Chapman & Hall.

DAVIS, D. 1994. "Partnerships Pay off," *Manufacturing Systems,* Vol. 12, No. 11, pp. 4–10.

DEMING, W. E. 1993. *The New Economics,* Cambridge, MA, MIT Center for Advanced Engineering Study, (November).

GALBRAITH, J. R. 1974. "Organization Design: An Information Processing View," *Interfaces,* Vol. 4, pp. 28–36.

GOLDMAN. S. L., R. N. NAGEL and K. PREISS. 1995. *Agile Competitors and Virtual Organizations.* New York: Van Nostrand Reinhold.

HALES, C. 1993. *Managing Engineering Design,* Longman Scientific and Technical, England, UK., Copublished by John Wiley & Sons, New York., USA.

KAUFFMAN, D. L. 1980. *System 1: An Introduction to Systems Thinking,* Cambridge, MA: Future Systems Inc., pp. 1–41.

KUSIAK, A., and J. WANG. 1993. "Decomposition of the Design Process," *Journal of Mechanical Design,* Transactions of the ASME, Vol. 115, No. 4 (December), pp. 687–695.

M'PHERSON, P. K. 1986. "Systems Engineering: A Proposed Definition," *IEEE Proceedings,* Vol. 133, Pt. A, No. 6 (September), pp. 348–Sr–47 to 348–SR–48.

PAWAR, K. S., and J. C. K. H. RIEDEL. 1994. "Achieving Integration through Managing Concurrent Engineering," *International Journal of Production Economics,* Vol. 34, No. 3 (June), pp. 329–345.

PRASAD, B. 1991. "Systems Automation—An Introduction." *International Journal of Systems Automation: Research & Applications.* Vol. 1, No. 1, pp. 1–5.

PRASAD, B. 1992. "On Systems Paradigms." *International Journal of Systems Automation: Research & Applications.* Vol. 2, No.1, pp. 1–3.

SAWYER, C. A. 1994. "The Systems View," *Automotive Industries,* Vol. 174, No. 9, pp. 82–87.

SIMON, H. A. 1981. *The Science of the Artificial,* 2nd Ed. Cambridge, MA: MIT Press.

SPURGEON, W. M. 1994. "System Engineering and Quality," *Automotive Engineering,* Vol. 102, No. 6 (June), pp. 17–19.

TAYLOR, L. E., and M. R. HENDERSON. 1994. "The Roles of Features and Abstractions in Mechanical Design." Proceedings of the ASME Design Technical Conferences—6th International Conference on Design Theory and Methodology, Minneapolis, MN, Sept 11–14, 1994, *Design Theory and Methodology DTM '94,* Edited by Hight and Mistree, DE-Vol. 68, pp. 131–140.

WHITNEY, D. E. 1992. "State of the Art in Japanese Computer-aided Design Methodologies for Mechanical Products—Industrial Practice and University Research," *Scientific Information Bulletin.* U.S. Navy Office of Naval Research, NAVSO P-3580, Vol. 17, No. 1 (Jan–March).

TEST PROBLEMS: SYSTEM ENGINEERING

6.1. How does system throughput limit individual productivity?

6.2. What role does virtual manufacturing play in manufacturing competitiveness? Is it a necessary trait or can a company maintain manufacturing competitiveness without converting to a virtual manufacturing?

6.3. Compare the elements of logistics integration among the five major types of manufacturing traits: craft, mass, lean, agile, and virtual. How do you describe plug-in capability?

6.4. How does flattening of a manufacturing operation help the time-to-market aspect?

6.5. How many elements are there in logistics integration? Discuss the difference between vertical and horizontal integration.

6.6. Give examples from your work life, when the whole system function was overlooked, the system as a whole could not be optimal.

6.7. Describe some pitfalls for the conventional design and development process. What is the core concept of IPPD? Why is system approach becoming a necessity in product development?

6.8. What are the differences in parts belonging to a system and parts belonging to a collection?

6.9. Give an example of top-down decomposition for an automotive wire harness product.

6.10. Where are the concurrencies in an Aircraft top-down decomposition? Is there anything called bottom-up decomposition? How does it differ from the top-down one?

6.11. Are current companies becoming more and more virtual? If so, what does this imply?

6.12. What is an agile virtual company? How does this differ from the "mass manufacturing" or a "virtual manufacturing" company?

6.13. What are the so-called fifteen organizational factors that differentiate various forms of manufacturing traits? Do you know of any new ones affecting your organization not contained in this list?

6.14. When we decompose a product system, why is it important to include interface requirements and constraints (RCs)? Choose those RCs that customers see as critical quality characteristics.

6.15. How can one apply QFD for designing software applications? What are its benefits?

6.16. Explain how a QFD can help in promoting "a constancy of purpose" philosophy in a company when it has several conflicting organizational goals and objectives.

6.17. What are the risks of decomposition without considering adequate interface definitions? Does decomposition increase or decrease chances of locating the mistakes?

6.18. Why is it unwise to use several unproven processes in manufacturing a new product? How would you separate the essential problems from the trivial ones in your product development process?

6.19. What are the different forms of knowledge sharing and collaboration that are possible between teams, computers, and processes?

6.20. Why are the effects of conducting QFD and CEs similar? Are the goals similar as well?

6.21. What are the differences between a system-centered organization and a goal-centered organization? What vital SE factors must you take into account when setting CE objectives?

6.22. What do you understand by the term system engineering? What are some of the main advantages of system approach to problem solving? How would you analyze and assemble common challenges among projects?

6.23. Even after the stage of layout design, the Japanese are more effective at making design changes. Why?

6.24. What are the steps required to complete a typical SE project? Are requirements related to constraints? If so how?

6.25. Why does the profitability curve often start with a negative value? Can this be changed? How?

6.26. What can a company do to reach (and surpass) its break-even point quickly so as to increase its range of profitability?

6.27. Describe some of the benefits that can be derived from using decomposition in a design project? Provide another example of class hierarchy using Figure 6.6 as an example.

6.28. List the various types of automotive engine families by using a class hierarchy structure (Figure 6.6).

6.29. Develop a basic representation of an automotive crankshaft production system (based on Figure 6.6).

6.30. How important are interface specifications in a system-centered approach? What type of metrics do we need in order to measure the systems' effectiveness?

6.31. If a chair can be regarded as a functional system, what are its parts? Diagram a class hierarchy of a typical chair.

6.32. How many ways can a product be decomposed? Describe some examples of typical decomposition scenarios of a physical-based description (PhD).

6.33. What is the right amount of decomposition to manage both the complexity and the communication efforts?

6.34. How has the modular design concept helped automotive industries realize benefits? How can organization make CE work? Did any auto company grow into modular subunits? Did they achieve the expected results?

6.35. Describe how a traditional organization differs from a CE organization. How does a company measure its progress in a CE case?

6.36. What are the elements of a part's definition class? Draw a process flow diagram showing a typical system level organization of a part you are familiar with.

6.37. Are the physical, functional, and geometrical aspects independent hierarchical components? If so, explain why.

6.38. What is a smart approach to setting clear and unequivocal goals? How would you balance new goals in today's globally competitive environment, and how would you avoid setting unrealistic targets?

6.39. What entails an integrated approach to manufacturing? Why is it important to ensure "constancy of purpose" between goals and objectives at various organizational levels?

6.40. What are the different ways of organizing a traditional organizational setup into a cross-functional basis?

CHAPTER 7

INFORMATION MODELING

INTRODUCTION

The integrated product development (IPD) process, as represented by the concept of concurrent engineering, is the foundation of product realization: This is the basic operating principle of an industrial entity, whereby it designs and develops its products or services. This foundation involves numerous management and technical disciplines in its functions, as well as indirect interfaces to nearly all other functions within an enterprise. Indirectly, it involves all, or a significant portion, of the management, administrative, contracting, pricing, and estimating functions of a corporation, related to the planning and marketing of a product. In the development area, IPD involves the design, analysis, evaluation, planning, product support, quality control, data and configuration management, and procurement functions within the company. The difficulties that one faces, as a developer of products or services, is that there is no real control over what the customers or potential users would like in a product. Manufacturers often do not know how consumers view their product or services and what they consider as value-added. Is cost an important factor? Is time an important consideration? The level of satisfaction of the consumer does drive his/her decisions to buy or not buy a product. A successful IPD operation requires a sufficient understanding of these processes. One way to achieve this understanding is to construct a series of reliable procedures or models for planning, designing, optimizing, and controlling each unit of the IPD process. Constructing models requires consideration and integration of an immense amount of diverse and complex information. Such consideration includes not only information about product requirements, constraints, form features, function, fabrication, marketing, and service, but also organizational and administrative procedures for the enterprise. The latter includes the manufacturing or production method/equipment that underlies the product realization process, and the human psychology and cognitive aspects that make it possible.

A model is a representation of some important characteristic of a subject system. The subject system could be physical (such as a part), conceptual (such as a view, or an abstract sketch), or analytical (such as an equation). The original definition of a model is from Murphy [1950, page 57]: "A model is a device which is so related to a physical system that observations of the model may be used to predict accurately the performance of the physical system in the desired respect." This definition applies more to a subject system that exists physically. The physical system, for which the predictions are to be made, is usually called a prototype. The conceptual model is a scheme or a method, which closely describes the behavior of the system at least in a particular domain. The analytical model predicts those behaviors by mathematical means.

The purpose of modeling is to simulate the various options in order to make informed decisions early in the process. Informed decisions (such as what happens in a practical sense) go a long way. The models act as road maps for predicting or simulating what should be done next, given the knowledge of what has already happened. It also helps relate one phase to the other, and to guide the CE teams along a consistent path of product design and development.

This chapter addresses the issues of model generation, identification of performance metrics, types of model creation, extension, and upkeep, and their verification.

7.1 INFORMATION MODELING

Most products have several years of useful life during which they must be maintained and supported. The product development cycle itself takes several years from conception to market. During this period, the product goes through a number of phases, each involving a number of processes, data conversions, specification transformations, product realization loops, etc. In addition, numerous sets of skills, policies, practices, procedures (3P's), and management philosophies are called upon, depending on the complexity and the level of organizational needs. For example, the conceptual design phase is dependent on the human designers' experience, intuition and abstract reasoning, and manipulative power to deal with geometric objects. Another requirement imposed by product and process complexity is the volume of information that must be considered explicitly.

Traditionally, a design team conceives an idea, which invariably is a part, an assembly, a process, a method, or a system. There could be more than one class of models, depending on the subjects' definitions and the aspects considered. Each model class describes the subjects in terms of their features (graphical or nongraphical), attributes (symbolic or nonsymbolic), and other technical or nontechnical information. These classes of models serve as a basis for investigating the product's behavior or seeking relationships between its constituents. One such traditional form of representation is a "conceptual (or imaginary) object." Such objects can be represented on paper in the form of an engineering drawing (figures, sketches, views, and isometrics), or electronically on a

CAD tube using computer graphics. Such intermediate representations generated during the design phase merely serve as a means of communication to augment memory and processing. These forms of representation, however, do not guarantee that the object represented is physically realizable. There are many ways in which information can be represented or captured into models. Some schemes provide more flexibility and are more efficient than others. Since ancient times, modeling in one form or another has been used to design and to develop buildings, machines, tools, and other manufacturable objects. During this modeling process, three types of schemes are often employed:

1. Physical Model
2. Conceptual Model
3. Analytical Model

The various methods of representing objects are illustrated in Figure 7.1. Characterizing and standardizing these schemes is an important step in ensuring smooth communication between the different stages of a concurrent design process. Good models are intrinsically aimed at representing a physically realizable object.

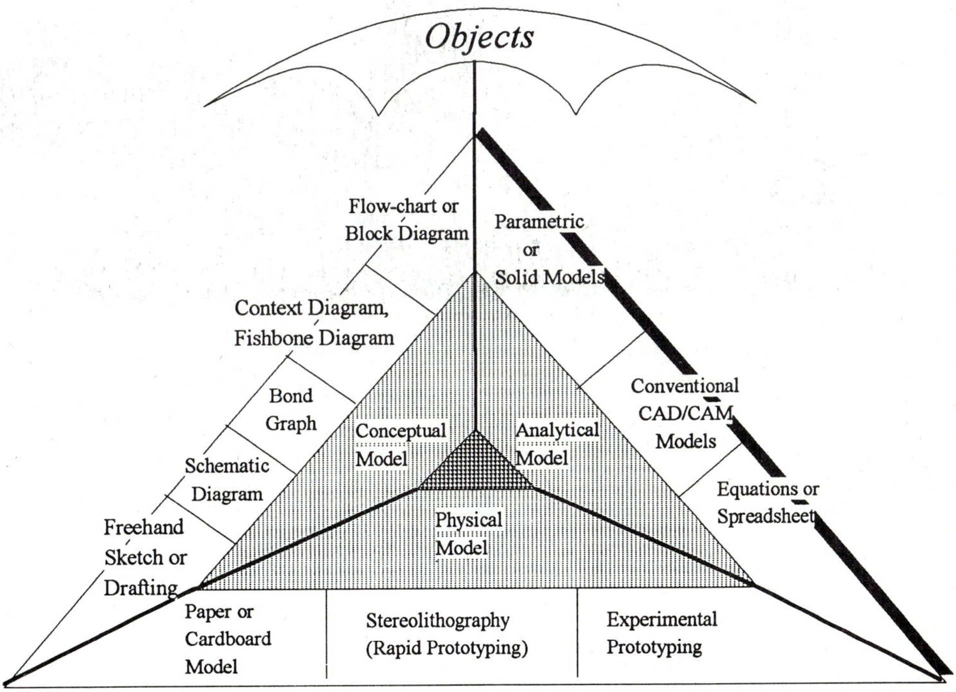

FIGURE 7.1 Various Methods of Representing Objects

7.1.1 Physical Model

In the early days, paper or cardboard models were used extensively. Later, as technology developed, it became easier to develop hardware prototypes made from a variety of sources, such as clay, wood, styrofoam, and even metals. People in industry still favor physical prototypes, since design is often considered an art rather than a science. Such hardware prototypes take the form of either experimental prototype or rapid prototyping systems, called stereolithography. The latter is discussed in Chapter 9, Section 9.5.4. However, with the advent of knowledge-based computer tools and cognitive methods of capturing human intelligence, hardware prototypes are now being replaced by software prototypes.

7.1.2 Conceptual Model

Another primitive form of the traditional design approach is a freehand sketch with lingual description to aid explanation. This approach may be rudimentary, but it serves a vital purpose—representing a conceptual view of something by someone from some point of view. Drafting a diagram is a more formalized form of freehand sketching, where drafting instruments are used to capture the design on paper. Such diagrams are called conceptual models (see Figure 7.1). Conceptual models are more prevalent during a traditional process particularly when dealing with the conceptual design phase of product realization. Teams need to determine what a product or process will roughly look like when done. There is a need to schematically capture the function and behavior of the system without any physical model so that the relationships can be studied further. During conceptual design with a CE approach, the teams also need to determine how the product and the related process will function at the system level or at the subsystem level. This is often referred to as a "system thinking" aid or intermediate diagram. Intermediate aids are useful in the group design environment to serve as records of agreements and social exchange. The idea is to quickly capture and structure the different object worlds using different forms of symbolic representations whether or not there is an underlying theory. Although each object world may contain information extracted from a particular angle or perspective, objects can be used in combinations to gain deeper insights into a particular phenomenon or subsystem behavior.

7.1.2.1 Forms of Representations

Conceptual models are effective visual techniques for displaying outcomes. Often, conceptual models are used to extract ideas from a group or are employed as a brainstorming tool. Conceptual models can be represented in five major ways: (see Figure 7.2):

- *Affinity Diagram:* The purpose of an affinity diagram is to categorize or group items from a brainstorming session or a random list of past practices. It groups items together that have some affinity in some way. Often, grouping is obtained as the result of a "group silent exercise". The participants are required to stand up of their own will (one at a time) and silently place the post-it labels along with other

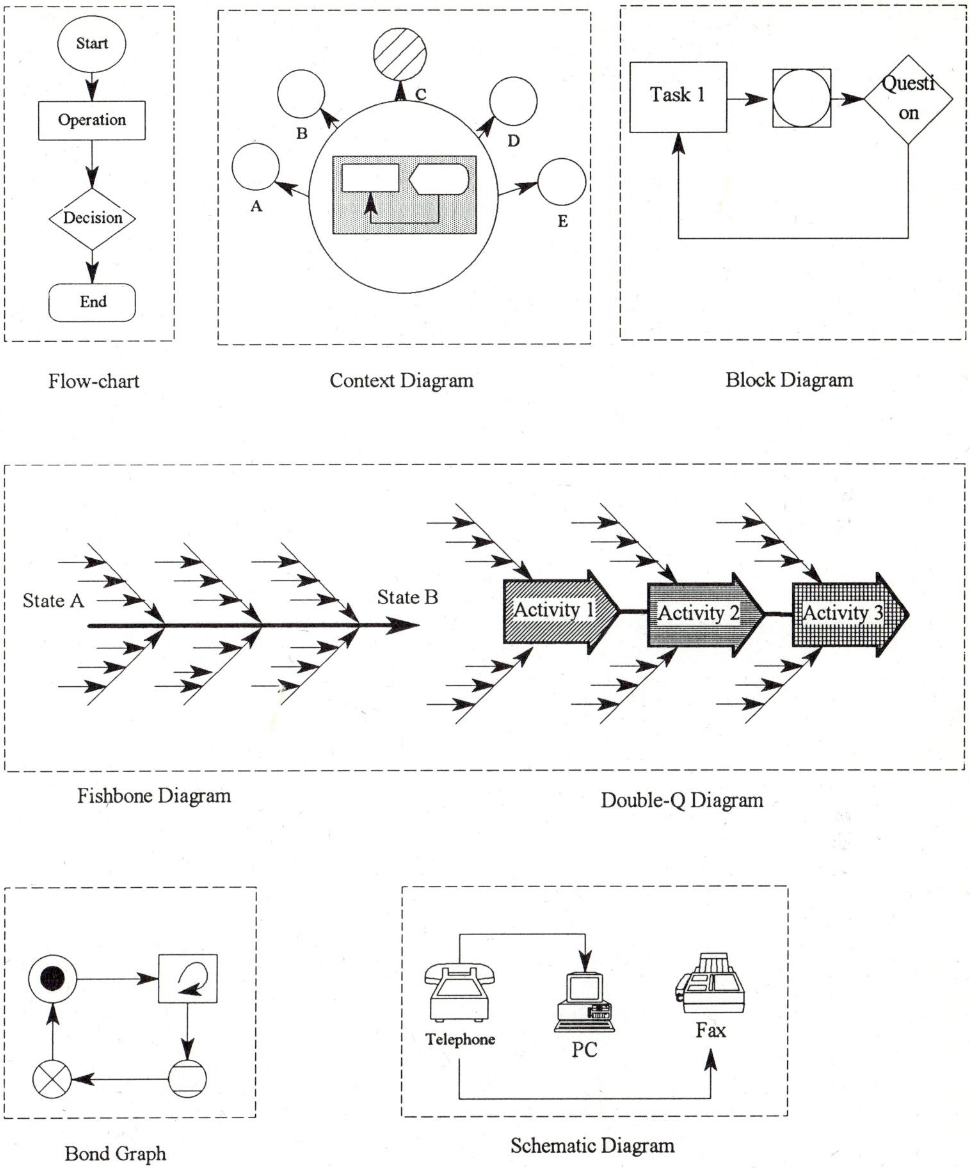

FIGURE 7.2 Various Methods of Representing Conceptual Models

labels that appear to show some affinity and commonality among themselves. Each participant takes a turn. The purpose of this exercise is to identify silently and simultaneously natural groupings of items, with no discussion. Once the group is finished moving around the post-it labels, a name is assigned to each identified category.

- ***Flow-chart or Block Diagram:*** Flow-charts or block diagrams are most suited for representing the function of a product or a process, that is, a detailed description of its purpose. These methods use inputs, outputs, and/or a set of decision blocks (optional) to describe a function. This is very similar to how a human being thinks in terms of cause and effects. Here, the blocks are connected through directed lines. This representation is suitable for analyzing interesting aspects of the system such as information or data flow (inputs, outputs, decisions), requirements (strategies, controls), constraints (mechanisms, feedback, dependencies, logistics, environments), etc. Integrated definitions (IDEFO and IDEFIX) are examples of such a tool [U.S. Air Force, 1981]; [U.S. Air Force, 1985]; and [Yang, 1989].
- ***Context Diagram:*** This is also referred to as interrelationship digraphs. It provides a method of determining one-to-one, one-to-many, or many-to-one relationships between a large number of items. Context diagrams are like block diagrams, where the relationship between "a child of" or "a part of" is represented by encircling the child or part information within a parent activity. Each activity "is made out of" or "requires" information from another activity and is captured by a directed arrow. This is useful for symptom analysis, or root cause analysis (such as repeated questioning used in the five WHYs). One may go over the diagram and have the group assign an S or R next to each circle to indicate whether it is a *S*ymptom or a *R*oot cause. The group can go back and break down each item which is marked S. Then, the group can evaluate the potential causes until all Ss turn into Rs.
- ***Fishbone Diagram:*** This is also commonly referred to as a Double-Q (QQ), root cause diagram, cause and effect diagram, and Ishikawa diagram: The two Qs stand for Qualitative and Quantitative factors; the name fishbone, because of the way they look; the name "cause and effect," because of the relationship they display. Kaoru Ishikawa claims to be the first one who described this approach in the 1970s in Japan [Ishikawa, 1986]. The diagram is made out of a heavy line or a box, slanted arrows, and parallel lines. Each line represents an activity and the arrow points to the issue being addressed. Such issues may include, for example, factors influencing the value of the products and services to the customers. There are three parts to a fishbone diagram. The first part is a solid line (or a combination of a box and a solid line)—a heavy horizontal arrow that runs straight from one state (tail) to another (head). There is a family of directed arrows that are placed on top of the horizontal line, representing major, "hard," (quantitative) factors or a line of functions. The "soft" (qualitative) factors, such as computer tools that incorporate these functions, run along the bottom. Arrows leading off the major factors represent subfactors. The QQ diagram provides a visual map of key factors at one glance. It often serves as a useful tool for brainstorming since both sides of the issues (hard and soft) remain equally visible.
- ***Bond Graph:*** This is most suited to capture the behavior of the physical systems instead of their functions. Most behavior is initiated by the energetic effects that take place across the multiple domains of the process. Bond graphs capture the flow paths for energy within an artifact using a directed graph technique [Finger and

Rinderle, 1989]. The vertices denote the energetic effects initiated by the subsystems, and the edges indicate the energy transfer between the subsystems. This graph forms a good basis for deriving a set of differential and algebraic equations for analytical or other forms of representation. Use of such multiple forms give both qualitative (such as comparision) and quantitative (such as sensitivity) insights into its energetic and dynamic behavior.

- ***Schematic Diagram:*** This is referred to as the most primitive form of physical modeling. It primarily shows a form of abstraction that tends to be near-complete with regard to domain-specific information and is very rudimentary with respect to the functional details of its components or parts. This diagram is useful in determining the missing elements and improving the quality of abstractions.

There are many ways information about a process can be captured or represented. Some schemes provide more flexibility than others.

7.1.3 Analytical Model

The typical formulas that we often find in standard reference books represent elements of the most simple analytical model. When they are used in conjunction with a spreadsheet program to satisfy a series of requirements, then they take the true form of a model. Each set of requirements may represent a subset of a CE design process to be captured. After adding modules, analogous with adding ready-made scripts to a spreadsheet program, the user can plug in the set of requirements and obtain results. The analytical model takes the form of an underlying "calculator" with a built-in "equation solver." With the advent of computer technologies, analytical and computer modeling techniques are gaining popularity. Analytical modeling is becoming today's dominating design tool in all aspects of concurrent design. There are several ways an analytical model can be represented (see Figure 7.3).

7.1.3.1 Primitive Forms

In the beginning, analytical models were mostly in primitive form. Many were based on standard reference book information, such as the structural design equations found in *Formulas for Stress, Strain, and Structural Matrices* [Pilkey, 1994]. Some primitive forms of analytical models are graphics-based (see Figure 7.3):

- *Spreadsheet:* A spreadsheet has a set format for user inputs—an array of predetermined equations, parameters (variables), and outputs. Parameters are variable lists containing data settings, outputs are where results are tabulated, and plot sheets define how results are graphed.
- *Function Plot, Run Chart, or Graph:* A function plot shows graphically how one variable affects another over the desired range of relevant values. A run chart shows how function changes with time (time is one of the parameters in the function plot).
- *Scatter Diagram:* This shows a mean curve drawn through a set of experimental or measured points.

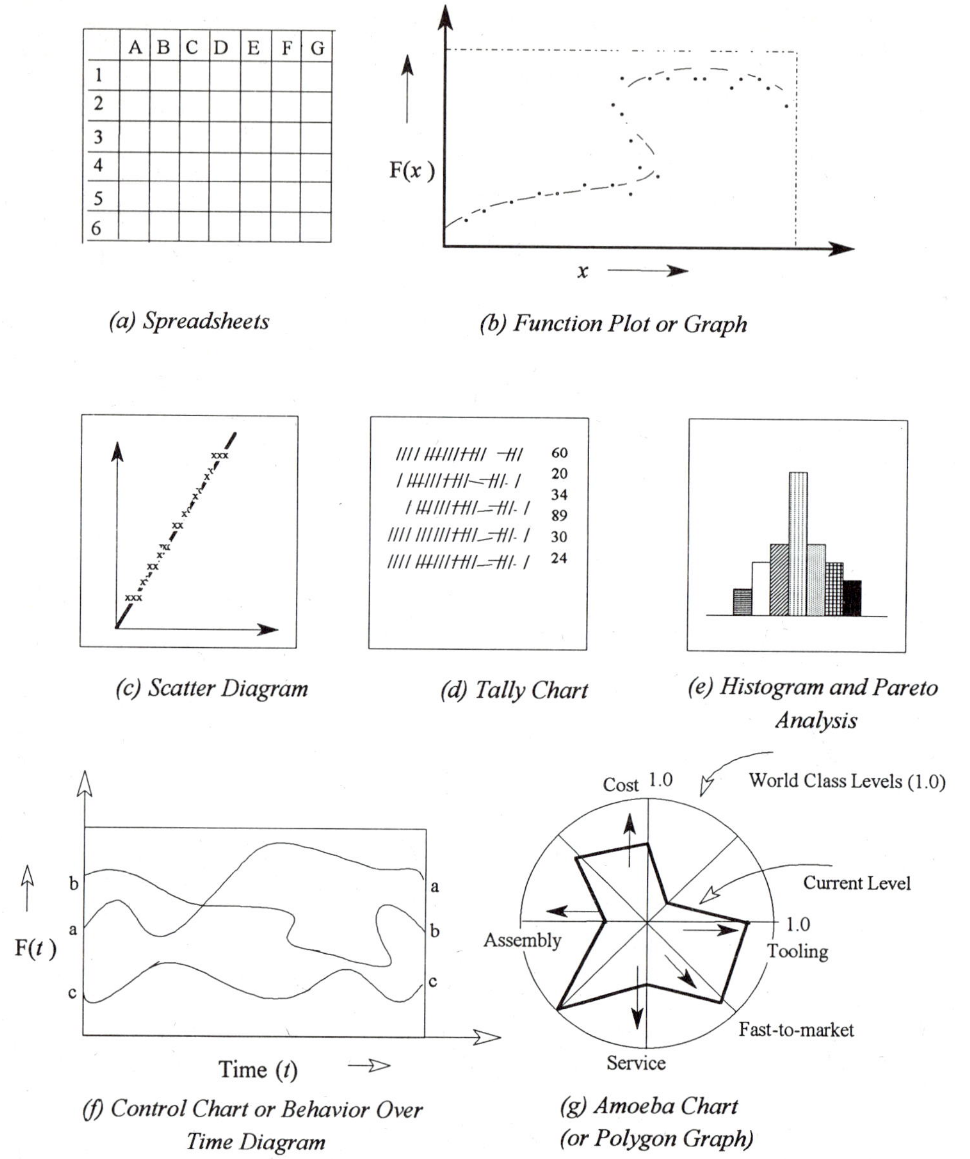

FIGURE 7.3 Graphics-based Forms of Analytical Models

- *Tally Chart:* This is also referred to as a check sheet and used for gathering data in an organized way (such as in tabular form). This is a convenient way to compare results (in whole numbers), such as voting, when the counts are functions of time. Tally charts are also used to show frequencies or percentages. Frequencies and percentages convey the occurrence of values in the data set. The two are proportional—the percentage is relative to the total number of scores in the data set.

- *Histogram and Pareto Analyses:* These are forms of bar charts graphically showing the frequency's distribution with respect to some chosen variable or class interval. The histogram not only indicates the frequency of each value, but also roughly indicates the range of data (lowest to highest value) and the shape of the distribution. There are many forms of distributions, such as a bimodal (two-humped) distribution and unimodal (one-humped) distribution. A normal distribution is a symmetric, bell-shaped distribution. Histograms can also be thought of as a crude form of a distribution. Given a distribution, we can calculate the probabilities of occurrences for various events.
- *Control Charts and Behavior Over Time Diagram:* The use of histograms and distributions is one way to summarize data graphically. Another way is to summarize data sets with the use of summary indices and control charts. A control chart is used to display how a parameter, measurement, or event progresses with time. If a number of functions are related to a single variable—say time—the behavior of each one can be graphed on a single chart. Two examples of summary indices are mean (central tendency) and variability (range). The control charts also show interdependence in their behavior. They are discussed in Chapter 2, Volume 2.
- *Amoeba Chart (or Polygon Graph):* This is also commonly referred to as a spider chart because of its final shape. It is a schematic way of plotting characteristic features relative to what is found in (or what makes up) a world class product. The points, which lie on a world-class level are plotted along the circumference of a circle. The points along the radial axis show the level of attainment of those features relative to their world-class level.

In the early days, CAD was primarily used to create drawings. The initial attempts to automate the activities of a production cycle were focused on reducing human involvement in preparing and interpreting two-dimensional (2-D) drawings. This effort led to the development of geometric modeling techniques in their most primitive form. In the beginning, CAD was (in its most primitive form) a tool for analytical modeling. However, it was soon realized that this "drawing through CAD" had limited usefulness since the information was entered manually through a computer at first and later output in paper form. All the effort put in to create the drawings, therefore, did not do much benefit except to automate the drawing creation process. Thus, the use of CAD had limited utility in the functional design process. Then came the two and a half CAD. It conveyed design features to clients and customers more rapidly than the customary 2-D drawings. The 3-D CAD system that followed, enabled the detection of layout problems, such as interference, that could not be detected from a 2-D CAD or a drawing. It also allowed determination of the physical properties of a part (such as volume, weight) without creating a physical model.

7.1.3.2 3-D CAD System

Analytical design has always been a process of thinking in three dimensions and then capturing these thoughts as two-dimensional drawings or isometric pictures. In general, the more removed a representation of a mechanical system is from its geometric form, the harder it is for an engineer to understand it intuitively. Solid modeling technology has become the core enabler of analytical modeling. Analysis of three-dimensional shapes is made simpler with solid models. With three-dimensional solids, electronic models of the product can simulate the design actions and manufacturing processes required to machine the part. Solid models can simulate digital preassembly and electronic mockup, and can perform three-dimensional tradeoff analysis (utilizing software prototyping concepts). Boeing eliminated the need for a full-size mock-up of the new 777 by using solid modeling. Solids are also quite useful in generating complete data for rapid prototyping systems to quickly obtain physical models of concepts and designs. Such mechanisms are key facilitators of the CE design process. These developments have brought analytical models closer to the realization of the physical objects than at anytime before. Still, it is not always practical to capture all the minute details. For example, most three-dimensional CAD models do not represent the geometric structure explicitly—in terms of variable dimensions. This requires recreation of the design even if only a small subset of parameters needs to be changed ("same as in . . . except . . . situation"). Also, specifications of tolerances and surface finish are difficult to define in most three-dimensional models (Requicha and Voelcker, 1982).

7.1.3.3 Virtual Analytical Models

The growing use of three-dimensional modeling in design stems from its convenience once the geometry is captured analytically. Designers are able to build and modify a part on a tube much faster and with greater ease than with the use of only a physical model. Moreover, the three-dimensional models created on computers are reuseable and modifiable with far less time and cost than models using lines and surface representation.

In recent years, software for three-dimensional geometric modeling has added techniques such as parametric modeling, variational geometry, variable dimension modeling, knowledge-based engineering, and other virtual techniques. It is no longer necessary to start over from the beginning to modify a solid model. Developers of solid models have linked up with suppliers of specialized applications to provide integrated functionality. Solid or three-dimensional design is becoming a preferred method of capturing analytical geometry. It allows the designers to communicate in the same manner as a common person thinks. After all, one cannot launch a newly designed jet or missile and just hope that it works. A look at the complexity of the aircraft and weapon systems used in the Persian Gulf reveals the magnitude of the role played by multi-level interactions. New aircraft designs, the landings on the moon, satellite networks, and space probes would never have been possible without a systematic approach to information modeling. How can this be done until all the data is available, all the parts designed, and all the bill-of-materials (BOMs) established? The answer lies in capturing information in an electronic surrogate knowledge-base and in developing generic models:

- ***Technical Memory:*** Technical memory is a form of surrogate database. It provides the basis for developing information models at the enterprise and specification levels. It can be used to develop realistic model alternatives, which can then be analyzed. The database, when used as the basis for these studies, can provide meaningful forecasts. It can predict the effects of the implementation of alternative modeling strategies on manufacturing facilities and operations.

- ***Specifications for Physical Realization:*** In order for analytical models to represent functional design as closely as a physically feasible object would, models must provide the following:
  - Mechanisms to explicitly represent the artifact's functional requirements and link them to the associated geometrical elements
  - Means of checking the internal consistency of the physical features of the object
  - Ways to adequately define the topology of the object (feature recognition)
  - Means of evaluating the design for assembly and disassembly (DFX) violations
  - Consistency and adequacy tests for selective assembly, part mating, and tolerance stack-up
  - Criterion to compute assembly clearances under working (loaded or stressed) conditions
  - Ways to specify tolerances and surface finish
  - Means of checking against under- or over-dimensioning.

7.2 MODELING METHODOLOGY

Analytical modeling provides a better modeling alternative than options such as physical modeling. It is very similar to ad hoc coding in software development. Most modeling methods formalize and expose errors and inefficiencies that can be remedied easily. The chances that such problems may be overlooked are slim with analytical modeling, as compared to the confusing complexity of the implicit nature of the alternate approaches. Newer tools which capture a product design as a computer model, and streamline the design process are available. These tools perform accurate modeling and can dramatically reduce the design time. In addition, tools allow engineers to evaluate alternate designs and transfer product data directly to other computer-based engineering systems. Modeling, based on systems thinking (discussed in Chapter 6, Section 6.3) or an enterprise integration (EI) metric, Section 6.6, provides new opportunities for system optimization. System complexity is often handled through a systematic breakdown of a product tree structure (discussed in Chapter 6, Section 6.4) followed by decomposing the tasks in smaller-sized chunks so that it can be easily managed.

It is difficult to cover all aspects of interest during a product design and development process with a limited set of model representations. As discussed in Section 7.1.3, each of the analytical representations within a model is better suited to a specific set of application areas (for modeling and analysis) than others. In CE, the ability to process information from multiple perspectives and levels of abstractions is critical. The multiple model representation supports the idea that one can enter and edit one model in several representations. Therefore, during product development, a variety of interconnected representations need to be considered. The applicability of a model with multiple representations tends to stretch over a large domain of applications, giving a better medium to analyze the design from multiple perspectives.

7.3 FOUNDATION OF INFORMATION MODELING

Systems such as computers, spreadsheet tools, numerically controlled (NC) machine tools, NC programming languages (APT), post processors, parametric programming, etc., have been all around for decades. Tools for modeling the physics of complex engineering devices and for simulation of their behavior are widely available. But, despite many advances in requisite hardware and software, their use in CE has not become widespread. On a critical level, the missing ingredient is the computational environment to model the invariant aspects of the problem or the product realization process. To support the latter, an organization requires an array of models to adequately capture the variability of the problems and their environment. A model is an abstract specification for domain functions that perform operations. There is a name-space behavior between models and applications that binds the abstractions to executables. The binding of behaviors is essentially a transformation where modeling concepts are converted into "real-world" applications. One of the important aspects in any product realization process is the existence of "real-world knowledge." This knowledge exists in many forms and structures: domain knowledge with its theory; abstract conceptual knowledge; and knowledge patterns. Abstract knowledge may be fragmentary without a consistent underlying theory of worldly experience or behavior models. The fundamental problem is the lack of abstract understanding of design and manufacturing processes to create math models that can capture or bind their behaviors. Figure 7.4 suggests a useful way of looking at the problem of information modeling. The objects in the real world are replaced by mathematical models, that is, by a series of abstract mappings defined in the appropriate domains of mathematics. The domain of knowledge has been broken up into three transformation name-spaces mapped on each other. The first transformation is from a physical object to a set of operational models, each providing a specific perspective of the "problem domain." This perspective addresses the problems associated with the "performance subspace." The second set of transformations represents the functional models of the perspectives. They belong to the "constraint subspace." Next, the notion of representation schema (entities) is defined. The mathematically defined entities, for example, attributes, symbols, and features, are shown to belong to the "parameter subspace." The functional models are expressed in this lowest level of representation for the mathematical modeling space. The representation schema underlying the defined entities of a computational system is important for ease in constructing and manipulating the functional models.

The transformation labeled "representation schema" in Figure 7.4 takes the functional models and produces mathematically defined "symbol structures;" for example, character strings, attributes, etc. Eventually, such symbol structures must be further mapped into "storage structures" in the programming environments; such as objects and classes in an object-oriented programming system (OOPS). This step is not shown in Figure 7.4. The information modeling scheme is concerned with the issues and approaches that provide this binding (for example, executable modules, foreign model importation, translation of model linkages to application linkages, and federated name-spaces). This type of information modeling scheme is based on some basic templates. Figure 7.5 shows a four-tier approach to information modeling. The top tier (floor 4) provides a mathematical

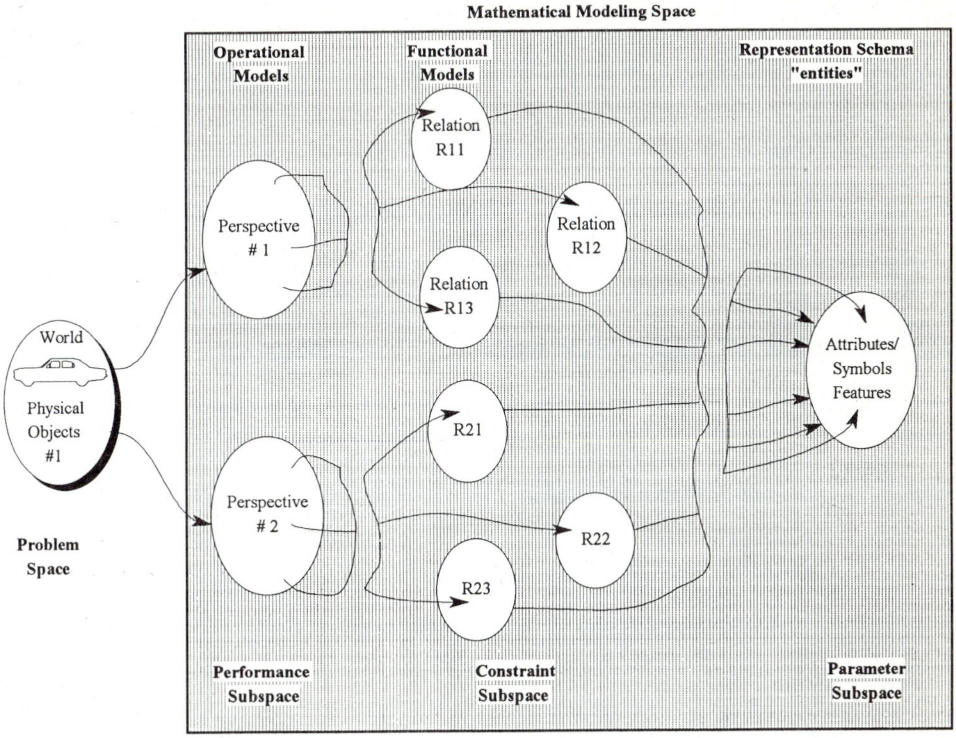

FIGURE 7.4 An Abstract View of Information Modeling

foundation to express the functions in math entities. The second tier is the uniform modeling scheme, which sits on the common process template tier. It provides representation-free mathematics. The third tier is a common process template, which furnishes a representational independent computational procedure. Finally, the fourth tier consists of common representation forms, which provide the foundation for building sharable models. In the following list, each building block is further explained in greater depth:

- ***Mathematical Foundation:*** This is the most fundamental tier. In order to be successful in all downstream operations, each model or function should be mathematically based. Each mathematical form should be capable of dealing with multiple aspects of information modeling, such as mechanical design, numerical control, process planning, manufacturing, assembly, inspection, etc.
- ***Uniform Modeling Schemes:*** This is the second tier. In CE, modeling schemes should be uniform across all aspects of product and process designs. Schemes must contain a uniform structure to accommodate various product data regarding its geometry, material, dimensions, tolerances, topology, form features, references, administrative, and technical notes. The algorithms should follow representation-free mathematics.

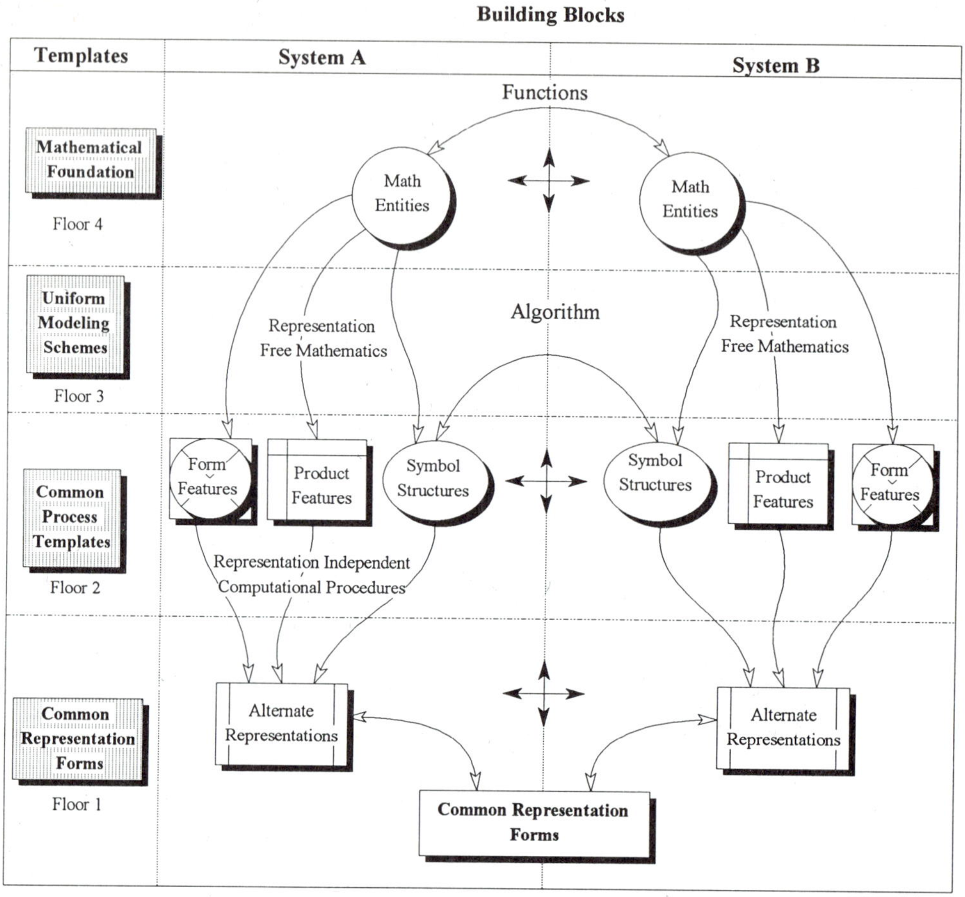

FIGURE 7.5 Building Blocks of Information Modeling (When Dealing with Multiple Systems)

- ***Common Process Templates:*** This is the third tier. Common process templates follow a consistent modeling approach to product design and development using the best industry practices across the entire organization. A common methodology enables integration and optimization. Uncommon processes are driven by multiple units performing similar functions but under different acronyms. Common process requirement templates, such as form features, product features, and symbol structures can provide the framework needed to enable commonality, consistency, cross-functional communication, and elimination of waste.
- ***Common Representation Forms:*** This constitutes the last tier. To build a good model, one needs a unified data representation scheme to organize all product data to support life-cycle product developments, such as product design, process planning, production planning, marketing, service, and administration. Parts, assemblies, stocks (raw materials), etc., need to be described through a common set of

representations. The forms of the entities are the same but they may yield different instances of the parts by taking different values. Each instance provides an alternate design. The same common forms can also be used to describe manufacturing, assembly, and inspection processes.

In a CE setup, the management staff needs to be more proactive in driving commonality and cooperation within the operating units. This results in a major change of the functional role of a common representation form (such as geometric model—from a mere drafting tool to a common information model) that is distributed across the networks geographically. Evidently, the lower level geometric abstractions, alone, are not sufficient to support the various collaborative activities of CE in modern manufacturing. It would be highly desirable to capture the natural and logical links among individual activities of the product cycle and then distribute them so that each is an add-on and supports a complex network of reasoning throughout the life-cycle of a product.

7.3.1 Mathematical Foundation

Representing the models of the product's assembly, its subassembly, and its components in some mathematical form provides a necessary basis for studying product behavior and possible interactions between them. The design engineers can use such tools to check whether the assembly, component, or the subassembly design perform in synchronization with all other product subcomponents and subsystems. Mathematical forms provide the basis to effectively handle the growing complexity of products today and to optimize each component relative to the overall product. Also, by virtue of a large knowledge-base, the pitfalls of traditional "cut and try" engineering are avoided by following this approach.

Moreover, computer models and knowledge-base eliminate the need for continuous retraining or for making up lost knowledge as people retire, quit, or are moved to other projects. The proprietary knowledge-base continually grows in capability and, when properly mined, can prevent repeating design errors during the IPD process. This technique has been effectively used for many years by the aerospace industry. Due to the extensive complexity of the products that one manufactures today, computer models are necessary to verify the design for X-ability (DFX) requirements before actual production can begin.

7.3.2 Uniform Modeling Schemes

Capturing mechanical objects in terms of CAD geometry or solid models that compute useful geometric properties, such as volume, moment of inertia, appearance, etc., is not sufficient for functional manipulations. Mechanical objects are not completely defined until physical behavior and functional aspects are also captured (in addition to the usual geometry, as shown in Figure 7.6). The manipulation of mechanical objects requires capturing information in the following three aspects:

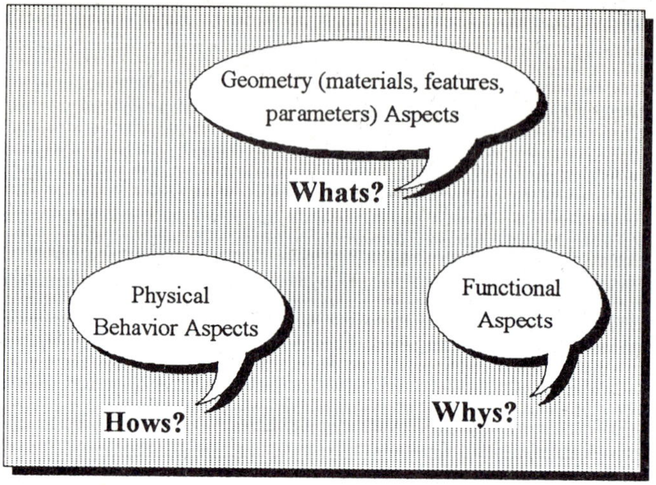

FIGURE 7.6 Definition of a Complete Product Model

- ***The geometrical aspect*** includes a three-dimensional spatial representation of parts, assemblies, joints, features, articulations, etc. A geometrical model answers the question "what are the elements of the artifacts?" In Chapter 6, Figure 6.3 shows that an artifact or a mechanical object may be decomposed into a hierarchy of elements: subsystem, components, parts, materials, features, parameters, etc.
- ***The physical behavior aspect*** includes the description of the artifact based on the physical laws of behavior. A behavior model answers the question "how do the elements behave individually or collectively as a system?"
- ***The functional aspect*** captures the design intent, purpose, a set of functions, and features or goals of the artifacts. It answers "why does the artifact function the way it does?" Unlike, geometrical features (such as line, arcs, or surfaces), functional aspects are difficult to capture programmatically. However, without the functional knowledge about the elements of design (such as an existence of a feature), it is difficult to make rational design suggestions. For example, a program might suggest removing features that are required for the proper operation of the device or it might suggest other similar impractical changes. It might suggest replacing a feature with one that is cheaper to produce but has the wrong shape or function. Without functional knowledge, it is difficult to determine which changes are good.

We need a mathematical definition of a product model connecting all such procedures and aspects, mapped across physical domains (product model is one of the five CE invariants described in Section 7.4 later). One way to achieve a product model is to assign a high level functional description with features, such as esthetics, weight reduction, clearance, slide, fit, etc. to the basic geometry. The following is a list of examples of some functional descriptions assigned to features:

- A large concave radius has an attached "clearance" function.
- Through holes are the "bolt" holes.
- A convex radius and the angles are for "weight reduction."
- A slot is for a "slide fit."

A physical aspect is represented by imposing a set of constraints on the part's functions. For instance, in the slide fit slot example, the top, bottom, and front sides must be open so that the bar can slide into it. Similarly, in the bolt hole example, the top and bottom sides must be open so that the bolt can pass through them. Having an understanding of the functional aspect and the physical constraints (which a functional aspect puts on the part), it is easy to determine what design changes are permissible [Hays, Desa, and Wright, 1989].

Among the three aspects outlined for a product model, geometrical definition is the most generically developed. Algorithms, for computing geometric properties of objects, depend strongly on how the objects are represented. Algorithms to operate on objects represented by their boundaries (that is, bounding surfaces) and those represented as compositions of solids, have been developed by computer graphic, computer-aided design, and numeric control communities [Requicha and Voelcker, 1982]. Probably the biggest difficulty in exchanging data between CE work-groups is due to the fact that the underlying geometric representations differ between the various CAD systems. This results in instabilities, ripples, and a lot of inaccuracies. There are also significant differences in the degree of completeness with which these models represent object geometry. Requicha identified six modeling schemes for unambiguous representation of solids [Requicha, 1980]. Among them, three-dimensional wireframe, the boundary representation (B-rep), and the constructive solid geometry (CSG) are the most popular. They are commonly found as internal representation schemes in most commercial CAD systems. A progression of geometrical models from the viewpoints of photorealism, compute power, and picture display and refresh is shown in Figure 7.7. Three-dimensional wireframe, which is the simplest form of a geometrical model, is at the bottom. It offers poor photorealism, but it takes less compute power, and is faster in picture display and refresh. As we move away from the wireframe model to the right, the degree of measures changes. For example, the CSG model offers sharper photorealism than wireframe; three-dimensional solids are even better than CSG; rendering and scientific visualization are still better; and interactive photorealistic rendering (IPR) is the best. A similar trend occurs for the remaining two cases—compute power and picture display and refresh. The IPR model offers the sharpest degree of photorealism, takes most compute power, and is slowest in picture display and refresh.

7.3.2.1 Three-dimensional Wireframe

A three-dimensional wireframe is the simplest way of creating surface geometry. This method models the object by traversing the trajectory with wireframes—recording primarily the edges and vertices in the database. The object is recognized by a series of u-v lines, called a wireframe, enclosing the object. Wireframe models are ambiguous. An area surrounded by a wireframe may represent a surface or a hole.

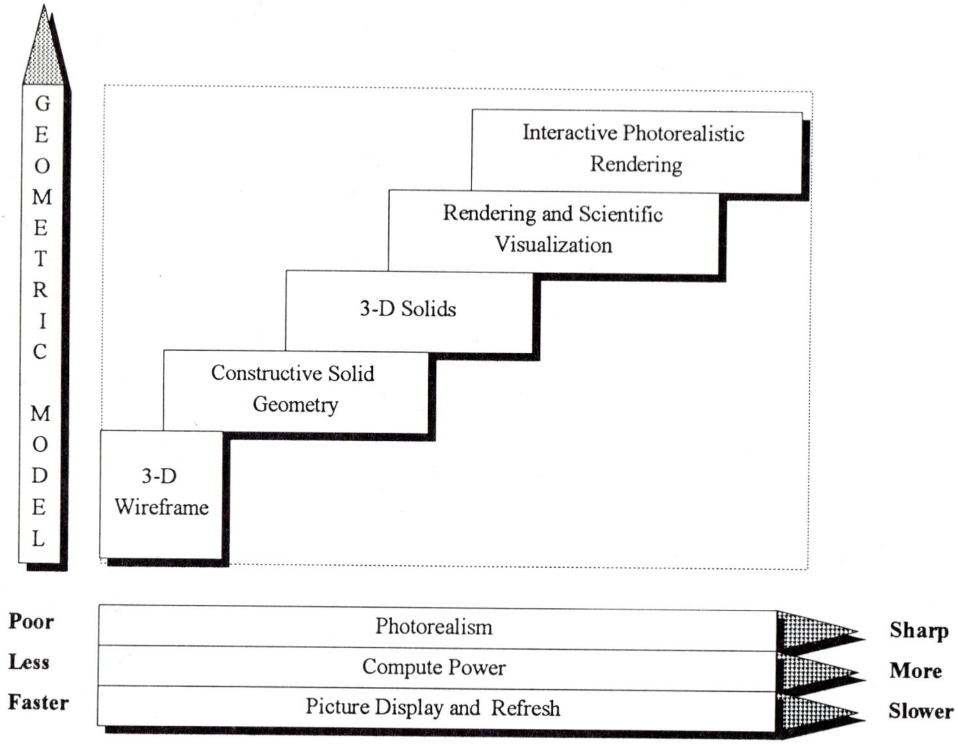

FIGURE 7.7 Common Progression in Geometric Modeling

7.3.2.2 Constructive Solid Geometry (CSG)

Figure 7.8 shows two schemes of representing solids. The selection of a representation can be crucial for ease of modeling, integration, and controlling the design through the same set of parameters. A CSG-based product model is represented by a set, theoretic Boolean expression of solid primitives. Commonly used solid primitives are block, sphere, cylinder, cone, torus, etc. New, complex, three-dimensional objects are constructed by allowing addition and subtraction—geometric manipulations of objects using Boolean operations (see Figure 7.8). For example, a user can create a new three-dimensional object, such as a void, by subtracting one shape from another. As a by-product of the Boolean operations, some CSG systems can automatically compute the edges and surfaces of the resulting object. However, it is harder to produce complex three-dimensional solids using Boolean primitives.

7.3.2.3 Boundary Representation (B-rep) Solid

In B-rep, the geometry of the object is represented by the surfaces. The object is recognized by the surfaces binding the faces of the object. Usually a three-dimensional wireframe is required to construct the bounding surfaces. Unlike the three-dimensional wireframe, B-rep can distinguish between a two-dimensional slit and an actual form feature on

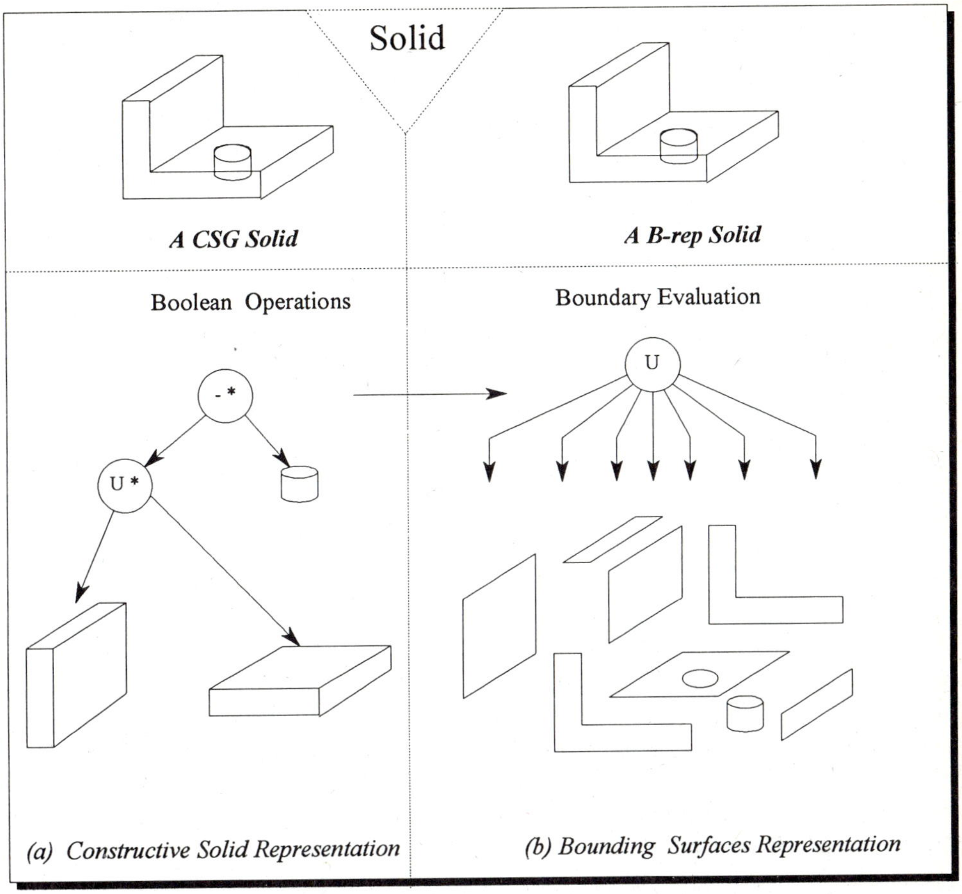

FIGURE 7.8 Two Methods of Representing Solids

the object's surface. However, the volume enclosed by a B-rep model may be empty or void. The creator of a B-rep surface can add intelligence or features (slots, holes, flat sheets, or volumes intended to represent material, etc.). As a result, the model can be interpreted by other programs that use the model as input. By contrast, a CSG solid model leaves little room to add any intelligence [Kinnucan, 1989].

Basic solid modeling operations include local modifications, global operations, and undoing operations. It is easier to implement local modifications in B-rep and global operations in CSG. With regard to undoing, a global operation on CSG is easy to undo, but difficult in a B-rep. On the other hand, a local modification is simpler to undo in a B-rep than in a CSG. The development of an extensible set of general three-dimensional concepts capable of representing a variety of design objects and constraints is also crucial. Three-dimensional concepts may overlap, that is, there may be more than one set of concepts to describe a portion of a design. Aspects, and views of aspects, can be described in terms of concepts, and the commonality of concepts between aspects can be used to build constancy of purpose.

The relative complexities of B-rep models do offer some significant advantages. Their judicious use allows automation of tasks that require identifying surfaces from holes. Furthermore, they offer faster response than CSG models. The responsiveness is further enhanced by representing surfaces via polygonal facets. Most B-rep systems use a hybrid approach. A faceted representation is used where speed, rather than accuracy, is of prime importance, such as for model display. Exact analytical representation is used where accuracy is of major consideration, such as surface intersection computations.

7.3.2.4 Three-Dimensional Systems

Today, most three-dimensional CAD/CAM systems use a hybrid approach. This approach employs either a CSG or a B-rep scheme as the primary representation option, but the three-dimensional system converts one to the other as the needs arise. Conversion of a constructive representation (CSG) into a boundary representation (B-rep), for example, is important in generating displays efficiently because an opaque object's appearance depends on its topological boundary rather than its interior.

Three-dimensional models, with their inherent capability to differentiate between void and solid volumes, offer some secondary benefits. They allow automatic computation of mass and inertia properties, and a method of easily verifying interference and clearances between objects. Three-dimensional systems automate many of the constructs that would have to be done manually with a B-rep or a CSG system. They are, therefore, better suited for boosting user productivity. Most three-dimensional systems do possess most of the underlying features of the B-rep and the CSG systems.

7.3.2.5 Rendering and Scientific Visualization

One of the biggest advantages of all geometry modeling systems is the ease with which a rendering or scientific visualization can be performed on objects. This includes producing color-shaded, three-dimensional renderings and the computation of surface intersections and measurements such as curvature, area, and volume. Wireframe based models are faster at picture display and refresh. In three-dimensional solids, color-shaded display and analysis operations are computer intensive. As such, three-dimensional models require more compute power than a wireframe system to achieve results in a comparable response time. To accelerate the display and refresh time, many systems utilize a faceted approach to surface representation. Internally, the surface is represented using a full and accurate mathematical form, but for display purposes, an approximated form is used. The mathematical representation and database remain unchanged. Now, with more compute power available per dollar, the emphasis has changed more to visual comfort than actual response time.

7.3.2.6 Interactive Photorealistic Rendering

Interactive photorealistic rendering (IPR) is a rendering methodology that combines interactivity with increased photorealism. IPR offers instant visual feedback for fine-tuning operations such as coloring, lighting, and texturing photorealistic images. It also provides increased image realism through high quality anti-aliasing, selective ray tracing, automatic motion blur, diverse shading effects, multiple light sources shadowing effects, and other rendering capabilities.

7.3.3 Common Process Template

A common process template follows a consistent modeling approach to product design and development using the best industry practices across the entire organization. This means interface procedures and standards should be common but decentralized. Common, here, refers to a single standard way of defining the interfaces linking various platforms, equipment, software, and applications. Decentralized means distributed across the networks and accessible through remote workstations. There are three possible methods of linking equipment with respect to time and place:

- ***Common centralized:*** A single method is used for defining the interfaces and the equipment, which are physically collocated (for example, one way, one place).
- ***Common decentralized:*** Here, the various interfaces, equipment, and applications are electronically linked to each other, but they can be geographically separated (for example, one way and a number of places).
- ***Unique decentralized:*** This is the opposite of "common centralized." Both the location and the interface for each location are different (for example, five ways, ten places).

7.3.4 Common Representation Forms

Representation formalisms, in which models are expressed, are not powerful enough unless (a) they have a common robust structure, and (b) they address a spectrum of aspects from generic to specific. For example, a geometry model can be adequately described using a common data description. A data description is used to facilitate consensus on the terminology necessary for multifunction design teams to communicate and operate effectively [Prasad, Morenc, and Rangan, 1993]. Common data representation is needed to design data structures to handle both immediate CE needs and the needs that will evolve as additional CE opportunities are pursued. Two forms of data representation are shown in Figures 7.9 and 7.10:

- ***Data description is local to each application:*** In this case (see Figure 7.9), different representation forms are used for each individual application. In such situations, interface definitions are required between shared applications to communicate entity, such as an attribute, a feature, or a relationship. This means that if there is a need for sharing common "entity," there would be a need for data integration across applications.
- ***Data description is global:*** Figure 7.10 shows a data description form, which is uniform (has common representation) across all applications. Notice the dividing lines (in Figure 7.9) separating the attributes/relationships from its applications have

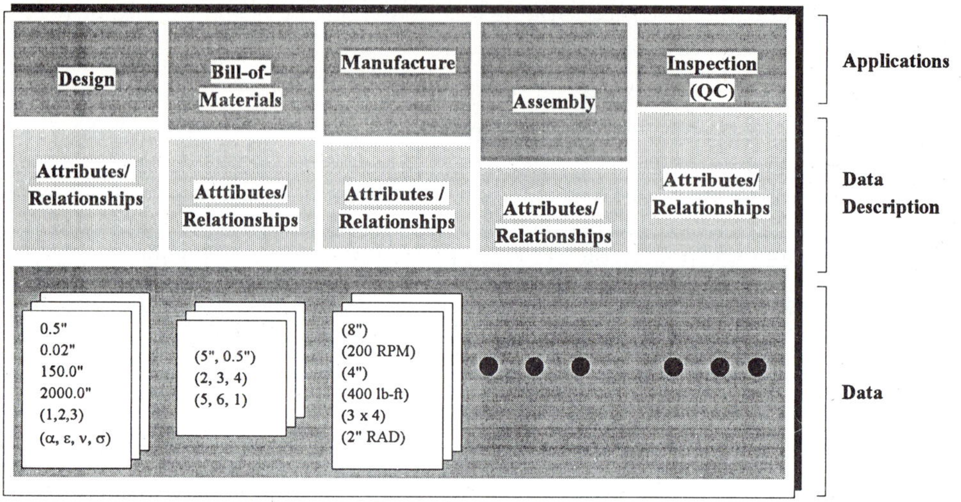

FIGURE 7.9 Data Descriptions (Attributes/Relationships) Local to Each Application

been eliminated in Figure 7.10. In this case, no translation of data is necessary for applications to communicate. Access to the data description entities (attributes/relationships) is available through database utilities for all applications to use.

The one way to model a part for a specific process is to design it using a library of generic representations of features [Curran, 1994]. If, for example, the designer needs in-

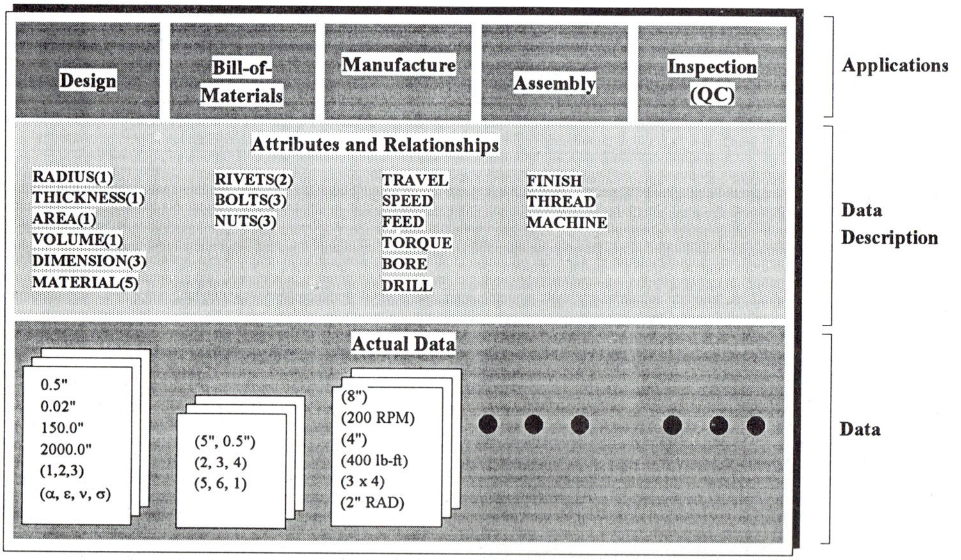

FIGURE 7.10 Data Descriptions (Attributes/Relationships) Common to All Applications

formation about key ways, the system should be able to provide the designer with information about different possible key ways and the processes which can be used to make them. Some systems include commands to create geometric feature families of manufactured objects, such as bosses, chamfers, holes, slots, countersinks, and counter-holes. They are often called "form features." A form feature is a local geometrical characteristic of a product at a higher data level then those of faces, edges, and vertices. "Form features" are the common thread between design and manufacturing. The use of form features on parts is entirely compatible with several new CAD, CAPP, CAM, and MRP systems [Hummel and Brown, 1989], which form the core of the manufacturing system.

A single monolithic information model environment would simplify system and applications development. However, a sufficiently powerful data definition scheme may take five to ten years to become standardized. Until that happens, the computational environment will most probably remain a heterogeneous collection of flat files, disk databases, and object-oriented databases of various sizes and degrees of maturity. The major objective of information modeling is, therefore, to define the semantics of data definitions independent of any particular application program which uses that data. In particular, one would like to do the following [Prasad, Morenc and Rangan, 1993]:

- Define a common generic scheme expressing parameters, constraints, form features, and tradeoff relationships
- Maintain a unified scheme for building models and defining their enterprise interrelationships with the product process and enterprise models
- Publish and distribute conversions (for example, data translation or interpretation to those models) among the various representations and platforms.

7.4 CONCURRENT ENGINEERING PROCESS INVARIANT

Concurrent engineering is still evolving into a cohesive methodology where diverse technologies of design, material science, material processing, manufacturing, computer networks, computer software, database systems, decision support tools, and man-machine interactions must be integrated. In order to provide all such variations in product, process, and tools, it is essential that a CE process focus on capturing the key aspects of the product realization definition and its environments, which are invariants [Prasad, 1996]. The role of identifying the various invariants of the CE process is crucial in accomplishing CE benefits. One way to accomplish this is to divide or classify the entire domain of CE into a set of patterns or models, which capture a distinct aspect of the CE phenomena. Such classification leads to a common set of CE process templates. The characterization and representation of models in mechanical products and systems for CE can be grouped into five model-classes as shown in Figure 7.11:

- Enterprise model-class
- Specification model-class

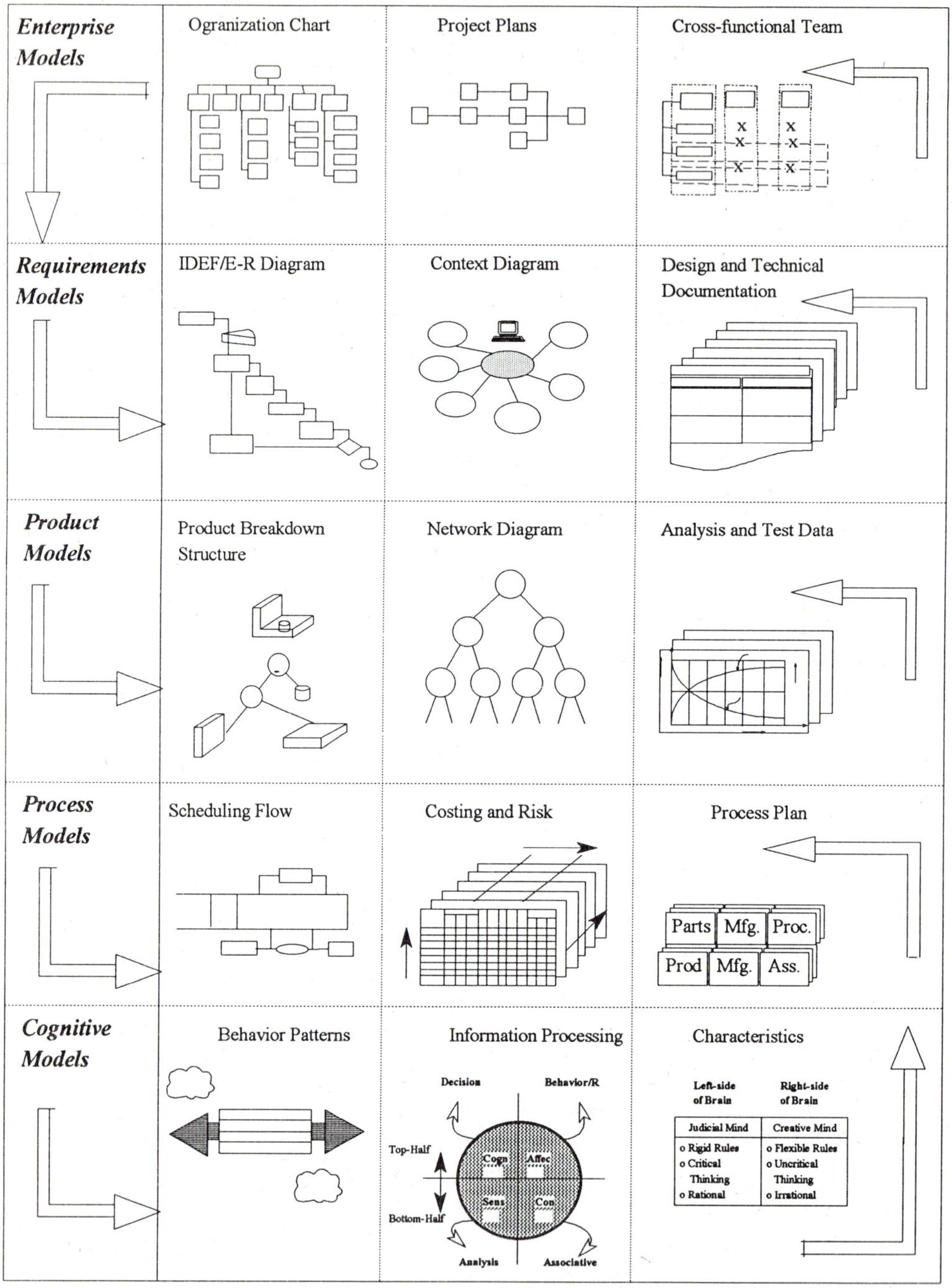

FIGURE 7.11 Classification of Information Modeling: Some Key Examples

- Product model-class
- Process model-class
- Cognitive model-class

Typical examples of models belonging to each model-class (invariant) are schematically shown in Figure 7.11. For example, organization charts, project plans, and cross-functional teams fall into the enterprise model-class. The corresponding structure of a model-oriented concurrent engineering system is shown in Figure 7.12. It employs the aforementioned five model-classes to build its foundation. The specification model is a pipeline that is extended (or connected) to all remaining four model-classes. An integration environment, in turn, links them to a number of application environments. Examples of an application environment might include design for X-ability (where X-ability can be manufacturability, assembly, flexibility, reliability, simplicity, quality, etc.), product feasibility and quality assessment, analysis or simulation, and process quality assessment. The engineering moderator, or user, shown in the middle (see Figure 7.12), ensures that

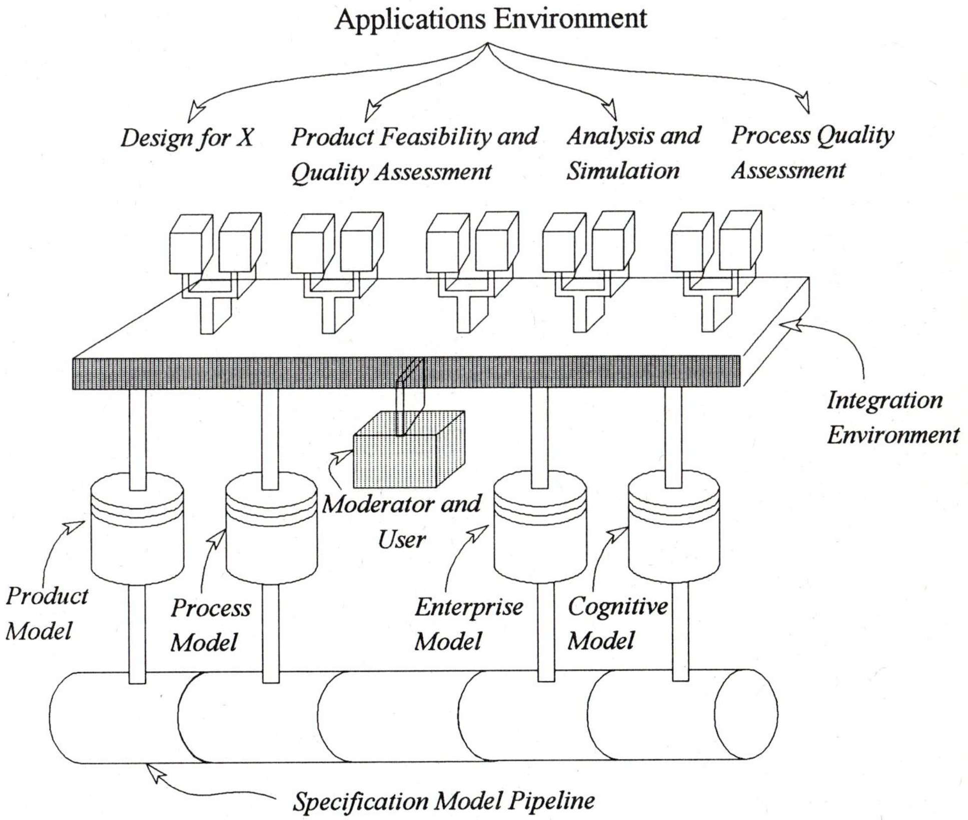

FIGURE 7.12 Structure of a Model-oriented Concurrent Engineering System

the evolving product realization process considers the different life-cycle values and assessments, represented by the application environment. The structure is such that any number of application environments could be virtually linked and supported. The five CE invariants emphasize the central role that modeling plays in the product realization process. CE engineers can continually use this approach to capture, reason, and structure the different types of information they require in realizing a product. The allocations of information into the five model-classes and their contents are discussed individually.

Figure 7.13 shows the relationships between these five model-classes:

- **The specifications model-class** provides the inputs and requirements pertaining to parts, manufacturing processes, process planning, tools' library, production scheduling, and plans for manufacturing, assembly, and inspections.
- **The enterprise model-class** provides the constraints for waste management, reliability, inspection, pallets, material handling, temporary storage, stations, and load/unload.
- **The product model-class** provides a transformation from inputs (baseline state) to an intermediate state.
- **The process model-class** provides the transformation from this intermediate state to a modified baseline state (output). Typical inputs to the product model-class are knowledge, raw materials, parts, work contents, priorities, process, etc. Typical outputs are production rate, machine utilization, efficiency, performance measures, etc.
- **The cognitive model-class** provides the intelligence or power of decision making.

This constitutes a functional design of a class-oriented concurrent information modeling system. This class-oriented functional design transforms the enterprise resources into a profitable product or service so as to serve as a link between enterprise, specifications, product design, process design (production, manufacturing, and inspection, etc.), and teams (cognitive behavior—see Figure 7.12). In the subsequent sections, each model-class is discussed individually.

7.5 ENTERPRISE MODEL-CLASS

The enterprise model-class provides a precise, accurate description of the activities, data connectivity, and communication network of an enterprise. Through the acts of flowcharting, the enterprise model-class captures the information flow through the units that are involved in the design and development of the product. A data flow model of an enterprise is shown in Figure 7.14. The model-class reflects the information needs of both the manual and automated units of an enterprise. It consists of four data bases: engineering data, manufacturing engineering data, manufacturing control data, and master schedules data. There are eight major activities that run in parallel: product definitions, process definitions, production operations and control, manufacturing and assembly, quality control,

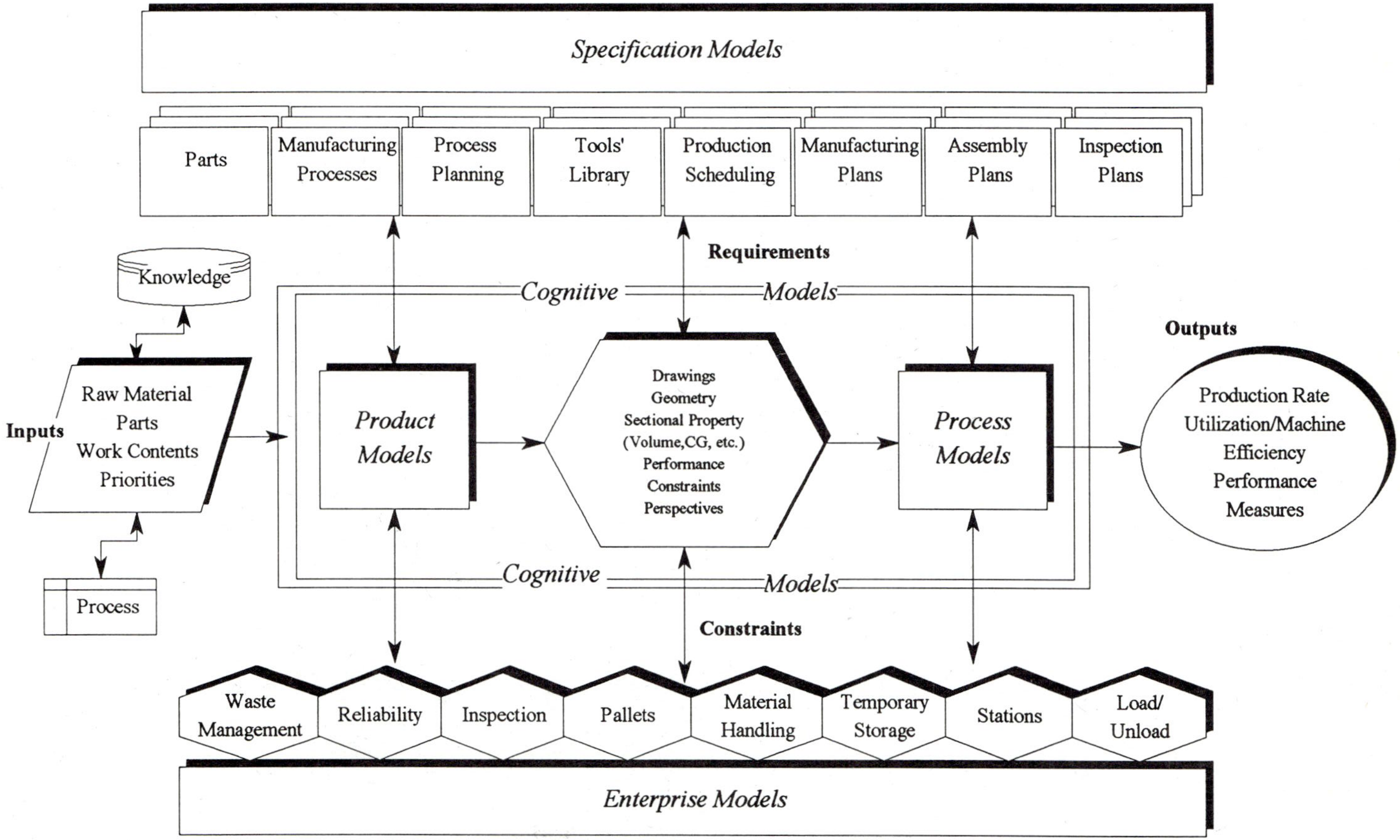

FIGURE 7.13 Functional Design of an Intelligent Modeling System

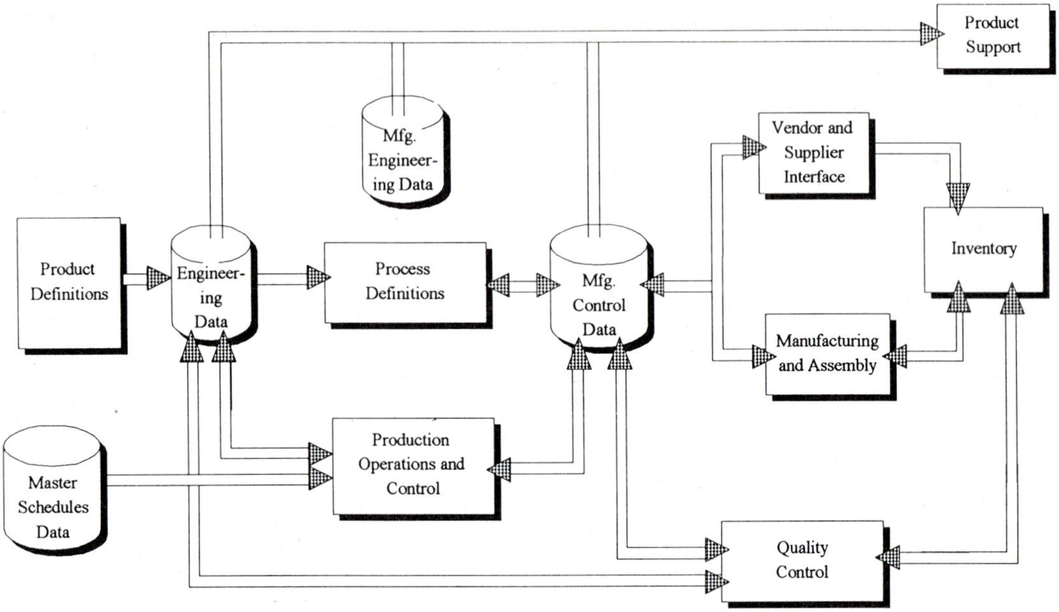

FIGURE 7.14 An Enterprise Dataflow Model

vendor and supplier interface, product support, and inventory. The data flow is shown through arrows and double lines.

7.5.3 Essentials for Enterprise Modeling

Enterprise modeling is a key tool for implementing concurrent engineering. It aids in planning what additional investments in information technology are needed to effectively manage complex configuration controls. An enterprise model serves as a guide for assessing information needs and for determining what impact concurrent engineering will have on the enterprise business. To be successful, an enterprise model must have many features. It should

- Contain a clear picture of the organization—who does what, how to do things, when to start, and why to do things a certain way, as well as the process flow one follows normally
- Include interfaces to assist a team or a work-group in defining the information needs of a project to another team or work-group
- Be a part of the communication and control information management system
- Include functional overlap, expressing situations when information ought to be shared across unclear or hazy organizational boundaries
- Be flexible and expandable.

Members from all CE units ought to be involved in setting up the "to be" process and in conducting facilitation sessions so that new product or process methods or technologies, which are available or applicable, can be harnessed effectively.

7.5.3.1 Surrogate Data as a Basis for Enterprise Modeling

The surrogate data (which is an electronic form of a technical memory) is a vast knowledge source for establishing the key product and process items. The surrogate data can define the requirements for developing the enterprise data flow model and for determining the needed space, staff, capital cost, etc. The surrogate data can also be used on a factual basis for determining what the assembly system should look like and what type of logistics support, within the factory, is necessary to support its production line rate. An enterprise model with surrogate data provides a strong base for conducting the following studies:

- Evaluate the impact of various production line rates and of various inventory levels on the facility for all, or a specific family of, components.
- Study the process system layout by changing the number of manufacturing stations and line lengths for both main line and subassembly components.
- Study the amount of material, which must be received, processed, stored (back-up and line stocks), and handled (through what means—material transport system).
- Determine material handling and line-feed methods for manual and automated assembly operations.

7.6 SPECIFICATION MODEL-CLASS

Specification is the organization or allocation of the incoming information into a set of inputs, requirements, and constraints. A constraint is the limitation about the environment the artifact will be subjected to. The elements of specification identify the flow of information from parts, manufacturing processes, process planning, tools' library, production scheduling, to many different plans—manufacturing, assembly, and inspection, as shown in Figure 7.13. The elements focus on processes and techniques for obtaining and translating the voice of the customer into producing a hierarchical set of "design to" functional specifications and modeling.

The specification model-class provides a planning documentation of the product development steps that an enterprise would undertake in actual practice. The process starts with a top-down approach. For instance, system specifications are first outlined and then broken down into further steps, generating a telescopic set of specifications. This process is repeated in increasing detail until the bottom level of substructuring is reached. This method ensures functional compatibility and logical integration of product breakdown into subsystems, components, parts, etc. There are five levels of specification model-classes:

1. System modeling
2. Subsystem modeling

3. Components modeling
4. Parts modeling
5. Materials/features modeling

These specifications are typically shown as "characteristic-type" information that must be addressed (that is, satisfied, scheduled, sequenced, prioritized, set values, relationships, etc.).

$$\text{Specification Modeling} \equiv \bigcup [\{\text{system modeling}\}, \{\text{subsystem modeling}\}, \{\text{components modeling}\}, \{\text{parts modeling}\}, \{\text{materials/features modeling}\}] \quad (7.1)$$

Where, the symbol $\bigcup$ denotes the union of.

7.7 PRODUCT MODEL-CLASS

The main purpose of building a product model-class is to provide information to the user in an unambiguous and complete form. A product model-class structures information about features of an artifact at multiple levels of abstraction [Solomons, Van Houten and Kels, 1993]. This is not the same as the product data modeling (e.g., IGES, PDES, STEP), which structures data elements in a common or uniform way. The latter is useful, however, for building consistent product models [Curran, 1994]. The product model-class serves many useful functions depending on the objectives of the CE teams. For example, design teams can use the models to gain insight into design problems and to analyze various solutions. The business office may use the model for marketing and pricing purposes. The engineering team may use it for analysis and performance evaluation. One of the important tasks in engineering design is performance modeling and validation. In general, performance modeling includes analysis or simulation of the product behavior, computing performance based on design for X-abilities, and other perspectives. These model-classes capture the life-cycle intent of the product perspective such as customer requirements, performance, geometry, loads, etc., which affect the product's behavior in some fashion. There are many ways in which one can capture the product life-cycle intent. Parametric, feature-based, and knowledge-based are examples of such methods. Some product model-classes do employ well-structured problem solving techniques with different levels of predictive and simulation capabilities.

A core aim of long-term design research in Japan has been to flush out the idea of what characterizes a good product model. Most product model-classes capture specific market and engineering specifications for a design object and link them to corporate design and manufacturing knowledge and capabilities. Knowledge in this context can be present in the following forms:

- ***Product knowledge:*** Such as past designs, form features, field performance, etc.
- ***Specifications:*** Such as test data, catalog data, inputs, requirements, constraints, etc.

- ***Process knowledge:*** Such as deep engineering understanding, manufacturing semantics, government regulations, etc.
- ***Standards:*** Company policy, practices, procedures (3Ps), and design rules.

Following these steps in succession as we refine the product, the product information can be captured in feature or other similar forms. These forms can be embedded directly into one of the product's database using an object-based or a knowledge-based scheme. The product model-class enables a CE user to understand the interactions of product parameters such as the effects of the features or materials on performance. There is an array of model classes from conceptual design to detailed design. In the conceptual stage the first level of business is to reduce the product's complexity, and then study impact of other parameters on performance indicators such as production cost. Most frequently, the greatest gain in cost reduction can be made by reducing the product complexity.

Product modeling for supporting multiple disciplines is a central topic in CE [Krause, Kimura, Iwata, Suh, Tipnis, Weck and Wolf, 1993]. Several approaches are available to view a design from different perspectives. Examples of such perspectives are manufacturability, reliability, assemblability, testability, and performance. In one case, bond graphs and structural models are used to analyze design from different perspectives [Finger and Rinderle, 1989]. Bond graphs show the energy aspects of a design, while the topological characteristics are presented by the structural models.

7.8 PROCESS MODEL-CLASS

Typically, the knowledge (so called process knowledge) necessary to get a design through the major life-cycle of the product is not well understood, especially in complex organizations. How problems are solved, the methods of creating or modifying an artifact, the sequence of events, the manufacturing issues, and the collected sets of information envelopes make up a large portion of this "process knowledge" bucket. In addition, the knowledge contained in procedure manuals, memoranda, textbooks, test results, and experience are not well-integrated toward the design and development of a complex product. Thus, the concept of a "knowledge-driven" product or process model becomes very powerful. The process model-class has the ability to structure state information or process knowledge. By using a process model-class to represent the knowledge required in each situation, the organization, the CE teams, and the employees all gain benefits and access for future "reuse." Practical developments of a computational support system for CE must also incorporate the process model-class and tools from the point of view of managing the "process knowledge." The primary process model-class includes process planning, production, and manufacturing process models. The manufacturing process model-class includes other process model-classes such as machining, forging, stamping, wire harnessing, assembling, routing and queuing, and other release management process models. Each process model-class must capture the activities related to process representation (modeling), process characterization (metrics) of a product, and process performance measurements (measures), such as NC machining assembly, packaging, production, manufacturing, inspection, quality control, etc. These three parts of a process model are shown in Figure 7.15. Three vertices of a triangle represent 3Ms: models, metrics, and measures. The three opposite sides represent the corresponding process elements: process representation, characteristic metrics, and performance measurements. Nodes and the sides, thus, together constitute a set of key elements required to create a good process model-class (as shown in Figure 7.15).

7.8.1 Group Technology

Group technology (GT) is an approach to improve effectiveness to minimize the cost of producing similar parts in small batches. A classification or a coding scheme (such as form-features) is used to identify parts with the following commonalities:

- Design features,
- Manufacturing procedures (3Ps)
- Machine tools usage

The most significant advantage of coding schemes is the potential of adding additional manufacturing semantics to a typical product geometric information model. Parts

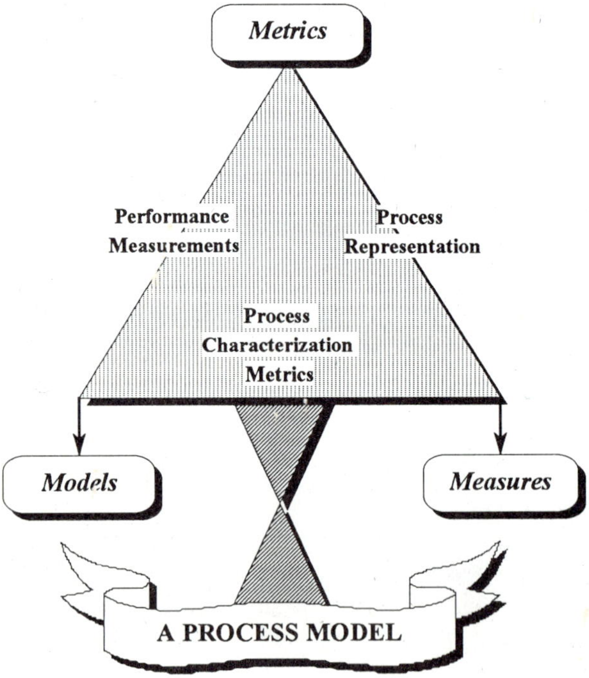

FIGURE 7.15 Three Parts to a Process Model

can then be scheduled batch-wise (all at once) to machine tools that are physically clustered together. The purpose of GT analysis is to give the user an automated means to help standardize the manufacturing processes and, thereby, minimize transportation and processing time. Group technology is useful both in product classification and process planning analysis. There are tremendous savings in using GT to identify groups of manufactured parts, which can utilize standardized machines and work flow.

GT analysis identifies groups of similar parts, simulates standardization production, and retrieves similar process plans. Common analysis outputs include machine utilization and work flow, cost, attribute, and layout of process plan. These analyses are designed to help standardize the manufacturing processes and group parts and machine tools to increase speed. GT analysis gives decision support information to help teams plan better. Once a group of parts is formed, a team needs to know if the cost is significant enough to warrant GT efforts for coding and analysis. Decision to use GT depends on the following factors:

- Value of the parts being produced
- Volume of the parts
- Significant machine requirements
- Current workflow
- Material handling requirements

For each group of parts, GT help teams standardize and optimize the manufacturing or assembly process. Teams can standardize the least cost processes, the process with the easiest work flow (least part movement and material handling requirements), or the processes that best utilize the machines. Some companies actually move machines or equipment physically together just to make a group of similar parts. This is called cellular manufacturing and has tremendous benefits in reducing setup time. This is similar to having a flexible machining center, but differs in that no new equipment needs to be bought. Lot size can actually be lowered for just-in-time manufacturing and still be very efficient because the setup and work-in-process times are dramatically reduced. Another benefit is the opportunity to lower costly inventories. There are a few commercial systems available that provide GT analysis tools to assist in the decision process.

7.8.1.1 GT Benefits for Manufacturers

Most of the benefits of utilizing GT are derived from the ability to RETRIEVE exact or similar information, and the ability to STANDARDIZE the manufacturing process for groups of similar parts. Retrieving information on similar parts implies a substantial reduction in duplication and proliferation of parts. For every new part that is introduced into manufacturing, a new process plan is created, followed by new NC tapes, new tools, quality instructions, time standards, etc. This is a very costly and time-intensive activity, which can be reduced or eliminated by retrieving information from a company's GT knowledge-base. The benefits of GT fall into three categories—one time (initial) benefits, repeated benefits that occur whenever parts are reproduced, and benefits due to the reorganization of production facilities:

- *One-time initial benefits:* By standardizing the manufacturing process for a group of parts, manufacturing engineering for new products, as well as machine setup times and work-in-process inventory are greatly reduced. GT standardization supports smaller lot sizes and just-in-time manufacturing. For many batch manufacturers the machine setup time is actually twice as long as the machine run time. If production methods are standardized, greater machine utilization can be achieved by making similar parts one after the other on dedicated machines. Standardization is the key to improving part quality and reducing scrap.
- *Repeated benefits due to reuse or regeneration:* If the design or features can be reused as they are, manufacturing documents such as process plans, shops and assembly drawings, NC programs, and machine instructions can also be reused rather than created from scratch. Similarly, if an existing design is modified, the corresponding manufacturing process plan can be regenerated rather than created anew. Further savings can be realized by scheduling longer production runs by group of parts rather than by scheduling individual parts.
- *Additional benefits due to reorganization:* This occurs when a company goes further and reorganizes its production facilities. This step substantially reduces material handling problems. The company reorganizes its production facilities to create GT work cells or work lines, which are a dedicated series of machines totally capable of producing a group of parts. A GT work cell can be called an "inexpensive flexible machining center." Often this entails a more complex long-term step towards product manufacturing.

Group technology classification and coding schemes also provide a basis for a "design retrieval system." A group technology retrieval system can provide manufacturing with many benefits similar to those provided by design standardization to design groups.

7.8.2 Computer-aided Process Planning

Figure 7.16 shows CAPP as the common region between CAD/CAM, the shop floor, and MRP II (Manufacturing Resource Planning).

$$\text{CAPP} = \cap\,[\{\text{CAD/CAM}\}, \{\text{Shop Floor}\}, \{\text{MRP II}\}] \tag{7.2}$$

Where $\cap$ denotes the intersection of the sets contained in the square bracket. MRP systems do not understand the similarity of part concepts or part standardization, CAD/CAM does. MRP only controls the time when the part will be made on the shop floor. CAPP determines the machine selection and machine sequencing; that is, how the part will be made. CAPP controls the development, maintenance, validity, and release control of manufacturing documentation (commonly called process plans, work instruc-

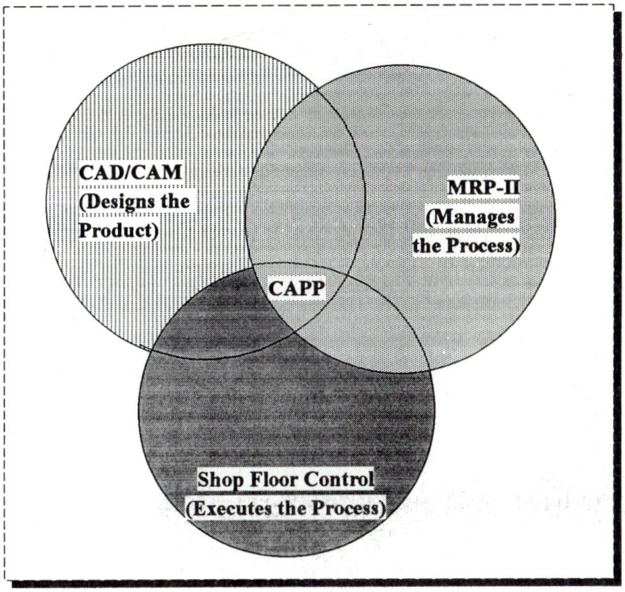

LEGEND:

CAD: Computer-aided Design
CAM: Computer-aided Manufacturing
MRP: Manufacturing Resource Planning
CAPP: Computer-aided Process Planning

FIGURE 7.16 Computer-aided Process Planning (CAPP)

tions, or method sheets). CAPP not only helps increase the speed of work, but also, and more importantly, helps improve manufacturing processes. In companies where manufacturing engineers have little or no automation tools, CAPP increases productivity by giving them a decision support tool to validate, standardize, and improve their decision process. Although many companies are using CAPP, many more spend more capital on automation for their office automation system than they do for their manufacturing engineers.

Good document management is very important for a process planning system. Other key planning features of a good CAPP system are security and release control, queue management, messages, revision history, mass updating, archival, and graphic illustrations from CAD/CAM drawings, scanned/digitized images, color options, highlighting, manufacturing word processing, special process planning edit features (operation insert and delete), standard reference notebooks with common operational and quality control instructions, flexible process-sheet generator, GT classification, GT analysis, report writer, customized printing, time standards, cost estimating, and open architectures for ease of integration with other CIM applications. Such applications could extend beyond CAD/CAM, MRP, and Shop Floor Control.

7.8.3 Computer Modeling and Simulation

Another method for increasing the performance of a process model-class is through computer modeling and simulation.

7.8.3.1 NC Machining Models

Manufacturing engineers are often confronted with the tasks of transforming the CAD geometry into NC output files to machine parts (to drive numerically controlled lathes and mills). The process is also suitable to form complex patterns, which can be molded in casting blanks. The problem is how to accomplish this in the least possible time and with the greatest accuracy. The NC model transforms a CAD design into a series of surfaces for CAM processing and gives instructions to actually machine the parts. There are many model parameters that control the machining process.

Some NC programs provide a better machined finish than others. Accuracy largely depends upon how the model surface has been represented. A two-dimensional machined surface mesh, for example, is faceted, that is, it is made out of a series of tiny flat patches, which approximates the actual intended three-dimensional shape. Two parameters that commonly effect the finished surface quality are:

- *Surface Approximation:* This entails defining points on the surface and defining how intermediate points are generated. Use of higher order forms, such as parametric or nonuniform rational B-spline (NURBS) rather than surface meshes, provides a better option for CAM processing. Using these forms, the surface can be described as a series of polynomial equations. The intermediate points between the mesh points trace a curve rather than a straight line. The more accurately we are able to represent the described surface, the smoother the machined part is likely to be to the actual surface.
- *Tolerance :* The accuracy of the finished surface is also affected by the machining tolerance. The smaller the size of the faceted meshes, the greater the accuracy of the machined part.

There are other model parameters that affect machining:

- *Types of Cut:* This describes the tool path generation criteria, such as two-axis, three-axis or five-axis machines. Two-axis features include contour and pockets with multiple islands. Three-axis adds the depth to the two-axis. A five-axis tool path can position the tool at a particular angle relative to the surface curvature.
- *Cutting Boundary:* The cutting boundaries form a series of elements bounding the surface. The selection process, referred to as "chaining," determines whether the cuts are inside or outside the boundary.
- *Tool Selection:* Usually, standard tools are defined as part of a tool library. The library provides a set of handy means to select a tool. Most programs allow provisions for adding new tools or modifying the ones already present in the tool library.

- *Material Parameters:* Material parameters include feed rate, plunge rate, and spindle speed.

Other model parameters include specifications for side clearance on islands, spacing between the rough passes, and the number and spacing of the finished passes.

Process models, once captured through computer simulation, provide a framework for taking an integrated look at the organization structure including the business and engineering processes in the entire organization. Simulation allows many CE process models to be tested and analyzed before implementation. Enterprise modeling, described in Section 7.5, is an example of computer simulation for analyzing organizational structure and studying the performance of the overall manufacturing system.

7.9 COGNITIVE MODELS

Information models described so far (namely the specifications, enterprise, product, and process models) do not solve the product realization problem. They describe the product and process behavior, but the problem is not solved unless a human brain or, more precisely, a virtual brain acts on it. The human brain, and more precisely the neuron morphology of the human central nervous system (CNS), is a complex phenomenon [Newell and Simon, 1972]. Recent studies have found the CNS of humans performing four important functions in information processing and decision making (Figure 7.17): cognitive (thought process), affective (emotional process), sensory (recognition process), and control (messaging associative process). The mental and cognitive functions of the brain, unlike the computational function of the computer, act on relative grades of information. Our sensory system's function is to acquire information in the form of relative grades rather than absolute values. Information may be viewed in a numerical form (for example the outside temperature is 70°F), but during the process of cognition, we perceive the temperature as being close to pleasant. The sensory function is essentially the brain's left cerebral hemisphere function (CHF). This kind of perception is important mainly for the judicial recognition of the problem's parameters such as its complexity or properties. Control, on the other hand, is essentially a right CHF and plays a major role in creative synthesis of design.

There are two sides of the brain that help each other in problem solving—the left side, called the judicial mind, and the right side, called the creative mind [Nicolai, 1992]. The judicial mind has two parts—a decision making part and an analysis part. Consequently, the judicial function is often called a deductive, analytical mental activity. Like the judicial side, the creative function also has two cooperative parts—a behavior part and a synthesis part. As a result, the creative mind is also referred to as the associative, behavior mental activity. The use of different parts of the brain in solving a typical design problem is cyclic in nature. There are five interactions (steps) that take place during a cycle completion:

1. ***Sensory and Cognitive Aspects:*** At the beginning of the cycle, the judicial mind (left-side of the brain) is engaged, as shown in Figure 7.17. Here the engineers

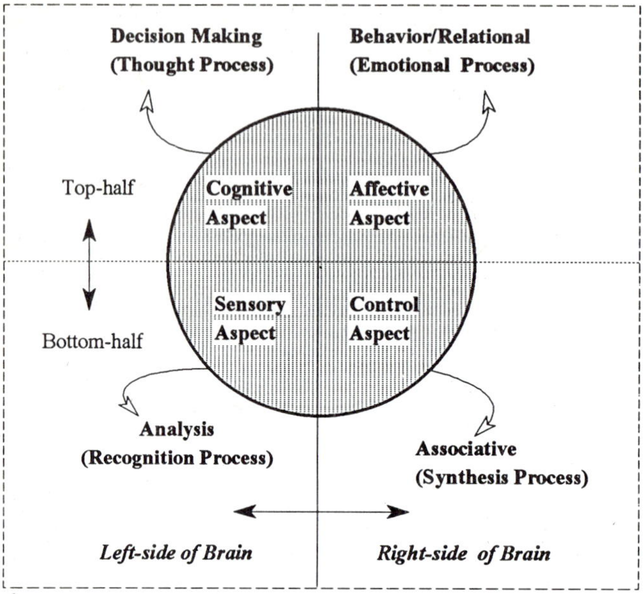

Judicial Mind	*Creative Mind*
• Rigid Rules • Critical Thinking • Rational • Logical • Definite Answer (One Solution) • Deductive or Analytical	• Flexible Rules • Uncritical Thinking • Irrational • Illogical • Multiple/Alternate Solutions/Paths • Associative or Creative

FIGURE 7.17 Cognitive Aspects of Information Processing and Decision Making

(through sensory aspect) set the rules by defining the problem, for example, problem or product decomposition—what parts of the product are to be designed, what specifications ought to be used? They must also decide (through the cognitive aspect) the relative importance of the specification attributes (inputs, requirements, and constraints), and measurement of merits (MOMs) such as performance, cost safety, simplicity, environment, etc. They must also decide the types of processes (process decomposition) required to transform the specifications into a physical output.

2. ***Control and Affective Aspects:*** In this case, the creative mind (right-side of the brain) is engaged, as shown in Figure 7.17. Here, the engineers (through the control aspect) ignore the norms that may inhibit creative thinking. Decision making is performed through some form of association (such as assembly, systems engineering, system theory, conferencing, group dynamics, or a collection of brainstorming ac-

tivities). Just as numbers are important in a brainstorming session, the associativity is more important here. The interest here is in the number of candidate designs or possibilities rather than in trying to find those which meet all the rules and constraints. Engineers (through behavior or relational parts) identify the boundaries (rules, constraints, assumptions, etc.), which support the various design candidates.

3. ***Sensory and Cognitive Aspects:*** After this creative function, the judicial mind acts once again (left side of the brain as shown in Figure 7.17). Each candidate system is analyzed in detail to identify the use of critical technologies and to determine their system specifications. During its decision making part, the team (or a design engineer) applies engineering principles to each design candidate to analyze or assess its feasibility and to quantify its MOMs. This completes one full cycle.

During the next cycle, interactions take place diagonally between the creative and judicial mind (see Figure 7.17). First, between the sensory and affective aspect and then between the control and cognitive aspects:

4. ***Sensory and Affective Aspects:*** During the second cycle, the person (through a combination of the sensory and affective aspects) conducts behavior sensitivity analysis of the boundary attributes discussed earlier. Each sensitivity analysis determines the impact of the boundary attributes (such as wheel-base, steer-angle, and so forth) on the MOMs.
5. ***Control and Cognitive Aspects:*** The second leg of the cycle ends with the control aspect of the creative mind interacting with the cognitive aspect of the judicial mind. During the synthesis process, the team (or engineer) applies the MOMs to all the feasible solutions provided by the associative part of the creative mind. The decision making part of the judicial mind selects the best design, which is usually a compromise between conflicting MOMs. The type of mental activity changes as the team moves from one design aspect to another, from sensory to cognitive, to control, to affective—and the cycle continues.

The process of teaming (sharing of cognitive minds) and common understanding (cross-pollination of cognitive minds) is the least understood phenomenon. Most managers consider cross-functional teaming primarily a management issue. Though management is a big catalyst to team-building and motivation, "effective teaming" is a natural process of informal human networking and the creation of a mind-set (deep common understanding) among the members. It has been proven that convergence (a collective mind-set) can do miracles in problem solving, which are impossible with just management control. Convergence involves taking specialized knowledge from multiple sources and forming a new collective mind-set that allows teams to work together at a much deeper level.

The cognitive model captures human behavior in teams and collaborative situations. Most CE environments involve a carefully orchestrated interplay between CE teams and machines. Information models are incomplete without a common framework of cognitive understanding. There are no definite cognitive models for accommodating human behavior such as different points of view, capturing thinking rationale, and various mindsets in a product realization process.

7.9.1 Modeling Collaborative Behavior

To develop goal-oriented or "constancy of purpose" solutions, a successful collaborative environment between the participating teams and their members is required. However, it is not easy to model the cognitive behavior of a collaborative and motivated goal-oriented work-group. Empowerment and reward systems help in motivating the teams and in removing some common barriers of communication. Implementing such tasks may not be difficult. However, developing collaborative behavior in a goal-oriented work-group is a more challenging task, since these are more related to the social and cultural roots of team-members, which may be difficult to change. There are some behavior patterns that help in collaboration more than others. Knowing and understanding the differences in behavior patterns can be helpful in organizing a better work-group. The behavior patterns that affect collaboration are as follows:

- Cognitive Framework
- Occupational Culture
- Education and Background
- Value System

This is shown in Figure 7.18. Cognitive framework, or nature, is how a team member thinks and acts during decision making. In individuals, certain cognitive skills tend to be more developed than others with one half of the brain (hemisphere) being utilized more than the other half. Individuals with creative minds are primarily right brained, while those with judicial/analytical minds are primarily left brained. Rarely are people exclusively right or left brained. More often they are in-between, perhaps leaning towards one side more than the other. For example, engineers generally have a more developed left brain, while designers have a more developed right brain. Human culture belongs to the affective aspect of one's creative mind. Occupational and organizational culture refers to a common tradition of functionally oriented departments in which the work is often like "deeply ingrained routine activities." When teams of work-groups are brought together in a CE setting, having initially different occupational and divisional cultures, they have a limited, and often restricted, view. Often, this is inherited with "native-view" paradigms that create organizational cultural conflicts. Differences in educational background of members of the work-group also result in differing "thought-views," creating further possibilities of conflict. Each teammember may have his/her own value system, depending upon what each team believes in. Some may have a stronger belief in one area than others.

It is important to realize that the above behavior foundations are the force behind creating strong cognitive minds. Cognition is a useful human resource since it contains the intelligence to differentiate good from bad. In the above context, the cognitive mind is mainly discussed in terms of its use at a personal level. It would be desirable to develop

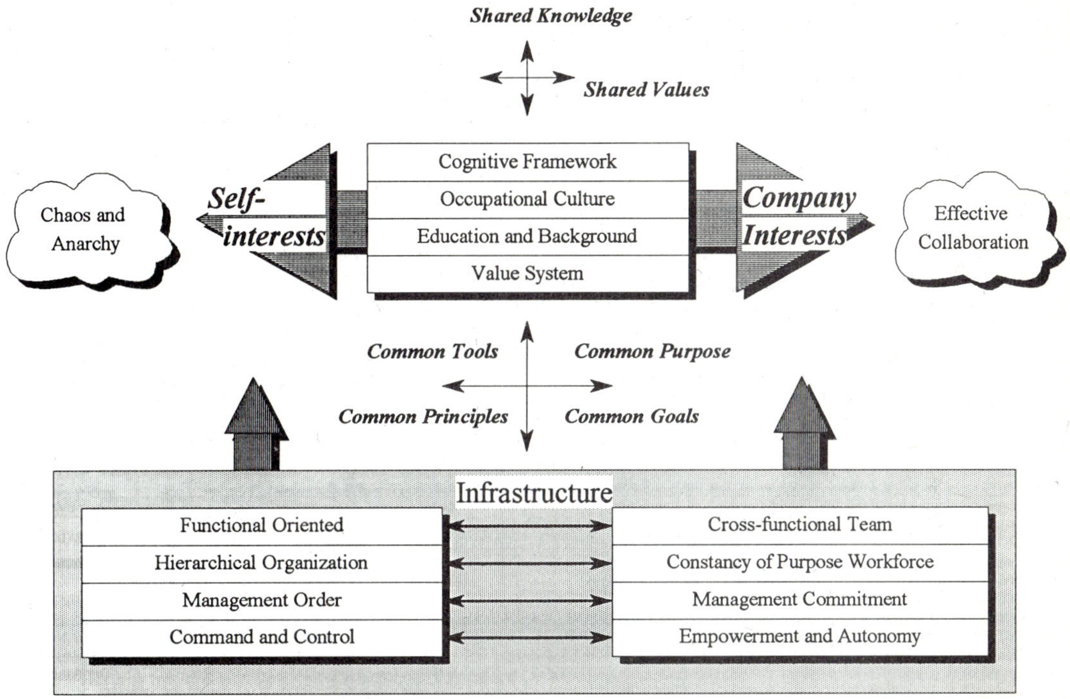

FIGURE 7.18 Behavior Patterns Affecting Collaboration

the same level of cognitive intelligence working at the level of the company instead of just at the personal level. This way, the teams would be able to deliver the same level of enthusiasm and dedication that existed on a personal level. All of the above behavior frameworks are important ingredients in making a company successful. For example, a value system that is based on the interests of the employees, customers, and the company would be a nice thing to have. Similarly, a company culture that is fully oriented toward building on the strengths of each member, group, department, and division of the company would be a perfectly right thing to have. Cultural views of the teams that are based on "local or native needs" are detrimental, but developing a cultural view that is conducive to global or company interests would be a good use of culture. This would require a culture in which work-groups would bring their own occupational or divisional "native-views" to define an "outcome" that is based on a "global sense of priorities" or "constancy of purpose."

7.9.2 Virtual Intelligence

Over the last couple of decades, several parallel advances have been made in two distinct disciplines: fuzzy logic and neural networks. Fuzzy logic, as the name implies, provides an analytical basis for the emulation of certain perceptual and linguistic attributes associated with the human brain. Neural networks, on the other hand, provide new computing morphology with learning capabilities. Virtual intelligence is considered to be one of the offspring of the marriage between these two disciplines. Other offsprings that result from the integration of these two fields are robust sensors and robust control mechanisms.

7.10 MERITS AND DEMERITS

The main objective of information modeling is to provide the concurrent teams with a comprehensive set of organizational tools and a class of models on which to base decisions. There are significant advantages for an organization to clearly define and maintain up-to-date model-classes for enterprise, specifications, product, process, and cognitive systems together with flow of data and material exchange between them. A list of these advantages are contained in the following:

- Through modeling of the information into invariants, diverse CE teams within an organization can gain valuable insight into the organization structure, specifications, product and process details.
- A model-class highlights the difficulties of some of the decisions by including them into the models early in the game.
- A model-class highlights alternatives and provides a framework for further specific design action on the part of the various CE teams.
- The development of a surrogate product manufacturing database as a part of the process model-class facilitates the setting up of a variety of realistic (strategic) manufacturing studies.
- Another advantage of the model-class definition is the ability of the system to identify and treat interactions explicitly among the various elements of the product realization process.
- A "customer focus" becomes the main thread throughout the program and forms the basis of the mission definition.
- A total system perspective in terms of a model-class leads to a common, consistent understanding of the product and its characteristics.
- Most importantly, it clearly identifies the desired result of the product realization steps before execution begins. This automatically forces participation and buy-in from all affected concurrent parties to agree on a previously identified set of CE objectives while also defining a monitoring and validation plan to meet these objectives.

This leads to opportunities for continuous process improvement by providing the ability to predict how changes in the CE structure or changes within the steps of the model-class information system affect the product or process yields. It is a useful philosophical approach for reducing the time-to-market for new products. The ultimate result of this process invariant approach is to reduce the duration and cost of the product development and produce a quality product that meets or exceeds the customer's requirements.

Although the potential benefits of the approach are substantial in terms of a better product at a reduced overall cost, there are a few disadvantages that cannot be overlooked. The largest impediment to the use of computer models of information and their subsystems is the investment that must be made in qualified people, facilities, and most importantly, the time. The creation of effective dynamic model-classes also requires a lot of patience. It requires close interaction between the system analyst and the development engineer to continually improve the accuracy of the model. Implementing the process invariant approach requires an investment in up-front analysis as well as resources to develop or acquire computer-aided tools. The formality of the process invariants might be considered restrictive and inhibiting, especially if improperly implemented. Finally, since it represents a change in product realization philosophy, significant training is required for implementation. However, in the long run, its benefits, by far, outweigh any shortcomings.

7.11 SUMMARY

In this chapter, we presented five model-classes as a common set (invariants) of the CE process template. These model-classes are not independent. They are interconnected through a primary set of parameters that are common to these classes. A secondary set of parameters is used to define the interrelationships between models of the same class. Primary parameters influence the behavior of models of one or more classes. This classification of parameters provides the basis for understanding the interactions between the models within a class (intraclass) and between classes (interclass). This level of distribution for the models is essential while balancing product requirements and the complex interdependency associated with the subsystems and the overall product. Together, these parts enable product developers to understand the interactions and produce a quality product. Moreover, if a parameter is altered, for example, if an old plan or a requirement changes, these models provide a basis to rework various tradeoffs between models (interclass or intraclass). Of particular importance to concurrent engineering is the means of unification so that the models are linked automatically. Examples of such attempts are unified parametric and pseudoparametric models [Kulkarni, Prasad, and Emerson, 1981].

Each model-class can have two interconnected components:

- *Product Process Organization (PPO):* The detailed descriptions of each perspective on the PPO can be captured in files or databases to be used by the relevant processors.
- *Auxiliary Model:* An auxiliary model can relate key parameters in multiple perspectives through tradeoffs and constraints (probably captured in compatible object-oriented databases).

Using prior CE process template and prior associated models makes the consideration of a new, similar process quicker and its implementation straightforward in a CE design process. Systems, of the type shown in Figure 7.13, for handling complex sculptured mechanical parts lie at least a few years into the future. The technology needed for representing complex parts (for example, ICAD-like languages [Saxena and Irani, 1994]) is slowly emerging. Design and development of algorithms to tie-in such representations and yield the outputs shown in Figure 7.4 are in their infancy. Process algorithms (for manufacturing, assembly, and inspection) are beginning to be used more often. Success will depend strongly on the development of appropriate process invariants and models for analyzing the various life-cycle options that effect manufacturing, assembly, and inspection. The general absence of measurements for systems manufacturing options, such as cost of manufacturing or design for manufacturability, attests to the work left to be done in this area.

REFERENCES

CURRAN, L. 1994. "STEP Bridges the Way to Better Product Modeling," *Machine Design,* Vol. 66, No. 5 (Mar. 7), pp. 137–142.

FINGER, S., and J. R. RINDERLE. 1989. "A Transformational Approach to Mechanical Design Using A Bond Graph Grammar." *Design Theory and Methodology -DTM '89,* Proceedings of the Ist Int'l Conference on Design Theory and Methodology ASME, Edited by Elmaraghy, Seering and Ullman, pp. 107–116. Montreal, Quebec, Canada. Sept. 17–21, ASME DE-Vol. 17.

HAYES, C. C, S. DESA, and P. K. WRIGHT. 1989. "Using Process Planning Knowledge to Make Design Suggestions Concurrently." *Concurrent Product and Process Design.* edited by Chao and Lu, Proceedings of the Winter Annual Meeting of the ASME, San Francisco, CA, Dec. 10–15, DE-Vol. 21. PED-Vol. 36, pp. 87–92.

HUMMEL, K. E., and C. W. BROWN. 1989. "The Role of Features in the Implementation of Concurrent Product and Process Design." *Concurrent Product and Process Design,* Edited by Chao and Lu, Proceedings of the Winter Annual Meeting of the ASME, San Francisco, CA, Dec. 10–15, DE-Vol. 21, PED-Vol. 36, pp. 1–8.

ISHIKAWA, K. 1986. *Guide to Quality Control.* Tokyo: Asian Productivity Organization, Japan.

KINNUCAN, P. 1989. "3-D Modeling Systems." *Computer Graphics Review,* (Sept.), pp. 16–26.

KRAUSE, F. L., F. KIMURA, K. IWATA, N. P. SUH, V. A. TIPNIS, M. WECK, and A. C. H. VON DER WOLF. 1993. "Product Modeling," *Annals of the CIRP,* Vol. 42.

KULKARNI, H. T., B. PRASAD, and J. F. EMERSON. 1981. "Generic Modeling Procedure for Complex Component Design." *SAE Paper 811320.* Proceedings of the Fourth International Conference on Vehicle Structural Mechanics, p. 99. Detroit, MI. (November), pp. 195–204.

MURPHY, G. 1950. *Similitude in Engineering.* New York: The Ronald Press Company, p. 57.

NEWELL A., and H. A. SIMON. 1972. *Human Problem Solving.* Engelwood Cliffs, New Jersey: Prentice-Hall.

NICOLAI, L. M. 1992. "Designing a Better Engineer," *Aerospace America,* Vol. 30, No. 4 (April). pp. 30–33.

PILKEY, W. D. 1994. *Formulas for Stress, Strain, and Structural Matrices,* New York: John Wiley.

PRASAD, B., R. S. MORENC, and R. M. RANGAN. 1993. "Information Management for Concurrent

Engineering: Research Issues." *Concurrent Engineering: Research & Applications.* An International Journal, Vol. 1, No. 1, pp. 1–19.

PRASAD, B. 1996. "Towards A Functional Design of a Concurrent Information Modeling System," *Computer Modeling and Simulation in Engineering*—An International Journal, Vol 1, No. 1, (in press).

REQUICHA, A. A. G. 1980. "Representations for Rigid Solids: Theory, Methods, and Systems." *ACM Computing Surveys.* Vol. 12, No. 4, pp. 437–464.

REQUICHA, A., and H. VOELCKER. 1982. "Solid Modeling: A Historical Summary and Contemporary Assessment." *IEEE Computer Graphics and Application.* Vol. 2, No. 2 (March), pp. 9–24.

SOLOMONS, O. W., F. J. A. M. VAN HOUTEN, and H. J. J. KELS. 1993. "Review of Research in Feature-based Design," *Journal of Manufacturing Systems,* Vol. 12, No. 2, pp. 113–132.

SAXENA, M., and R. K. IRANI. 1994. "A Knowledge-based Engineering Environment for Automated Analysis of Nozzles," *Concurrent Engineering: Research & Applications*—An International J., Vol. 2, No. 1, pp. 45–58.

U.S. Air Force, 1981. "Integrated Computer Aided Manufacture (ICAM) Functional Modelling Manual (IDEFO)," Air Force Materials Laboratory, Wright-Patterson AFB, Ohio, Report #AFWAL-TR-81-4023, (June).

U.S. Air Force, 1985. "Integrated Computer Aided Manufacture (ICAM) Architecture Part II, Information Modelling Manual (IDEFIX)." Air Force Materials Laboratory, Wright-Patterson AFB, Ohio, Report #AFWAL-TR-85-4023.

YANG, Y. 1989. IDEF1X Style Guide For PDES Users, NIST TN 1295. Draft, Washington, DC. August 31.

TEST PROBLEMS: INFORMATION MODELING

7.1. Why would you look at the system, the product, and the process requirements separately? List the salient points for the system, the product, and the process requirements.

7.2. What is contained in specification modeling? Why is this information separated from product or process modeling? Why is it critical for the design team to concentrate on the specifications?

7.3. What roles do abstraction and precise specifications play in system modeling?

7.4. What are the various methods of representing conceptual models? Why do we prefer structure flow charts to describe a process instead of context diagrams?

7.5. What are the differences between the icon-based modeling described in Chapter 3 and the other forms of conceptual models described herein? Which model is more intuitive and why?

7.6. What is meant by modeling? Does an information model possess a tree structure? Why is it considered different from a database or a technical memory?

7.7. Describe each of the classifications of information modeling (IM) and provide some examples of each. Is there any relationship among the five classes of information modeling? Why is IM called invariants?

7.8. What is information modeling? What purpose does information modeling play in IPPD? What significant advantages does it bring to an organization?

7.9. What are the elements of product modeling? How many ways can an object be described? Describe these methods.

7.10. Define the term geometric modeling. What are some of the common progressions of geometric models? What are the methods to represent solids?

7.11. Today CAD/CAM systems are widely used for geometric modeling. What are some of the advantages of using the CAD/CAM system for product modeling?

7.12. What are the schemes for modeling a process? Give a typical breakdown structure (a decomposition tree) for product and process modeling.

7.13. What are the characteristics of an analytical model? What are some of the merits of analytic modeling over physical modeling? What techniques are involved in transferring geometric aspects into physical aspects?

7.14. What is cognitive modeling? Describe its four important functions? How can the fundamental differences of cognitive minds be applied to the company's benefits?

7.15. Why are self-interests often directly opposite to company interests? What can be done to reverse that?

7.16. Is it possible to draw a process model for a virtual company? What are the benefits of virtual modeling over physical modeling?

7.17. How does information flow contribute and support DFM? List some of the information technology requirements for DFM?

7.18. Does information modeling provide an understandable view of the manufacturing world? If so, how? Explain the importance of modeling in manufacturing.

7.19. What defines a complete product model? Describe the product modeling steps using a fish bone diagram.

7.20. What is the relationship of information modeling with respect to internal or external customers? How do behavior patterns of employees influence productivity and quality?

7.21. How can information modeling assist an enterprise to be agile (help achieve manufacturing flexibility)?

7.22. Most problems that commonly occur during product design and development result from discontinuities among the three model types: physical, conceptual, and analytical. How can these three types of models be linked continuously in order to facilitate quick updates and reduce development time?

7.23. Many companies are in the process of "right-sizing," which means down-sizing. How can companies use information modeling techniques to determine the "right" process, the right number of employees, and maximize productivity from those employees?

7.24. Describe how you could utilize the various types of models (conceptual to physical) to build a kitchen cabinet?

7.25. Why isn't a mechanical object completely defined until its physical behavior and functional aspects are also captured?

7.26. Explain the modeling process as it relates to process planning. What model parameters control the machining process in NC?

7.27. How would you exploit group technology to simplify information modeling complexity? Define and discuss GT as it relates to CE. How does GT improve effectiveness and minimize the cost of producing parts?

7.28. The brain works in many diverse situations. Describe what interactions occur during a typical problem solving session.

7.29. Use an automotive camshaft as an example to describe the templates for information modeling schemes.

7.30. Describe the significance of enterprise modeling. Give some common reasons for requiring a new set of enterprise models even though an enterprise might be operating just fine.

7.31. Explain the concept of a multifloor template to information modeling. What are the advantages associated with this concept?

7.32. What are the common benefits derived from a formulation of common process templates?

7.33. What should management do before engaging into an enterprise model?

7.34. Describe the structure of a model-oriented engineering system. How does a functional design of a model-oriented CE system will look, if it is applied to a Pizza chain?

7.35. What are the building blocks of information modeling when dealing with multiple systems? What are the different methods of linking equipment with respect to time and place?

7.36. Does information modeling mean centralizing all of the product and process information into one system/database? What are the advantages of setting data descriptions common to all applications?

7.37. What are the three parts of a process model? What role does information modeling play in CIM?

7.38. Explain the behavior patterns that affect collaboration. Explain why they are essential to a successful work-group.

7.39. Describe three possible methods for linking equipment/facilities/computers. Present your ideas using a Double-Q diagram.

CHAPTER 8

THE WHOLE SYSTEM

INTRODUCTION

Often while designing a product, teams forget that the product is a system. It consists of a number of subassemblies, each fulfilling a different but distinct function. A subassembly is a group of two or more units that can be brought together (say for instance, share common axes). Subassembly information includes subsystems, components, parts, and features (materials, attributes, parameters, and geometric features). Process information includes layout, machines and equipment, control and scheduling, inspection and fabrication, and material handling and form features (see Figure 8.1). A feature is a representation of form, intent, and relevance to some aspect of part design of interest to either functionality (part features or so-called form features) or manufacturability (for example, DFX) [Nielsen, 1990]. *Form* signifies the attributes of the features present, their connectivity (topology) and geometry. *Relevancy* points to the reason a feature is included in the design (for example, issues related to functionality or DFX rules). *Intent* represents the imposed constraints by the CE teams on the freedom of the parts' function or rules associated with a DFX principle. All these needs have to be considered simultaneously in order to achieve a truly integrated product and process design (IPPD). The core concept of IPPD is the system definition. Consider the statement by Szucs [1980, pp. 42]:

> The term *system* is used to denote an object composed of certain elements that are linked by well-defined (but not necessarily known) relationships. It is stipulated that the system may interact with its environment (receive external stimuli) and that its behavior comprises responses that are useful or profitable to the operator. Objects completely isolated from their environment (completely closed system) and phenomena that cannot be influenced by persons are not regarded as systems in the technological sense.

Break-down	Enterprise	Product	System	Work	Process	
					Modeling Steps	Graphical Representation
HIERARCHY	Strategic Business Unit (SBU)	System	Complete Assembly (assembly of PPO -- product process and organization)	Project Vision and Mission	Process Layout	
	Subunit	Subsystems	Subassembly (assembly of PPO subsystems)	Values, Objectives and Strategies	Subprocess Layout	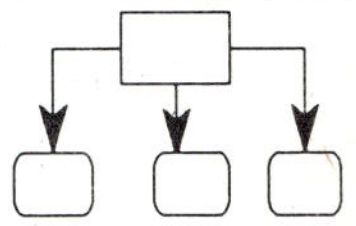
	Product Development Team (PDT)	Components	Sub-subassembly (assembly of PPO components)	Project Initiatives or Tasks	Activities or Function-chain or Sub-subprocess	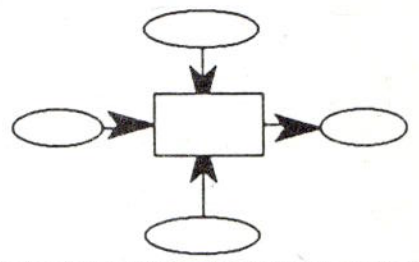
	Work-group	Parts	Sub-sub-subassembly (assembly of PPO parts)	Project Teams	Steps or Subfunction or Sub-sub-subprocess	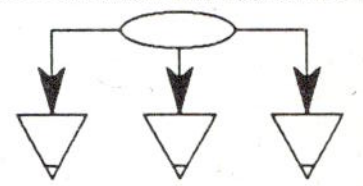
	CE Teams Experts Designers	Materials/Attributes/ Features/Parameters (Geometric and Nongeometric)	Mating Specifications and Sequences	Teammember, Schedule, and Deliverables	Actions or Methods (Sequential, Parallel, Alternative, Join, Loop)	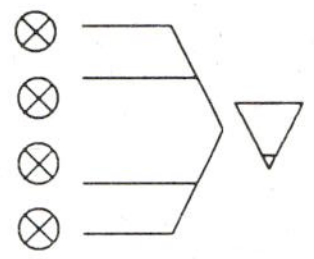
Guiding Principles	7Cs	Common Representation and Standards	Constancy-of-purpose Structure Whole-of-the-parts, Parts-of-the-whole	7Ts Project Plan	3Ps Policies, Procedures, and Practices (tacit or explicit)	

FIGURE 8.1 Enterprise Product, Assembly, Work, and Process Hierarchy

Before we take a close look at the requirements of a full IPPD system, let us review what constitutes a system. A system has three parts:

1. ***Class Hierarchy (a structure definition):*** Class hierarchy is a very efficient mechanism for system definition because one can use method and variable definitions in more than one subclass (such as subsystems, components, etc.) without duplicating their definitions. For example, consider a system that represents various kinds of human operated vehicles (see Figure 8.2). This system would contain a generic class for vehicles, with subclasses for all the specialized types. Five major subclasses include: auto, truck, bus, aircraft, and land transport devices. Their vehicle class would contain the methods and variables that are pertinent to all vehicle features, such as identification numbers, passenger loads, fuel capacity, etc. The subclasses, in turn, would contain any additional methods and variables that are specific to the specialized cases [Taylor, 1993]. Trucks may consist of pick-up, van, or tractor-trailer. Land transport devices may include subclasses such as motorcycles, bicycles, skateboards, and unicycles.

$$\left.\begin{array}{l}\{\text{motorcycles}\}\\ \{\text{bicycles}\}\\ \{\text{skateboards}\}\\ \{\text{unicycles}\}\end{array}\right\rangle \in [\text{Land transport devices}] \tag{8.1}$$

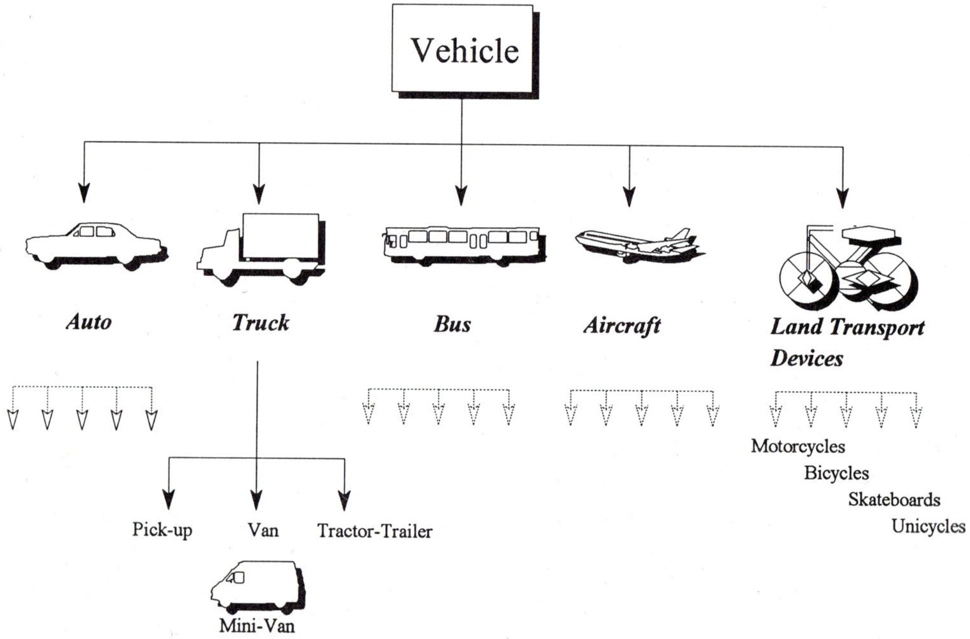

FIGURE 8.2 Class of Human Operated Vehicles

Where $\in$ denotes an element of a set or member of the class hierarchy.

2. ***Integration Hierarchy (an assembly definition):*** Creation of the integration hierarchy begins by using the previous class hierarchy structure to generate an assembly definition of the required information flow, assembly sequence, and precedence relationships. Resolving the assembly problem in complex product development is a two step process. The first step is to identify the assembly sequence, orientation, and alignment needs of parts to minimize the assembly time. The second step is to develop a sequencing plan into the assembly process to facilitate this integration.

$$\begin{array}{c}\text{[Land transport devices]} \supset \{\text{motorcycles, bicycles,} \\ \text{skateboards, and unicycles}\}\end{array} \tag{8.2}$$

Where $\supset$ denotes a superclass set. The elements in the curly bracket are its member classes.

3. ***Constancy of purpose or goals:*** This identifies the key functions or characteristics of the system that are also common to its elements (subclasses). For example, general company goals are more important than components' (for example, department) objectives. Constancy means carrying the same definition and the same sets of goals and management process throughout the company. Teams must agree on a common set of purposes that relate to the company goals and customer needs and which provide a common set of ground rules for mutual cooperation and communication throughout. If a product development process is "left to itself in the Western world, components become selfish, competitive, independent profit centers" [Deming, 1993].

$$\begin{array}{c}\text{Goals}_{\text{parts}} \exists \text{ Goals}_{\text{components}} \exists \text{ Goals}_{\text{subsystems}} \exists \text{ Goals}_{\text{system}} \exists \\ \text{Goals}_{\text{department}} \exists \text{ Goals}_{\text{SBUs}} \exists \text{ Goals}_{\text{enterprise}}\end{array} \tag{8.3}$$

Where $\exists$ denotes an element of.

Having a definition of what constitutes a system is the first requirement of being able to manage an IPPD system. Other requirement descriptions include the following:

- ***Associative:*** A portion (for instance a part or a process) should not be designed in isolation from the rest of the product or process. This is derived from the proposition that if each of the components are suboptimized in isolation from the rest, the whole will not be optimized.
- ***Cooperative:*** The design of a typical automobile, aircraft, helicopter, or computer involves thousands of engineers making millions of design decisions over several years. The activities for product design and development should not be carried out by a single individual or a group acting in isolation [Harkins and Dubreuil, 1993]. The ultimate measure of cooperation is the accomplishment of organization goals (constancy of purpose), which requires 7Cs (collaboration, commitment, communications, compromise, consensus, continuous improvement, and coordination) to drive actions. Transformation of information into actions occurs as shown in Figure

5.3, Chapter 5. The New United Motor Manufacturing Inc. (NUMMI) operation in Fremont, California developed a cooperation planning matrix that matched priority topics with appropriate media. A great deal is now known about such factors that contribute to successful team efforts. This is discussed in Chapter 5.

- ***Characteristics or Viewpoints:*** The product should not be designed from a single viewpoint or characteristic [Gutwald, 1988]. A characteristic, or viewpoint, defines a consideration of each life-cycle concern and a method for generating or criticizing design decisions. A single characteristic design often undergoes major changes when other life-cycle concerns, such as assembly and serviceability, are not considered. DFX methodologies (in which DF stands for design-for, and X stands for any life-cycle concern) such as quality, assembly, manufacturability, serviceability, and maintainability, are important elements of CE. They should all form a part of a complete evaluation.
- ***Synthesize:*** It is not enough to have the product subjected to a series of characteristics (such as DFX) or aspects one at a time giving specific guidance or assurances in each case. How designs are generated depends on which characteristics are weighted and what portion of the artifact has been synthesized at any stage. The more weight given to a characteristic, the greater role it can play in synthesizing a design. Any characteristic or aspect may become more or less important as the design evolves. The functional characteristics tend to dominate at the outset but later subside when trade-off between the conflicting characteristics is summed up. A trade-off across different DFX principles is likely to provide a well-rounded outcome.

8.1 CONVENTIONAL DESIGN AND DEVELOPMENT PROCESS

Conventional product design and development practice often concentrates solely on constructing "end-use" descriptions of the product, often called layout designs. Most layout descriptions fall in two categories: concept descriptions and math-based descriptions (see Figure 8.3). Examples of concept descriptions are a concept paper, a product schematic sketch, a blueprint or drawing, and/or a physical model (mock-up). Many math-based descriptions, in the layout stage, are an explicit representation of geometry in either "flat files" (such as CAD file/2-D parts geometry) or in "digitized 2-D/3-D forms" (such as in popular CAD entities, arcs, lines, surfaces, etc.) In the preliminary stages of layout design, some of the tangible deliverables are functional decomposition, device models, back of the envelope calculations, digitized geometry and drawings. When a layout is "locked-in" in digitized form, it is easy to create tool-path or NC data, post processing data (machine inspection), and prototype machining (streolithography), etc. This is shown by the blowup portion of a shaded block in Figure 8.3. It is also easy to create downstream results, such as bill-of-materials (BOM), process plan, products manual, users manual, service plan, etc. This is shown by a set of additional blocks following the shaded block in Figure 8.3. Some of the intermediate deliverables at this stage include design, manufacturing specifications, tooling specifications, and even physical prototypes. These deliverables are for

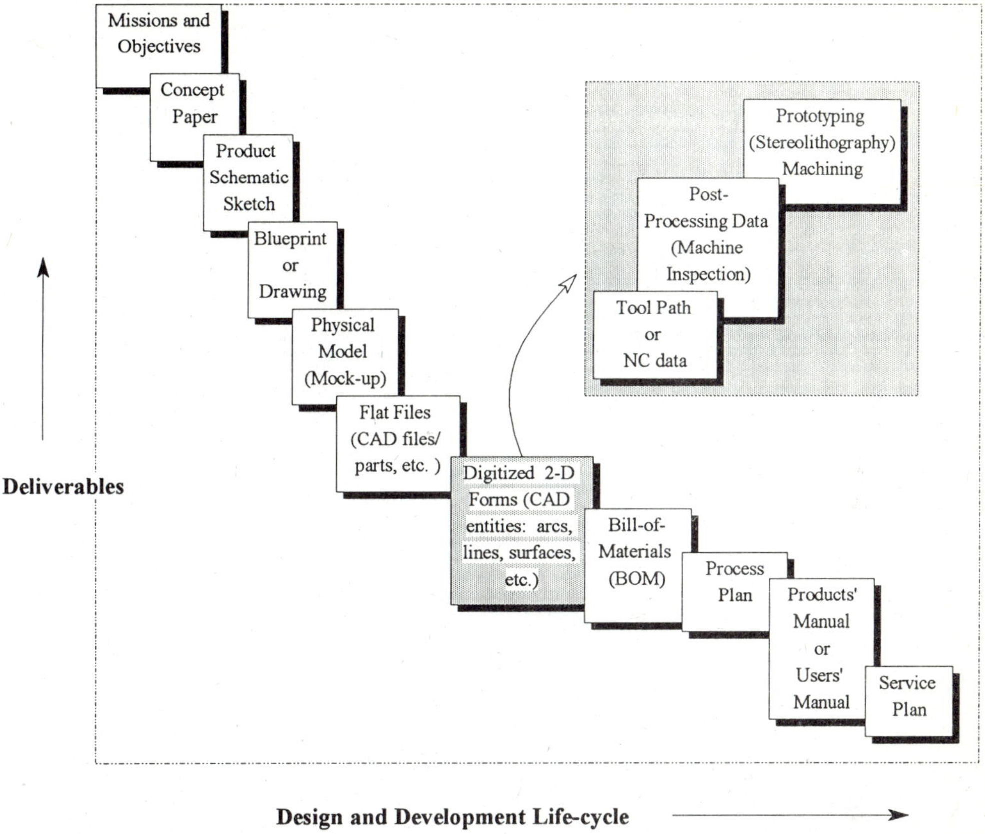

FIGURE 8.3 List of Typical Deliverables in a Conventional Design and Development Process

the "locked-in" form of the CAD data. The text and point data in such cases present no problem, if the original intent of the design is not altered. However, via digitized data, it is very difficult to transfer or reformat the intent of a design to any other form or object. For example, recognizing a circle from its digitized form (series of points) can present a big problem. Intermediate deliverables are not the end product of the design process. Using a symbol or a dimensional-driven system, one can generate these intermediate deliverables easily for a variety of new or altered situations. Parametric and knowledge-based CAD representations are examples of such systems since they are based on capturing design intent.

It is also generally said that 70% of the product cost is committed or locked-in after the layout design is complete [Whitney, 1988]. This presents with three dilemmas:

- ***First,*** the layout design (locked-in in digitized form) is not flexible, so engineers are incapable of incrementally modifying it as they move into other iterative life-cycle phases.
- ***Second,*** even if such flexibility exists, no matter how many engineers try, the ex-

pected gain will only be a fraction of the remaining 30%. The reason—70% of the life-cycle cost was already frozen or committed at this point [Nevins and Whitney, 1989].

- ***Third,*** even if description of the design process is captured parametrically or through similar means (of representing design rationale), it will not be of much use since most communications in the current process are carried out in explicit forms (such as blueprint or drawings, CAD geometry, etc., as shown Figure 8.3).

Japanese face the same dilemmas. However, their process differs from the American way in that they periodically employ systematic design reviews to "life-cycle optimize" the layout design right from the start. This works for them since their units are more coherent and they possess a very strong, culturally inherent communication link among different product teams. In the United States this type of cultural link is very difficult to implant. There is also a cultural disadvantage in the way American industries have learned to work together. Take, for example, a typical product manufacturing setup. Each process is designed serially and specifications are transmitted from one department to another [Harkins and Dubreuil, 1993].

8.2 A TRANSFORMATION MODEL FOR A MANUFACTURING SYSTEM

At the heart of every discrete (parts) manufacturing system—whether traditional or flexible, serial or concurrent, manned or lights-out—there exists a series of process engines that convert raw stock materials into a useful product. One simple example of such a process is a "design process," where a designer transforms an abstract functional specification into a list of design issues or design criteria. In this sense, product realization is a series of transformations through a number of states of an artifact—from a highly abstract form of specification to a detailed specification and finally to a null form, that is, leading into a physical domain. Figure 8.4 shows a semantic representation of a typical manufacturing process that transforms a set of specifications into a functional, producible finished good or a tangible service:

- ***Inputs*** include raw materials, machines, equipment, and capital.
- ***Requirements*** include labor, manpower (direct or indirect), energy management, organization, suppliers, and tools.
- ***Constraints*** include cultural or human factors, inventory, expenditures, company policy, regulatory/ethical/and environmental obligations, experimental factors, technology, information, natural resources, methods, and systems.

One way to mathematically describe such a production system (in Figure 8.4) is in terms of a linear programming approach. The system of equations is as follows:

$$\text{Maximize: } \{V_i\}\{O_i\}^T \quad \text{[Objective Function]} \tag{8.4}$$

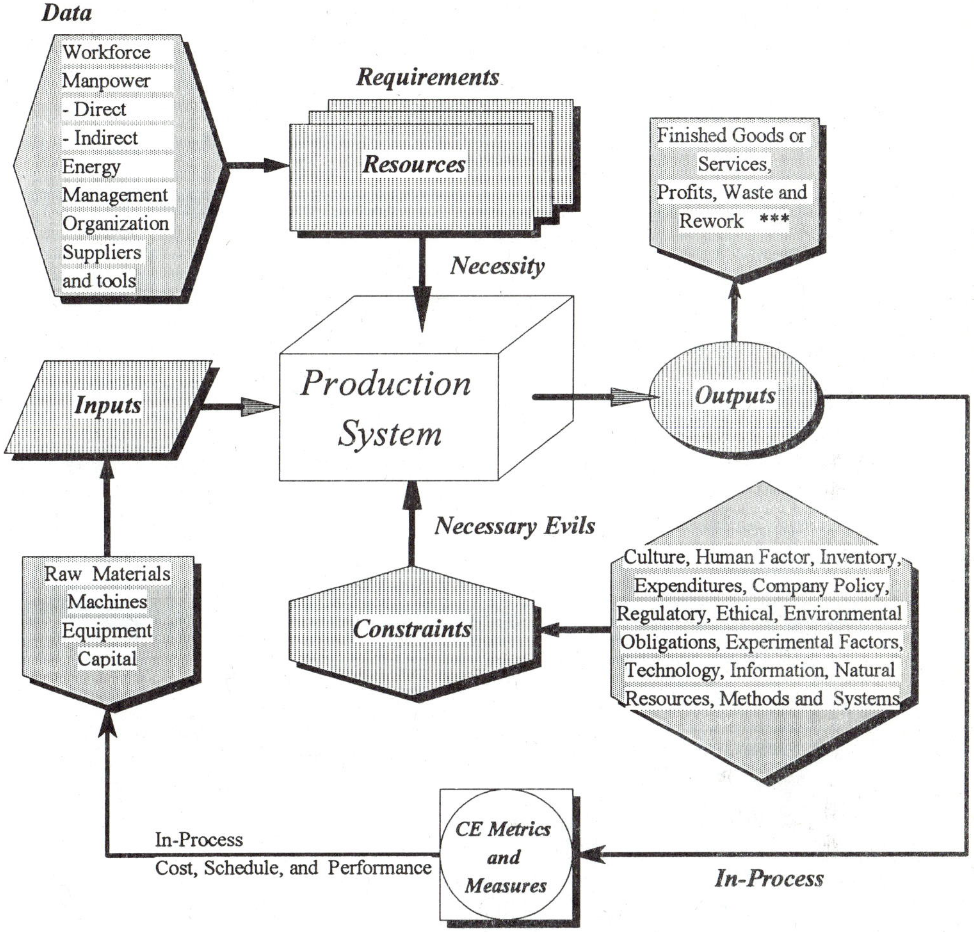

FIGURE 8.4 A Typical Production System

Subject to the following:

$$[T_i]\ \{S_i\} = \{O_i\} \quad \text{[For an } i\text{th state Transformation]} \tag{8.5}$$

The $\{V_i\}$ is the weighting factor, and $\{O_i\}$ is the output of each transformation. The product $\{V_i\}\{O_i\}^T$ indicates how much benefit (incremental gain or value) accrues from each transformation.

T_i is a transformation matrix of an ith stage. S_i is the specification set that is transformed into useful outputs; O_i. The constraints and requirements are represented by separate equation sets. Requirements $\{R_i\}$ are generally imposed in the form of equalities. These are necessary evils or minimum requirements.

Where, index i represents an ith stage of transformation.

$$[A]\ \{S_i\} = \{R_i\} \quad \text{[For an } i\text{th state Requirement Set]} \tag{8.6}$$

Where A is a matrix relating the specification set $\{S_i\}$ to the requirement set $\{R_i\}$.

The constraints are imposed in the form of inequality (greater-than or less-than sign). This is further discussed in Chapter 9.

$$[B]\ \{S_i\} \geq \{C_i\} \quad \text{[For an } i\text{th state Constraint Set]} \tag{8.7}$$

Where B is a matrix relating the specification set $\{S_i\}$ to the constraint set $\{C_i\}$.

Where $\{S_i\}$ is defined as a union of the ith state inputs, $\{I_i\}$, requirements $\{R_i\}$, and constraints $\{C_i\}$;

$$\{S_i\} \equiv \cup\ [\{I_i\}, \{R_i\}, \{C_i\}] \tag{8.8}$$

and

$$\{S_{i+1}\} = \{O_i\} \tag{8.9}$$

The quality of finished output (product or service) depends on the quality of this transformation. The quality of each transformation is represented by the VCs (value characteristics) associated with each transformation and the quality characteristics associated with each intermediate outputs. The input I_i is represented in the specification set S_i.

The initial specification, S_1 is a full specification. It consists of a set of inputs, requirements, and constraints. The final specification is a null or an empty set.

$$\{S_1\} \Rightarrow \{\bullet\} \text{ a Full State} \tag{8.10}$$

$$\{S_{n+1}\} \Rightarrow \{\emptyset\} \text{ a Null or an Empty State} \tag{8.11}$$

Figure 8.4 also shows a feedback loop consisting of a set of CE metrics and measures. CE metrics provide a measure of merits (MOMs), such as cost, schedule, and performance. The basis of defining and selecting a series of appropriate transformations has not been well-understood or well-characterized for mechanical systems [Finger and Rinderle, 1989]. Two chapters of the book are therefore devoted to these two important topics. Chapter 9 deals with taxonomy of transformation and Chapter 2, Volume II deals with CE metrics and measures. A taxonomic form is employed here as a method of characterizing a transformation state and for understanding its success factors.

8.2.1 Key Success Factors

The success of a transformation system can be measured by the following degrees:

- ***Degree of compliance*** in meeting the specifications, such as performance, schedule, production rate, and cost target following a set taxonomy of product realization
- ***Degree of adaptability or flexibility*** in meeting the dynamic demands of product mix (or customer specification)
- ***Degree of automation*** in getting the product to meet its stated purpose in a minimum number of transformations

- ***Degree of reengineering*** in changing the inherent task distribution of the product development process (such as task scheduling, paralleling the activities, creating concurrent process plans) with least manpower and lead time and still able to follow a set taxonomy for product realization.

8.3 CE ENTERPRISE SYSTEM TAXONOMY

A model of a CE transformation system is shown in Figure 8.5. It is a variant of the production system shown in Figure 8.4. Though the parameters entering and leaving the two systems are the same, the task distributions within the system are different. A bulk of inputs consists of

- Stock (raw materials)
- Time and energy
- Methods and processes
- Capital and labor

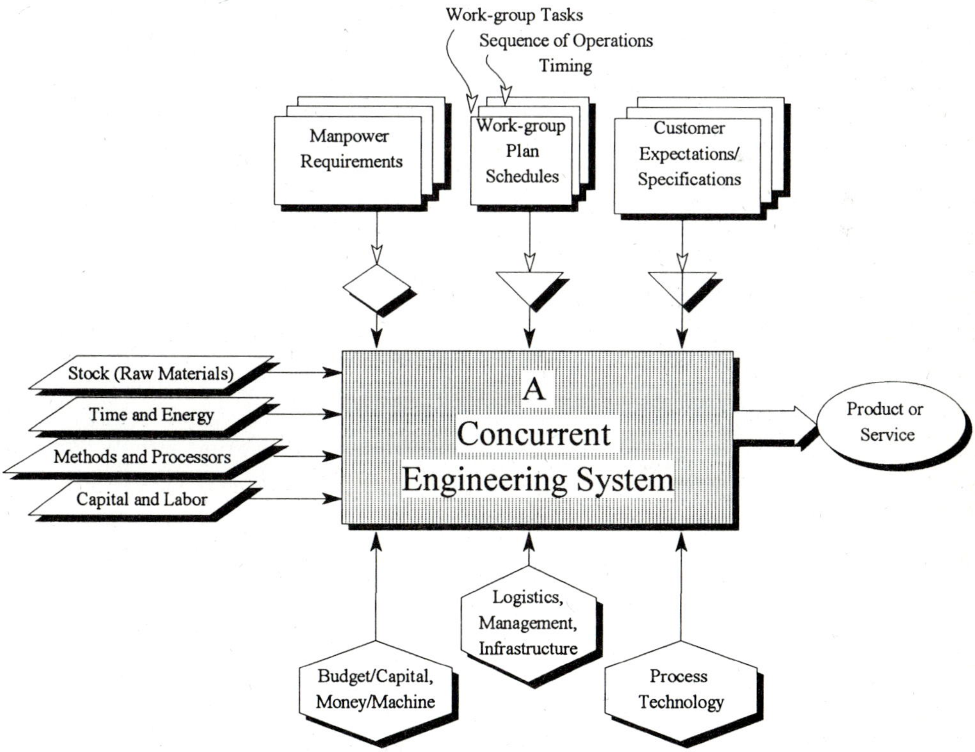

FIGURE 8.5 A Model of a CE Transformation System

Specifications are transformed and converted into products or services. The central difference between a CE transformation system and any other manufacturing system, such as serial engineering, is the manner in which the task distribution is stated and specifications accomplished [Simms, 1993]. In a CE transformation system, the purpose of every process step is not just to achieve a transformation but to accomplish this in an optimal way. A transformation imposes constraints on each "state definition":

- Budget, capital, and money place restraints on teams' activities or functions, e.g., amount of available capital to buy equipment.
- Logistics management infrastructure is a source of constraints: what's the right time to start an activity; who are the people doing it; how is this activity to be done and, will it be done in relations to others.
- The physical proximity or layout of the process technology limits how close a manufacturing function is placed with respect to the next function, and how the activities within a process model are accomplished.

Actually, many variants of a transportation system are possible. In each transformation case, processes require specifications and activities (tasks) create a series of evolving states (for example, definition of geometry, definitions of assembly, process plan, production plan, etc.). The specifications in these transformation cases may not change very much from one variant to the other. The differences in the system transformation models result from how specifications are met. A transformation imposes requirements on the state definitions, for example:

- Manpower needs
- Work-group plan schedules {which include sequence of operations, work-group tasks, and timings}
- Customer expectations/needs

The serial and concurrent transformation models represent two forms of a production system, differing in how the teams' plans and schedules have been specified in each case:

- In the ***serial*** transformation system model, the process runs serially. Activities may be performed independently at different times or in sequential order.
- In the ***concurrent*** transformation system model, the activities within a plan or schedule are arranged in chronological order [Pourbabai and Pecht, 1994]. Timing relationships between the work-group's plan activities are expressed using concurrent logic (such as decomposition, overlap, or temporal logic). Process taxonomy accounts for the parallel scheduling of tasks to realize drastic reduction in time and cost, and to support life-cycle operations.

Ideally, a transformation system that are adaptive (flexible), accurate, and automatic—is most suitable. A model of an efficient CE system would be the one which meets these goals. This means, for instance, selecting an optimal combination of the ten items as shown in Figure 8.5 (four inputs, three constraints, and three requirements). Together they minimize the life-cycle time to efficiently develop a product.

8.4 INTEGRATED PRODUCT AND PROCESS DEVELOPMENT

An important distinction in integrated product and process design (IPPD) is to recognize that the artifact being designed is an assembly or a system (that is, something which is composed of a hierarchy of subsystems, components, parts, materials, and features). A system level organization of product development activities is shown in Figure 8.6. The time has come to understand and manage IPPD as a system. At this level, the system for IPPD consists of two redesign paths:

- ***System Development:*** During system development, the constructs of product and manufacturing systems proceed concurrently with integrated system planning and scheduling on one end and comparison to goals and objectives on the other. If the goals are met, the design enters into system construction, otherwise it goes through an iterative redesign path as shown in Figure 8.6.
- ***System Construction:*** During system construction, the manufacturing system is constructed and installed, and later its operation is validated. Value engineering and continuous process improvement are considered part of this validation. If no improvement is desired, the artifact is produced, otherwise, necessary improvements are incorporated through a secondary redesign bypass loop.

During the highlighted development and construction processes, the system involves employees, CE work-groups, customers, competitors, and the organization as a whole. These components are interdependent. The product information associated with a particular development or construction process is overlapped, that is, none of the product and process information is kept on hold until a phase is completed or a set of criteria have been fully satisfied. The Japanese seem to do a better job in "optimizing the artifact" for the life-cycle with the help of people's knowledge and experience. However, as international pressure mounts for more and more downsizing (that is, cutting down the lead time), more and more companies around the world will be taking a balanced approach: First, acquire the product and process knowledge and then use the power of the intelligent systems to guide decision making. Second, supplement it with the multidisciplinary teams' expertise to further narrow down the choices.

8.4.1 Broad System Model

In order to understand the "artifact's intent" definitions, it is useful to start with a broad concept of a system model. This broad concept model is identified in Figure 8.7 using a five-block schema:

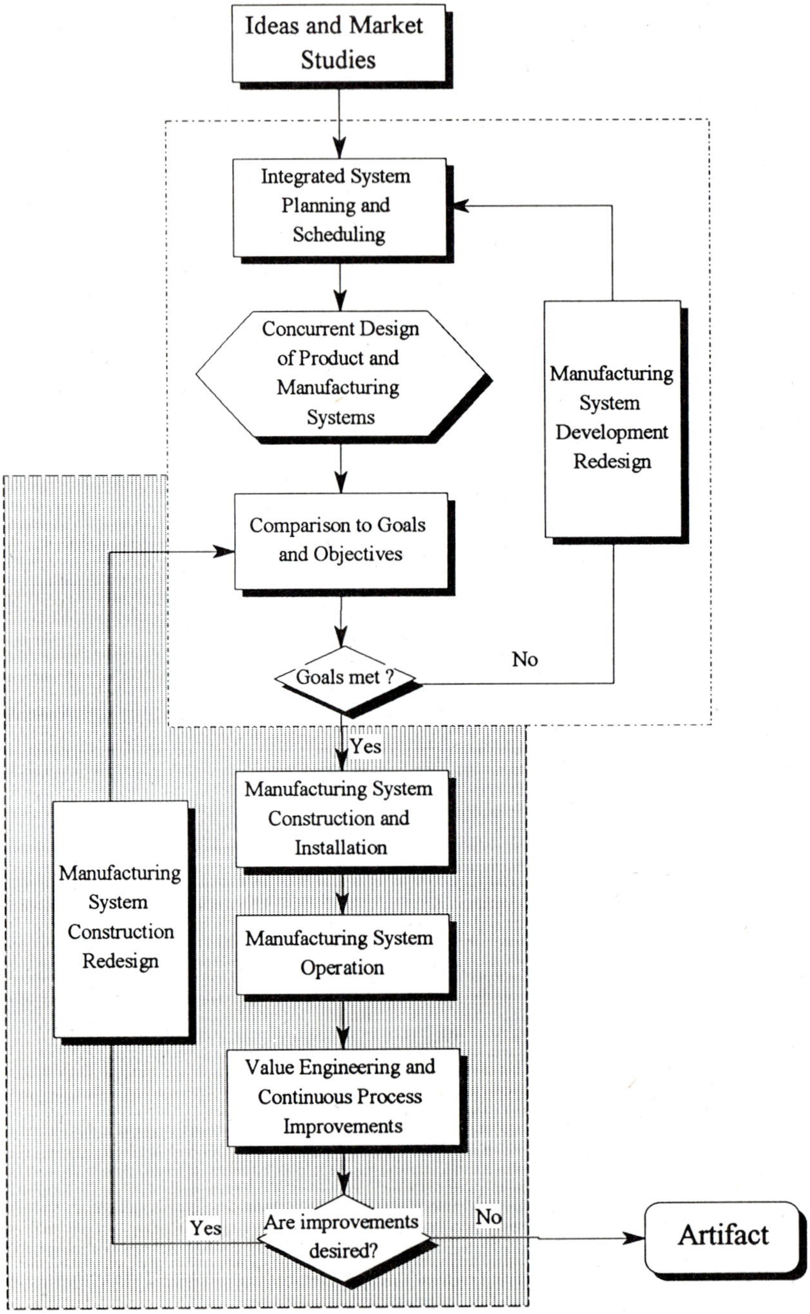

FIGURE 8.6 A System Level Organization of Integrated Product and Process Development Activity

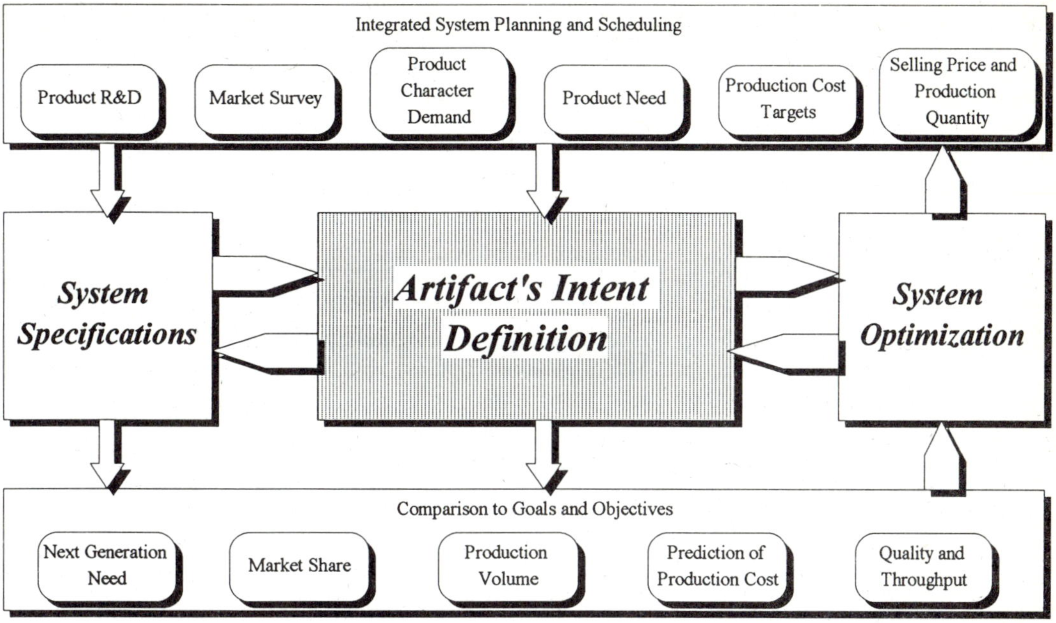

FIGURE 8.7 Broad System Model for Integrated Product and Process Development

- ***Requirements:*** Integrated system planning and scheduling comprises the requirements block.
- ***Inputs:*** A set of system specifications makes up the input block.
- ***Transformation:*** Artifact's intent definition represents the transformation block—converts inputs into outputs.
- ***Constraints:*** Comparison to goals and objectives forms the constraint block.
- ***Outputs:*** System optimization is an output block.

These five components are interrelated, and can be used to explore and define the artifact's intent according to the perceived customer needs. Using this approach, the system model broadly defines the necessary means, the constraint boundaries, and the overall goals of a CE enterprise. Each of these components is discussed next.

8.4.2 Integrated System Planning and Scheduling

Integrated system planning and scheduling provides the system specifications to help identify the artifacts' intent definitions. It consists of teams from product R&D, sales and marketing, design planning and manufacturing planning, and system and management staff who recognize the needs of a product through research and development, market study, customer survey, etc. The product needs and requirements are determined from what is collectively called voice of customer (VOC). VOC provides information about the would-be demand of the product character, functions, and features (see Figure 8.7). Sales and marketing specify the demand information in the form of product need, production cost targets, selling price and production quantity, and other similar parameters. Relevant information, for system specification needs, is then extracted. Design planners convert the set of perceived needs and market constraints into high level "product specifications" for use during the artifact's intent definitions. Manufacturing planners convert the set of form features and manufacturing functions into high level "process specifications."

One of the persistent problems in a traditional production system is scheduling work and assigning resources. The production environment is somewhat dynamic and loads change with the results of previous activities. One cannot know definitely what the task is, how long, or who will be involved, until it is actually begun. Because of these problems, the traditional production system is often described as serial engineering or "design as you go" type. Each step consumes valuable time: Parts sit waiting to move; parts move; then parts change—fixture changes; tools are brought in; parts are waiting to be processed next, etc. Products usually gain value for only 20% to 40% of the time they are in the factory.

8.4.3 System Specifications

This is the input block for the broad system model (see Figure 8.7). A system specification is of two types:

- ***Behavioral:*** Those specifications that describe the desired behavior of the overall system at an abstract level are called behavioral specifications. Examples of behavioral specifications include global behaviors, such as limits on overall frequency, stiffness, etc. Note that strength is not a system behavior because it is generally governed by local characteristics such as presence of notch, crack, etc.
- ***Physical:*** Physical specifications place physical restrictions on the outputs, such as geometry, allowable sizes and weights, amount of scraps, etc.

The behavior and physical specifications are often related. The relationship may exist in an explicit form, like an analytical form (such as design equations.) A physical form (for example, the weight of a helical coil spring) is analytically expressed as the product of stiffness and the square of the allowable deflection, which are behavioral terms. Other forms in which this relationship exists are database entries, like a spreadsheet, or an implicit form (such as FEM or CAD geometry) which relate geometric characteristics (such as weight) to behavior characteristics (such as stiffness).

Each node in the specification tree may contain the following:

- ***Class Definitions:*** This states the membership requirements of the product (PtBS) or process (PsBS) tree for inheritance purposes. This is useful for building a specification hierarchy since high level specifications may be a concatenation of low level specifications.

- ***"Children of" or "Made of" List:*** This is useful for identifying relationships, dependencies of objects, or systematically organizing the information as a network tree (see Figure 8.1).
- ***Requirements and Constraints (RCs) List:*** It asserts the allocation of RCs to various nodes of a tree depending on the product and process functional needs. The detail of specifications depend on the degree to which a team would like to place restrictions on the nodes of the product or the process tree.

Functional requirements are characterized into three groups:

- System Requirements
- Product Requirements
- Process Requirements

8.4.3.1 System Requirements

CE teams from different domains and disciplines view the design with different interpretations. Depending upon which of the team is making decisions, or what requirements of the design is considered important, the design outcome (or preference) changes. System requirements alleviate these uncertainties. System requirements are "system centered" (such as improved system reliability, better overall performance, and lower total price) as opposed to "component centered" (such as improved component reliability, better component performance, or lower unit price). System requirements are transformed from customer needs or perceived requirements. Most customer needs are generally not related to the geometrical form of the product. When key factors of customer needs must be addressed, benefits (such as "more" of "*x*" and "less" of "*y*") lead to difficulties in interpretation and implementation. Decision making is based on past experience and knowledge, or planning tools such as QFD, rather than being purely subjective. Using QFD system requirements can be satisfied by a set of well-posed engineering characteristics. (See Section 2.8.2.2)

8.4.3.2 Product Requirements

Product planners develop requirements for the design of the product. Product requirements include information such as reliability, safety, performance, and cost targets. Product requirements are of three types:

1. *Technical Requirements:* Technical requirements are the equivalent of the customer requirements that are stated in meaningful and quantifiable terms. "Meaningful" in the sense that they are understandable to the engineering community. Quantifiable indicates that they can be measured in physical parameters like force, distance, torque, acceleration, rates of change, and so forth. The customer requirement for "an engine that accelerates fast" might be translated into a technical requirement of "the time it takes to pick up a speed of 100 miles per hour under idle conditions." The customer requirement for "a door that opens easily" can be translated into technical requirements for "the amount of force required for a given push or rotation"

depending upon the type of handles used. Technical requirements can also be established directly using the technical expertise of experienced engineers. Requirements can also be taken from product acceptance standards if the work-group can intelligently reflect the customer needs or expectations.

2. *Geometrical Requirements:* Most of the activities in the product development cycle of mechanical products are centered around generating and designing geometric shapes that perform some specific functions. Geometrical requirements may include generation of geometry, topology, dimensions, and process tolerances. There are two types of geometrical requirements (GRs) that are usually needed to model the product geometry adequately: quantitative GRs and abstract GRs. A complete design of a product requires not only the design parameters, but also the geometric configurations that realize the required functions. Although quantitative GRs are very important to the determination of the design parameters, the quantitative GRs alone cannot create an embodiment of design. Under normal conditions, it is possible that more than one type of final geometrical form may satisfy the required quantitative functions. At other times, some of the GRs of design are also difficult to represent by quantitative variables. Those GRs which cannot be directly represented by a quantitative value or simple logical computation are characterized as abstract GRs.
3. *Interface Requirements:* An assembly contains more interface information since planning for assembly deals with a number of individual parts, symmetry, and spatial reasoning. Types of product specifications captured during the modeling of the assembly are the subset of the interface specifications. Interface specifications are not complete until all forms of spatial and mating relationships are contained in it. Types of mating relationships could be symmetry, parallel-offset, aligned, incline-offset, include-angle, etc. The spatial aspect may include criteria such as orientation (angular rotation in axial, vertical, or longitudinal direction) or positioning keywords (distance measured from top, bottom, front, rear, left, right, etc.)

8.4.3.3 Process Requirements

Manufacturing planners produce process requirements for the product's manufacture. Process specifications include information such as materials, manufacturing process, tooling, costs, environmental conditions, etc. Product or process specifications are described hierarchically. The first level nodes in the product hierarchy define the specifications of the desired product: subsystems, components, parts, materials, and form features. Similarly, the process hierarchy defines the specifications for the desired process: layout, machines and equipment, control and scheduling, inspection and fabrication, material handling, and process features.

8.4.4 Comparison to Goals and Objectives

The key to gaining an insight into any system is to develop its breakdown structure tree and define a set of specifications and goals as a part of this system tree. Comparisons to goals and objectives are in fact the constraints for the broad system model (see Figure

8.7). Objectives are related to goals—a goal is broken up into several objectives. Typical examples of goals and objectives are next generation need, market share, production volume, prediction of product cost or quality, and throughput. The triad—tree structure, specifications, and goals—provides a basis for system optimization. As Deming said, "optimization is a process of orchestrating the efforts of all components towards achievement of the stated goals." Without a structure there is nothing for the system to stand on. Without specifications, there is no unique system. Without a set of consistent goals, there is no system. The task of simultaneously redefining all pertinent components of a complete product development in a concurrent engineering environment so as to be mutually supportive of each other's needs (RCs), both in content and timing, is a mind boggling task. The simple contention that if the components of the tree would be optimally designed it will, in turn, yield an optimized system does not hold true. Reciprocally, an optimized system at the end does not guarantee that its constituents will also be at their optimal. This is because each of the constituents of the system is dependent on each other. The system, therefore, cannot be optimized assuming its constituents to be completely independent. Besides, there exists differing nature of specifications (inputs and RCs) and goals for each constituent. In order to create common corporate goals, product values should be considered, and the IPPD process should be utilized. Product values must be established by weighing company and customer interests (see Figure 8.8). Other important factors for defining a consistent set of goals include the following:

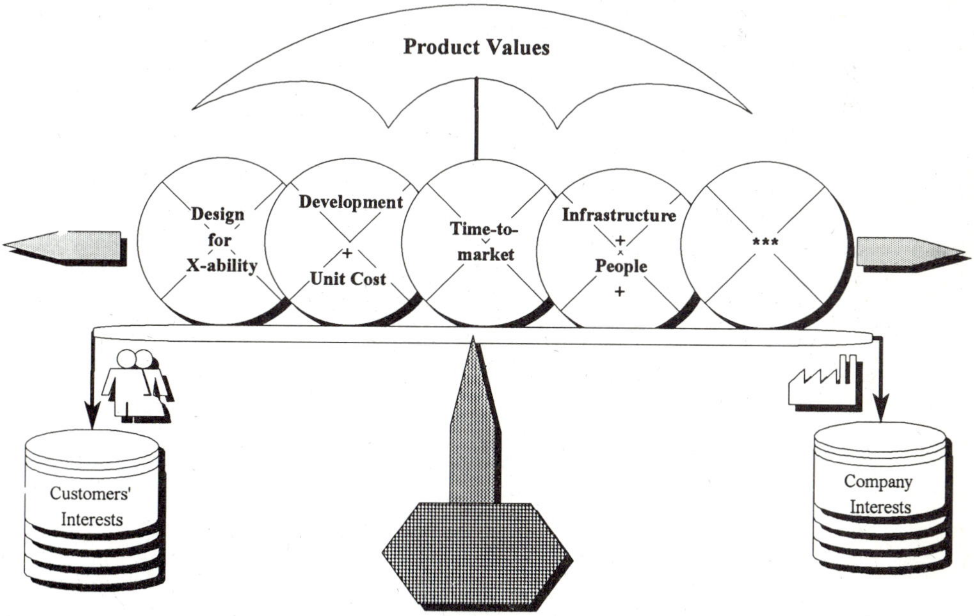

FIGURE 8.8 Balancing the Product Values Set

- Corporate or SBU goals should be based on artifact's intent definitions. (Note, artifact should be defined in terms of both customer interests and company interests.)
- Department or unit goals should be subordinated to corporate-based goals. (Note, this is similar to the enterprise hierarchy shown in Figure 8.1.) Goal optimization at the SBU level does not necessarily equate to optimization of subunits or departmental goals.
- Feedback should be provided to ensure that subunit goals compliment and help achieve the SBU goals.

Furthermore, conflicts between subunits are less likely to occur as a result of the characteristics of the IPPD system: associative, cooperative, characteristics or aspects, and synthesize. This is discussed in section 8.0

8.4.5 System Optimization

This represents an output block of the broad system model, as shown in Figure 8.7. In concurrent engineering, the multifunctional teams have to consider goals and objectives that provide a balance between the customer and company interests. In Figure 8.8, several aspects, like functionality, behavior (design for X-ability), cost (development + unit cost), time-to-market, and infrastructure (form, organization, people, etc.), are displayed on a balance. At one end of the fulcrum, the balance is loaded with company interests and at the other end of the fulcrum with customers' interests. The selection of a mix of the product values-set and its location across the balance provide the relative weights of their (company and customer) contributions. The challenge is to appropriately place these weights on the lever so that their contributions are balanced. As mentioned earlier, the system optimization is a process of orchestrating the contributions of all of the value-sets toward achievement of the stated objective, which is, in this case, to balance the customer and company interests.

8.5 TRANSFORMATION SYSTEM FOR PRODUCT REALIZATION

Organization of a process plays an important role in ensuring the effectiveness of the resulting production system. CE is a problem of high complexity involving a variety of Ts (talents, tasks, teams, techniques, technology, time, and tools). Organization of CE can be approached by carefully splitting the system level problem into its mutually separable transformation states, followed by modeling of each state, then the reconstruction of a system definition from the aggregation of the definitions of its constituent states. Figure 8.9 shows a mapping of the functional description of the product through a series of transformations leading to a physical description. The series exhibits a precedence relationship among its transformed states. Each state is driven by a set of specifications and produces a set of outputs.

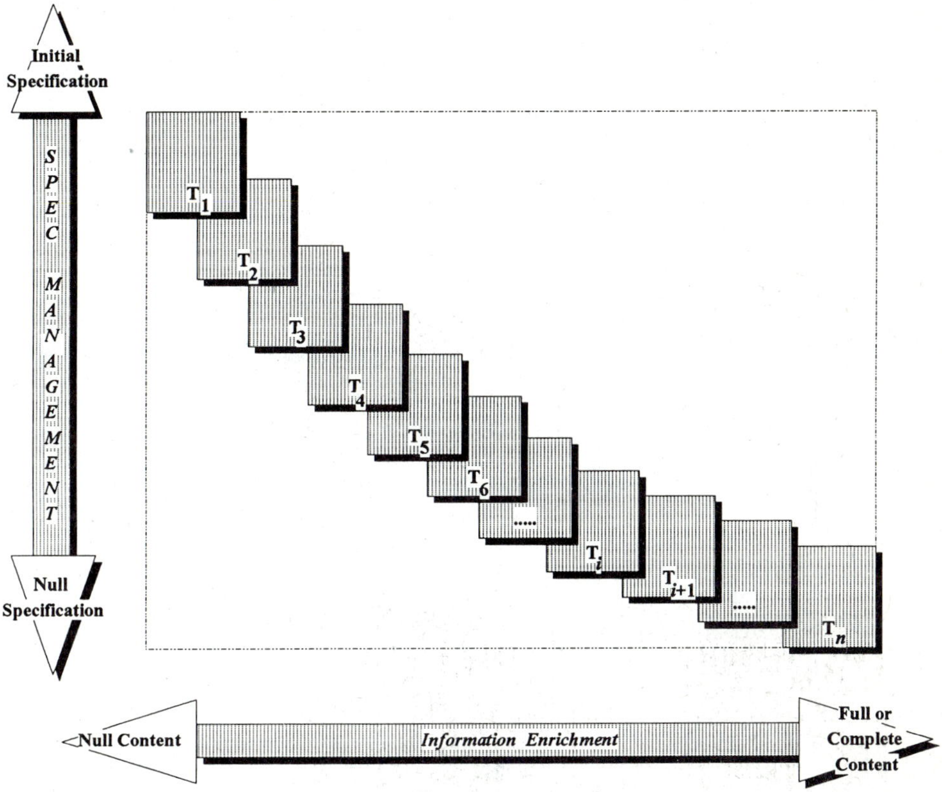

FIGURE 8.9 Mapping of a Functional Description Into a Physical Description

The modified output of the previous state becomes the specification data for the new state:

$$\{S_{i+1}\} = \{O_i\} \tag{8.12}$$

From the definition of transformation, T_i transforms a set of specifications at state i from the S_i to O_i:

$$[T_i]\ \{S_i\} = \{O_i\} = \{S_{i+1}\} \tag{8.13}$$

In accordance with the precedence relationships, the transformation matrix T_i describes the relation between the specifications of any two adjacent taxonomic states. The transformation equation for a state $i = 1, 2, 3, \ldots, n$ can be written as follows:

Stage 1 ($i = 1$): $$[T_1]\ \{S_1\} = \{O_1\} = \{S_2\} \tag{8.14}$$

Stage 2 ($i = 2$): $$[T_2]\ \{S_2\} = \{O_2\} = \{S_3\} \tag{8.15}$$

Stage 3 ($i = 3$): $$[T_3]\ \{S_3\} = \{O_3\} = \{S_4\} \tag{8.16}$$

$$\text{Stage } i\ (i = I): \quad [T_i]\{S_i\} = \{O_i\} = \{S_{i+1}\} \tag{8.17}$$

$$\text{Stage } n-1\ (i = n-1): \quad [T_{n-1}]\{S_{n-1}\} = \{O_{n-1}\} = \{S_n\} \tag{8.18}$$

$$\text{Stage } n\ (i = n): \quad [T_n]\{S_n\} = \{O_n\} = \{S_{n+1}\} \tag{8.19}$$

At the beginning of the transformation (stage 1), the design exists in pure specification form (see Figure 8.9).

$$\text{Stage 1:} \quad \{S_1\} \Rightarrow \{\bullet\} \text{ a Full State} \tag{8.20}$$

$$\{O_1\} \Rightarrow \{\emptyset\} \text{ a Null or an Empty Content} \tag{8.21}$$

As various tasks within this taxonomy are performed, designs (meaning product and/or process designs) begin to take shape. The set of specifications changes over time. At the end (nth stage) all specifications have been implemented. At that point, the designs are at the "full content" and specifications are at the "null state."

$$\text{Stage } n\text{:} \quad \{S_{n+1}\} \Rightarrow \{\emptyset\} \text{ a Null or Empty State} \tag{8.22}$$

$$\{O_n\} \quad \Rightarrow \{\bullet\} \text{ a Full Content} \tag{8.23}$$

Assuming that at the end of the nth state, the initial specifications have been converted into a physical artifact. The new specification is a null set. Substituting the equation in reverse order, we get

$$[T_n][T_{n-1}][T_{n-2}] \ldots \ldots [T_1]\{S_1\} = \{\emptyset\} \tag{8.24}$$

If we denote

$$[T] = [T_n\, T_{n-1}\, T_{n-2} \ldots \ldots T_1]$$

then the above equation can be written as

$$[T]\{S_1\} = \{\emptyset\} \tag{8.25}$$

Finally, the artifact contains all the information (enterprise, requirements, product, process, and cognitive) needed to function as a unit as initially desired. The process taxonomy includes both the specifications

$$[\{S_1\}, \{S_2\}, \{S_3\}, \{S_i\}, \{S_4\}, \ldots, \{S_i\}]$$

and the activities

$$([T_1], [T_2], [T_3], [T_i], [T_4], \ldots [T_i])$$

as the design evolves from its empty or null or $\{\emptyset\}$ content for a new design (or "a baseline content" for a revision or a change design) to a full or complete content.

The design process begins with the establishment of initial specifications $\{S_1\}$. The $\{S_1\}$ are defined in the functional domain to satisfy a given set of customer needs. This first step is subjective but it is clearly one of the most critical steps. In this step, it is important that $\{S_1\}$ should always be stated in the functional domain, that is, $\{S_1\}$ must be defined in the solution neutral environment.

The design can be thought of as having reached a full or complete content when the sets of specifications stabilize and when all the intermediate specifications have been satisfied. This is why this taxonomy is also referred to as a "pull system for product realization". The pull system serves as a standard taxonomy for problem solving. The aim of a good transformation strategy is to uncouple the system so that each transformation state affects only one set of outputs. This is very similar to Suh's first axiom [Suh, 1988] in the axiomatic design theory. Such creative design process requires rational prediction. The chances that such creative efforts will result in a finished artifact are more likely with this taxonomic transformation than otherwise. For CE management, it guarantees a paper trail for the finalized product or service.

8.5.1 Taxonomy for Product Realization

As discussed in Section 8.3, an evolutionary definition of artifact is a transformation process. Transformation means a change in form, shape, or appearance (*Oxford English Dictionary,* 2nd Edition, Vol. IX, Clarendon Press, Oxford, 1989). The product realization process can thus be viewed as a series of transformations converting the specifications into a useful product or service. Each transformation state represents a subset of the product realization space at different levels of abstraction and/or granularity. Frames are the attributes that characterize a design state from the viewpoint of any transformation level. These definitions establish a list of taxonomy terms to express a distinct activity of the product realization process, in view of standardization. A taxonomy is the theory, principle, or process of classifying organisms in established categories (*The American Heritage Dictionary, 1981*). Long-term success depends on the need to create a taxonomy of transformational strategy for product realization in which product specifications (frames) can be transformed into a description of a series of subspecifications (subframes) for the lower level transformations [Finger and Rinderle, 1989]. Interaction between elements of the product is important in our synthesis strategy; the following defines three major steps of a process taxonomy:

- ***Linking*:** Representations of the behavioral inputs and physical requirements must be linked to representations of their physical characteristics which dictate the constraints. Constraints at different levels of abstraction guide the design process. RCs are also used to maintain consistency and propagate design decisions. Thus, for the taxonomy, we are interested in capturing both the frames (inputs, outputs, requirements, and constraints) and the concepts—identifying the systems' hierarchy and their relationships. The concept includes modeling the relationship between form and functions of the transformation process (the predictor-corrector process model). Let us assume $[T_i]$ represents an activity plan for a series of transformations from $[T_1]$ to $[T_n]$. $\{S_i\}$ denotes a set of specifications at a transformation state i. Following the definitions of the transformation state Equation (8.5), we can write

$$T_i = f\,[\{I_i\}, \{R_i\}, \{C_i\}] \tag{8.26}$$

where

$$I_i = f\,[\{D_i\}, \{K_i\}, \{P_i\}] \tag{8.27}$$

Where, f denotes the function of, and $\{I_i\}$, $\{R_i\}$, $\{C_i\}$, $\{D_i\}$, $\{K_i\}$, $\{P_i\}$, $\{O_i\}$, represent the sets of inputs, requirements, constraints, data, knowledge, process, and outputs, respectively, for a state i.

- ***Mapping Process:*** A process taxonomy captures the system's behavior in some form of abstraction or formalism. To use this formalism, it transforms some crude description of the desired behavior (called specifications) into a physical description (called an artifact). The relations between the functional specifications $\{S_i\}$,—namely $[\{S_1\}, \{S_2\}, \{S_3\}, \ldots, \{S_n\}]$ and the output variables $[O_i]$—namely $[\{O_1\}, \{O_2\}, \{O_3\}, \ldots, \{O_n\}]$; are expressed through a transformation plan-set. There are a series of mutually separable matrices which describe the relations between sets of specifications. Matrix representation is amenable to mathematical treatment and some rules of calculus to evolve a rational conceptual basis for product realization. This was discussed in Section 8.2 Equation (8.5).

$$[T_i]\,\{S_i\} = \{O_i\} \tag{8.28}$$

In Section 8.4.3, specifications are characterized into three groups: systems requirements, product requirements, and process requirements. As a part of this activity plan, if a CE team uses the classifications of Section 8.4.3, there will be three sets of mapping:

1. A series of mapping which converts a set of system requirements, SRs (such as those to fulfill the customer requirements) into product requirements. Then development of the physical embodiment takes place first in product domain and in parallel to this in process domain.

$$[T_i]\,\{SRs_i\} = \{PtRs_i\} \tag{8.29}$$

2. A series of mapping for the product management cycle, transforming the product requirements (PtRs) into process requirements, PsRs is as follows:

$$[T_i]\,\{PtRs_i\} = \{PsRs_i\} \tag{8.30}$$

3. A series of mapping for the process management cycle transforming the process requirements (PsRs) into the physical domain (or the artifact itself) is as follows:

$$[T_i]\,\{PsRs_i\} = \{O_i\} \tag{8.31}$$

Where, $\{SRs_i\}$, $\{PtRs_i\}$ and $\{PsRs_i\}$ represent the vector of system requirements, product requirements, and process requirements, respectively. $[T_i]$ is the corresponding transformation matrix. The mapping process from the system to product, from product to process, and finally from process to the physical artifact is not unique. There can be an infinite number of plausible solution paths. However not all proposed solution paths are good solutions. The criteria for judging the good from the bad is through knowledge capture discussed in the next section.

- ***Knowledge Capture:*** Often, experienced designers can make decisions which far exceed the power of many computer programs. This is due to our inability to quantify the various types of creative knowledge that makes a design a good design. Any attempt to capture such creative knowledge should be based on an efficient and modern manufacturing process and 3Ps (policy, practice, and procedure). For instance, one of the major problems, in "functional tolerancing", is the lack of knowledge concerning the relations between tolerance types and functions, surface finish and functions, or geometric forms and functions [Chen and Menq, 1992].

The potential impact of this (linking, mapping process, and knowledge capture) taxonomy steps on product realization is comprehensive. Following these steps, future concurrent teams would be able to design the product right the first time.

The taxonomy of CE-based integrated product realization is different from that of serial engineering. The serial engineering taxonomy is based on linear network techniques (such as program evaluation and review technique (PERT), critical path method (CPM), and the precedence diagramming method (PDM)). Such network techniques assume that serial engineering is a one-way progression along linear paths, with no procedural feedback or iteration and no looping or back-chaining. In CE, the information flows bidirectionally to upstream and downstream tasks. The emphasis here is on the interactions between the interdependent components that must be facilitated. If a planning or design team does not know parts of a product's configuration, how can a team estimate the total product cost? Similarly, if a design team does not know the limitations of the manufacturing processes that are available, how can the team design the product in a cost-effective manner? The same can be said for the downstream tasks. How can a manufacturing group produce a part, if the manufacturing process is not accurately defined or the design is not well-understood? How can a product be supported in the field, if the accurate information is not made available to the product support team interfacing the customer?

The taxonomy is represented conceptually in this book using a dataflow engineering convention (as shown in Figure 8.10). It employs a holistic approach to product realization. The order in which the functions are described is the order in which these activities will usually be executed. Each function establishes the data necessary to support the succeeding function. One of the advantages of this representation is that it combines the information flow with the work-flow. The conceptual model covers "what is to be done" and sometimes identifies the interface requirements, possible overlaps, and constraints. It usually does not state how the processes are performed or whether they are performed "locally" or are "distributed." The relative timing and flow of the tasks, however, are included. This differentiates a process taxonomy that is concurrent from one that is not, say, a serial taxonomy. Concurrent taxonomy is characterized by overlap timing, parallel teams, and flow of overlapped tasks. In summary, the taxonomy represents a snapshot of the life-cycle transformations in terms of work-flow (specifications, activities, actions), and information flow. The taxonomy is useful for representing both a carryover and new product introduction situations.

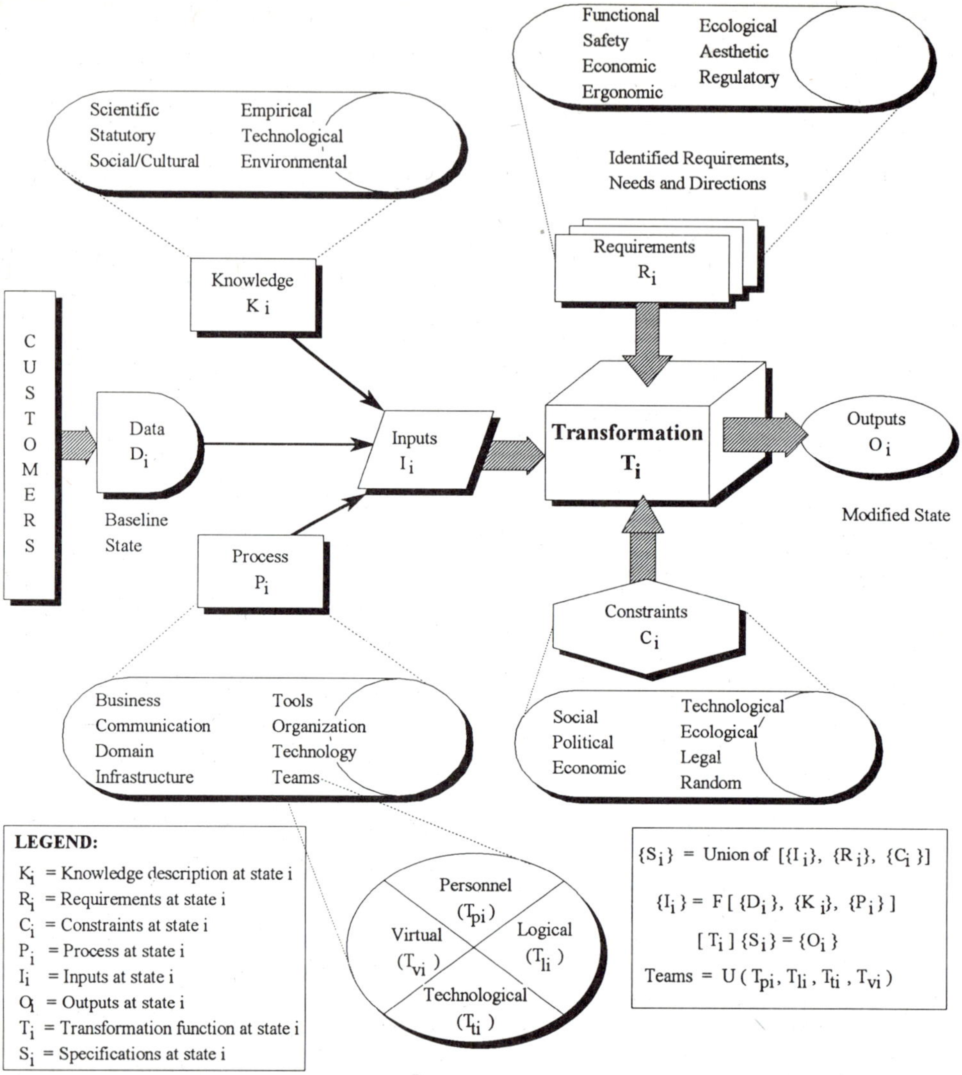

FIGURE 8.10 Basic Terminology in a Transformation Process

8.5.2 Transformation Process—A Process Taxonomy

One of the most important steps in integrated product development is the creation of a detailed taxonomy of the process which permits the concurrent production of an artifact. Process taxonomy is the set of functional building blocks which transform the specifications (inputs, requirements, and constraints) into an output (leading to a physical artifact). Taxonomy serves a very useful purpose since it embodies a basis of comparison of prediction with measurements. Rational prediction requires theory or criteria. Two common criteria exist for defining taxonomy [Stauffer and Slaughterbeck-Hyde, 1989]:

- ***Orthogonal:*** The elements of the taxonomy show orthogonal properties. The term orthogonal is used here in a qualitative sense rather than the usual mathematical sense (that is, vectors have a dot product of zero). That is, the series of taxonomies should be mutually independent (and thus separable) so that every category within the domain of transformation can be uniquely defined. Mutually separable means that each transformation state exhibits a minimum number of overlapping functions or specifications/outputs. This resembles Suh's first axiom [Suh, 1988] which states that good designs maintain independence of functional requirements. That is, to generate a good transformational strategy, the taxonomy must show independence. Each input, requirement, and constraint in the mutually separable series of transformation, affects only those functions of the product that are part of a single transformation.

$$\{T_i\} \cap \{T_j\} \cong [\text{A Diagonal or a Triangular Matrix}] \tag{8.32}$$

  The product of the two transformations, if they are orthogonal, should be either diagonal (which is called an uncoupled mapping) or triangular (which is called a half uncoupled mapping).

- ***Whole set:*** The taxonomy must be exhaustive over the domain of interest. That is, there would not be any need for recreating a new set of abstraction to accommodate a new design situation or a scenario. There are many alternate sets of abstractions that are possible over the life-cycle domain but some of them may be redundant. Most life-cycle situations can be accommodated with a "complete basic set" of abstractions. This idea is similar to Suh's second axiom, which states that the information content of the design should be minimized [Suh, 1988].

$$\cup [\{T_1\}, \{T_2\}, \{T_3\}, \{T_4\}, \ldots, \{T_i\}, \ldots \{T_n\}] \rightarrow \text{All possible outcomes} \tag{8.33}$$

The orthogonal criteria distinguishes between a bad (or unacceptable) mapping from a good (or acceptable) mapping. The "whole set" criteria is the method for selecting a basic set of mapping descriptions from among those which satisfy the "orthogonal" criterion. The two criteria, together, establish a uniform rational basis (theory) for classifying and characterizing the domains. By distributing specifications into independent categories, the information content (inputs, functional requirements, and constraints) can be minimized without compromising the independence of taxonomy. The goal of a rapid product realization taxonomy is to create a transformation strategy by which the product specifications for a manufacturing system can be transformed into three descriptions:

1. ***Concept Description:*** The behaviors and physical representations (models) of the transformation process form what is called "concept" descriptions.

2. ***Structure Description:*** Structure descriptions identify a class hierarchy for the concept representation.
3. ***Frame Description:*** The behavioral and physical specifications form what is called the "frame" description of a series of connected product and process units.

These descriptions must be included as a part of a "taxonomy" to come up with a transformational approach to product realization. The idea is to first systematically lay the specifications from its abstract descriptions to its detail descriptions. Then, at a particular transformation state, one must satisfy the specifications from the most important ones first, to the least important ones last.

8.5.2.1 Concept Descriptions

Concept description is for the purpose of representing the required functions of a product. This is a representation for the behavioral and physical models of each transformation state that this taxonomy represents [Eppinger, Whitney, Smith and Gebala, 1994]. There can be two types of concept descriptions:

- **Algorithmic:** The algorithmic concept description uses a set of analytical models to describe its behavior.
- **Abstract:** The abstract concept description uses set theory to capture life-cycle requirements such as specifications, design, manufacturing, marketing, etc.

The concept description is concerned with describing a wide range of functions of a product which include the following:

- The modeling of the inside representations for behaviors and physical characteristics
- The relationship between specifications and outputs of the transformation process
- The methods by which conflicts between the constraints can be detected and checked, which may encompass chaining—how the implications of a design decision are propagated to other parts of the system; the methods may detect violations among the constraints and conflicting decisions that may arise during product realization
- A conflict resolution strategy that will lead the design to a settlement—meaning how the detected conflicts are to be successfully resolved.

In summary, concept descriptions specify how a baseline system is transformed from its prior (input) state to a modified (output) state. It includes the physical principles to fulfill the functions as well as the corresponding concept or embodiment to capture its intent or knowledge. It includes methods by which conflicts can be detected and a resolution strategy to arrive at an amicable settlement.

8.5.2.2 Structure Descriptions

This views an assembly as a hierarchy of standard class structures and uses a strategy of known classes to devise a plan for the whole assembly. This identifies a class hierarchy within each concept representation (such as behavior or physical models). For example, instead of viewing an assembly as a set of parts, one can view it as composed of a set of class structures. Structure description could be a class representing standardized relationships among a set of parts. These descriptions typically form a class, which is made out of other classes. For example, a door structure is composed of a latch structure and two door hinge structures. A door hinge structure is composed of a hinge pin structure and two multiple screw plate structures. The multiple screw plate could be composed of an odd number of simple screws and a solid circular pin. The simple screw would be composed of a variety of contact (seating and positioning) structures.

8.5.2.3 Frame Descriptions

This is a method to capture the behaviors and physical specifications that a particular transformation state contains. While the concept descriptions of a product are characterizations of the transformation state at an abstract level (such as models), the frame descriptions define the boundaries of the concept descriptions within which it can operate. Along with the concept description of a product, a frame description imposes a set of behavioral characteristics so that product functions the way it does. When a team develops a product, it specifies three types of information, inputs a given baseline system, requires certain behaviors from the existing system, and imposes some changes or restrictions on the future behaviors of the system. In the nomenclature for taxonomy they are characterized as inputs, requirements, and constraints, respectively (see Figure 8.10). The frame representation provides a clear picture of engineering and manufacturing needs at each structure description level.

The transformation activities which form the core of the system are shown in Figure 8.10. It includes various inputs (data, process and knowledge), requirements, and constraints. The following list defines the basic frame items:

- *Inputs:* Inputs are everything needed to carry out a work-group process or a function. Inputs, $\{I_i\}$, include data $\{D_i\}$, knowledge $\{K_i\}$, and process $\{P_i\}$ at state i. The inputs at an intermediate state, assume the values of the previous state during transformation. Initial data is usually derived from environmental conditions, customers' requirements, or voice of the customers. Data may also be present as capital investments, people, money, materials or equipment. Another element of input is the process P_i at state i. The process includes teams, skill, and leadership, as well as the state of domain, infrastructure, tools, organization, information, communications, technology, and other business processes. K_i is the product or process knowledge that could be scientific, technological, empirical, statutory, social, cultural, or environmental (see Figure 8.10). The functions of data, knowledge, and process result in 7Ts (talent, tasks, teams, techniques, technology, time, and tools).

$$\text{Inputs} = \text{Function-of (Data, Knowledge, and Process)}$$

or

$$I_i = f\,[\{D_i\}, \{K_i\}, \{P_i\}] \tag{8.34}$$

- *Requirements:* These are defined as "necessary" statements of the characteristics of a domain object that are required to describe the domain, its features, or its multiple behaviors. The simplest form of requirements, R_i at state *i,* is a form or function features described in Section 7.4.3. Requirements may take the form of industry standards, materials characteristics, aesthetic considerations, human factors, safety, ergonomics, regulatory, ecological, or environmental conditions among others.
- *Constraints:* Constraints are the restrictions on the inputs and the requirements. Constraints, C_i at state *i,* restrict the behaviors of the domain object to function in a particular or specific way. Collectively, the constraints define what will be an acceptable outcome. The restrictions range from social, political, economic, technological, ecological, and legal, to random behaviors. The simplest form of constraint would be like assigning a value to a form or function feature. For example, if a countersunk feature of a product is defined as a requirement, the limits on tolerance define the constraints. Constraints can also be specified in the form of a parametric relationship, the feasibility limits of a set of design alternatives, and elimination strategy for unfeasible designs. In the context of engineering design, a constraint can be thought of as a required relationship among design attributes and features. Constraint may embody a design criterion (for example, stress, deflection, stiffness, etc.), a physical law (for example, $F = ma$), geometric compatibility (for example, mating of parts), production requirements (for example, no blind holes), or any other limitation set by one subsystem onto another related subsystem. Thus, in a coupled system, some constraints are part of "coupling functions" that exist between the units and are derived from "concept" descriptions. Constraints keep the team focused and enable them to make tradeoff analyses.

  The three discussed frame items are collectively called, herein, as a specifications set $\{S_i\}$:

$$\{S_i\} \equiv \cup\,[\{I_i\}, \{R_i\}, \{C_i\}] \tag{8.35}$$

- *Outputs:* Output (at any point) is an instance of an artifact object (at that point) with specific values for all parameters, functional and form features, and domain knowledge. Outputs, O_i at state *i,* contain the information of updated (transformed state) design. Intermediate outputs are intermediate states of products or services that are created by the work-group. The most mature form of an output is a finished good (artifact) or service, which is handed off to the customer. For any state *i:*

$$[T_i]\,\{S_i\} = \{O_i\} \tag{8.36}$$

The three descriptions (concept description, structure description, and frame description) define a transformation model within a state *i*. There would be a series of models, one for each transformation state, and a "broad system model" representing the top-level transformation. This is discussed in great length in Section 8.4 (Figure 8.7). A broad system model in this case, represents a conceptual semantic form of a total manufacturing system. Under a given set of constraints on technology and resource limits on the 4Ms (money, machines, manpower, and management), often the objective of the enterprise is to find an efficient transformation system that will maintain a constancy of purpose. This may be specified in objectives such as: (a) meet a given plan or delivery target; (b) comply with product and process specifications; (c) please the customers, etc. These objectives do not state the process that will be used for each intermediate transformation state. The global model represents a taxonomy for a manufacturing enterprise at the highest conceptual level. The final outputs are in the form of finished goods, services, waste, and corporate profits.

8.6 KEY DIMENSIONS OF A CE SPECIFICATION SET

The specification set (inputs, requirements, and constraints) is usually a part of a transformation (see Figure 8.11) which characterizes the state of the refinement of the object being designed. A synthesis loop is defined by the initial state of knowledge that exists about a product (say a baseline design) at the outset, analysis or evaluation of the current

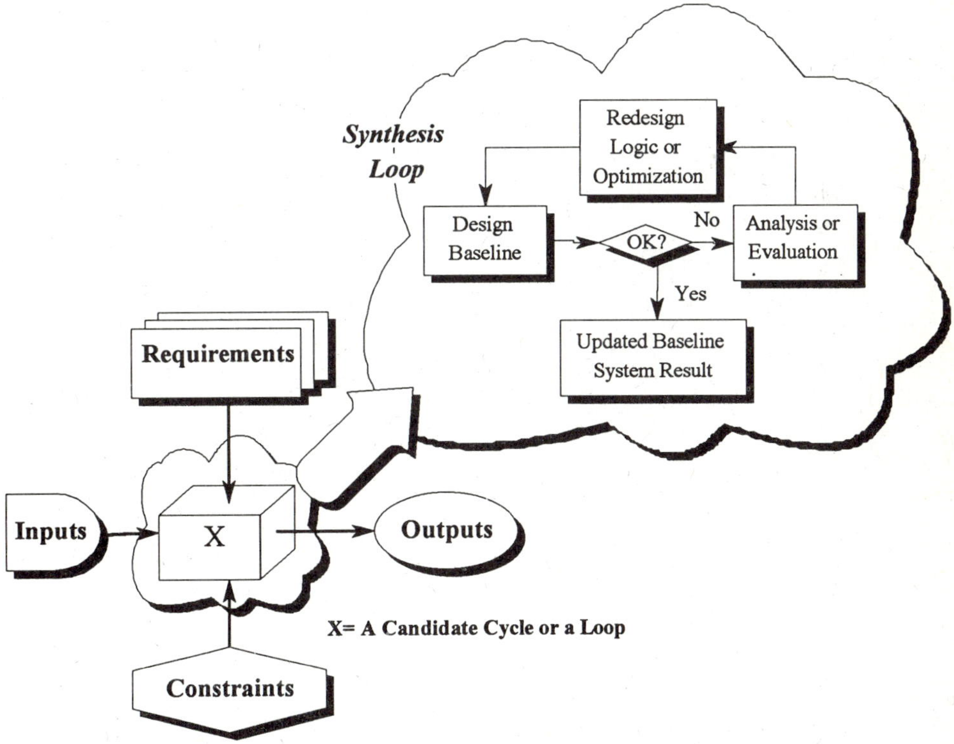

FIGURE 8.11 Typical Legend for Representing a Transformation Loop or a Cycle

state, proposed redesign logic or optimization, and an updated state of knowledge (updated baseline model). Specifications are not independently specified, but are governed by information, which not only describes the inputs (data, "domain knowledge," and the state of the object's process), but also its environment (constraints) and the value system (requirements) as shown in Figure 8.12.

- ***Environment:*** This includes domain (for example, design office, network tools, H/W, S/W, applications), organization, communication, and the infrastructure (the information surrounding the object in which transformation is occurring) as shown in Figure 8.12.
- ***Data and Knowledge:*** This is characterized by what state the object is in at the beginning of transformation. Domain knowledge and data partly define the scope of the object being transformed from its baseline state to a modified or improved state. The process completes the scope definition.
- ***Process:*** The output is dependent upon the PDD process of transformation, and its limitations on how the change in status (from specifications to outputs) is brought about. There are seven distinctive elements of the PDD process—7 Ts (talents, tasks, teams, techniques, technology, time, and tools) that need to be tied together

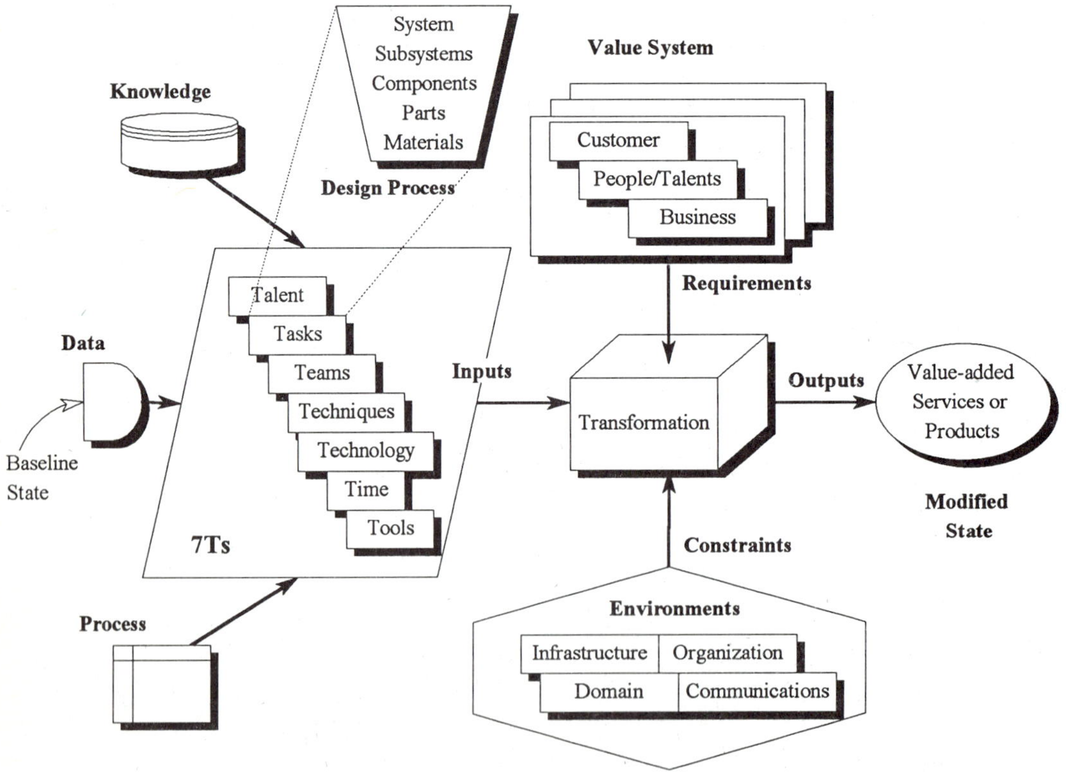

FIGURE 8.12 Key Dimensions of a CE Specification Set

and reshaped as inputs for change. Ts have a great impact on shaping the product outcome and laying down its strategies. Ts create opportunities and pressure to change and improve products, and strategies determine the success of product positioning in the global marketplace. Tasks include the product breakdown structures (PtBS) and process breakdown structures (PcBS). PtBS includes systems, subsystems, components, parts, materials, etc. PcBS includes methods, approaches, and the 3Ps (policies, practices, and procedures) that are used to accomplish activities in work breakdown structure (WBS). WBS activities, such as tasks, teams, and time are the center of this process definition. Teams include the participants shown in Figure 8.10—logical (Tl_i); personal (Tp_i); technological (Tt_i); and virtual teams Tv_i. That is:

$$\text{CE Teams} \equiv \cup\, ((Tl_i, Tp_i, Tt_i, Tv_i) \tag{8.37}$$

- ***Value System:*** The value system provides a cultural foundation to create a business based on the requirements obtained from the people/talents and the customers. The value system is like a three-legged stool. The three complementing legs of the value system are: customer, people/talents, and the business (see Figure 8.12). They support each other to provide overall stability.

$$\text{Value System} \equiv \cup\, (\text{Customers, People/talents, and Business}) \tag{8.38}$$

As discussed in Section 8.3, taxonomy is established through a series of transformations leading to the realization of a physical artifact (see Figure 8.9). The outputs of the transformation depend on the specific key dimensions of the CE specification set shown in Figure 8.12. The creation of a formal taxonomy using a finite set of transformations and a corresponding description of concepts, structures, and frames, can generate the descriptions of a physical artifact that maintain useful interactions between its multidimensional design space (such as subsystems or components).

8.6.1 Requirements and Constraints (RCs)

Product and process taxonomy is based on transforming the sources of information (called specifications) into outputs.

$$\text{RCs} \equiv \cup\, (\text{Requirements, Constraints})$$

or

$$\text{RCs}_i \equiv \cup\, (R_i, C_i) \tag{8.39}$$

Where, $\cup$ denotes union-of. Requirements and constraints are related hierarchically; that is, as one proceeds from one state of transformation to another, new RCs (which are local to a transformation domain) are derived from those found in earlier domains. Taxonomy uses RCs decomposition and hierarchy analysis to guide the generation of design alternatives to ensure that life-cycle concerns are explicitly incorporated in the product realization process. Furthermore, at each state, the relationships between the inputs, requirements, and constraints are complex, and, therefore, the transformation of specifications into outputs is an iterative process. A systematic and hierarchical representation of RCs is the key to capturing the design rationale and the design intent [Thompson and Lu, 1989].

8.6.1.1 Attributes of RCs

Characterizations of RCs, through identifying attributes, facilitate their hierarchical decomposition and analysis. There are three major attributes that characterize RCs:

- ***Dependencies:*** Dependencies are the directed pointers to the RCs which participate in forming relationships. A pointer identifies the RC (or its attributes) and a direction of dependency. In a two-object linear relationship, direction of dependency is either "input" or "output." In other situations it can be "low", "medium", or "high." In a single plane situation, it can be governed by angular attributes.
- ***Expression:*** Expressions are the enunciation of relationships between the RCs. A wide variety of forms are employed for expressing the relationship during product design development. They range from a natural form suitable for expressing "indefinite narration" (such as textual narratives, rules, goals, activities, etc.) to a more definite form (such as algorithmic, temporal relationship, or mathematical values). Definite forms can further be classified into "fixed" (or "equality") and "relaxable" (or "inequality") forms. Fixed RCs cannot be relaxed; they typically represent criteria that can make the design solution unacceptable if they are not satisfied. Generally, fixed RCs are equality types of relationships or expressions. Inequality forms, on the other hand, are more flexible. The design can be changed by adjusting RCs until the conflicts between the RCs are resolved. In a complex situation, expression may include the execution of a procedure or a program to form the basis of such relationships.
- ***Assumptions:*** Assumptions are the "limitations" on the validity of the expressions or the dependencies. Most assumptions are stated in implicit or logical type expressions, specifying the conditions or the circumstances in which a set of expressions is valid or invalid. In some sense, assumptions can be considered an extension of the "expressions" that were discussed.

8.6.1.2 Forms of RCs

Through the above attributes, the initial customer needs are transformed into a RCs hierarchy. A set of RCs during product development can exist in explicit, implicit, or logical forms (see Figure 8.13):

- ***Explicit Form:*** In most cases, customers desire several quantitative properties or specifications from their anticipated products, such as power, velocity, speed ratio, life, etc. These are called quantitative functional requirements (FRs). In the beginning of iteration, these FRs are usually assigned certain explicit quantitative values or ranges. This indirectly sets up an initial trial design configuration. Other design parameters are then governed by the magnitudes of these FRs. Actually, most traditional designs are determined from a set of quantitative FRs. In particular, mechani-

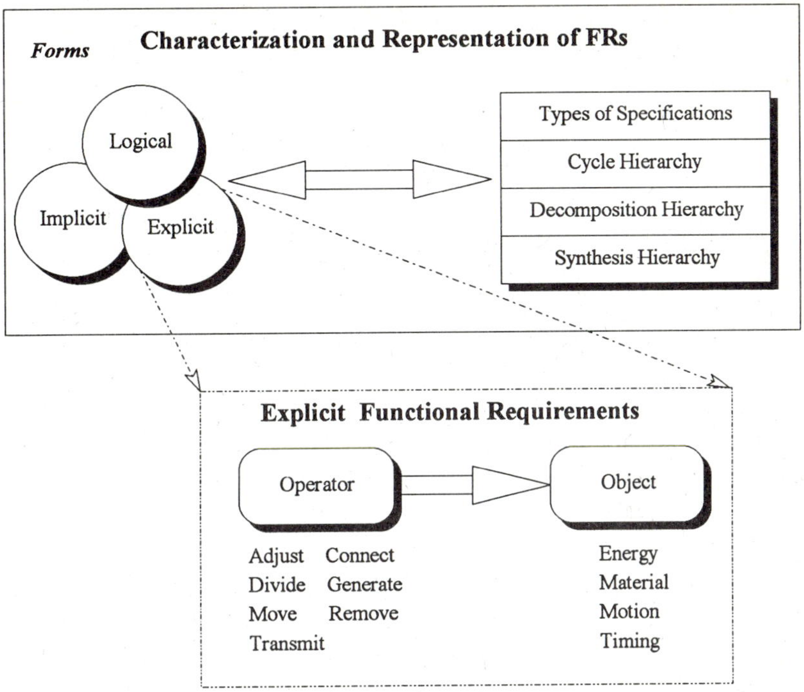

FIGURE 8.13 Characterization and Representation of FRs

cal engineering designers are mainly concerned with static, fatigue, dynamic, and thermal characteristics, and these characteristics are directly related to the quantitative FRs. Determining specification of products has long been recognized as weak links of any development process—concurrent or traditional. There is a recent CALS standard in the United States and a British standard aimed at addressing this issue: BS 7373 [BSI, 1991]. Pugh recognized this fact and identified a number of product design specifications to be considered prior to commencement of, or as a part of, the initial design process [Pugh, 1991]. Many examples of such product design specifications are listed in Figure 8.14. A quantitative FR can be represented by an operator applied to an object, as shown in Figure 8.13. The object can be energy, material, motion, and timing, and the operator can be adjust, connect, divide, generate, move, remove and transmit [Chen and Menq, 1992]. Each of these quantitative FRs can be satisfied by one or more design parameters. In other words, most design parameters are determined by these FRs from static, fatigue, dynamic, or thermal considerations. This process can be characterized as a mapping from a functional space (functional requirements set) to a physically designed space (design parameters set).

- ***Implicit Form***: Typically, implicit FRs come from initial statements of the objective that are often in abstract form—qualitative descriptions. Descriptions may be

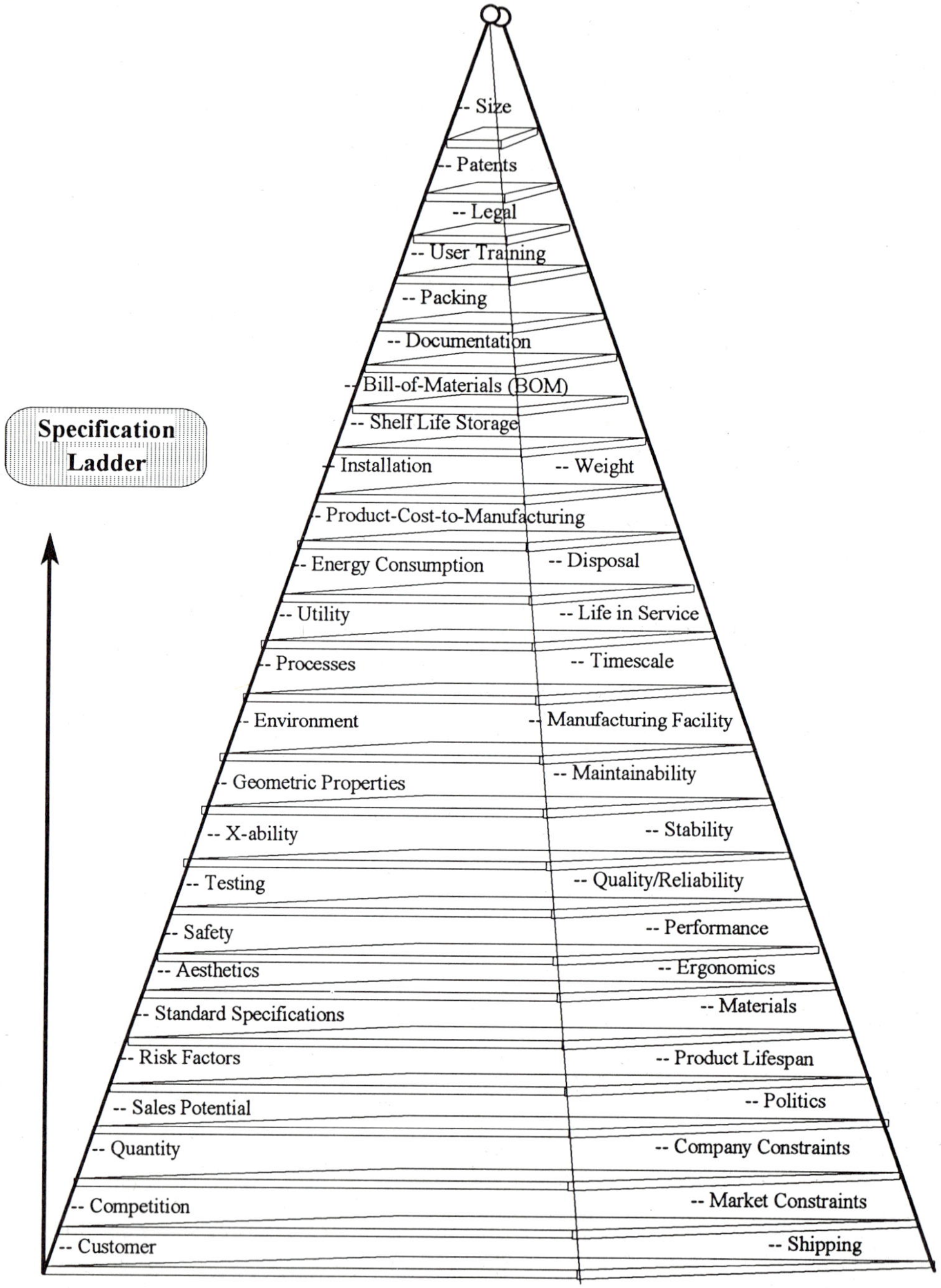

FIGURE 8.14 Examples of Product Design Specifications

brief implicit statements, such as, "the product x must be safe and reliable and be used by y for the purpose of z." In order to support engineering-related decision making, teams need to infer, from the abstract FRs, more precise objectives that lead to engineering-related activities. Most system constraints are simple logical computations, and are represented logically [Chen and Menq, 1992].

- ***Logical Form:*** Logical forms are those system requirements that specify the spatial relationship of the various components of a product or the interactions between them. For example, a logical form of RCs can be expressed by different assembly relations. They can also be generated from what the product or system intends to perform. For example, "surface A should contact with surface B" in an assembly. Usually logical forms are generated during the synthesis process of the product. Logical forms are useful for eliminating infeasible design alternatives.

As design evolves, RCs are added or retracted dynamically by different perspectives. Each RC or its form can be further decomposed into independent sets or subforms that can better serve different perspectives of the product cycle. Each perspective may have its own design model or limited understanding of models of other perspectives.

8.7 ARTIFACT'S INTENT DEFINITIONS

This forms the core of the broad system model. The CE system, in its entirety, promotes a complete change in the way products are designed and released. CE requires a complete change or restructuring of the processes of PDD process. These changes require the following steps:

- ***First,*** an organization must have an understanding of how to reconstruct enterprise functions and subsystems into a taxonomy of transformation for product realization. This is discussed in Chapter 9.
- ***Second,*** an organization must have an understanding of the meaning and value of high level/system level/end object metrics of measurement, called value characteristics matrices (VCM), and how to incorporate these VCMs into product taxonomy. This is discussed further in Chapter 2, Volume II. There are many types of metrics that are commonly employed during concurrent product and process design. They range from team building, project management, assessment types (such as reliability engineering, competitive analysis, quality characteristics (QC) matrix), to measurement types, such as SPC, to other related management and engineering metrics and methodology.
- ***Third,*** an organization must have an understanding of an integrated product development (IPD) methodology. This is discussed in Chapter 4, Volume II. An IPD methodology includes mechanisms for capturing purpose, boundaries, and goals into a set of VCMs as a part of the PtBS, PcBS, and WBS (product, process, or work breakdown structures, Figure 8.1) so that they are appropriately linked with

the product, process, or work specifications at that level. The methodology of realizing specifications and RCs in a transformation history is discussed in Section 8.5.2.

8.7.1 Product and Process Management Cycles

In CE, it is required that the processes be organized in hierarchical activity sets so that they can be executed concurrently. The middle activity for artifact's intent definitions, shown in Figure 8.7, is a large single block. It may have to be branched into a tree structure hierarchy, so that each can be adequately dealt with at its own level. This process of branching is a part of a methodology which is commonly referred to as "branch and bound" methodology. A second level branching of this block is shown in Figure 8.15. The artifact's intent definition block has been branched into two coupled cycles: product management cycle and production management cycle. The two cycles are shown in Figure 8.15 in a shaded cylinder drum. Similar to the setup in Figure 8.7, the drum is still flanked by the same type of blocks around all four sides. On its left side it is flanked by system specifications, on its right by system optimization, on the top by system planning and scheduling, and on the bottom by comparison to goals and objectives. Two new blocks have been added on the top: functional requirements and proposed modification. Functional requirements provide the needed classification when one considers size, complexity, degree of innovation, and management of risk. The decision block in the bottom of Figure 8.15, ensures that all the goals and objectives are satisfied, otherwise, a new design modification is proposed and the process is repeated. The proposed modification block provides the change that must be introduced in the product and process to comply with system specifications and to meet the required goals and objectives.

In order to maintain a "constancy of purpose" it is essential that the two artifact cycles be further branched down into smaller pieces. The redefinition of cycles is governed by the same "branch-and-bound" terminology as previously discussed in Figure 8.11. This results in multiple loops of the product and process design (see Figure 8.16). The loops are actually a part of a larger system called product realization taxonomy which is discussed in greater detail in Chapter 9. The existence of loops is mentioned here to point out their dependence on the "whole system" concept. Interactions take place between the teams and the loops, and the loops themselves. The two key characteristics of this loop methodology are "iteration" and "integration". The process of iteration is a good classical feature of many implicit-type problem formulation and solving methodologies. Iteration through the loop enables concurrent teams to try imposing different sets of metrics or measures at each pass, which may in turn reveal problems not seen before. This may trigger applying additional measures to use in the next pass making the design solutions even more robust. Teams begin to gain confidence in the approach and the solution. Loops also improve cooperation between the CE teams since each loop is specific to a work-group and has shared interfaces which support communication between different CE teams.

The bounding of the loops in a product development process are shown in Figure 8.17. The CE team can perform each of the five loops in an ordered sequence of overlaps

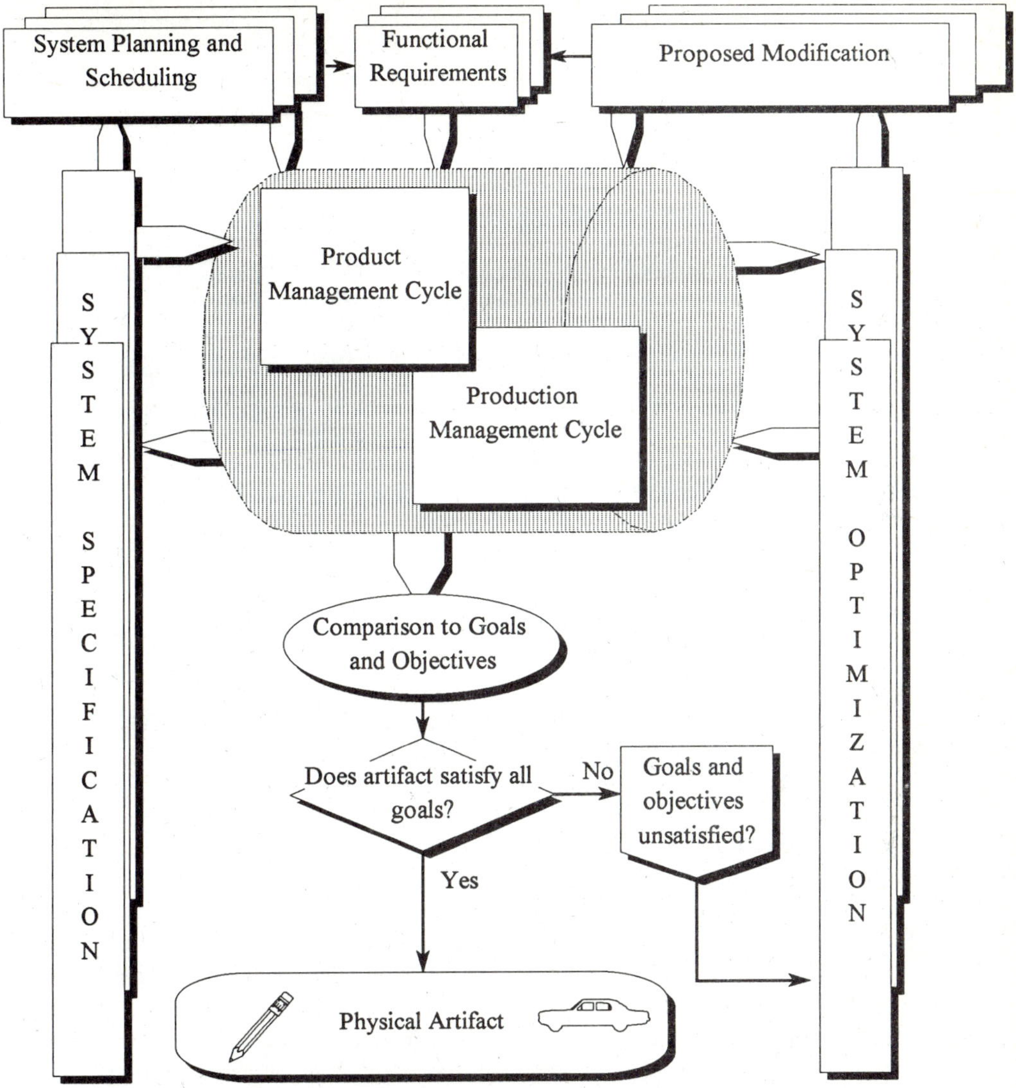

FIGURE 8.15 A System Level Redefinition of a Product Realization Process

as shown. A partial output for one loop becomes input for the next. Each loop is shown to follow a previous five block schema (Figure 8.11) with core in the middle surrounded by requirements to its left, constraints to its right, inputs on its top, and outputs on its bottom. The core is represented by a predictor-corrector logic. It consists of three interconnected elements: a model, a predictor, and a corrector. The baseline state (shown by a parallelogram icon) represents the model. A decision-logic (shown by a rhombus icon) serves as a corrector. A set of supporting analysis or assessment tools (shown by a matrix icon)

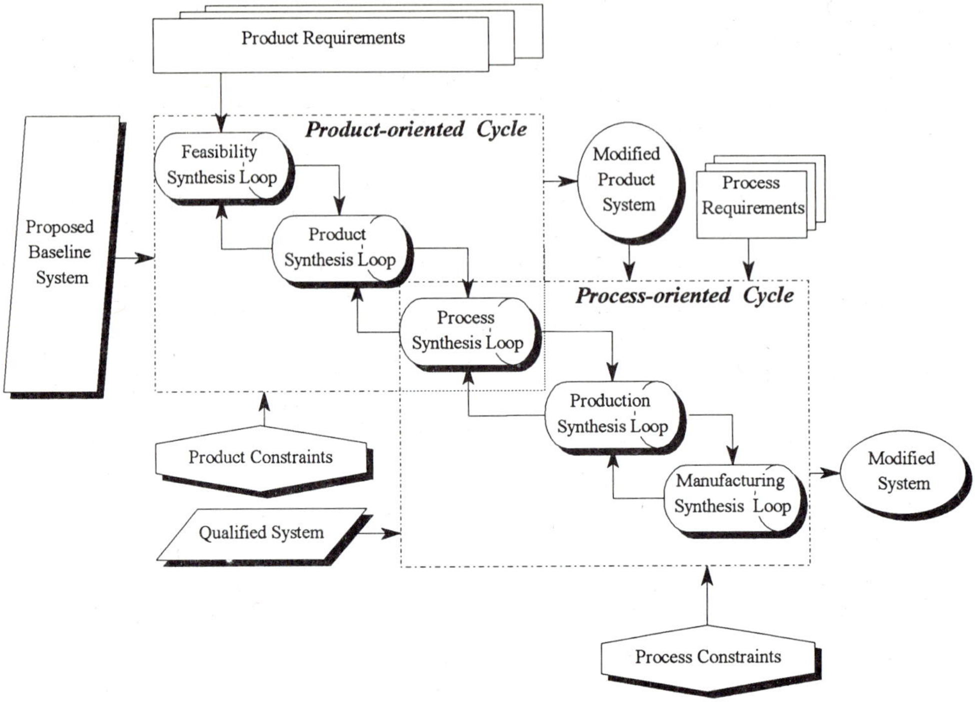

FIGURE 8.16 Branching of Product and Process Management Cycles into Loops

serves as a predictor. The requirements are identified as a D-shaped icon and constraints by an ellipse icon, as shown in Figure 8.17. In the product synthesis loop, for example, "design requirements" represent the requirements block and "user insight, vendor capability" as a constraint block. The input to the product synthesis loop is the output coming from the decision block belonging to the feasibility synthesis loop. During each loop, the team feeds in the requirements and the recurring output from the last loop to the core system. The predictor-corrector logic of the core system carries out a series of required analyses, and with the help of the built-in model logic, recommends a new set of parameters to be tried in the consecutive cycle. These output parameters serve as a modifier to the new baseline model and, together with requirements and constraints, they form a trial design for the next cycle. Information flow can be affected by many factors, such as scrap, rework, absenteeism, or machine breakdowns. Consequently, planning adequate capacity through the loops is a necessary activity, but not enough in itself. Metrics of measurements for actual transformation efficiency throughput, and reaction to the inevitable discrepancies must be imposed to maintain a balanced throughput, and capacity content throughout the loops. This measurement and reaction activity are represented by the VCM metrics, linked to the loops. Chapter 2, Volume II discusses the VCM metrics and measures of merit in a greater detail. There are two levels of bounding in this system:

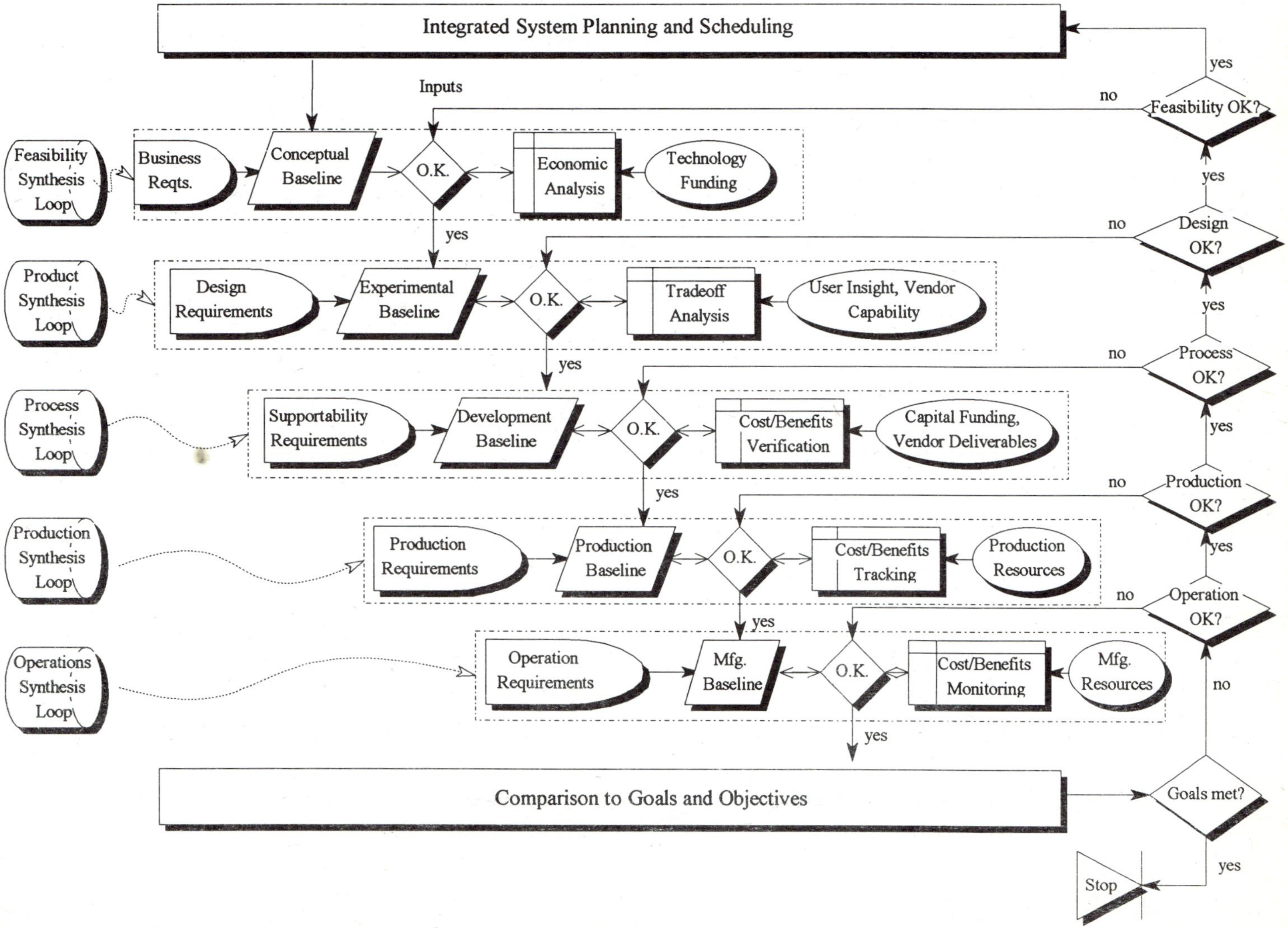

FIGURE 8.17 Bounding of the Branched Loops

- The first level of bounding is caused by the interloop iteration—decision block in the predictor-corrector logic. This interloop iteration is devoted to feeding the new inputs to the representative baseline model of the cycle, analyzing the baseline concepts, and determining their validity with respect to the unsatisfied goals and objectives.
- The second level of bounding is caused by the series of intraloop iterations. There is a set of decision-logic shown to the right of Figure 8.17 which carries the intraloop iterations. Design updates from any loop may nest through one or more of the five loops depending upon whether outputs (a set of goals at a particular loop level) are met or not. If all the goals from the feasibility loop to the operation loop are met, the resulting system is considered converged.

If the goals and objectives are all met, the state of transformation of the two mangement cycles and the loops at this point should yield the physical artifact. Several additional steps are required to further reduce design and development time:

- **Minimize** the number of product and process interfaces, as discussed in Chapter 4, so that activities can be initiated simultaneously.
- **Consider** the partial and/or complete overlapping of activities wherever possible.
- **Exploit** the overlapping so that a synthesis loop, which requires more lead time, can be initiated prior to the synthesis loop from which it requires some inputs.
- **Reduce** the amount of feedback by constantly comparing synthesis loop outputs to goals and objectives. If the output of a synthesis loop is not consistent with goals and objectives at that level, then it might be desirable to take care of the unmet goals locally before moving on to the next synthesis loop.

8.7.1.1 Loops of Product-oriented Cycle

The three loops of product-oriented cycle, shown in Figure 8.16, deal primarily with such things as estimated weights, costs, materials, inefficiencies, sizes, and performances:

- The ***feasibility loop*** iteration takes place on its own with the business and performance requirements and technology and funding constraints placed on a candidate (conceptual baseline) system. The outputs from this loop determine a set of design parameters for an experimental baseline design. If advanced technologies are required, plans are made to ensure that the next two loops are carried out to minimize risk.
- During the ***product synthesis loop,*** experimental baseline is iterated rapidly until the team emerges with a set of feasible technology that meets product synthesis loop requirements.
- During the ***process planning of synthesis loop,*** developmental baselines are iterated rapidly until the team emerges with a feasible concept that meets process synthesis loop requirements. The relation between the functional requirements and process variables is an example of a product planning QC matrix. The process planning QC

matrix describes the relations between process variables of applicable manufacturing processes.

This completes the first cycle of the iterative two-cycle optimization process. The following is a list of fundamental issues for this product-oriented cycle:

- **Reduce** product complexity before worrying about cost. This may involve combining parts and processes and evaluating alternatives.
- **Study** the design alternatives at this stage, when design experimentation is easy.
- **Focus** on multiple objectives simultaneously, like unit costs, number of parts, material alternatives, and alternative manufacturing methods, while evaluating design options at this stage. Focusing on a single objective, say, reduction in parts' costs, may dictate a design change from sheet metal stamping to plastic components. Consideration of other critical performance (for example, durability) may drive it in the opposite direction (see Figure 8.18). A combination of the two may drive it to a new alternative manufacturing method.

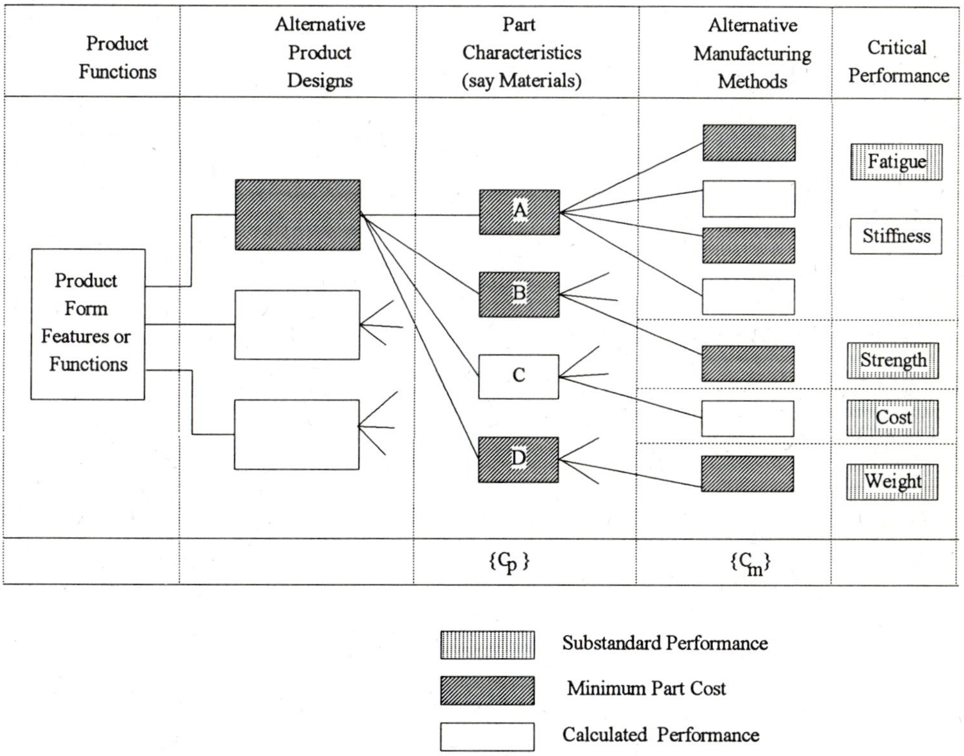

FIGURE 8.18 Examples of Information Generated During Product-oriented Cycle

- **Consider** net savings through the entire product-oriented cycle, not the unit cost savings from each loop.
- **Utilize** analytical techniques, such as crude "tradeoff analysis" or "cost/benefit verification," if applicable, to aid in decision making.

8.7.1.2 Loops of Process-oriented Cycle

There are three additional loops in a process-oriented cycle:

- The **process execution loop** is actually an implementation part of the process synthesis loop. The planning part was described as a part of the product-oriented cycle. At the end of this loop a production baseline or a trial design is obtained.
- The **production synthesis loop** begins by feeding the production requirements to this production baseline and tracking the cost/benefits outputs for different candidate processes. A manufacturing baseline design, which meets the goals, is approved for operations checking.
- During the **operation synthesis loop,** the operation team feeds the operation requirements to the manufacturing baseline and monitors the cost benefits. The loop ends if the series of goals and objectives applied at each loop (which are derivatives of the system goals) are met.

Any product development system seeking to provide process compatible design will be necessarily bound by constraints from the process subdomain. It should satisfy two types of constraints: product constraints (such as topological and geometrical description constraints on part geometry) and process constraints. Generation of process constraints for real world parts, in an activity such as process plan, requires extensive reasoning, both heuristic and analytical. The product design must reflect the needs of process design. However, once the product is modified, a new process plan must be generated. Each modification introduced in product design has the potential for requiring a complete regeneration of process design.

A concurrent design would require that the design team consider process constraints while the process planners consider design constraints. A designer waiting for a new process plan after each change made to a product design is no good. This is like reverting to a serial mode of operation. Unfortunately, time, costs, and ergonomics place severe restrictions on the amount of such planning and cooperation that can occur. In the initial stages, therefore, the cycle must be based on high-level heuristics information that is captured in previous designs or is available from technical memory.

The activities and steps associated with an IPPD process are shown in Figure 8.19. Familiarity and knowledge of the system, its constituents, and their interactions, are necessary to have the greatest impact. There is no definite and absolutely correct way of defining the steps in each loop. Steps will be developed iteratively over an extended period of time. They will also have to change over time as product lines change, product functionality changes, design and manufacturing technologies change, and so forth. In other words, the process of redefinition should never be thought of as numbered. It is a

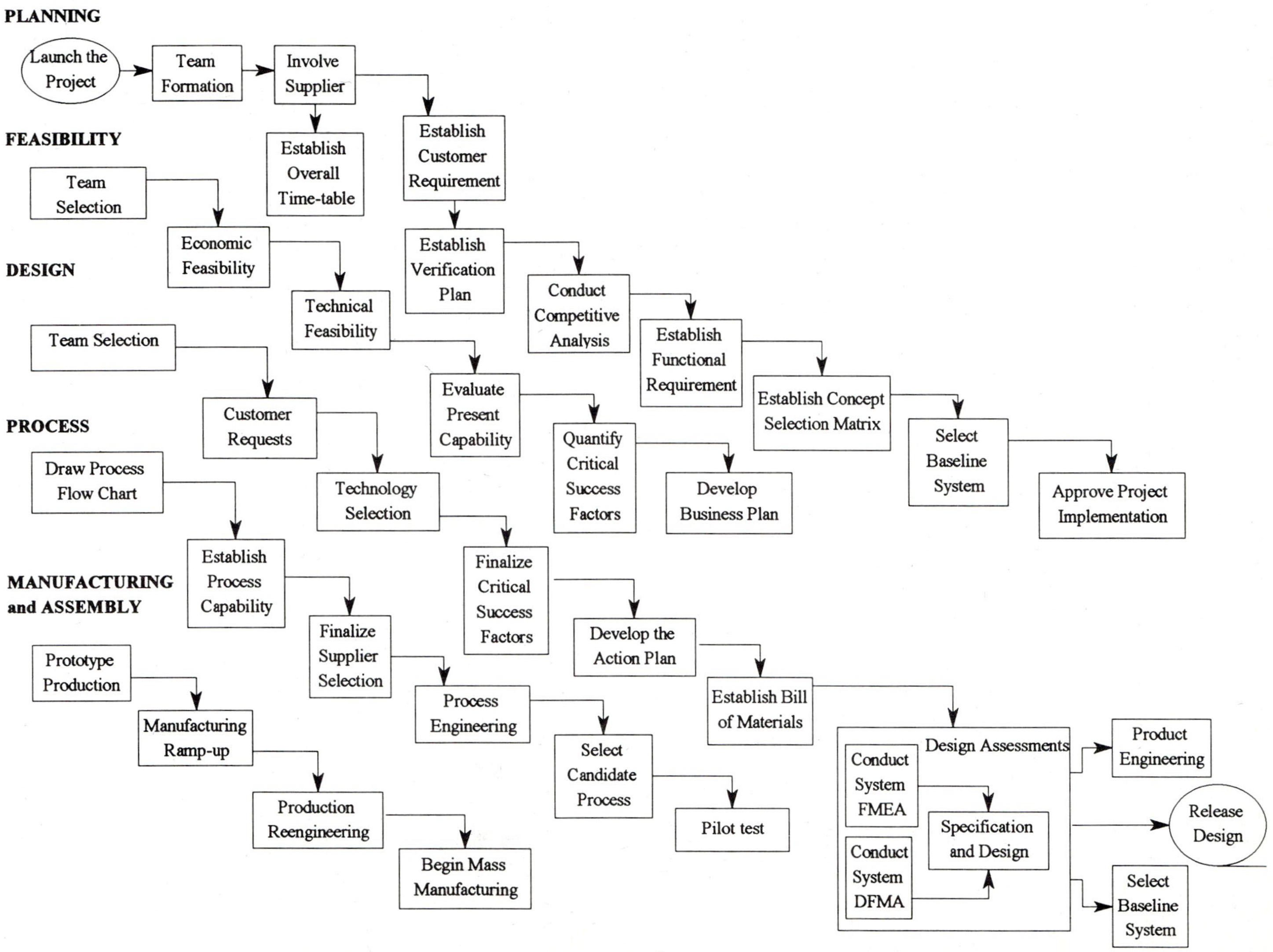

FIGURE 8.19 Activities and Metrics Associated with Each Loop of the IPPD Process

case for system optimization and should be treated accordingly by its participants and those who are managing it.

REFERENCES

BSI. 1991. *BS 7373—Guide to The Preparation of Specifications.* Milton Keynes, UK: BSI.

CHEN, B., and C.-H. MENQ. 1992. Initial Attempts On the Characterization of Functional Requirements of Mechanical Products. *Concurrent Engineering,* Proceedings of the Winter Annual Meeting of the ASME, Edited by D. Dulta, et al. PED-Vol. 59, Anaheim, CA., pp. 315–329.

DEMING, W. E. 1993. *The New Economics*, Cambridge, MA: MIT Center for Advanced Engineering Study.

EPPINGER, S. D., D. E. WHITNEY, R. P. SMITH, and D. A. GEBALA. 1994. "A Model-based Method for Organizing Tasks in Product Development," Research in Engineering Design, Vol. 6, No. 1, pp. 1–13.

FINGER, S., and J. R. RINDERLE. 1989. "A Transformational Approach to Mechanical Design Using A Bond Graph Grammar", *ASME DE-Vol. 17,* Edited by Elmaraghy, Seering and Ullman, Proceedings of the ASME Ist Int'l Conference on Design Theory and Methodology (DTM '89), Montreal, Quebec, Canada, Sept. 17–21, pp. 107–116.

GUTWALD, M. J. 1988. "Concurrent Product/Process Development—A Key to Competitiveness," *SAE Transactions,* Vol. 97, Section 5.

HARKINS, J. R., and M. P. DUBREUIL. 1993. "Concurrent Engineering in Product Design/Development," *Plastics Engineering,* Vol. 49, No. 8 (Aug.), pp. 27–31.

NEVINS, J. L. and D. E. WHITNEY. 1989. *Concurrent Design of Products and Processes,* New York: McGraw-Hill, pp. 2–3.

NIELSEN, E. 1990. Designing Mechanical Components with Features. Ph. D. Thesis, University of Massachusetts, Amherst, MA.

POURBABAI, B., and M. PECHT. 1994. "Management of Design Activities in a Concurrent Engineering Environment," *International Journal of Production Research,* Vol. 32, No. 4 (Apr.), pp. 821–832.

PUGH, S. 1991. *Total Design.* Wokingham, UK: Addison-Wesley Publishers.

SIMMS, R. 1993. "CE: Engineering a Change in the Design Process," *Aerospace America,* Vol. 31, No. 4 (April), pp. 18–22.

STAUFFER, L. A., and SLAUGHTERBECK-HYDE. 1989. "The Nature of Constraints and Their Effect on Quality and Satisficing." *ASME DE-Vol. 17, Design Theory and Methodology—DTM '89,* Edited by Elmaraghy, Seering and Ullman, Proceedings of the ASME Ist Int'l Conference on Design Theory and Methodology, Montreal, Quebec, Canada, Sept. 17–21, pp. 1–8.

SUH, N. P. 1988. *The Principles of Design* Oxford, UK: Oxford University Press.

SZUCS, E. 1980. *Similitude and Modeling.* Elsevier Scientific Publishing Company.

TAYLOR, D. A. 1993. *Object-Oriented Technology—A Manager's Guide.* Reading, MA: Addison-Wesley Publishing Company.

THOMPSON, J. B., and S. C.-Y. LU. 1989. "Representing and Using Design Rationale in Concurrent Product and Process Design." *Concurrent Product and Process Design.* Proceedings of the Winter Annual Meeting of the ASME, San Francisco, California, DE-Vol. 21. PED-Vol. 36, Dec. 10–15, pp. 109–116.

WHITNEY, D. E. 1988. "Manufacture by Design," Harvard Business Review, Vol. 66, No. 4, pp. 83–91.

TEST PROBLEMS: THE WHOLE SYSTEM

8.1. Explain the systems thinking behind a virtual product development process. How does government conduct these synthesis loops for their own process?

8.2. What is the system level definition of the product realization process? What are the benefits of incorporating synthesis loops into the product development cycle?

8.3. What do you understand by "system"? What constitutes a whole IPPD system? Give examples from manufacturing and nonmanufacturing environments.

8.4. Illustrate a scenario for a whole CE system at your workplace. Explain the differences between interactions (for example, coupled, dependency, etc.) and constraints.

8.5. List a set of typical deliverables in a conventional design and development process.

8.6. What are the key success factors for a good production transformation system? Why is it essential to look at the problem as a whole rather than only at its subsets?

8.7. State the main differences between a CE transformation system and a serial engineering system.

8.8 How does a CE enterprise system taxonomy differ from a typical production system? What are its inputs and outputs? What does concurrency do to the specifications?

8.9. How do you describe a basic transformation process? What is the common terminology used?

8.10. Explain branching-and-bounding methodology. Why are cycles branched into loops? Can we achieve the same results without branching and bounding? If so, how?

8.11. Show an example to illustrate the branching of cycles into loops and how they are bounded back into a whole system.

8.12. What factors influence the product management cycle? Why it is essential to decompose, work, product, process, system, and enterprise into hierarchical sets?

8.13. Figure 8.4 shows a typical production system. Draw a similar chart that would correspond to (a) mass manufacturing case, (b) agile manufacturing case, and (c) virtual manufacturing case.

8.14. Why do you require a broad system model to understand the artifact's intent definitions? Describe its five major components. How are the goals of the CE enterprise and the constraint boundaries defined?

8.15. Describe a model that can optimize the life-cycle of an artifact. Identify a set of goals and objectives for realizing your company's vision with regard to product development. Why is it essential to define a set of consistent goals?

8.16. What is system optimization? Why is it essential to define product values that weigh and balance both company and customer interests?

8.17. Looking at Figure 8.6, where do you see goals that are accommodated or changed?

8.18. Looking at Figure 8.4, explain why feedback is essential. Can feedback go back to change the requirements or the constraint types?

8.19. On the basis of Figure 8.18, generate a design cycle loop for an automotive cylinder head. Identify and list a set of CE specifications for an engine component.

8.20. What is the core of a broad system model? How does the loop (subset of a cycle) methodology help support this broad CE concept? Describe the major loops for product-oriented and process-oriented cycles. Does this change from one product case to another?

8.21. Provide a list of potential gap examples which best illustrate a lack of common goals or lack of constancy of propose in an organization.

8.22. Describe two major types of system specifications. If system specification is considered as a tree, enlist some of its nodes. What are the major attributes that characterize requirements and constraints (RCs)?

8.23. Describe the characteristics of the transformation process in product realization taxonomy? How is the functional description mapped into a physical artifact? What is the basis for choosing the number of transformations? What ensures that the intermediate states would be mutually separable? What defines a good transformation strategy?

8.24. If system specification is considered a high level input for this transformation, describe the various types of requirements that one may encounter in IPPD.

8.25. What are the key dimensions of a CE specification set? During product development, what are the various forms in which RCs can be present? Give some typical examples of product design specifications.

8.26. Are product and process requirements related to system requirements? Describe the relationship. Why are such relationships beneficial from a system optimization perspective?

8.27. Does the content referred in Figure 8.9 ever reach the enrichment level of a full or complete content set? What is the result when that happens?

8.28. What are the different parts of knowledge required during IPPD?

8.29. Describe three key strategies for synthesizing the taxonomy of product realization. Why is knowledge capture considered one of its strategies?

8.30. During the transformation process, why are concept, structure, and frame description models needed? What are these descriptions? How does the use of the description model help in meeting rapid product realization goals?

8.31. What is a process taxonomy? Describe two criteria that are commonly used for defining taxonomy.

8.32. Does each disciplinary team need to understand the product realization process?

8.33. Why is it necessary for the enterprise team to understand requirements, specifications, interdependencies, and priorities of the product?

8.34. List a set of activities and metrics associated with each loop of an IPPD process.

CHAPTER 9

PRODUCT REALIZATION TAXONOMY

INTRODUCTION

Product or process development is a complex activity. The complexity arises from many sources, including the multiplicity of functions it usually provides. A mechanical product or an artifact, for example, serves many different functions, such as transportation, consumer needs, control and measurements, generation or conversion of energy, etc. By mechanical artifacts we mean engineering systems, which include mechanical, hydraulic, pneumatic, electrical, electronic, magnetic, and thermal units and effects, as well as control subsystems and other feedback mechanisms. Complexity also results from the fact that most mechanical and electrical products are composed of highly integrated (tightly coupled) units. Similar to what is shown for tasks in Figure 4.6, the units are interdependent. Each unit may contribute to several required behaviors, and a single required subsystem behavior may contribute to several required subunit behaviors [Finger and Rinderle, 1989]. Also, there exists strong coupling between independently specified objectives of the product (such as weight, cost, quality, performance, throughput, etc.). In these cases, design decisions are not governed by a single consideration. One-to-one translation from an independent set of organization goals to the goals of its operating units (with the constancy of purpose) is often difficult, if not impractical. Complexity is also introduced by the manufacturing process used and the manner in which one designs and develops a product. Even with a simple design, the success in realizing a product requires a very large, focused, labor-intensive workforce. Potentially, engineers from various disciplines are called upon during complex decision making. In new product development, most designs evolve—from a set of qualitative and incomplete specifications to a more quantitative and exact description—through a number of stages, each with a definite focus

[Larsen and Alting, 1992]. Successful, rapid, product realization requires decision making early in the cycle. Information at early stages often lacks detail and complete specifications. This means many critical design decisions, for the product or process, have to be made when only incomplete information is available. Identification of steps in the process, when such decisions ought to be made, and their corresponding level of incompleteness can serve many useful purposes. Many design process models have been developed in the past as block diagrams to characterize the design process and to provide the design engineers with a somewhat defined procedure for applying available design techniques [Hales, 1993]. In the 1990s, there was renewed interest in the human activity of engineering design. Pugh proposed design activity models representing a set of engineering activities within an overall management framework [Pugh, 1990]. The approach provided a useful model dimension toward development of conceptual means for product realization. However, until recently such models and associated techniques have been all but ignored in practice. Engineering design has tended to be something that was just done by somebody in the drawing office [Hales, 1993], without regard to any well-structured pattern. Each designer in the drawing office held different views of the design and each approached the work in different ways.

A taxonomy-dimension phrase is introduced in Chapter 8 to mean the abstractions and formalisms used to express product realization knowledge, in an implementation independent manner. Taxonomy describes a pattern based on CE invariants, structured techniques for problem decomposition and integration, tools, methods, and successful 3Ps [Dixon, Duffey, Rohinton, Meunier, and Orlup, 1988]. A taxonomy dimension advances a systematic approach for achieving a consistent set of product realization goals. The scope and content of the taxonomy for product realization are implicit in the following:

- ***Types of Information and Sources:*** In a complex process such as integrated product and process design (IPPD), the information base is heterogeneous [Urich and Eppinger, 1994]. Information differs in form, content, representation, relation, and structure. Data is fragmented. For example, a bill-of-materials (BOM) comes from product description. Information (as in data, process, and knowledge forms) comes from various related sources and domains. There is seldom a record showing where they are originated, or why they are important. It is difficult to verify if necessary requirements have been considered and whether or not there are constraint redundancies or requirement omissions. Documentation comes from multiple sources: CAD files, dimensional component drawings, three-dimensional solid models, sketches, process charts, etc. It is difficult to maintain consistency.
- ***Complexity of Tasks:*** Many design tasks are interdependent and cannot be separated from each other. Parallel scheduling of tasks is particularly difficult since tasks cannot always be separated. For instance, the bill-of-materials (BOM) is common to both product design and MRP (materials resource planning) implementation. If MRP starts when product description is not complete, the launch of MRP has to be delayed until some surrogate BOM data is made available or the product description is completed. Solutions to such problems require more convergence-

time since many iterations of the information transfer have to take place back and forth.

- ***Degree of Incompleteness and Ambiguity:*** The third dimension is the degree of in completeness of a task while the information is being processed. Task ambiguity refers to the degree of uncertainty—whether a process or the content of the task can be reliably anticipated or whether the task behaves in uncertain or unpredictable ways. Questions asked might include the following:
  - Is it better to use an approximate analysis to predict the behavior early in the game or wait until information is complete?
  - Is it better to proceed with the analyses with a lot of information taken from the last design iteration?
  - Is it better to delay the information-rich analyses until the design is somewhat matured or stationary?

  The most effectively designed product can still suffer from cost overruns, loss of market share, quality problems, etc. These problems are frequently caused by the inaccuracy, ambiguity, impreciseness, inconsistency, or incompleteness of the available information, resulting in unnecessary iterations.
- ***Timing of Decision Making:*** Normally, decisions that are made during initial stages commit about 70% of the total cost of production. At the same time, the flexibility or the degree of control to activate a change decreases as one moves away from the initial stages. For example, when produceability and supportability issues are deferred until the manufacturing stage, the design change could be very costly. This is discussed in Chapter 2 in great length.
- ***Order of Decision Making:*** This issue becomes relevant during parallel scheduling of decomposed tasks. In parallel scheduling it is important to specify when critical decisions have to be made and to what level of completeness. In CAD/CAM-based design, there is a heavy pressure on teams to produce drawings quickly. This forces designers to create a depth-first detail design ignoring the measures of merits. Designers bypass the essential scheduled DFX checks. This often does not yield a rigorous design. If critical aspects of the work content and its scheduling order are accurately followed, design intent or life-cycle intent can be preserved.
- ***Communication Mechanism:*** The sixth dimension is to identify and communicate the different types of information even earlier in the life-cycle since, in CE, many different types of information must be considered in parallel. Unfortunately, however, there is no good tool to estimate product costs early in the design process. When costs are ultimately calculated, it is often too late to introduce major design changes. Even with the most effectively designed product, cost overruns and quality problems occur when information is not communicated early enough to the work-groups requiring it.

The level of complexities of the IPPD process differs depending on the types of information to be processed and its standing with respect to the above six dimensions.

Today, a serial engineering (SE) process, to a certain extent, can handle incomplete and uncertain cases (as enumerated in the first three dimensions). This is because a human is capable of working with incomplete information. Teams are capable of extracting meaningful concepts from the available data and ignoring unrelated details. However, the latter three dimensions add additional demands to product and process development (PPD), which cannot be easily accommodated using SE. The CE taxonomy has six dimensions: types of information and sources, complexity of tasks, degree of incompleteness and ambiguity, timing of decision making, order of decision making, and the underlying communication mechanism. In CE, information (embodied in data, process, and knowledge forms) occurs with different levels of abstraction and granularity. The ability to handle, coordinate, and communicate effectively—heterogeneous, qualitative, and incomplete information and their dependencies—is a major difference between a CE and SE concept. The taxonomy of concurrent product realization defines a set of systems and processes that have the ability to handle incomplete and partial information, and the ability to communicate it effectively to both upstream and downstream organizations. This is accommodated using a series of computer abstractions (models); this is similar to how the human cognition process works when looking at an incomplete design. The human cognition process does not arrive at the final design in one shot. We, as humans do this in several stages. At each stage we look at an abstraction of the design working with data that are necessary at that stage. Two conclusions can be drawn from this observation:

1. *Completeness of a design depends on its level of abstraction:* A design is incomplete when viewed at a lower level of abstraction or granularity, but may be complete when viewed at a higher level of abstraction.
2. *Not all data are meaningful at all levels:* A design data is complete when viewed at a certain angle but may be incomplete when viewed from a different angle. In other words, a design data is "incomplete" only with respect to the richness or level of granularity.

Concurrent product realization taxonomy (CPRT) can use the above observations in defining concept descriptions (see Section 8.5.2.1). In the fuzziness of the those six dimensions, coupled with the heterogeneity of the information base, it is challenging to define a well-connected CE process taxonomy for product realization. This is discussed next.

9.1 DEVELOPMENT METHODOLOGY FOR CPRT

Inputs (I_i), requirements (R_i), and constraints (C_i) are collectively named, herein as specifications:

$$\{S_i\} \equiv \cup\,[\{I_i\}, \{R_i\}, \{C_i\}] \tag{9.1}$$

Specifications, S_i belongs to a reference frame description of a transformation state, which is represented using concept descriptions. At a point when the product design is first conceived, specifications are usually present as an incomplete set.

$$\{S_i\} \exists \{S_i^*\};\ \text{for}\ \forall\ \{S_i\} \tag{9.2}$$

where i ranges from 1,2, . . . n and where $\{S_i\}$ is the set of specifications at ith state and $\{S_i^*\}$ represents the complete set at ith state.

It is not possible to identify a set of all specifications, since relevant specifications (at a transformation level) depend on the state of the concept descriptions, which is evolutionary. If the design is under the conceptual design stage, then a "hole feature" may consist of a hole diameter and a depth, and constraint may be present as tolerances. If the design is under the process planning stage, then machining constraints on the bore may be present as surface quality and finish. Constraints depend on the frame descriptions, which in turn depend on the concept description. The latter relates to things like level of abstractions, relationships, methods, and resolution strategy. Specifications are required relationships among frame description parameters (such as inputs, requirements, and constraints) that may be retracted, augmented, or refined as the product moves from one transformation state to the other.

The direct (one-to-one) transformation of specifications into physical components may result in undesirable or infeasible solutions. One-to-one matching does not ensure satisfaction of constraints that arise from assembling these components together. Requirements and constraints (RCs) are often difficult to assign independently, since they are often coupled, incomplete, or qualitative. Further, the satisfaction of specifications at the components' level does not guarantee that an artifact will function as an integrated system and not just as a collection of these components. The physical components may satisfy these specifications but may also exhibit behavioral and physical characteristics that are in conflict with each other. It is essential, therefore, that each subsystem functions as a unit and supports the system level goals. Finger and Rinderle used a bond graph grammar to model this transformation sequence [Finger and Rinderle, 1989]. The modeling was an integral part of this description. The modeling of product and process behavior is separated from the taxonomy in this book. Modeling topics are discussed in Chapter 7 and taxonomy is described in this chapter.

It is one thing to model the information or make it available to all members of the PDT (product development team). It is another thing to make sure it is used at the right time and place in the development process. This section describes a multilevel process taxonomy (breakdown of the life-cycle specifications) called the "spiral approach" for designing and developing a competitive product. In addition to the taxonomy, this section describes what models to use in the PPD process and when to use them. Concurrent product realization taxonomy has three parts:

1. Concurrent product realization methodology
2. Track and loop methodology
3. CE measures of merit

Development methodology for concurrent product realization is discussed in Section 9.2. Track and loop methodology is presented in Sections 9.3 through 9.5 and CE measures of merit are described in Chapter 2 of Volume II.

9.2 CONCURRENT PRODUCT REALIZATION TAXONOMY

In concurrent product realization taxonomy setup, the IPPD process is divided into seven concurrent tracks that may have staggered starting points but which run in parallel. The concurrent tracks include the following:

- Mission
- Planning (also called product planning)
- Design (product design)
- Process (that is, production process design)
- Production
- Manufacturing and assembly
- Delivery and service

This is shown in Figure 9.1. If each track is "completely inter-dependent" on the outcome of the previous track, concurrent engineering would be no different from serial engineering. There would not be any need for overlap. The time-to-market will be identi-

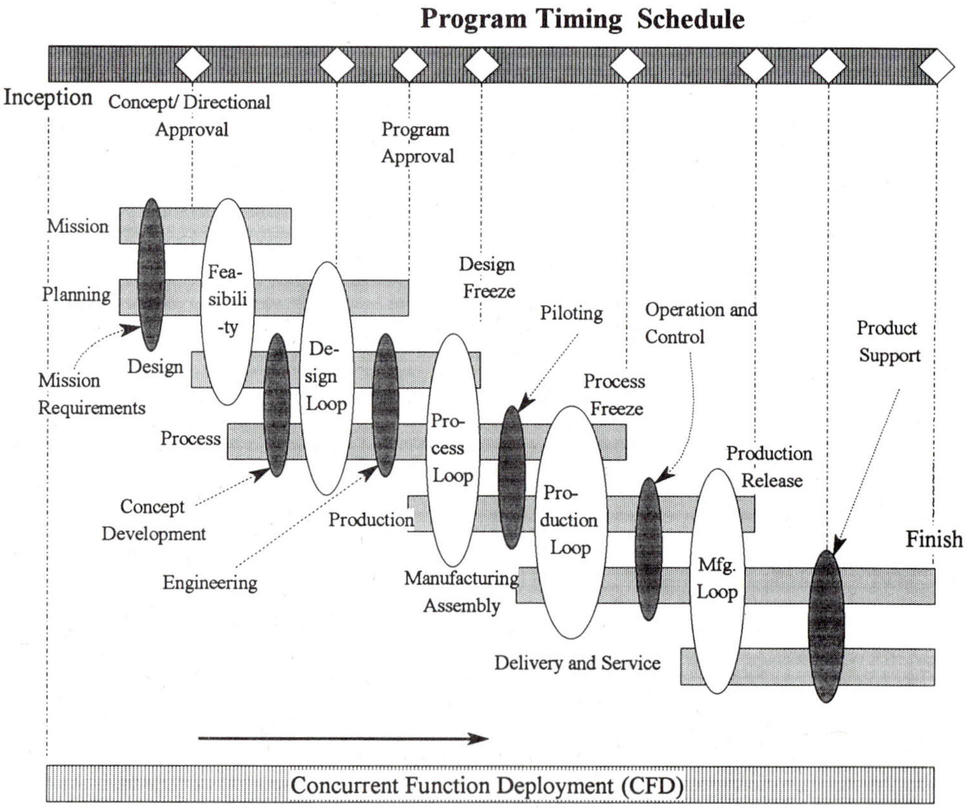

FIGURE 9.1 Overall Product Design and Development Process (PDD) Process

cal. Mutual dependencies and subsequent interactions between the tracks are useful mechanisms in satisfying the conflicting requirements that arise, especially in a complex product realization case (like automobiles or airplanes). For example, when a designer chooses a certain part configuration or feature as a part of its design, indirectly, such choices narrow down the manufacturing possibilities. The freedom for manufacturing that part is, thus, inadvertently constrained by that choice. To get an overall better design, it might be necessary to check whether the part designed can also be economically manufactured.

Paralleling creates overlaps. One of the advantages of overlap is that possible interaction between the tracks can be exploited. While overlap teams are actively performing tasks in real-time, the required information from one track can be dynamically moved to the other track requiring no additional effort. As the product moves through these tracks, a series of transformations occurs from its basic inception state to the finished state. This transformation framework can help carry out the tasks of satisfying the interfacing constraints simultaneously, and can offer an efficient product realization environment in which each team can concentrate on doing its part. At the finished state, the artifact converges on an interesting part of the design space. This is referred to in Section 8.5 as a concurrent product realization methodology.

9.2.1 Concurrent Product Realization Methodology

In order to realize this pull-type transformation, the following problems have to be solved:

- ***The transformation system must be demystified,*** so that specifications are steadily satisfied as the product contents start growing (product begins to take its physical form or shape).

$$\text{Determine a set of transformation} \Rightarrow \{[T_1], [T_2], [T_3], [\ldots], [T_i], [\ldots], [T_n]\} \tag{9.3}$$

Such as

$$\{S_{n+1}\} \rightarrow \text{Null Set } \{\emptyset\} \tag{9.4}$$

and

$$\{O_n\} \rightarrow \text{An Artifact} \tag{9.5}$$

- ***Interactions (refinements, iterations, recursions, etc.) must be handled*** between these tracks, so that the system at each level manages the corresponding constraints logically; that is, conflicts are resolved systematically, constraints are satisfied at each level, and results are consistent. A design conflict is reflected as a violation of a constraint (or a set of constraints) of one perspective that may be triggered by a decision(s) about some variable made by another perspective.

A process of taxonomy is employed to demystify the transformation system. Product realization usually takes place in a series of recognized stages [Larsen and Alting, 1992]. The stages are independent of (a) whether they are intuitive, (b) whether they follow a recommended systematic procedure, or (c) whether they are followed in a linear, cyclic, or iterative fashion [Eder, 1990]. The goal of this demystification is to develop a modeling framework methodology which is useful for allocating the specifications and ensuring that needed iterations take place at the appropriate time. There have been several attempts to give formal structure to the transformation process [Suh, 1990; Pahl & Beitz, 1988; Alexander, 1964]. The major stages can be recognized by uncoupling the definitions of the product from its abstract description to a more concrete description. It allows the design teams to strategically uncouple the major product and process subdomains into a taxonomy of transformation (partitions into sizable chunks). Such transformation gives CE teams a methodology to map the total processes required by their partitioned solutions.

9.2.2 Track and Loop Methodology

The track and loop methodology has been developed to facilitate quick and concurrent transformation of the initial specifications into desired end outputs. The methodology allows quick insertion of new technology in response to changes in market conditions. Each loop is represented using the conventions described in Figure 8.11 of Chapter 8.

Three sets of information are required, namely data, process, and the knowledge to form a baseline system for each loop.

$$\text{Baseline system at state } i \Leftrightarrow \cup\ [\{(\text{Data}), (\text{Process}), (\text{Knowledge})\} \text{at state } i] \quad (9.6)$$

This serves as an input to a synthesis loop. The output is a modified or updated baseline, which, in turn, takes the form of a new baseline system for the start of the next loop.

$$\text{Modified or updated baseline model at state } i \Leftrightarrow \text{Output at state } i$$

$$M_{i+1} \Leftarrow M_i \quad (9.7)$$

Where M_i is a modified or updated baseline model at state i, and M_{i+1} is a baseline model at state $(i + 1)$.

As usual, there are requirements and constraints (RCs) for each loop that completes the loop description. Requirements and constraints are hierarchically divided into system level RCs and loop level RCs. Clustering the RCs into system and loop level sets allows one to determine the degree of overlap, concurrency, or linearity (the order in which they must be satisfied to minimize any apparent couplings among them). Figure 8.14 of Chapter 8 shows some typical examples of RCs. RCs information changes from one loop to another. For example, during the feasibility loop, environmental effects and cost requirements are predominant. During design, this may change to quality and reliability. Redesign logic and optimization includes requirements for optimum system performance, improved manufacturing/operations economics, and waste minimization. Analyses or evaluation results (for example, economic analyses, design alternatives, etc.) are considered integral to each synthesis loop. Analyses provide inputs to redesign logic, rational choices among design alternatives, and optimization. It is desirable and inevitable for this track and loop methodology to spawn and affect all the stages of product design and development. This causes the stages to shift and change so that the product realization taxonomy adapts successfully to the ever-changing marketplace.

One of the simplest forms of a loop is "1-T loop," where, "1" denotes "one-dimension" and T denotes a track. An individual track of a loop represents the partition of the life-cycle activities. Figure 9.1 shows seven parallel tracks of a 1-T loop. Since 1-T is one-dimensional, there is no interaction between the tracks. Activities within each track are carried out independent of each other. For example, in an 1-T loop the following is true:

$$A_i \cap A_j = 0;\ \text{when } i \# j \text{ and } i = 1, 7;\ \text{and } j = 1, 7 \tag{9.8}$$

and

$$= 1 \text{ when } i = j \tag{9.9}$$

where A stands for an activity, and i may stand for a track—mission, planning, design process, production, manufacturing/assembly, etc.

In a 1-T loop, the overlap between the tracks, due to concurrency, also has no significance since possible advantages from their interactions cannot be realized. In real situations, interactions among two or more of these parallel tracks do take place. Many work-groups and organizations recognize these interactions by activities differing in the level of interactions between the tracks.

Each loop represents an interaction between one or more essential tracks of product development. A 1-T designation indicates interactions take place within the track itself. A 2-T designation indicates that looping or interactions take place between two parallel tracks. Similarly, a 3-T designation means that there are interactions between at least three parallel tracks. Depending on the product type, its complexity, and the organizational setup, a series of these 2-T or 3-T loops can be executed to reach the final product realization state. Though it is difficult to characterize every phase of transformation, a set of popular loops is shown in Figure 9.1 by their most recognized activity names. 2-T loops are shaded and 3-T loops are unshaded ellipses. From this diagram, it is clear that there could be many alternate traces of loops leading to the final state. In principle, it is true that the larger the number of interactions, the more balanced will be the output. However, this may not always be advantageous. The relationship is given by (Figure 9.2):

$$\text{Number of intergroup interactions} = L\,(L - 1)/2 \tag{9.10}$$

Where, L is the number of loops. For example, in a 2-T loop, there is only one intergroup interaction. In a 3-T loop, there are three intergroup relationships. In a 4-T loop, there are six intergroup relationships and in a 5-T case, there are ten intergroup relationships. The more loops, the more will be the number of intergroup interactions. With increasing interactions, it is possible that it may get progressively difficult for every group to agree.

The six shaded ellipses corresponding to 2-T loops (Table 9.1) are described in section 9.4:

$$\text{2-T loops} \equiv \{\text{Mission Requirements, Concept Development, Engineering, Piloting, Operations and Control, Product Support}\} \tag{9.11}$$

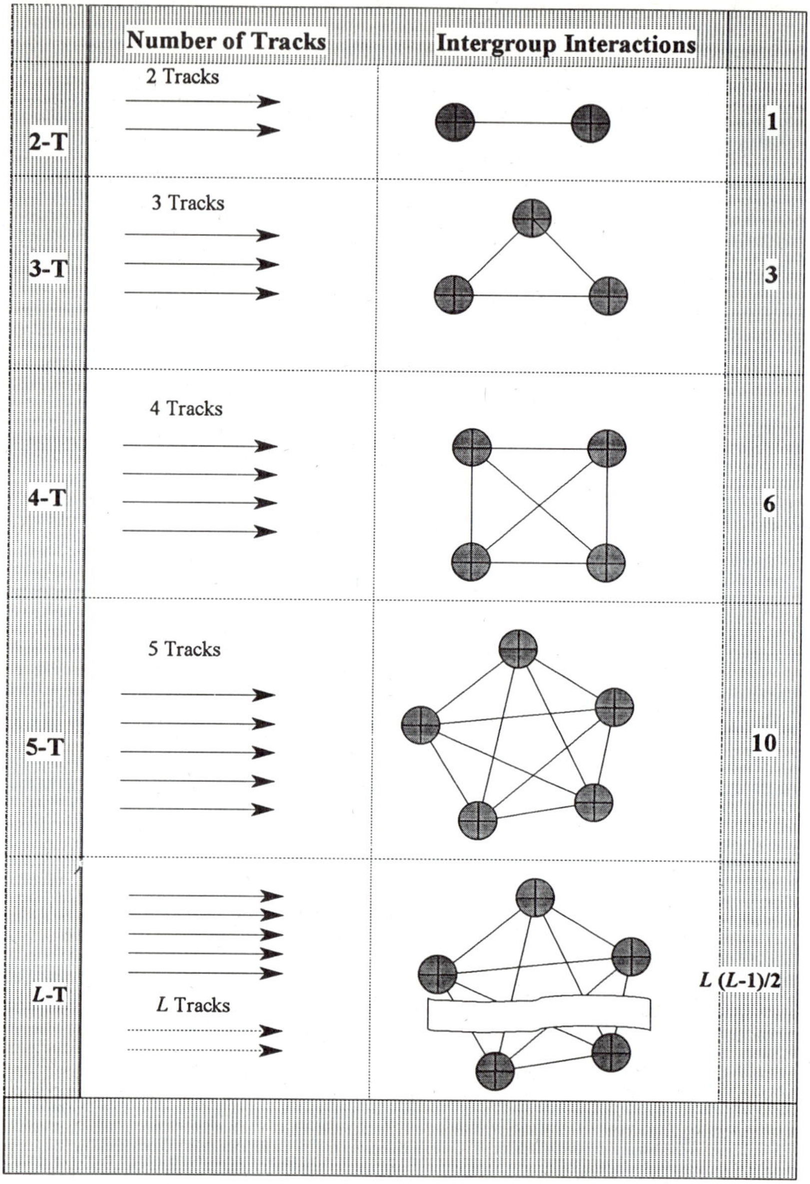

FIGURE 9.2 Number of Intergroup Interactions

TABLE 9.1 2-T Loops

Activity Names	Definition	2-T Interaction Elements
Mission Requirements	$\equiv$	$(A_{mission} \cap A_{planning})$
Concept Development	$\equiv$	$(A_{design} \cap A_{process})$
Engineering	$\equiv$	$(A_{design} \cap A_{process})$
Piloting	$\equiv$	$(A_{process} \cap A_{production})$
Operations and Control	$\equiv$	$(A_{production} \cap A_{manufacturing/assembly})$
Product support	$\equiv$	$(A_{manufacturing/assembly} \cap A_{delivery\text{-}and\text{-}service})$

The next set of five un-shaded ellipses, corresponding to 3-T loops (Table 9.2), is described in section 9.5:

$$\text{3-T loops} \equiv \{\text{Feasibility Loop, Design Loop, Process Loop, Production Loop, Manufacturing Loop}\} \quad (9.12)$$

Distribution of the product life-cycle into loops provides an opportunity to view the process of design and development in a systematic manner. Figure 8.15 of Chapter 8 shows the simultaneous satisfaction of the life-cycle values during two half-cycles, that is, product management half-cycle and production management half-cycle. During each half-cycle (see Figures 9.3 and 9.4) the design is checked or assessed against a number of "design for X-ability (DFX)" constraints. There is a definite set of constraints specified for each half-cycle derived from the corresponding specifications. They are lumped here as "product" and "process" constraints. Each specification represents the allocation of life-cycle values that need to be satisfied. The candidate's initial system and the qualified system represent the inputs for each half-cycle, respectively. The product realization or incorporation process is shown in Figure 9.5. Collaborative concurrent design considers the product realization process on a holistic basis [Lu, Subramanyam, and Thompson, 1988]. One may recognize from Figure 9.5 that DFX and other metrics of measures are not added steps, but they are an integral part of the product realization process. During these two (spiral) half-cycles, as DFX analyses and evaluation proceed, the specification envelope shrinks but the information-content envelope containing product and process knowledge grows. At the beginning (product half-cycle) stage, the specification envelope is full

TABLE 9.2 3-T Loops

Activity Names	Definition	3-T Interaction Elements
Feasibility Loop	$\equiv$	$(A_{mission} \cap A_{planning} \cap A_{design})$
Design Loop	$\equiv$	$(A_{planning} \cap A_{design} \cap A_{process})$
Process Loop	$\equiv$	$(A_{design} \cap A_{process} \cap A_{production})$
Production Loop	$\equiv$	$(A_{process} \cap A_{production} \cap A_{manufacturing/assembly})$
Manufacturing Loop	$\equiv$	$(A_{production} \cap A_{manufacturing/assembly} \cap A_{delivery\text{-}and\text{-}service})$

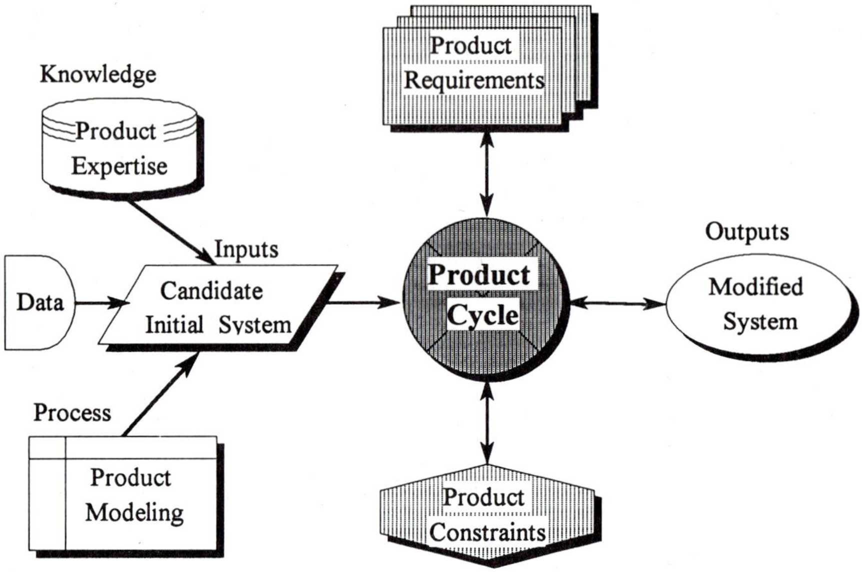

FIGURE 9.3 Product Design Management Half-cycle

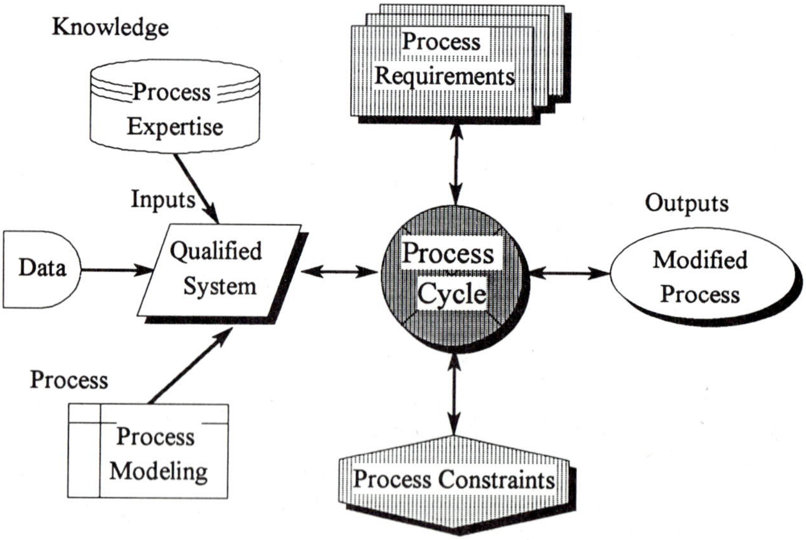

FIGURE 9.4 Process Design Management Half-cycle

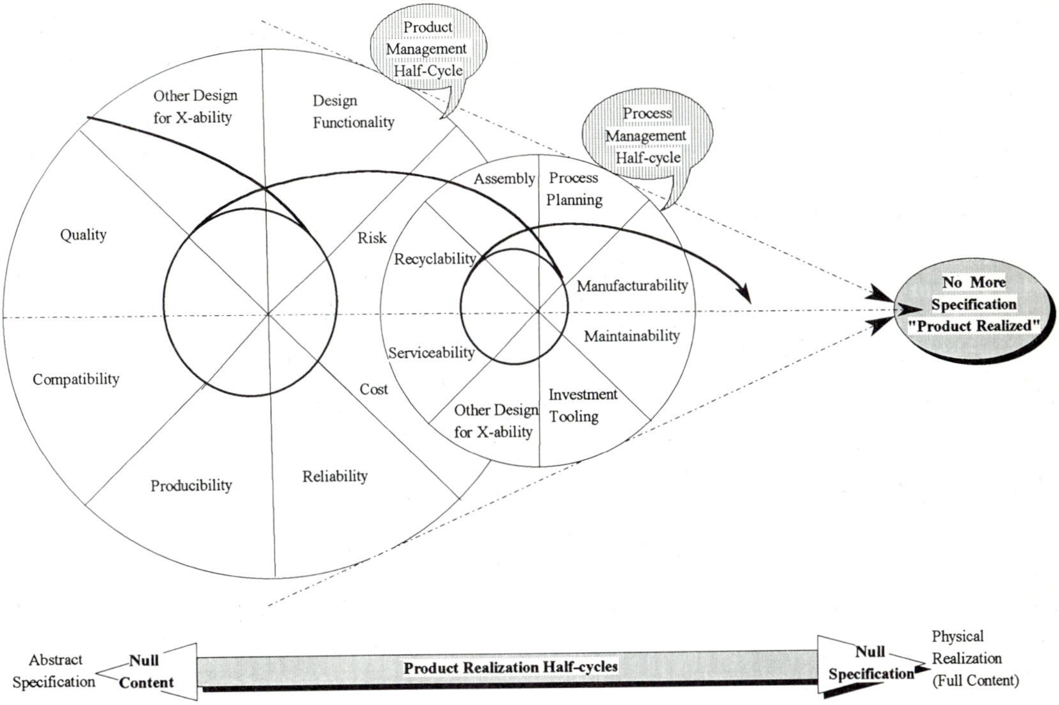

FIGURE 9.5 Product Realization or Incorporation

of RCs and the product might only contain a few conceptual semantics or information. By the time the product half-cycle incorporation is complete, the specification envelope is leveled-down and the product may contain a wide range of information about all aspects of design phases. The production incorporation starts when all the product loops are finished and the changes are blended into the master product schedule (MPS). During this half-cycle, as the process continues through the assessment of production loops incorporation, more and more process constraints are satisfied. The product realization process spirals forward from one half-cycle (product subdomain) to the other (production subdomain) and finally takes the form of a physical artifact. During this process, the specifications that existed in the beginning in pure raw form, are transformed into a physical form. Figure 9.5 shows important life-cycle requirements (key X-ability characteristics or measures of merits) of the specification envelope. Because of their relevance to CE, these measures of merits are separately discussed in Chapter 2, Volume II. The supporting methodology and computational tools are discussed in Chapter 4, Volume II.

A product realization process is described in terms of taxonomy of objects representing models of specifications, product, process, enterprise, and cognitive behaviors. They are also referred to as invariants of the taxonomy world. A CE participant generates and analyzes a series of well-connected models of the product realization alternatives within a taxonomy world. This has associated breakdown structures, a transformation system to convert specifications into outputs, convergent loops, symbol systems, solution techniques, and tools. During this taxonomy, a product goes through a transformation into two half-cycles, then into loops, followed by concatenation. This methodology achieves its constancy of purpose (a common set of consistent goals and objectives) using a branch-and-bound methodology that guides the teams as the product evolves. The branch-and-bound methodology is briefly discussed earlier in Section 8.7.1. Following this taxonomy, many ambiguities in the various activities of the life-cycle of a product can be eliminated. Taxonomy characterization is very appealing since it permits the following:

- ***Parallel Design:*** It allows many parts of the products to be designed in parallel. Design variants can be considered quickly, and life-cycle time, cost, and quality disadvantages of serial engineering can be largely eliminated. This can have a positive impact on enhancing concurrency and reducing the total workforce.
- ***Concurrent Interface Definitions:*** In addition, the "inside-out" decomposition of the product into product breakdown structure (PtBS) and "outside-in" integration of the decomposed elements back into the product ensures that the "interface definitions" occur concurrently.
- ***Convergence by Iterations:*** During this taxonomy, interactions are handled through "iterative loops." Since many decisions are coupled, loops provide a natural means of accomplishing convergence by "iterations" or "negotiations" among the diverse members of the CE work-groups.
- ***Natural Part of the Process:*** A number of business and technical activities within loops are conducted by the participating teams and organizations to develop products and services that meet the wants and needs of the customers.
- ***Identification of Coupled and Difficult Tasks:*** This may enable CE teams to spot potentially complex or coupled tasks or processes. This may also pinpoint the tasks that can create difficult convergence problems later in the product's life span. It generates confidence in the solutions and builds stronger commitment among the CE team members.
- ***Integrative Looping Process:*** This builds-up the convergence slowly during each pass. The decomposed product and process elements will need fewer and fewer changes on each iterative pass until it finally arrives back at the participating teams with no new redesign suggestions. At this point, the outcome of the transformed product is considered to be feasible and, in all probability, somewhat optimized. In a traditional design process, many iterations are required to adequately address all RCs.
- ***Cooperative Decision Making:*** The loop provides provisions for including management as a part of the negotiation team, thus gaining the early management buy-in, if needed. The method reduces the need for cooperative decision making and limits the size of the required CE work-groups.

The process taxonomy captures the specifications (inputs, requirements, and constraints), outputs, and their relationships in the form of a transformation matrix. Considering the direction of traversal, two types of analyses are possible for studying the sensitivity of taxonomy changes:

- ***Forward Traversal:*** This analysis is used to identify the subgraphs of the transformation outputs that are affected when a specification data (inputs, requirements, and constraints) is changed. The analysis uses dependency information and specification/output relationships for each transformation state to trace the behavioral patterns of the affected outputs.
- ***Backward Traversal:*** This analysis traverses backwards in a process taxonomy to identify inputs and RCs that could perform the stated outputs. It identifies a possible set of parameters (RCs) that need to be changed to predict a desired outcome. The back-computed RCs depend on the choice of the "loop methodology" selected for the transformation. Change analysis increases in importance with the specification complexities and RCs dependencies.

9.2.3 CE Measures of Merit

Along with the classification of product design and development (PDD) into loops, specifications are also classified into hierarchical networks. A unique set of evaluation criteria—called measures of merits (most commonly referred to as MOMs or metrics)—is associated with each loop. The methodology can be used to facilitate various CE activities arising primarily from product and production subdomains. At any loop level, the baseline design is subjected to an evaluation that is uniquely defined for that loop. Evaluation may consist of a set of analyses, simulation, and understanding of the life-cycle considerations (for example, X-abilities, heuristics, and other MOMs). This allows the teams to review the performance of the product, process, and related production schemes for the baseline design. Feasible concepts are presented at design review to check against the original intent's definition. If the concepts do not meet the original intent, the design work-group must decide whether to look into other new candidate concepts, or to return to the initial functional requirements and modify them. After a decision is made, the process continues. If the evaluation is passed, the resulting design is stored as a new baseline design and the control is returned to the next loop. If the evaluation is failed, a redesign or optimization procedure is activated to correct the constraint violations. The forthcoming changes are introduced to the baseline design, the redesigned baseline system is reevaluated, and the constraint satisfaction process continues. If all RCs for a loop are satisfied, the design is stored and passed on to the next loop. This becomes a new baseline design for this loop. This process is repeated for all the loops successively. If evaluations for all the loops are passed, the design is stored as a possible final design. If, during an evaluation, all constraints belonging to a loop cannot be adequately satisfied, the control is returned to the previous loop with "unsatisfied constraints" now becoming the "requirements" of the previous loop. The personnel team is the primary driver for this taxonomy. Computer programs may be used to evaluate and store all design iterations and modifications. The virtual team concept may be employed to complement the personnel team.

There will always be feedback loops between the steps described above. Normal program timing should allow for some of these to take place. If the team has done its work properly, the overall process should provide them with confidence. Teams should have a limited set of design alternatives that meet the design requirements. Teams should have a good understanding of cost, weight, investment, reliability, process capability, and other important factors.

Such a taxonomy helps the generation of highly accurate product and process task breakdowns for each parallel team. Tasks come as a natural by-product of the teams' control over many concurrent loops intrinsic to the taxonomy. Once these underlying loops of the system taxonomy are carried out by each concurrent team, the corresponding tasks are executed in the background. Better results (such as higher accuracy and consistency of design) come at little or no additional cost.

9.3 PULL SYSTEM OF PRODUCT REALIZATION

The pull system of product realization is a method of controlling the transformation of the product into an artifact by

1. *Following* a standard taxonomy of product realization (track and loop methodology)
2. *Substituting* only what has been changed or modified out of the taxonomy
3. *Minimizing* the dependency of change into the product realization process, so that the loops can be performed in parallel.

The characteristics of a concurrent product realization process (CPRP) are listed here:

- **Establish** one consistent taxonomy for controlling and balancing the flow of specifications (inputs, requirements, and constraints).
- **Organize** specifications into a hierarchical setup and development process into loops so as to minimize revisions or engineering change orders (ECOs)
- **Decompose** the product transformation process into a series of semi-independent loops, so that necessary iterations having intertrack communication of common constraints are performed locally.
- **Pass** on the controls of the product transformation process to the individual taxonomy-based loops. Loop methodology takes care of local satisfaction of constraints and essential interactions between the product tracks.
- **Minimize** the dependency of the decomposed parts in the product (PtBS) tree, tasks in the process (PsBS) tree, or work content in the work breakdown structure (WBS) tree.
- **Substitute** only what has been modified or changed out of the taxonomy.

- **Run** the decomposed independent tasks in parallel to each other. Run only the coupled or inter-dependent tasks serially.
- **Build-up** "what information is required" to start the next activity immediately.
- **Overlap** two or more processes so that the next process can start just after the first one is finished generating the required information (just-in-time).
- **Reduce** lead time for each decomposed tasks individually if possible, in the concurrent product realization process.

9.4 DESCRIPTION OF PARALLEL TRACKS

9.4.1 Planning Track

The mission is considered a part of the planning track. Due to the complexity of the products manufactured today, there is a growing complexity of requirements. Hierarchies of requirements are at every track and loop-level and must be considered during the product development process. Basic information required at product planning stage is derived from functions associated with integrated system planning and scheduling (see Figure 8.7, Chapter 8). At the conclusion of this track, pricing is set and the major characteristics of the product are decided. Integrated system planning and scheduling has two functions: (a) product planning; and (b) manufacturing planning.

9.4.1.1 Product Planning Requirements

Product planning starts with developing customer requirements (market analysis, voice of the customers), product data, and warranty field data (see Figure 2.22). Management recognizes the need of a product in conjunction with the marketing department. The marketing department does the market survey to determine this need. Once the product character demand and need have been formulated, the primary goals of a would be product are determined. Estimates are made of the production quantity and the target cost of manufacturing the product. At the same time, prices of competitive products that may appear in the marketplace with similar features are estimated (see Figure 8.7 of Chapter 8). On the basis of production cost and competitive analysis, the selling price (the price at which the product would be sold) is estimated so that the company can make a fair profit.

9.4.1.2 Program Approval

Program approval in the automotive industry ordinarily occurs two to three years before full production and involves the decision to expend large amounts of resources in workforce and tooling. It involves a commitment on the part of the product development community to simultaneously meet objectives for product quality, cost, weight, and investment within a time constraint. Program approval is the decision point of the product planning track (see Figure 9.1). The product development team must have agreed upon a plan for accomplishing product objectives at this point of program approval, or risk program failure. Product objectives are worth very little unless the following holds true:

- The objectives must be supported by a reasonable assurance from the work-groups or teams.
- Teams must have some clues to the feasibility of the design and process plans to back them up.
- The objectives, in addition to being challenging, must also be achievable.

9.4.2 Design Track

The design track integrates customer needs, functional requirements from integrated system planning and scheduling, and artifact's intent definitions into various activities so as to serve as a link to downstream operations, such as process planning, manufacturing, assembly, and inspection. A quick pass through the X-ability satisfaction process of the product and process subdomains is initially performed (Figure 9.5) to eliminate the conflicting RCs. This process is repeated for each design improvement and whenever a new design feature is added. The design track normally involves a continuous interplay between what organization wants to achieve and how teams can achieve it. The candidates initial design is usually specified in terms of desired functions, geometrical parameters, features, and materials. The end point of the design track is called the design freeze point (see Figure 9.1).

9.4.3 Process Track

The process track focuses on achieving the necessary process quality levels, while creatively searching for ways to do it at the lowest possible manufacturing cost, weight, and investment (such as new equipment or technology). It is an important part of the artifact's life-cycle, which is refined further as more detailed information emerges when the production or manufacturing and assembly tracks are set in motion (see Figure 9.1). As can be seen by overlapping functions of these tracks, the process track is actually a composite of manufacturing, quality control, resource planning, and inspection tools. Many of these steps, such as SPC, JIT, and QC are interwoven and they proceed concurrently. For example, an early task of this track may be to use JIT activities to reduce buffer stocks on the line, so that shift operators will have only the parts they require to complete that shift's production. The constraints imposed on this activity, in turn, requires that all the parts shipped to the operators are defect-free. For this to happen consistently, it is important that the SPC and QC programs are in place, so as to guarantee zero defects.

The process track contains knowledge that must be considered as the track evolves, such as stock selection, operation selection, machine cell selection, tool and fixture selection, operation sequence, etc. (See Figure 9.6). A sheet metal process, for example, has five steps: punching, shearing, bending, welding, and painting. The end point of the process track is called the process freeze point.

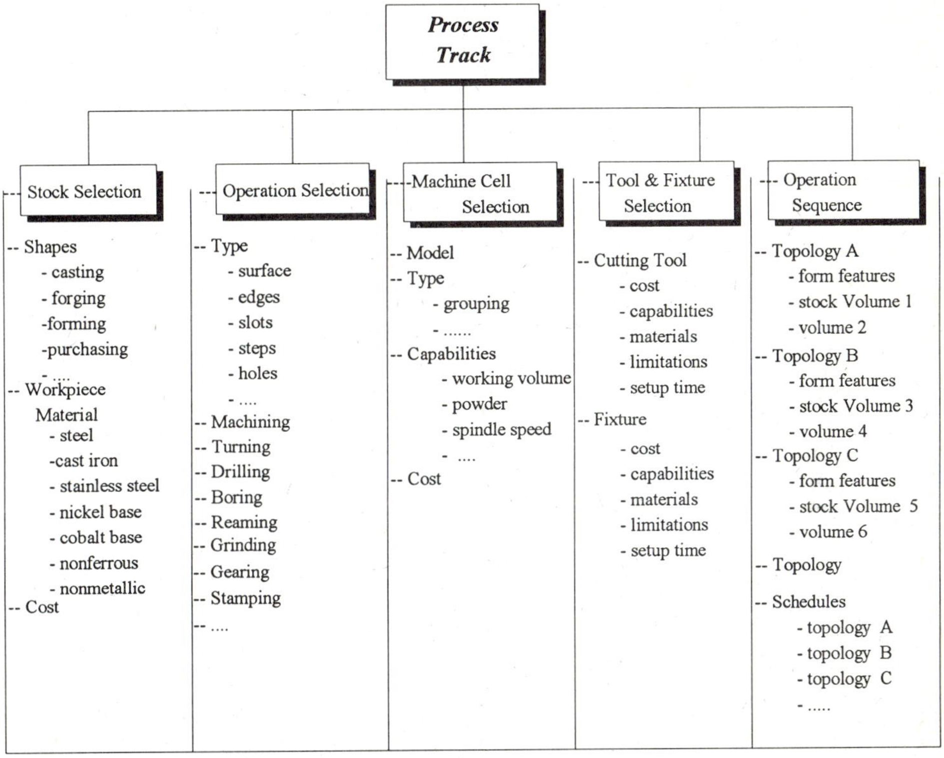

FIGURE 9.6 Types of Objects in a Process Track

9.4.4 Production Track

In the production track, one or more of the fabrication stations must be selected and then grouped. Station selection may be done on a case-by-case basis depending on schedule, priority, and workloads. There are advantages in selecting a group of stations rather than a single station. If it is possible to put more than one job on a station, two processes can be scheduled on the same station neck-in-neck. The wait time, machine setup time, changeover, etc. all cannot be eliminated. A flexible manufacturing system allows many such technologies and efficiency choices. This practice is increasingly common with multipurpose NC machine tools. If the different capabilities of NC machine stations can be quantified, several candidate sequences can be obtained. The determination of alternate sequences can be based on one or more of the following criteria:

- Minimize the number of setups.
- Minimize the cost of fixtures.
- Minimize the number of changes between machines.

- Minimize the number of orientation changes.
- Minimize the transportation needs or waiting time.

It is possible to use knowledge-based tools to select groups of machines, optimal sequence, and people suitable for performing groups of fabrication tasks. These can be used to meet production quantity targets on multiple models of a product. Less transportation need and less waiting time to carry parts from one station to the other (see Figure 9.7) minimizes time for unit production. The end of the production track is called the point of production release.

9.4.5 Manufacturing or Assembly Track

In the manufacturing assembly track, (see Figure 9.8) the stations must also be selected and then grouped. With increasing availability of different types of robots, there is an increased possibility of selecting groups of robot stations and assigning more than one task to a station. The suitability of a robot or a person for a task depends on its design, cost, speed, and quality of work. The assembly sequence is particularly important, as is the definition of subassemblies (see Figure 9.8). In an assembly, different schemes of accomplishing a task may have different chances of success. It depends on what mechanical accuracy or force levels are involved, how much care or clearance is required, and how sensitive the technique is to small errors.

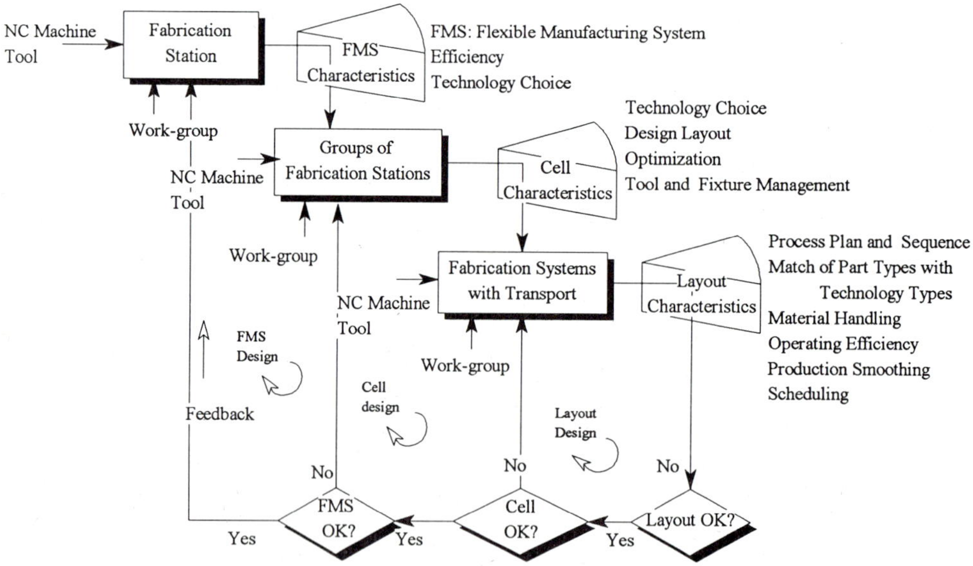

FIGURE 9.7 Information Flow in a Production Track

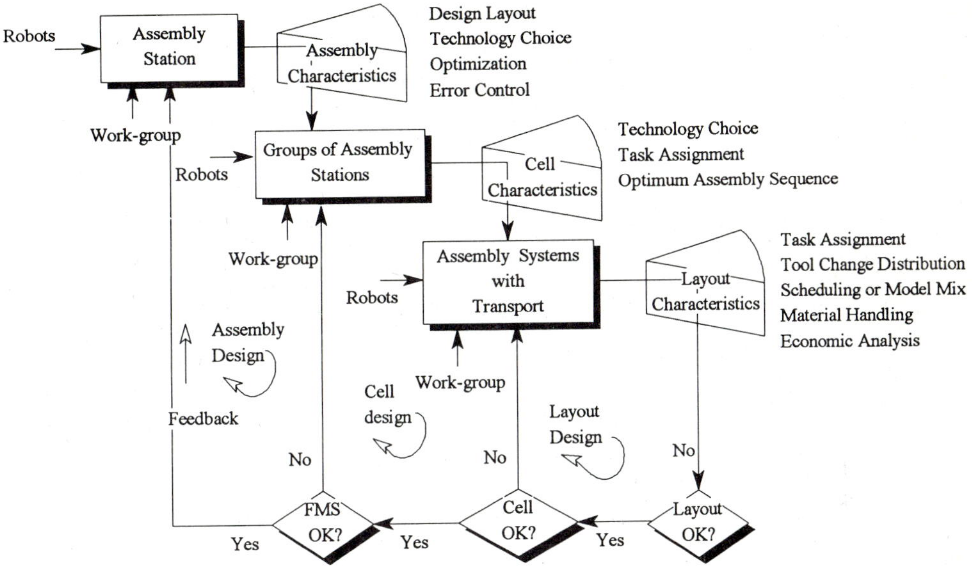

FIGURE 9.8 Information Flow through a Manufacturing or Assembly Track

9.4.6 Delivery and Service Track

With manufacturers achieving parity in a number of competitive tracks (mostly one-dimensional tracks (1-T)), product delivery is a place where manufacturers are now concentrating to gain a competitive edge. They are looking for ways to improve the service level and please the customer.

9.5 DESCRIPTION OF 2-T LOOPS

As stated earlier, the goal of any loop is primarily to match the individual specifications (desired behaviors and physical specifications at each loop level) directly to the units. This normally does not ensure an integration of behaviors into the units, or a more balanced collection of the units. An evaluation of the baseline design produces a set of unwanted incidental behaviors and characteristics. This usually occurs at the start of an iterative loop. The "goodness" of the product, its configurations, and the determination of the right mix of its units depends on successful satisfaction of these incidental behaviors. This is accomplished during subsequent iterations of the loops. The newly determined incidental behaviors and characteristics are imposed as constraints in the subsequent iterations. The iterative cycle continues until the resulting product outputs a consistent design with specifications that are consistent with a majority of desired behaviors and physical characteristics.

A decision or review is performed after each step to ensure that the requirements allocated at each loop level are met. This is usually proposed in the form of questions:

- *Mission requirements:* Are functional requirements satisfactory?
- *Concept development:* Is concept design satisfactory?
- *Product engineering:* Is preliminary design satisfactory?
- *Product design:* Is detail design satisfactory?

The step reviews are shown by a rhombus shaped icon in Figure 9.9. At the end, an overall design review is performed to determine whether or not the product is optimized with respect to all relevant RCs. The control is returned to the appropriate steps (as shown in Figure 9.9) if problems occur with parts/materials, components, subsystems, and systems. If the problems occur at the system level, the overall functional requirements and mission definition of the system are then reevaluated. If every step review and design review (at the end) is passed, it means the emerging baseline system at this stage met the allocated RCs. Then, the process moves on to the next loop.

9.5.1 Mission Requirements Loop

This is the first 2-T loop of product realization taxonomy. The purpose of the loop is to define the program requirements in terms of overall goals and the operational value of the product as is outlined in Figure 9.9. Mission requirements are due to interaction between the "mission" and the "planning" tracks. This brings out marketing and business aspects of the new product program and describes the specifications of the proposed products in QFD (customer) terms. During this loop, the product design team, first, translates the desired features (marketing and customer inputs) into a set of mission requirements. In most cases, mission requirements are of two types:

- ***Performance goals:*** These ought to be satisfied in the most economical way.
- ***Operational values:*** These are a preferred set of values that are nice to have, but can be traded off against each other (for example, fuel economy and 0–60 mph acceleration time).

The first step is to identify product characteristics, possible options, and market focus (type and size of the market, methods of distribution, likely competition, etc.). The next step is to determine the available technologies and limitations (such as government regulations, human factors, economic realities, and customer desires). Market research techniques and customer interviews help to determine customer requirements (Figure 2.22). Using available inputs, CE teams define a complete set of program mission requirements, including the desired product features, performance goals, and constraints, without specifying how these will be attained [Shilke, Rhode, Fruechte, and Rillings, 1989]. A quantitative and functional equivalent of the mission objectives is defined. The qualitative

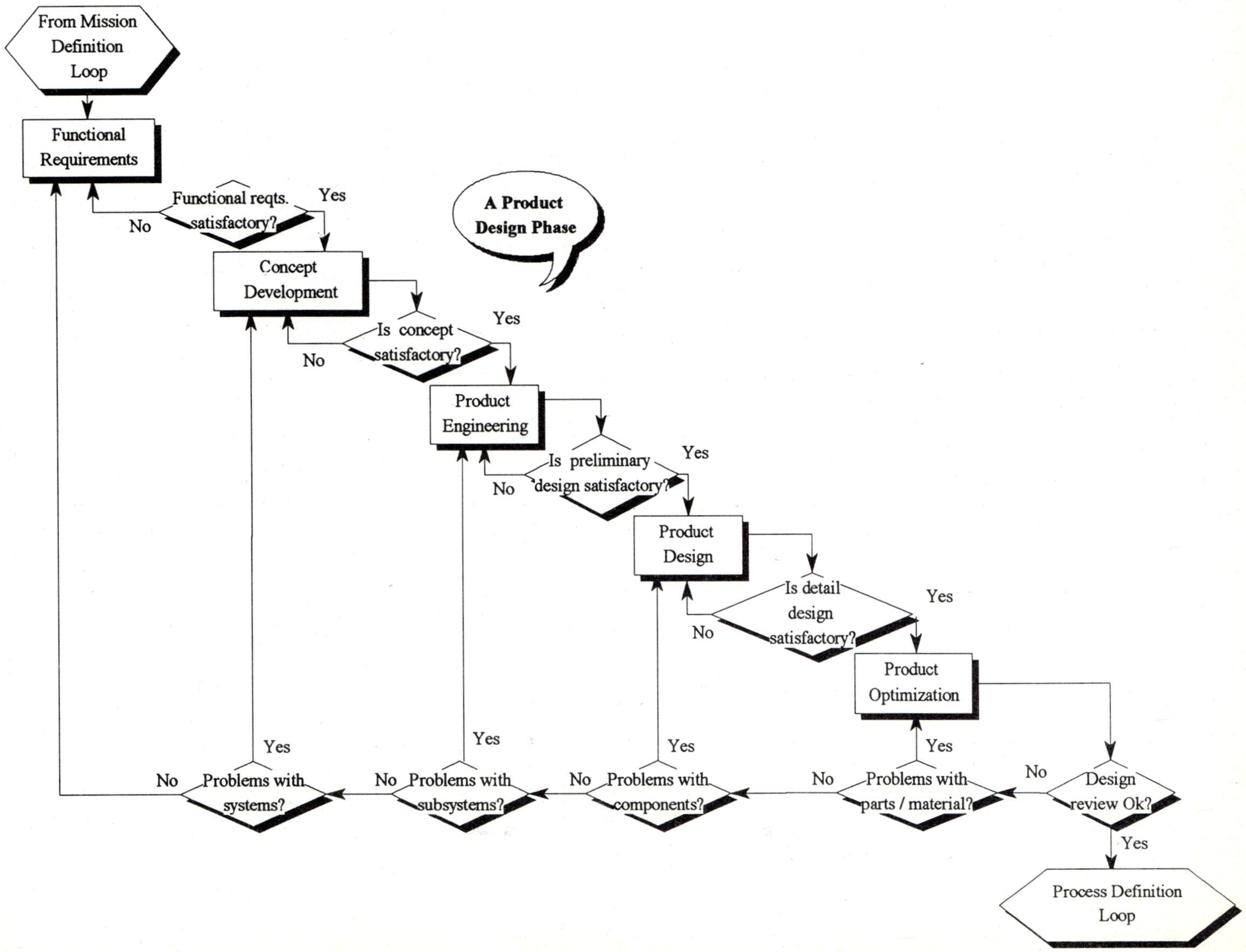

FIGURE 9.9 Information Flow Diagram for a Product Definition Phase

performance goals and the desired product features are translated into confidence factors or operational values, relative weights on constraints, and a set of (minimizing or maximizing) objective functions.

A feasibility validation plan is developed so that the functional performance goals of the product can be tested and evaluated. A planning matrix is set up to verify that the requirements stated through mission objectives are adequately satisfied. A QFD review is performed to verify the adequacy of the planning matrix. If the QFD specification does not meet the mission requirements (such as performance goals and operational values), objectives are changed, and the steps repeated. The iteration provides feedback information on the tradeoff between different markets and operational variables.

9.5.2 Concept Development Loop

Concept development is due to interaction between design and process tracks. Concept development takes place between the feasibility loop and the design cycle loop (see Figure 9.1). Concept development deals with the activities that happen early in product development. It includes activities such as conceptual design, CAD, parametric design, systems design, crude analysis, etc. The concept development process steps through these activities in a global way, over a relatively short timespan prior to program approval. Here, details are analyzed for alternative concepts with candidate systems modeled generically. Members of planning, product design, and process design participate in developing the concept. This is the period where the knowledge and experience of the work-groups have the greatest leverage, and where team effort has to come together. The activity is completed in a reasonably short time prior to program approval. As a result, only a small portion of the development resources is generally expended in this activity, though a large portion of the product costs is usually committed. Clearly, the directions established here have a profound impact on all the loops that are to follow. During this activity, a number of feasibility tests are carried out by an experienced team to develop some of the most attractive design concept alternatives [Shilke, Rhode, Fruechte, and Rillings, 1989]. Feasible tests may be comprised of analytical or experimental verifications of concepts, along with DFX evaluations of the design concepts. It results in a choice of suitable principles of operations and modes of internal action, and some concepts that start to define the future product as a solution to the original problem. Identification of a new concept is attractive since the major advances in technology are due to new concepts rather than improvements in embodiment. The concept development loop establishes a solid foundation by assuring that the best overall product design concept is selected for a detailed evaluation during product engineering loop.

9.5.3 Product Engineering Loop

This loop deals with activities that happen after the completion of the planning track and before the product design loop is completed (see Figure 9.1). Members of product design, process design, and production-ready teams participate in engineering the product. Beginning with a feasible product concept from the earlier loop, the detailed subsystem design requirements are specified in this loop. Parts, components, and subsystems are modeled generically. Optimization and design improvement are carried out in an iterative fashion with alternatives further refined through the process outlined in the concept development loop. Each functional requirement, including both product and manufacturing requirements, is satisfied one-by-one. The product engineering process has six major activities (see Figure 9.10):

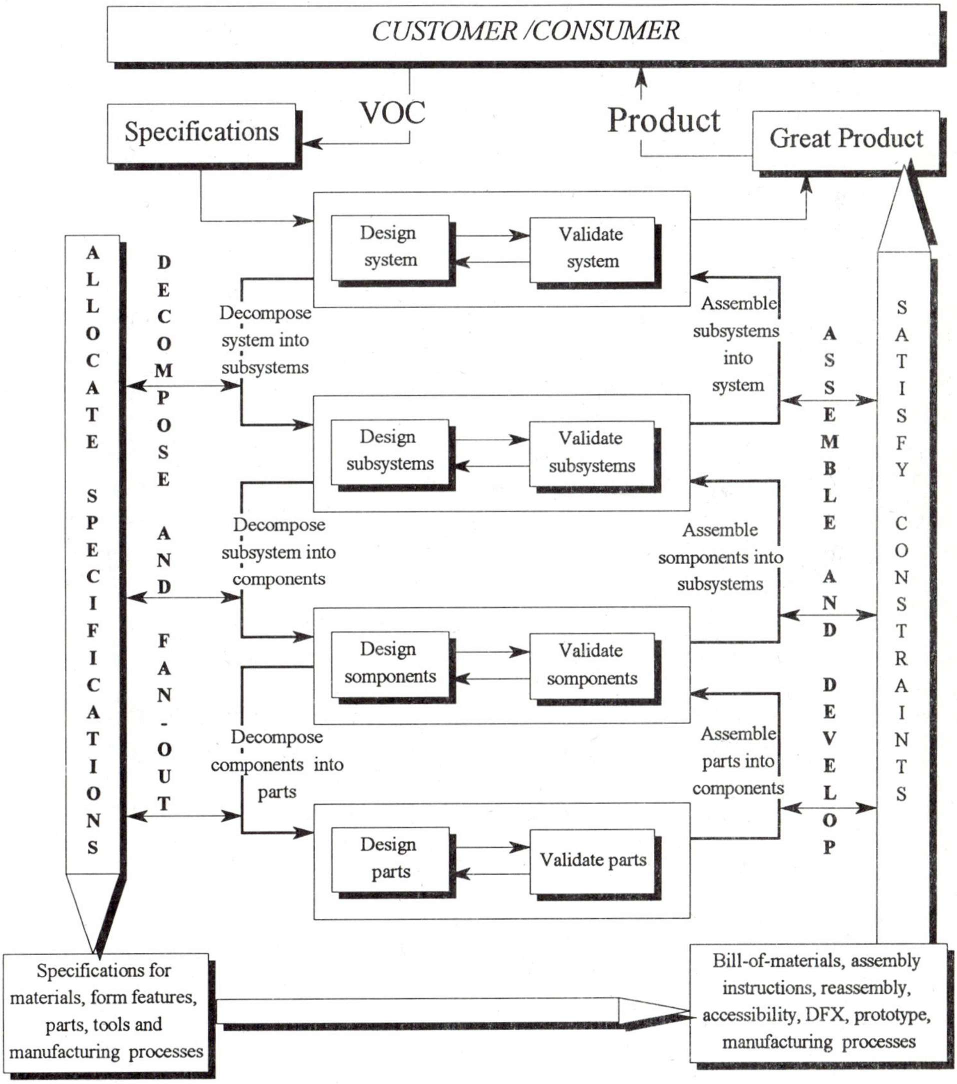

FIGURE 9.10: Steps in Product Engineering Loop

1. *Determination of the specification levels from VOC:* The voice of the customers help determine the specifications and to categorize them into different levels: systems, subsystems, components, parts, materials, and features, etc. Specifications come in forms such as materials, structures, features, parts, tools, and manufacturing processes, as shown in Figure 9.10.
2. *Allocation of specifications:* Parts can be initially designed using a feature-based product representation scheme, such as descriptive form features, etc. An assembled product model can be created using the interpart mating information specified by the designer or through the feature recognition aspects of the intelligent CAD system.
3. *Design and build of "top-down" PtBS tree:* An exploded view of the CAD model can be obtained from the initial assembled model of the product using methods, such as a rule-based or knowledge-based approach. An exploded view represents an ordered disassembled state of a product. The precedence described by the relative location of parts ensures the parts' compatibility.
4. *The assembly sequences and development (aggregation) of assembly (or system) plan:* The assembly process involves systematic introduction of parts or groups of parts to a fixture or to an assembled component. The precedence information implicit in an exploded view can be used to generate the selected assembly sequences. The generated assembly plan must ensure part mating conditions and a set of assembly criteria designed to avoid redundant and awkward assembly sequences.
5. *Validation of the assembled design against the allocated specifications:* An assembly operation is valid if, for example, it establishes planar or cylindrical contacts or does not enclose components. The enclosed situation is most commonly found at the axes' intersection point where the parts after the proposed operation are geometrically inaccessible. An assembly sequence is considered better if it requires the minimum number of changes in the direction of assembly or is geometrically accessible and does not involve awkward assembly operations.
6. *Great product:* The performance of the integrated system is optimized to derive a quantitative set of subsystem functional performances. The performance of each individual subsystem is then optimized by trading in system and subsystem activities. If the resulting feasible system design, subsystem design, component design, and part design outcomes are not satisfactory, the six-step procedure/activity, 1. to 6., is repeated. The procedure is continued until a reasonable model (for system, subsystems, and components) is obtained.

Figure 9.10 shows the interactions among the six-step procedure. Sometimes, a given function is achieved by a combination of several subsystems (for example, heating usually requires heating elements and a blower). In those cases, the CE team allocates the functional requirements to a set of candidate subsystems to engineer the performance of an integrated system, as shown in Figure 9.10. Product engineering activities include embodiment (subsystem design, assembly, prototype design, manufacturing processes), detail design (components, parts), and bill-of-materials (assembly instructions, reassembly, accessibility, DFX, etc.) [Shilke, Rhode, Fruechte, and Rillings, 1989].

9.5.4 Piloting or Prototype Fabrication Loop

This deals with activities that occur after design freeze and before process freeze as shown in Figure 9.1. Members of the process loop and the production-ready loop are involved in testing, verifying, and piloting. Rapid prototyping is a common technique to reduce the product's development time.

Rapid prototyping, in a generalized sense, can be any technique that allows for a quick and accurate physical or functional description of a product and process design presented in an understandable way [Houpt, 1990]. Figure 9.11 shows an overview of one particular prototyping methodology. The multifunctional team is shown as a focal point to integrate the activities from both product and process half-cycles. Rapid prototyping enables a lower life-cycle cost by allowing product and process information to be shared early in the design process. Rapid prototyping techniques range from soft prototypes (computer models of product/process design) to actual 3-D physical prototypes (hard prototypes) developed from CAD model data by a variety of techniques [Poindexter, 1991]:

- *Soft Prototypes:* In a soft prototype, early product/process prototyping is accomplished by using computer tools before any attempt of physical representation is made. Computer modeling, analysis, and simulation tools enable designers to experiment with several alternate strategies in order to assess their characteristics and to arrive at the most reasonable solution for a given set of product requirements.
- *Hard Prototypes:* Current "hard" rapid prototyping techniques (see Figure 9.11) include NC patterns and molds, stereolithography, selective laser sintering, laminate object manufacturing, and fused deposition modeling.

Prototyping, both analytical (soft) and physical (hard), is now commonly used to narrow down the producability, cost, X-ability, and other considerations. In the rapid prototyping when product and process synthesis are performed, the physical and functional interactions between subsystems must be accounted for explicitly. This is the basic tenet of the systems engineering process. Figure 9.12 shows subphases of a prototype system—soft and hard interactions for arriving at a feasible design. During the first three subphases, scientific visualization and the inference checks are performed on the soft prototypes. Thereafter, depending on the needs for evaluation, a hard prototype of the suspected part is built. The resulting mock-up (as built) is inspected and critiqued by a design review team (PDT). If the PDT team finds it unacceptable, it is sent back for further refinements/improvements to the "soft" prototype development team. If the PDT finds the mock-up design acceptable, the process planning subphase may begin. Since most systems are diverse in nature, any computer-based tool that attempts to simulate and track both these prototypes (soft and hard) must be multidisciplinary. A flexible three-dimensional

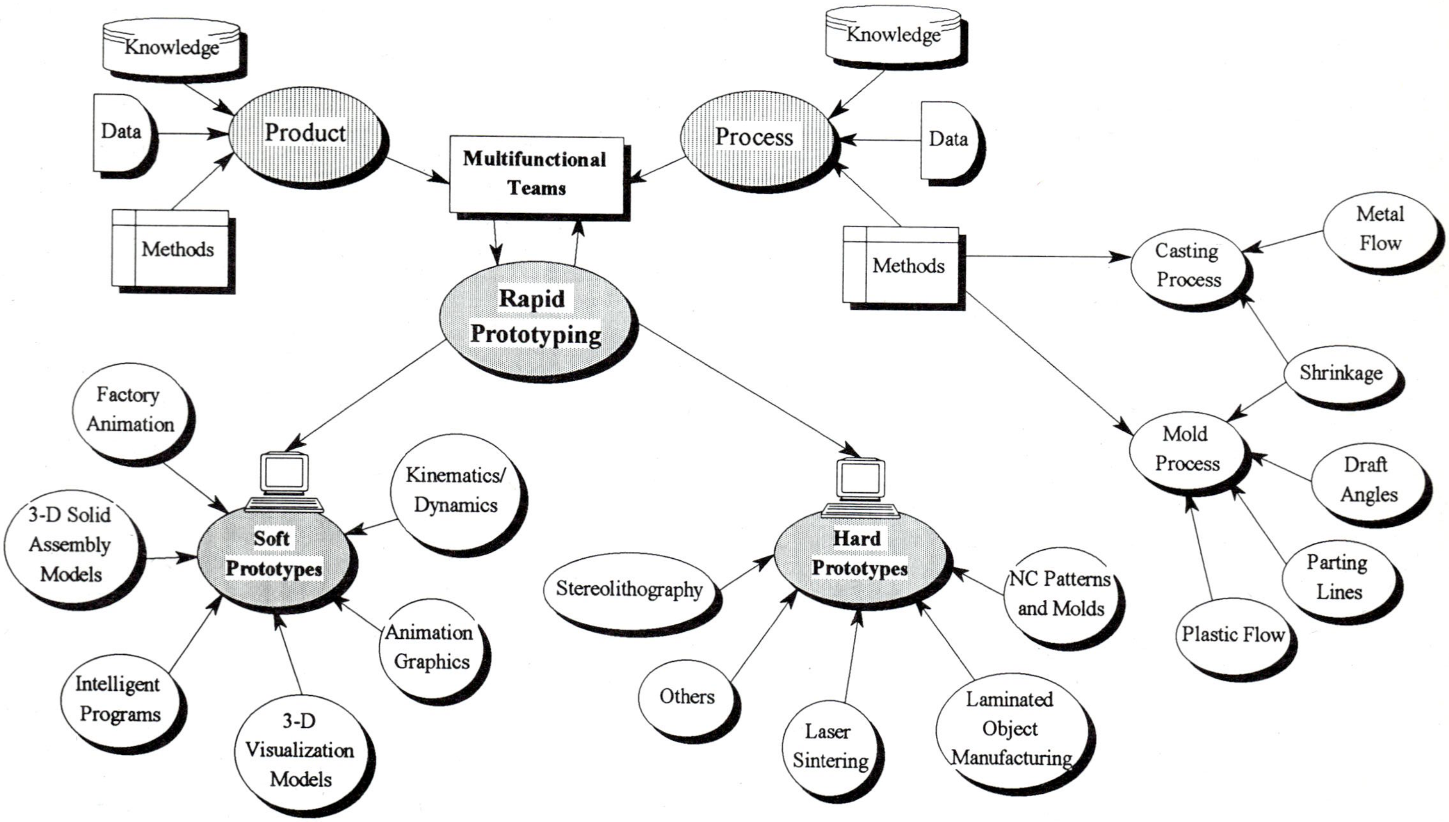

FIGURE 9.11 Rapid Prototype Techniques

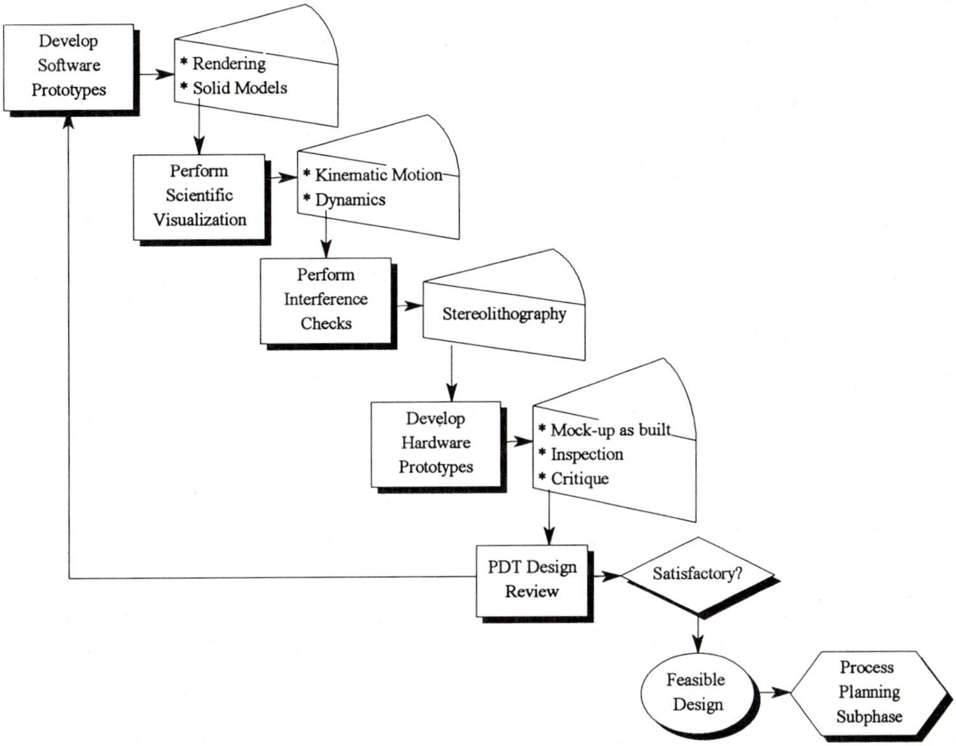

FIGURE 9.12 Prototype Subphase Information Flow

modeling environment, based on solid elements, is needed to allow design teams to rapidly simulate three-dimensional assemblies. The three-dimensional modeling environment is also needed to focus on obtaining quick hardware through an integrated stereolithography or a three-dimensional DNC (distributed numerical control) process.

9.5.5 Operations Control Loop

This deals with activities that occur after process freeze and before production release (see Figure 9.1). Members of teams from production-ready, production-implementation, manufacturing/assembly, and delivery and service are involved. It includes material handling, production of parts, assembly, fabrication, quality control, dies, etc. Many of the decisions on manufacturing have already been made at the embodiment (product synthesis) stage.

The operation synthesis loop serves as a basis for measuring functional tolerances, the intent specification, as well as quality evaluation, to form an assessment hierarchy [Menq, Yau and Wong, 1991]. An assessment hierarchy can also be generated from the relationship between the product-variation and the process-variation features list. Moreover, the deviation of a "product intended" function from its new "process intent" function becomes more evident by assessing and ranking the critical constraints of the hierarchy. It then becomes possible to feed back the result of this assessment to the design loop, and other upstream loops of a product development cycle, as shown in Figure 8.16 of Chapter 8. After critical assessment, a modification of part geometry is frequently made to make the product X-able (such as manufacturable) and, at the same time, keep the product's characteristics equal to or better, than what was achieved in previous steps.

During the process of limits checking, some modification of part geometry is required, which may or may not affect the initial design intent. It may violate the product constraints that were checked earlier or it may violate the process constraints. Thus, a strategy is required to determine when to go back and start repeating the cycle or when to continue with the in-progress checking. The following criteria can be used with the appropriate definitions of features to guide decisions during product developmental efforts:

- If the modification changes the process intent of the part and the modification itself is associated with features/parameters that are of the process-variation type, then change the value of those parameters and continue the assessment.
- If the modification does not change the process or the product intent of the part, change the value of the corresponding parameters in the next "features" list and continue the assessment.
- If the modification does not change the functional intent but changes the process intent of the part, and is associated with features that are of the product-variation type, then change the value of those parameters and continue the assessment.
- If the modification changes the functional intent of the part and is associated with features that are of the product-variation type, avoid making modification and preferably go back to the product subdomain. In a small case, if modification is definitely necessary for X-ability, only a small amount of variation in features can be allowed. On the other extreme, if the necessary modification causes variation in functional intent beyond its prejudged limit, a change of design concept or manufacturing process must be sought.

9.5.6 Product Support Loop

Product support deals with production implementation, delivery, and service aspects of the product, such as quality assurance, CPI, SPC, and maintenance issues. With quality and reliability (now being a set of requirements for an entry into the marketplace) reaching parity, product support and service is increasingly becoming an important competitive advantage tool. User requirements for product support that must be integrated into the product design process include the following:

- Serviceability, diagnostability, testability
- No trouble-found issues
- Partitioning of on-board versus off-board diagnostic capabilities
- Information system support

In Chapters 2 and 4, much has been written about the need for an overall implementation plan for concurrent engineering. It is assumed that people who implement a CE system will employ an architecture similar to that described in Chapter 5, Volume II and embrace the philosophy of CE as elucidated in Chapter 4. Regardless of the degree of success achieved in concurrent engineering implementation, the personnel who are involved in it are the ones who can improve it.

9.6 DESCRIPTION OF 3-T LOOPS

A concurrent design activity for product realization is shown in Figure 9.13. As indicated, CE domain is branched into two halves: product management and process management half-cycles.

$$\text{CE Domain} \equiv \cup\,[\text{Product management half-cycle} + \text{Process management half-cycle}] \qquad (9.13)$$

The process synthesis loop itself is also branched up into two parts: planning and execution. There are six loops which have been branched into two halves. The first three loops belong to the product management half-cycle and the next three loops belong to the process management half-cycle. The process planning part falls into the product management and the process execution part belongs to the process management subdomain.

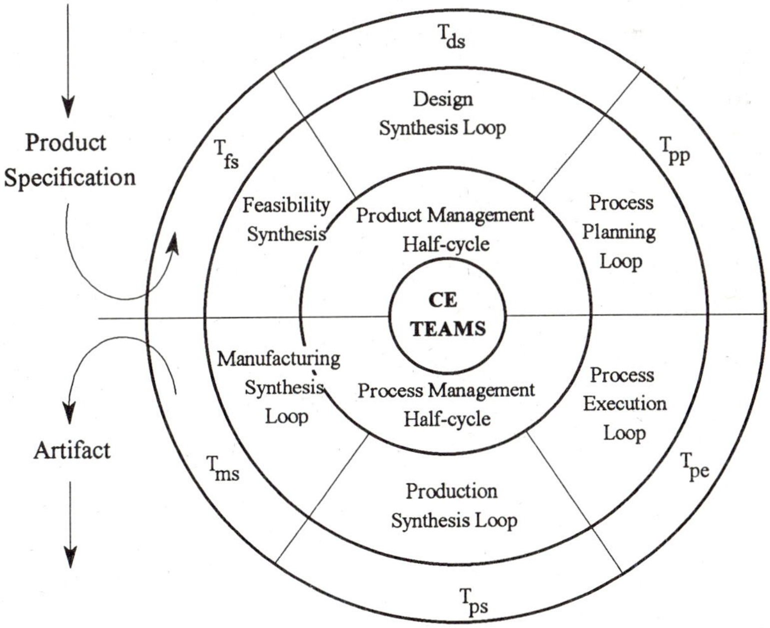

FIGURE 9.13 A State of Transformation Through 3-T Loops

$$\text{Product management half-cycle} \equiv \cup\,[\text{Feasibility synthesis loop, Product design synthesis loop, Process planning loop}] \quad (9.14)$$

$$\text{Process management half-cycle} \equiv \cup\,[\text{Process Execution loop, Production synthesis loop, Manufacturing synthesis loop}] \quad (9.15)$$

Each loop forms its own functional requirements, but the manner in which they are branched and later bounded into the "whole" system enables CE teams to systematically address the integration issues. Section 8.7.1 explains this methodology.

The goal of the CE team is to evolve the product through a series of branch-and-bound loops to create a manufacturing design intent that can yield a desired production quantity at the target production cost so that the company can make a fair profit. The 3-T loops, represented by unshaded ellipses, interact simultaneously with at least three tracks (see Figure 9.1). It is expected that this will result in a more integrated path for product realization.

9.6.1 Applying 3-T Loops for Product Development

A product development consists of a series of transformations (mapping between inputs into outputs) to provide its desired functions. Inputs represent the problem parameters that govern the statement of the problems and the environment in which they will function or operate. Functional requirements can be specified as form features, problem parameters, relations, or geometrical shapes. Constraints are imposed so that the product performs as expected. There are six steps to product development (see Figure 9.14):

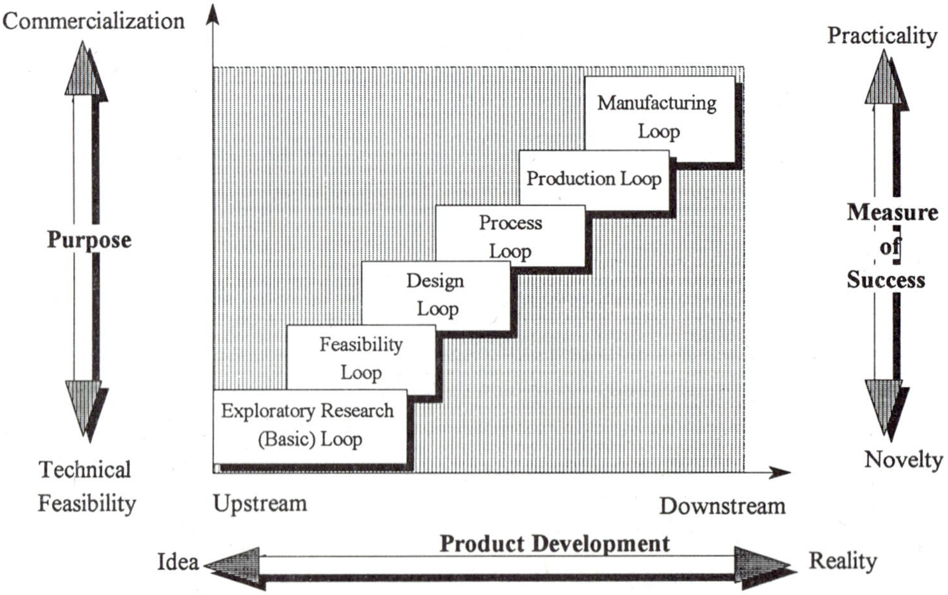

FIGURE 9.14 Major Aspects of a Product Development (Realization)

- Exploratory research (basic) loop
- Feasibility synthesis loop
- Design synthesis loop
- Process synthesis loop
- Production synthesis loop
- Manufacturing synthesis loop

A product definition starts with specifications (for each track, cycle, and loop) and a corresponding specification document. As in 1-T and 2-T loop cases, specifications (inputs, requirements, and constraints) can also be decomposed into 3-T loops—representing the six major aspects of product development.

The product requirements (PRs) are defined to satisfy a given set of customer needs including the basic functions, specifications, environment, and life expectancy, etc. Generally speaking, the product requirements are those essential engineering-related characteristics, which are transformed from perceived customer needs, and describe what the product is intended to do, as contrasted with purely physical properties. Several attempts at representing PRs and linking them to the design process have been made recently [Chen and Menq, 1992]. Nam Suh of MIT [Suh, 1990] employed an axiom-based approach where he developed specific methodologies for analysis and problem solving (see Section 4.3.1, Volume II for an additional description). Some researchers [Wolter and Chandrasekaran, 1991] used a perspective approach where they built a representation of function and geometry. PRs were transformed into perspectives such as dynamic, thermal, electrical, and kinematics considerations. Researchers [Sthanusubramonian, Rinderle, and Finger, 1991] used a decomposition-based approach to transform the design specifications for a mechanical system into a collection of mechanical components.

A decision or review is performed after each step to ensure that the requirements allocated at each level are met. For example, during product design definitions, a question is posed: "Are the product design definitions acceptable? During process definitions, a question is posed: "Are the process definitions acceptable?" During production definitions, a question is posed: "Are the production definitions acceptable?" The step reviews are shown by a rhombus shaped icon in Figure 9.15. At the end of the manufacturing definitions, an overall design review is performed to determine whether or not the product is manufacturable with respect to all relevant RCs. The control is returned to the appropriate steps, as shown in Figure 9.15. If the product is not competitive or profitable, problems may occur with human factors, with manufacturing allocations (RCs), with design allocations (RCs), and with customer expectations. Only when problems occur with customer expectations, the market needs, schedules, or overall product needs are modified. If every step review and the manufacturing review at the end is passed, it means the product has met all the allocated specifications and is ready for production.

The product/process development is the process of filtering or narrowing down a set of initial requirements and constraints (RCs) into a residual set (see Figure 9.16) congruous to a target system. At the start of the process, the product is at the state of the baseline system. The product subdomain feeds in product RCs while the production subdomain adds in process RCs. At the baseline state there may exist a number of such initial RCs

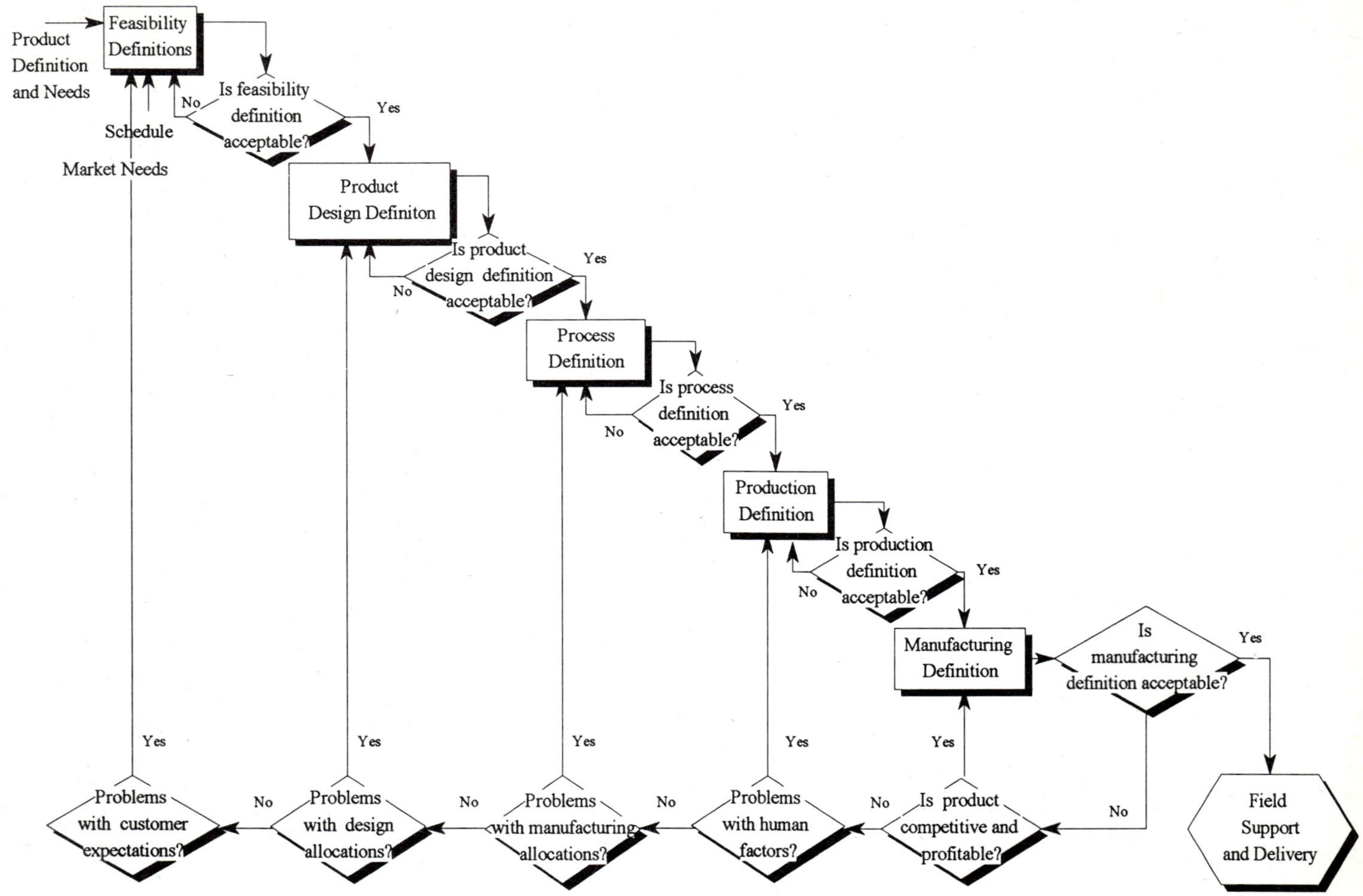

FIGURE 9.15 Information Flow Diagram for an Entire Product Life-cycle

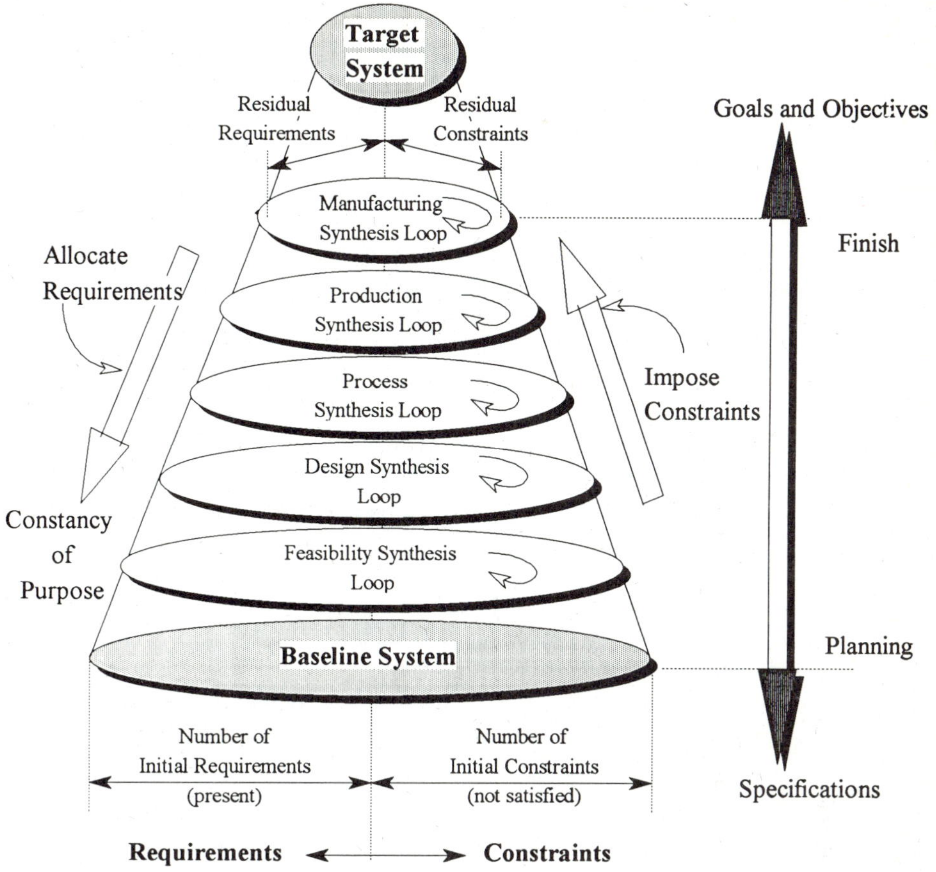

FIGURE 9.16 Narrowing of Requirements and Constraints as Product Development Continues

that may not be satisfied. There are requirements and constraints at all loop levels. For example, the functional requirements during an embodiment design (feasibility synthesis and design synthesis loops) produce design requirements. Similarly, the functional requirements during the process, production, and manufacturing synthesis loops produce process, production and manufacturing requirements, respectively (see Figure 9.16). To ensure compatibility among the loops in this framework, constraint violation techniques are employed, most of which are quantitative descriptions. These constraints may originate from the embodiment of a design (product feasibility synthesis and design synthesis loops) and may have influences on the subsequent production and manufacturing activities. Constraints can originate from downstream activities of the production subdomain, such as process planning, manufacturing, assembly, and inspection, and may have an influence over the design specifications. The satisfaction of RCs yields a target system—a functional, physical artifact. This process of narrowing the RCs as product development continues is not unique; the actual outcome depends on a designer's individual creative process. It is possible to obtain a number of competitive design solutions that meet the residual RCs objectives. In the following, each of the 3-T loops is discussed individually.

9.6.2 Feasibility Synthesis Loop

Having developed a mission and a concept (or several concepts) for the product, it is wise to assess feasibility before committing large sums of capital outlay on design and development. This step establishes the preliminary feasibility of the candidate system subject to the constraints imposed by technology limitations. A feasibility study is like conducting steps of the design loop in a miniature form. Supporting data includes published results, competitors' data, and results from key experiments. As shown in Figure 9.17, the goal of feasibility synthesis is to produce a detailed economic model description of the component's functions to aid in ascertaining the viability of the product. The economic interaction models describe basic behaviors that the characteristic system, subsystem, and components must possess to carry out the mission. Together they form a complete statement of feasibility requirements. Economic analyses (such as QFD or planning matrix) guide the design in meeting performance goals. The functional breakdown structure of product (PtBS), process (PsBS), and work (WBS) can be used as a basis for the allocations of RCs to the economic interaction models.

The steps in the feasibility subloop are shown in Figure 9.18. The feasibility loop is a nested iterative process. It has four subloops, starting with the project mission's definition and ending with specifications and substantiation of product requirements:

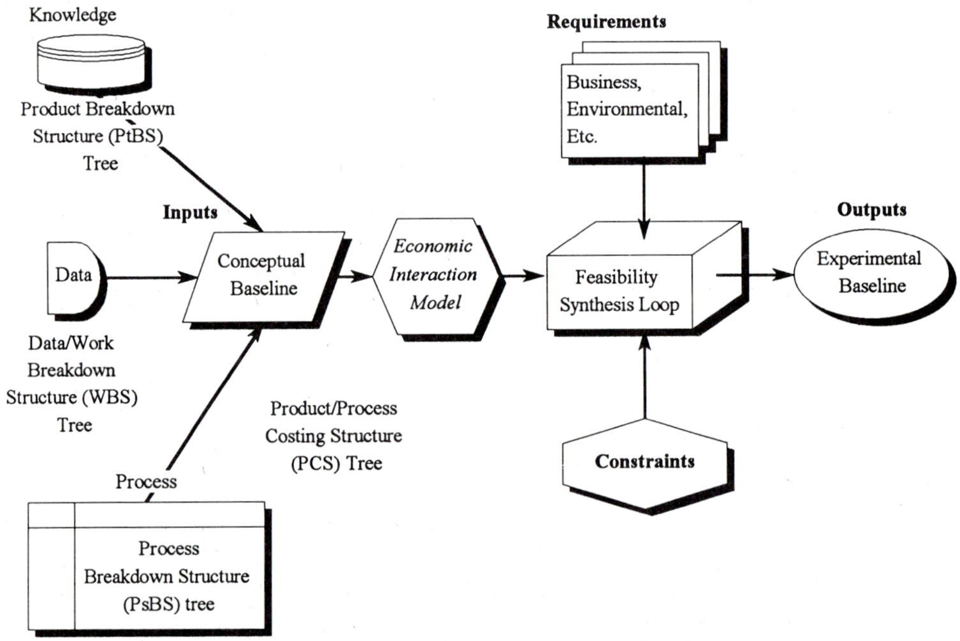

FIGURE 9.17 Feasibility Synthesis Loop Management

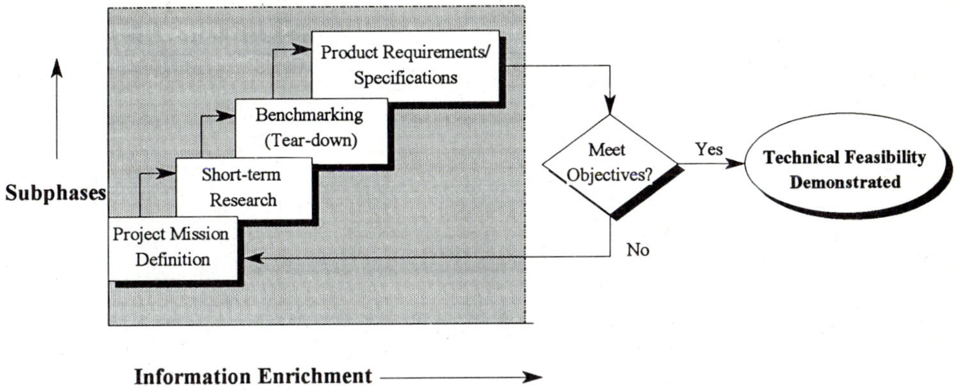

FIGURE 9.18 Subphases of a Feasibility Loop

- The mission definitions involve nonnumerical or nonalgorithmic information. The problem at this step (mission) is often ill-structured and is usually not amenable to algorithmic computation. Coming up with an initial feasible solution requires application of reasoning and decision rules, derived from the domain knowledge and the teams' experience.
- The second step of the feasibility subloop is a short-term research. Iteration is continued for the second subloop (short-term research on subsystems and component refinements).
- The third step of the feasibility subloop is a benchmark (or tear-down) function. Iteration is continued for the third subloop (benchmark) to verify that product concepts meet all functional requirements (objectives).
- The fourth step is the creation of product requirements.

When iteration through all four subloops is complete, the result is a detailed model containing the description of the desired functional performance of the product, and a set of models of functional interactions. Such models generally go through a continuous evaluation of improvements and refinements of the design as the product matures. However, many functional elements may remain unsatisfied at this stage of product realization.

9.6.3 Design Synthesis Loop

There are two stages of design synthesis loop (see Figure 9.19):

- Top-down half-loop
- Bottom-up half-loop

The top-down half-loop defines a breakdown structure of the work (WBS) tree, product (PtBS) tree, and process (PsBS) tree. At the lowest level, each part is initially de-

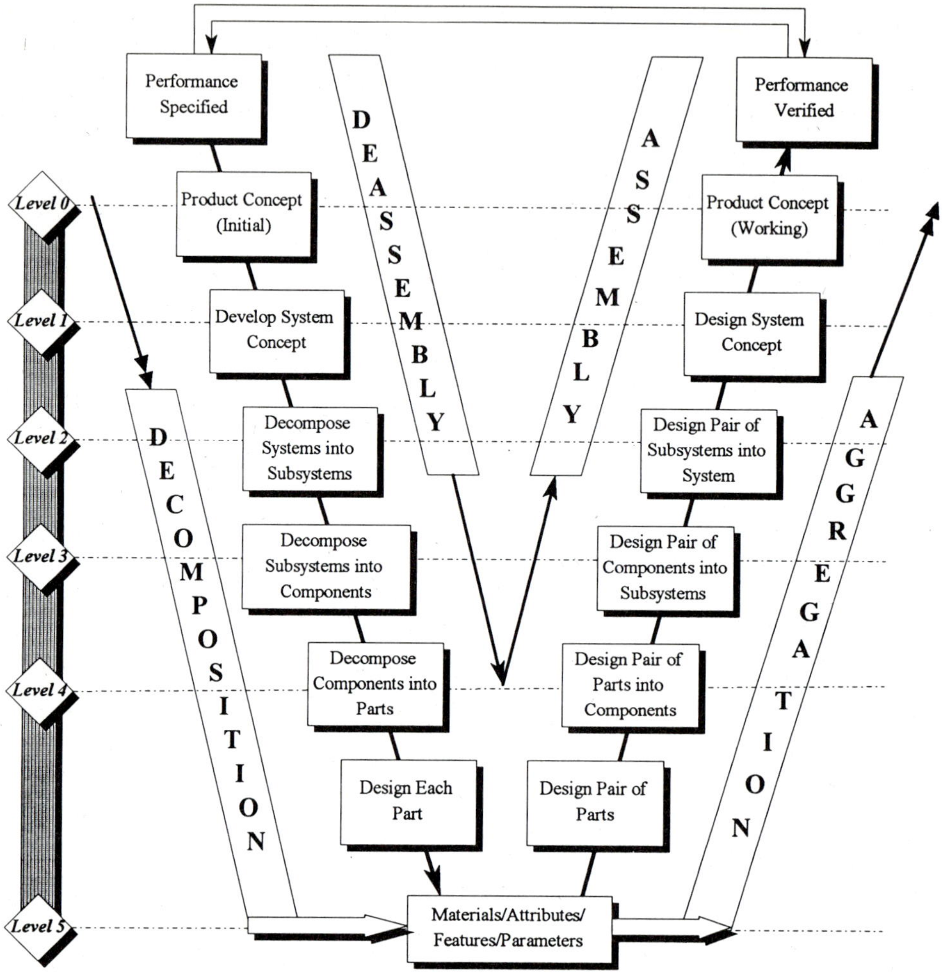

FIGURE 9.19 Steps in a Product Design Loop—Whole to the Parts and Parts to the Whole

signed in terms of descriptive form features, materials, etc. The bottom-up half-loop drives the aggregation or integration process. Concurrency is achieved by considering the design of multiple parts, multiple components, multiple subsystems, etc. The multiple parts' synthesis speeds up the integration process, but may create waste when alternate parts are eliminated.

9.6.3.1 Top-down Half-loop

The top-down structure defines the breakdown structure of the parts and the tasks to be designed. PtBS organizes a product structure by using a differentiation technique. The top-down synthesis begins with "inside-out"—starts from a set of initial total product concepts (this is level 0). They are used in developing a system concept. The differentiation technique is employed to decompose the system concept into smaller and smaller elements. A six-level breakdown structure tree is shown in Figure 9.19. In the top-down product design, half-loop decomposition produces a set of functional requirements to be imposed on the "parts of the whole." Systems are decomposed into subsystems, subsystems into components, components into parts, and finally parts into materials, attributes, features, and parameters. During this top-down process, different teams can be assigned to work concurrently on different tree levels (product networks).

9.6.3.2 Bottom-up Half-loop

The bottom-up synthesis begins with the "outside-in," where a pair of parts is designed from basic information (such as materials, features, etc.). Later, it is rolled up into larger and larger parts, components, subsystems, and then into a system via a design aggregation process (see Figure 9.19). During this process, the decomposed parts (in top-down half-loop) are assembled to form components. Components are brought together to form subsystems. Subsystems are assembled to form systems, and finally, systems are brought together to construct the artifact as a whole (see Figure 9.19). Most often, a product is composed of two or more constituents. Each component or part of the component has specified nominal design parameters and tolerance assignments. During assembly, when these parts are brought together to form the product, they may not function properly. In fact, many of the critical design functions that the product should possess are not functions of a single component, but are functions of the assemblage of components. In order for the parts to function well, the relationship between the assembled parts must be established.

The aggregation process (in the bottom-up approach) requires a set of constraints on the assemblage to function as a "whole-of-the-parts." As shown in Figure 9.19, a representation of the product design loop can be based on two half-loops:

- One half-loop to generate the possible disassembly sequence, based on the so-called product requirements
- Another half-loop for successfully generating the mechanical assembly, based on the so-called assemblage constraints.

The assembly sequence of a set of components plays a key role in determining the various constraint characteristics of the assembly tasks. Aspects, such as difficulty of assembly steps, needs for fixtures, potential for damage during assembly, and cost of assembly, are all affected by the assembly sequence. The spatial and functional relationships imposed on the parts, components, subsystems, etc., as a result of the various levels of aggregation, are characterized as assembly constraints. By identifying all the assembly relations among the parts, components, subsystems, etc., a full set of assembly constraints of a product can be established. These assembly constraints are typical of assembly modeling needs, such as the assembly relation graph or an assembly plan. An assembly plan may contain both assembly constraints and other geometrical considerations, including assembly sequences. The assembly plan must contain the part mating conditions and a set of assembly criteria designed to avoid redundant and awkward assembly sequences.

There are two ways in which a V-assembly or a pair-wise coupling can be achieved:

- ***Cumulative Aggregation:*** The assembly strategy first, starts with designing a pair of parts and then extending this partial assembly to other elements of a product decomposed network for coupling more and more parts (see Figure 9.20). By incorporating these pair-wise considerations of assembly constraints in the design loop, the functional requirements associated with disassembly (possible reverse assembly) sequences are satisfied. The complete product subdomain is thus capable of identifying the assembly requirements, incorporating constraints from assembly assessment, and satisfying them early in the functional design stage.
- ***Aggregation with Several Cores:*** Obviously, whenever there is more than one pair of parts whose assembly constraints are very much in conflict, this strategy can be improved by taking a pair-wise distance of the assembly plans from each other. An efficient way of generating total assembly is to generate the sets of larger and larger partial assemblies by giving preferences to the generation of assemblies with the most conflicting (that is, the nearest) plans.

Decision making can also be based on assessments. Examples include minimum number of changes made during assembly, minimum number of components in a subassembly, geometry compatibility, or topology validation, etc. Most of the assembly type process constraints and topology-related product constraints are checked during this bottom-up process. Also, the quality of a product can be better characterized with the help of the decomposition of requirements and the aggregation of constraints.

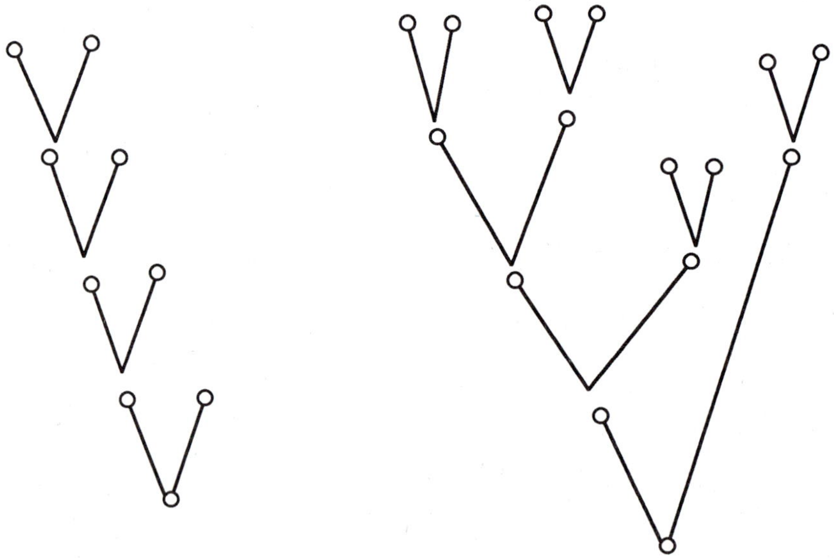

(a) Cumulative Aggregation *(b) Aggregation with Several Cores*

FIGURE 9.20 Aggregation Methods

Many researchers have investigated techniques for generating assembly sequences [Ko and Lee, 1987; Woo, 1987; DeMello and Sanderson, 1991; Baldwin, Abell, Lui, de Fazio, and Whitney, 1991], but research in assembly modeling which describes the interrelationship between the assembled components, is lacking. Recently, Turner established several types of assembly relationships as a part of his automatic tolerance research [Turner, 1991]. These assembly relations between part positions are useful since they can be transformed into product constraints.

The second-level breakdown of proposed taxonomy, for the design synthesis loop, is shown in Figure 9.21. It is shown to follow a previously defined five block schemata. The product synthesis loop is in the middle. It is surrounded by inputs to its left, outputs to its right, requirements on its top, and constraints on its bottom. Inputs to the right are compactly made out of two interconnected elements: a predictor, and a corrector. The experimental baseline state (shown by a parallelogram icon), along with product information (data, process, and knowledge) make up the predictor. The PtBS tree forms the knowledge; PsBS tree forms the process; and WBS tree forms the work. The product/process costing structure (PCS) tree is a source of cost data, which follows the structure of WBS. Figure 9.21 contains the expanded form of these abbreviations. The tradeoff analysis model serves as a corrector. A decision-logic (not shown in this Figure 9.21) is made an integral part of the corrector. Requirements are driven by performance, design geometry, and DFX (such as

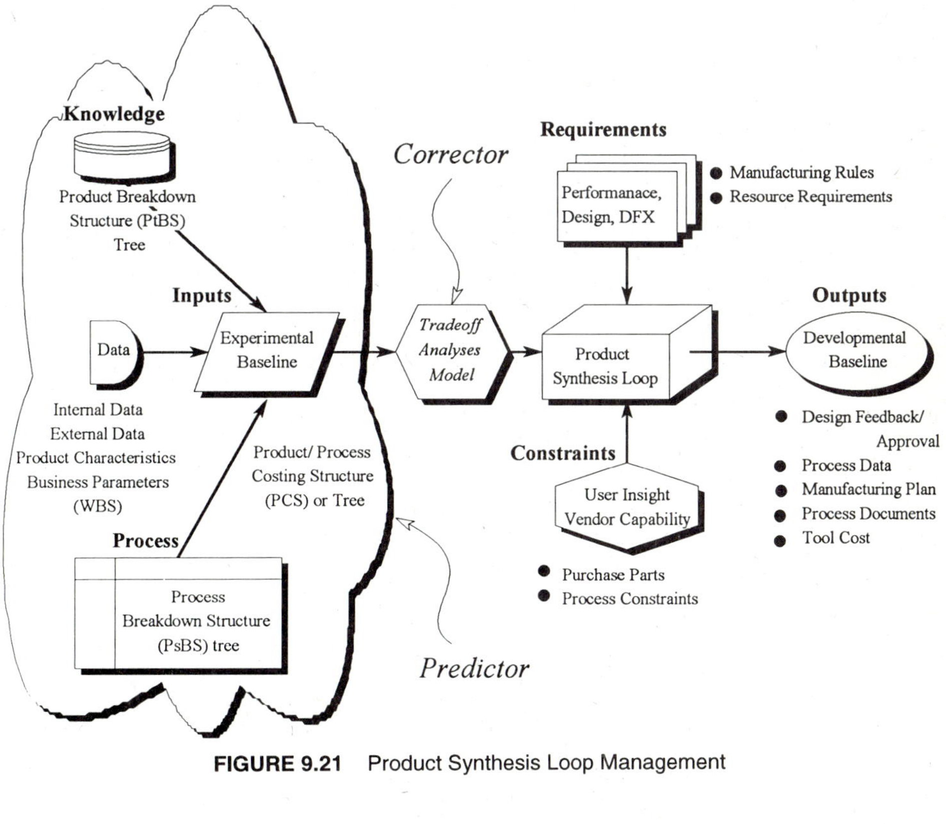

FIGURE 9.21 Product Synthesis Loop Management

manufacturing rules and resource requirements), and constraints are mostly imposed by user insight and vendor capability (such as JIT, purchase parts, process constraints, etc.). The output is a developmental baseline with other associated parameters, such as design feedback, process data, manufacturing plan, process documents, tool cost, etc.

The subphases of the design loop are shown in Figure 9.22. The loop consists of a nested iterative design process with four subloops: conceptual design, configuration or layout design, detail design, and manufacturing design. The design loop starts with problem needs and ends with an instance of an artifact.

Conceptual design is usually made out of three additional sub-subloops: functional intent design, engineering design (phenomenological), and embodiment design:

- The ***functional intent design*** determines the general specification, primary intent, important parameters and functions to be performed.
- The ***engineering design*** selects the primary structural form, alternate materials and components, and establishes phenomenological design rationale, engineering principles, etc.
- The ***embodiment design*** provides the components' design and performance measures, in both numerical and symbolic forms, to the configuration design subtasks.

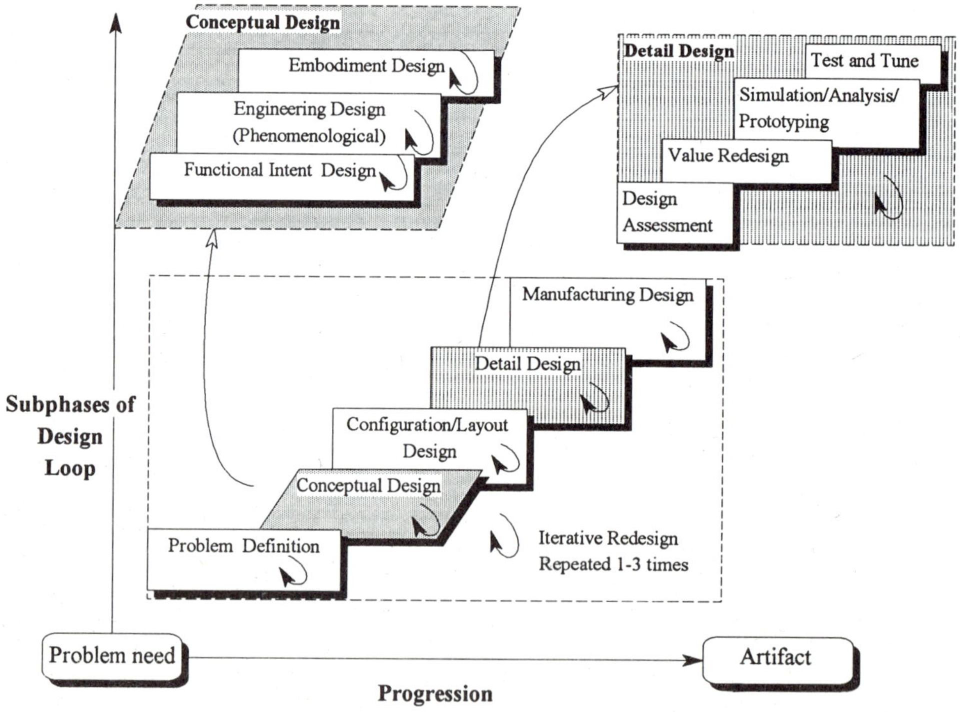

FIGURE 9.22 Subphases of the Design Loop

Conceptual design flushes out the functional intent into a series of layout alternatives. The steps in the conceptual design subloop are shown in Figure 9.23. The conceptual design is best captured as a "smart model" such as those created using parametric, feature-based, or knowledge-based concepts. The iterative redesign symbol is used here to indicate a synthesis loop with analysis or evaluation assessment of the baseline design, along with some form of redesign logic or optimization (see Figure 9.22). The idea is very similar to the branch-and-bound concept that was discussed in Section 8.7.1. This concept can be employed during this iterative process for zoning the problem at each iterative subloop and thereby finally arriving at an acceptable solution.

Configuration or Layout design determines the connections and relative positions of all components for a conceptually designed system. Designs are usually composed from an identified set of components determined during conceptual design, but for which allowable configurations are not fixed. The steps in the configuration layout design subloop are shown in Figure 9.24. Configuration design starts with the following:

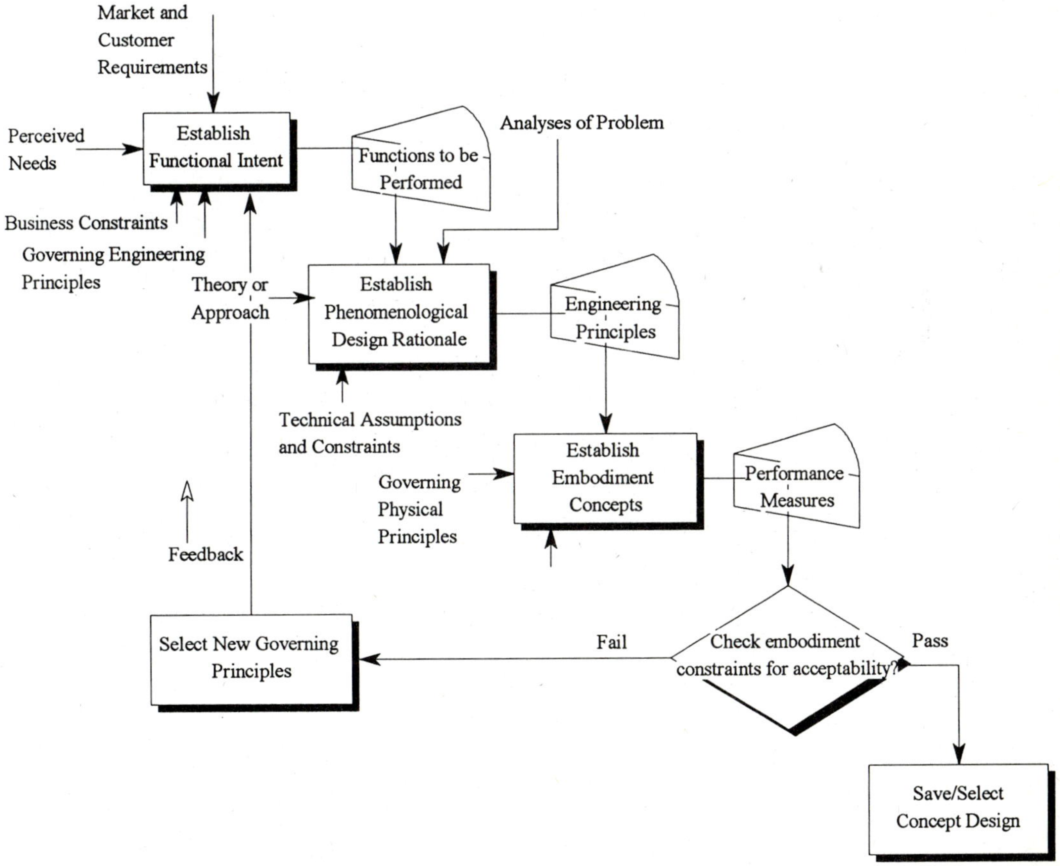

FIGURE 9.23 Steps in a Conceptual Design Subphase

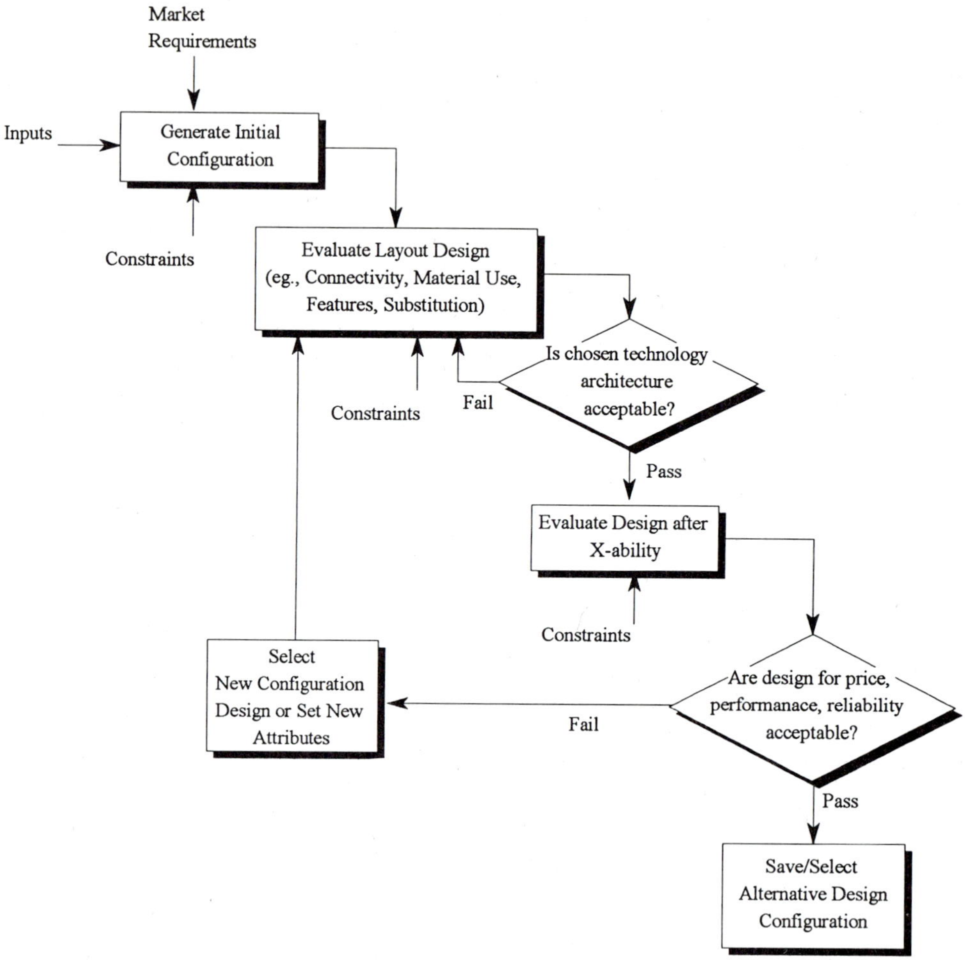

FIGURE 9.24 Steps in Configuration Design

- The conceptual design as its baseline
- User requirements of its layout (e.g., whether or not fasteners are to be used, or a chosen technology architecture is acceptable)
- Constraints limiting the domain of alternatives (such as X-ability, price, performance, reliability, etc.).

The output of the configuration subloop is a set of possible arrangements (layouts) for configuration design, including relative positioning and orientation information. At this point in the design process, layout of an artifact, in terms of its attributes and relationships, not values, is defined. During this subloop, configuration is evaluated based on criteria such as connectivity, load paths, manufacturability, material use, etc. (See Figure 9.24). Configuration design is a hybrid engineering problem of performing algorithmic numerical computation and symbolic inference simultaneously. Selection is made and a particular design layout is chosen from a set of possible candidates. Choices made during configuration design can influence such things as material choices, fabrication techniques, capital equipment decisions, and supplier contracts.

Detail design begins by obtaining an instance of the CAD geometry and topological layout passed down from the configuration subloop. The steps in the detail design subloop are shown in Figure 9.25. The four additional sub-subloops of the detail design are as follows:

- *Design Assessment:* The first step is to assess the design from the results of X-ability assessments.
- *Value Redesign:* Value redesign adds dimensions, notes, details, etc. to the CAD geometry.
- *Evaluation:* Simulation, analysis, or prototyping is part of evaluation.
- *Test and Tune:* Testing and tuning sub-subloop changes tolerance values to determine its acceptability.

The major attributes of the design at this stage are known and their numerical values are varied. Detail design fully defines the individual components so that they can be manufactured.

In manufacturing design, the alternatives are numerically evaluated for a set of criteria, such as variability reduction, manufacturing cost, tolerance, manufacturability, produceability, maintainability, etc. Once the design is deemed satisfactory, it is stored (see Figure 9.25).

9.6.4 Process Synthesis Loop (Planning Phase)

The process synthesis loop is an important activity, linking product design and manufacturing operations. At the beginning of the process synthesis loop, the geometric shapes and sizes of a part, along with its functional surfaces (that is, required finish) are identified. The initial assembled model of a product is inadequate to specify the required process considerations. An exploded view from its initial assembled model is required to determine the precedence information implicit with exploded views. The loop also needs a basis for the identification of the parts function within the product and allowable process-type variations. Examples of such variations include material and heat treatment requirements, geometric and dimensional tolerances, and surface finish, etc. In addition, the manufacturing tolerances and mating specifications are also needed to evaluate dimensional quality and to plan prevention [Menq, Yau and Wong, 1991]. This can be accomplished by extending the definition of three-dimensional CAD features to include process-type variations as well as assembly features. The types of assembly features include the following:

- Relationship between parts within a product—this indicates the nature of parts' connection such as free contact, tight fit, etc.
- Relative position of each individual part in a product—this is indicated by positioning information.

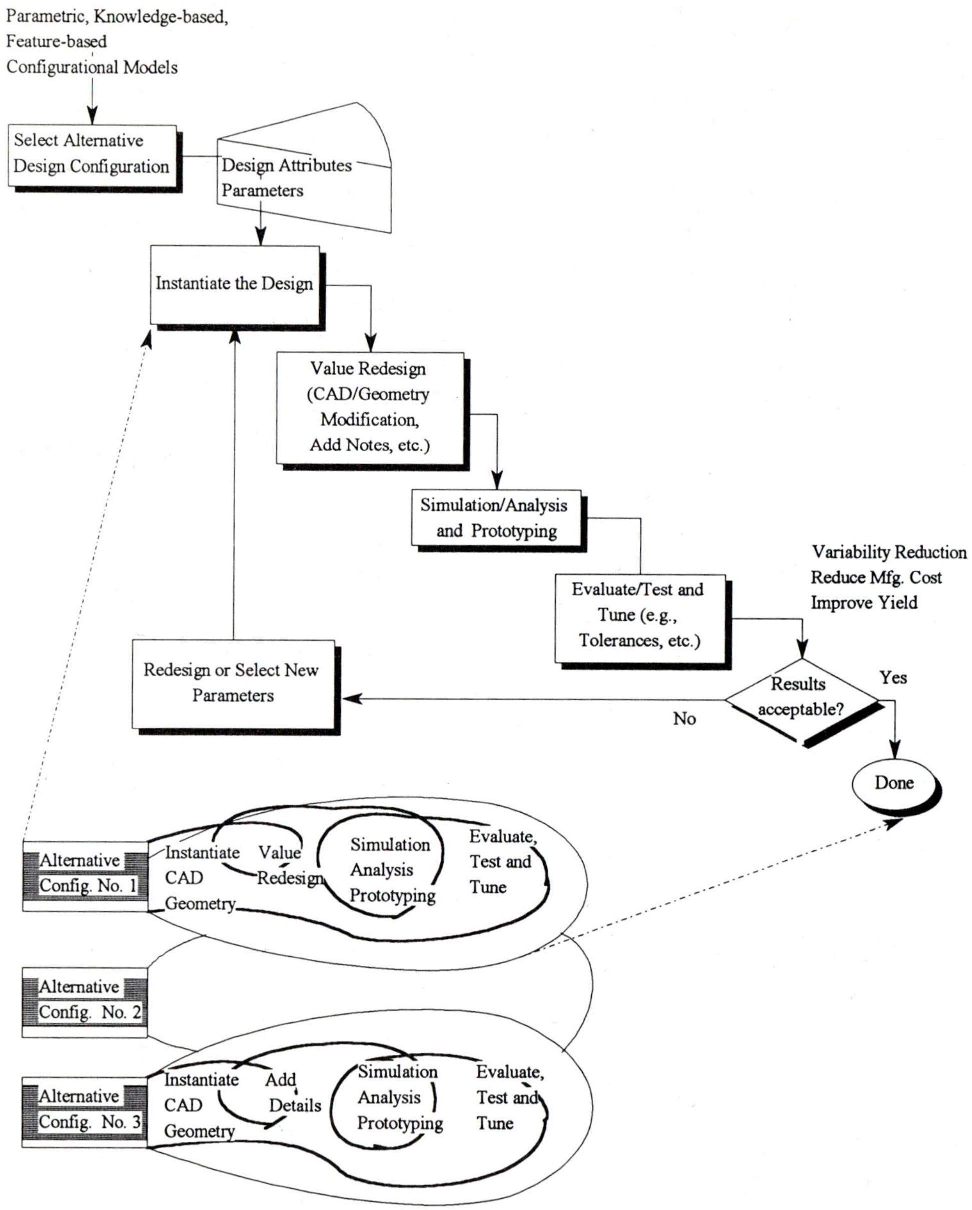

FIGURE 9.25 Steps in a Detailed Design

By using combined features' forms, along with the three-dimensional CAD model of a design, the part not only has a better description of its functional intent (such as geometric shape), but also has an explicit expression of how it can be processed. For example, if a "feature" definition contains manufacturing tolerances, features can be employed, along with a CAD model, to determine the best manufacturing plans and suitable assembly operation sequences according to the available manufacturing capability. The representation of RCs can be integrated with process knowledge to support the concurrent design activity, and perform functionality and other X-ability assessment. The product should be driven towards minimum human intervention. If a computer can be used to automate some product design subphases, it should be encouraged at every step.

9.6.5 Process Synthesis Loop (Execution Phase)

One of the current obstacles in X-ability assessment in the redesign process is loss of time in iterations to satisfy all specifications. There is a considerable gap between the time when designers and engineers are done with a part design and the time when actual production starts. Often, this time is spent in painstakingly evaluating a design, only to find out that some of the RCs cannot be satisfied. This leads to a return at the start of an iteration. With the hierarchical decomposition of specifications and systematic distributions to the appropriate loops (as proposed earlier in Section 8.7.1), a set of independent RCs is satisfied at each loop as the product develops from raw materials to manufacturing. Each loop is self-contained, that is, the specifications are well-defined (what the product intends to do). The RCs are well-posed (meaning most of what is included can be satisfied at a particular loop level).

Manufacturability is often assessed by experienced manufacturing engineers. A successful design should conform functionally while being acceptable (meaning X-able) and compatible with a selected manufacturing process. The second level taxonomy for the process synthesis loop is shown in Figure 9.26. It is shown to follow a previously defined five block schemata with the synthesis loop in the middle. It is surrounded by inputs to its left, outputs to its right, requirements on its top, and constraints on its bottom. The developmental baseline state (shown by a parallelogram icon) along with process information (data, process, and knowledge) makes up the predictor. The PtBS tree forms the knowledge. The PsBS tree forms the process, and the WBS and PCS trees form the data. The cost/benefit verification model serves as a corrector. Requirements are driven by processability (for example, reliability, maintainability, supportability) and constraints are mostly imposed by facilities, tooling, capital funding, and JIT parts deliverables. The output is a production baseline.

During the process synthesis (execution) loop (as shown in Figure 9.27) decision making is based on raw stock, tool selection, fixture, cutting data, etc. An assessment of time and cost standards are performed. This analysis involves the use of well-established process plan details, raw stock, and tooling information. By integrating processing (machining and routing) knowledge with machine, tool, and fixture capabilities of manufacturing plant operations, many process planning constraints can be satisfied. The decision

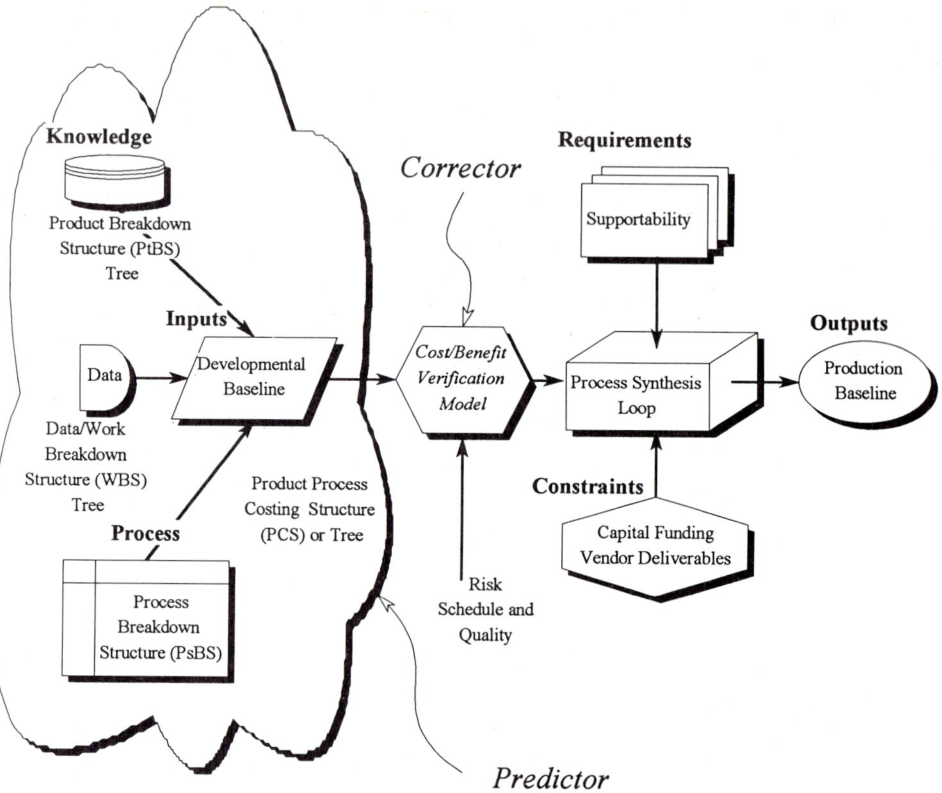

FIGURE 9.26 Process Synthesis Loop Management

basis can be implemented using a group-technology based process, or with a knowledge-based tool. The output of the process synthesis loop is the finalized process plan, master production schedule, manufacturing execution plan, and cost estimates. There are several research efforts relating to automating the manufacturability assessment process [Luby, Dixon, and Simmons, 1986; Chen, Miller, and Vemuri, 1991].

9.6.6 Production Synthesis Loop

The layout of a factory system is vital to the realization of a time-based competitive advantage. The production synthesis system (see Figure 9.28) is one in which tooling and machine equipment is organized and sequenced according to the production schedule specifications to produce the product. The second level taxonomy for the production synthesis loop is shown in Figure 9.28. It is shown to follow a previously defined five block scheme. Inputs to the right are compactly made out of two interconnected elements: a predictor, and a corrector. The production baseline state (shown by a parallelogram icon) along with production information (WBS, PsBS, PtBS, and PCS) makes up the predictor.

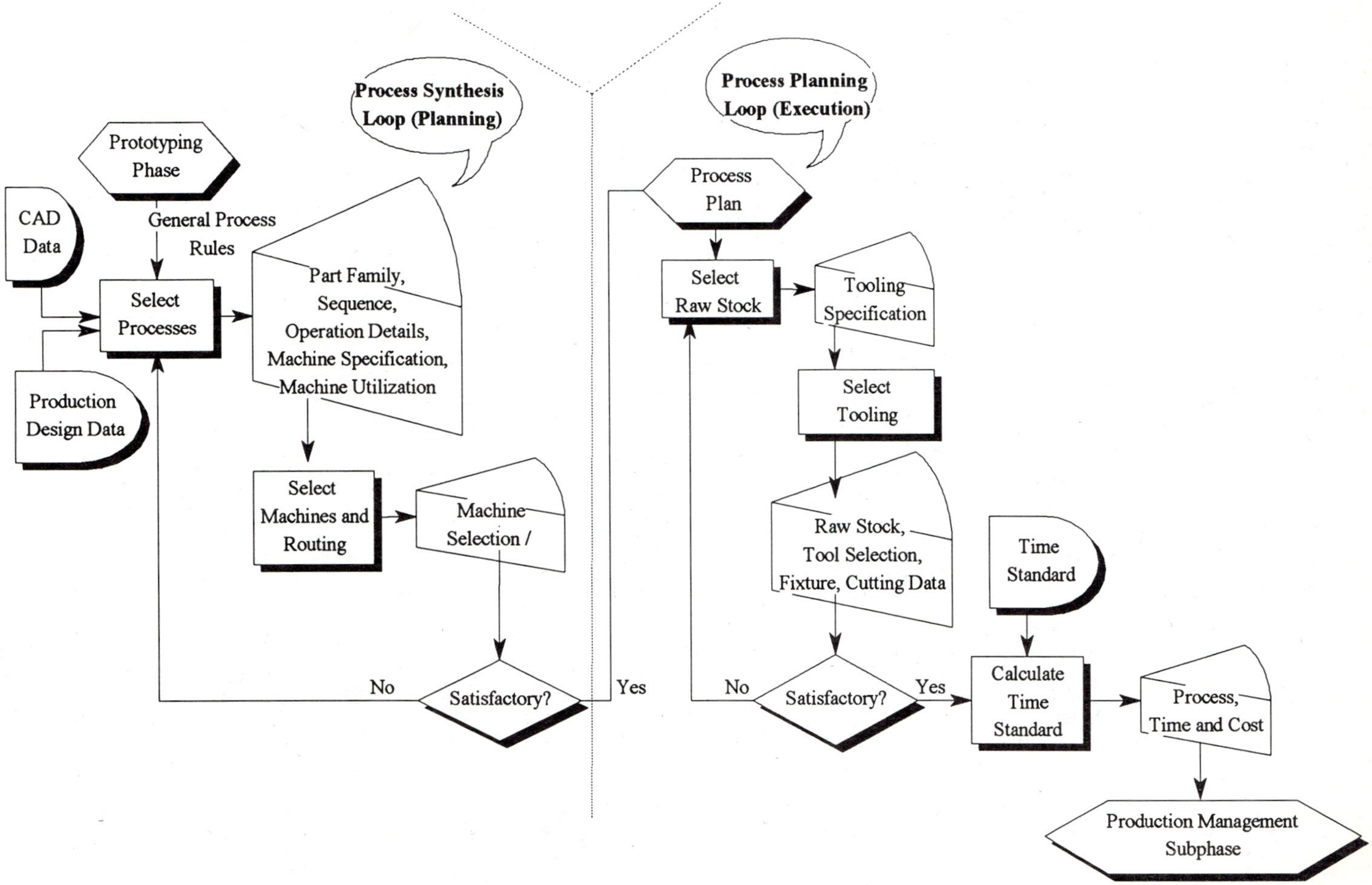

FIGURE 9.27 Flow of Information in a Process Synthesis Loop

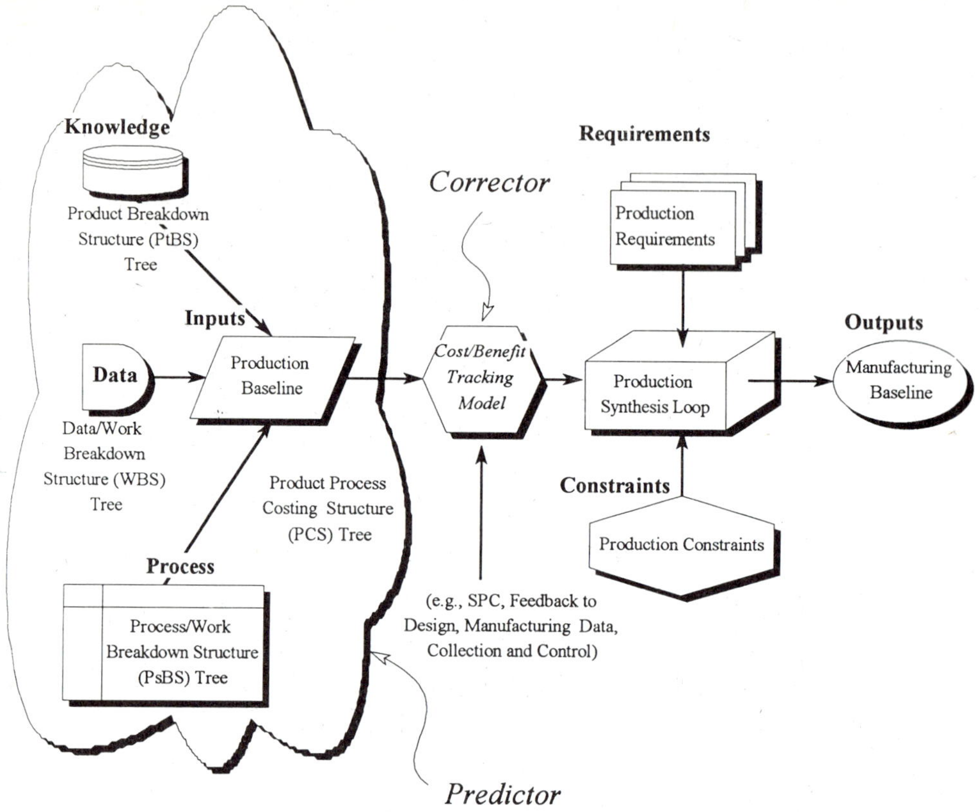

FIGURE 9.28 Production Synthesis Loop Management

The cost/benefit tracking model (for example, SPC, feedback to design, manufacturing data, and collection and control) serves as a corrector. RCs are driven by corresponding production limits. The output is a manufacturing baseline. The manufacturing baseline places the various machines in the exact sequence required to process a family of parts; the machines and the parts family are based on group technology analyses. Cells are arranged in a certain order—X, U, or L shapes to minimize the material handling time and motion between adjoining processes. Production synthesis includes a changeover reduction and quick setup strategies. Optimizing the production layout reduces the time required to convert raw stock to the finished part(s) and minimizes the time elapsed between the production of the last good piece of part-type "A" and the first good piece of part-type "B."

9.6.7 Manufacturing Loop

Traditional factories are usually organized by process technology (for example, cellular manufacturing) or machine centers (for example, flexible manufacturing). Parts move from one machine center to the next. Each step consumes valuable time: parts wait in the bin to be transported; then they are transported; then they wait for the next process to start. A time-based manufacturing loop is organized by product family. To minimize the handling and moving of parts, the packaging of a finished part and assembly optimizes the material flow process of the supplier (for example, JIT), manufacturing plant, dealer, or customer. This will provide a balance between manufacturing, transportation, and assembly needs, while protecting the quality of the part at the lowest overall system cost. The subphases of such a manufacturing loop are shown in Figure 9.29. It is broken up into six steps:

- The ***first step*** involves screening the initial process and manufacturing data using producibility analyses.
- The ***second step*** is component, prototype, and assembly testing. This step may utilize the process models depicting the generalized behavior (process-intent) of the manufacturing processes, DFM, and estimate variable costs associated with assembly and production.

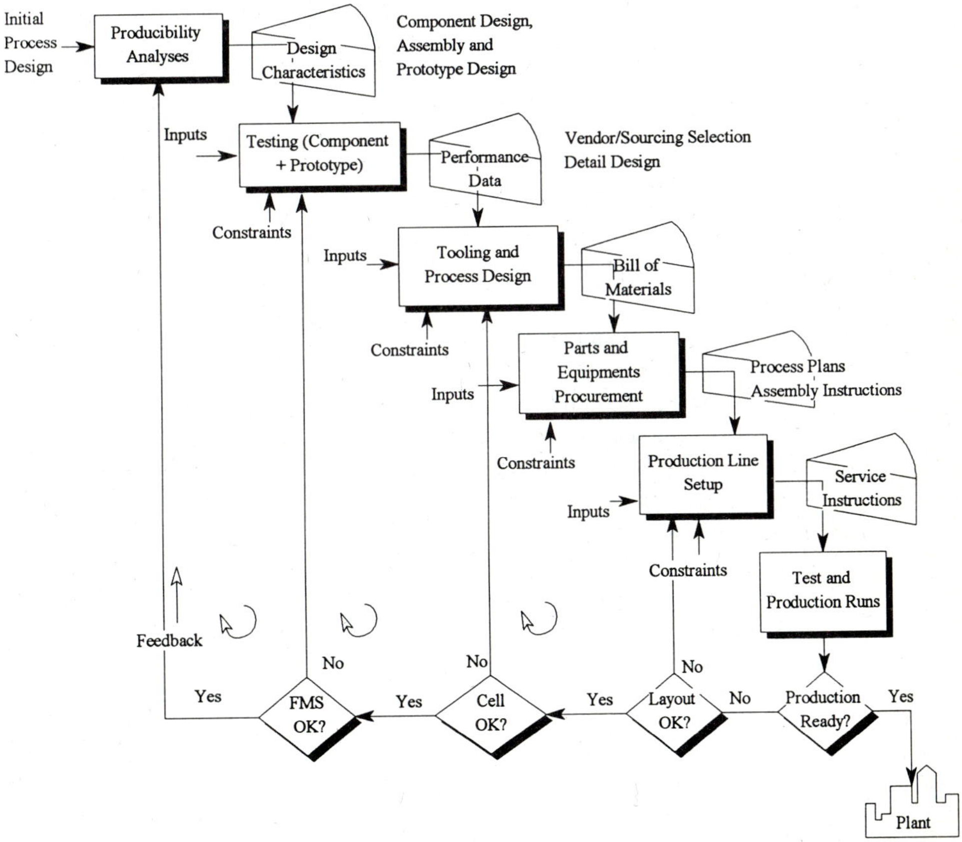

FIGURE 9.29 Information Flow Through a Manufacturing Loop

- The ***other steps*** are tooling and process design; parts and equipment procurement; production line setup; and test for production runs.

At the end of production runs, results are reviewed to determine whether or not the product is production worthy. The control is returned to the appropriate steps (as shown in Figure 9.15) if product is not production ready and problems occur with layout, with cell design, or with machine allocations (for example, flexible manufacturing system). If reviews, at every step and during the production-ready step are passed, the product is stamped for production.

9.7 MAJOR BENEFITS

Taxonomy has the following major benefits:

- ***Common Representations:*** The largest benefits occur in common representation of transformation models and specifications. In a taxonomy-based process, models are represented using abstractions. Representations are common across product lines, processes, systems, work-groups, cost, etc. Specifications are often governed by the level of abstraction. Thus, the taxonomy-based process provides a natural extension to assigning specifications. Depending on the level of abstraction associated with a transformation, specifications can be uniquely assigned to each transformation.
- ***Common Framework:*** It provides a framework of classifying the downstream operations (manufacturing) and their loops using the same class of definitions and nomenclatures as used upstream (such as planning and design).
- ***Common Method:*** The loops provide a systematic way of meeting specifications, while isolating coupling effects temporarily. It can generate the description of a physical system that is coupled, that is, where units are interacting with each other. A well documented and thought out taxonomy (design plan) and methodology can streamline the propagation of inputs, requirements, and constraints. It can enable early detection of design inconsistencies and allow more efficient revisions to eliminate the detected inconsistencies.
- ***Consistent Basis:*** It provides a consistent basis for analyzing all problems, big or small. The specifications are dynamically linked, that is, they change from one state to another as the transformation proceeds. However, the environment, domain, and process provide checks and balances so that specifications are consistent with the overall company goals, policies, and value systems. Parts flow quickly and efficiently through the factory.
- ***Assures Completeness:*** When a taxonomy-based process is followed, the organization assures completeness of all necessary considerations of metrics, measurements, and perspectives.
- ***Applicability Across All Product Lines:*** The taxonomy is not constrained for a par-

ticular line of products (such as a hair dryer or an automobile). Thus, many opportunities, in terms of possible products and services, exist for applying this process.

- ***Manageable Size:*** The size of the problem and the number of iterations is minimal since the problem is decomposed into a series of semi-independent loops that can be solved somewhat independently of each other.

In engineering fields other than product realization, the taxonomies are usually not so stringent. For example, in software or circuit designs, and in some architectural applications, there is very little or no interaction between hierarchical sets [Finger and Rinderle, 1989]. Each set more or less fulfills a single function with minimal interaction. In those cases, taxonomies are simple, since independence requirements are already satisfied.

REFERENCES

ALEXANDER, C. 1964. *Notes on the Synthesis of Form.* Cambridge: Harvard University Press.

BALDWIN, D. F., T. E. ABELL, M-C. LUI, T. L. DE FAZIO, and D. E. WHITNEY. 1991. "An Integrated Computer Aid for Generating and Evaluating Assembly Sequences for Mechanical Products." *IEEE Transaction on Robotics and Automation.* Vol. 7, No. 1 (February), pp. 78–94.

CHEN, B., and C.-H. MENQ. 1992. "Initial Attempts On the Characterization of Functional Requirements of Mechanical Products. PED-Vol. 59, *Concurrent Engineering,* ASME, edited by D. Dutta et al., Proceedings of the Winter Annual Meeting of ASME, Nov. 8–13, 1992 Anaheim, CA., pp. 315–329.

CHEN, Y-M., R. A. MILLER, K. R. VEMURI. 1991. "On Implementing an Integrated Design Manufacturability Assessment Environment." *Computers in Engineering.* Proceedings of the International Computers in Engineering Conference, ASME, Vol. 1, pp. 407–413.

DEMELLO, L., A. C. SANDERSON. 1991. "A Correct and Complete Algorithm for the Generation of Mechanical Assembly Sequences." *IEEE Transaction on Robotics and Automation.* Vol. 7, No. 2 (April), pp. 228–240.

DIXON, J. R., M. R. DUFFEY, K. I. ROHINTON, K. L. MEUNIER, and M. F. ORLUP. 1988. "A Proposed Taxonomy of Mechanical Design Problems." Proceeding of the International Computers in Engineering Conference ASME, pp. 41–46 (Aug.), San Francisco, CA.

EDER, W. E. 1990. "Design Science—Meta-Science to Engineering Design." *Design Theory and Methodology—DTM '90,* Edited by J. Rinderle, Proceedings of the 2nd International Conference on Design Theory and Methodology, ASME, DE-Vol. 27, Sept. 16–19, Chicago, IL, pp. 327–335.

FINGER, S., and J. R. RINDERLE. 1989. "A Transformational Approach to Mechanical Design Using A Bond Graph Grammar," *ASME DE-Vol. 17, Design Theory and Methodology—DTM '89,* Edited by Elmaraphy, Seering and Ullman, Proceedings of the Ist Int'l Conference on Design Theory and Methodology, ASME, Montreal, Quebec, Canada, Sept. 17–21, pp. 107–116.

HALES, C. 1993. *Managing Engineering Design.* Co-published by Longman Scientific & Technical, and John Wiley & Sons, New York.

HOUPT, T. J. 1990. "The Importance of Rapid Prototyping to Manufacturing and Integrated Product

Development." Concurrent Engineering Office, Manufacturing Technology Directorate, Wright Research and Development Center, Wright-Patterson Air Force Base. Ohio, (August).

KO, H., and K. LEE. 1987. "Automatic Assembling Procedure Generation From Mating Conditions." *Computer Aided Design.* Vol. 19, No. 1, pp. 3–10.

LARSEN, N. E., and L. ALTING, 1992. "Dynamic Planning Enriches Concurrent Process Production Planning," *International Journal of Production Research,* Vol. 30, No. 8 (Aug.), pp. 1861–1876.

LU, S.C.-Y, S. SUBRAMANYAM, and J. B. THOMPSON. 1988. "A Cooperative Product Development Environment to Realize the Simultaneous Engineering Concept," *Proceedings of the International Computers in Engineering Conference,* San Francisco, CA. (August), ASME, pp. 9–18.

LUBY, S. C., J. R. DIXON, M. K. SIMMONS. 1986. "Designing with Features: Creating and Using Features Database for Evaluation of Manufacturability of Casting." *ASME Computers in Engineering.* Proceedings of the International Computers in Engineering Conference, ASME, Vol. 2, pp. 285–292.

MENQ, C-H., H-T. YAU, and J. L. WONG. 1991. "An Intelligent Planning Environment for Automated Dimensional Inspection Using Coordinate Measuring Machines." *Transaction of the ASME, Journal of Engineering for Industry.* Vol. 114 (May), pp. 222–230.

PAHL, G., and W. BEITZ. 1988. *Engineering Design: A Systematic Approach.* New York: Springer Verlag.

POINDEXTER, J. W. 1991. "Rapid Prototyping in an Integrated Product Development Environment." *SAE Paper # 911113,* Presented at the SAE Aerospace Atlantic Conference. April 22–26, Dayton, Ohio.

PUGH, S. 1990. *Total Design—Integrated Methods for Successful Product Engineering.* Reading, MA: Addison-Wesley Publishing Company.

SHILKE, N. A., and S. M. RHODE, R. D. FRUECHTE, and J. H. RILLINGS. 1989. "GM Systems Engineering Center," GM Project Trilby, *Automotive Engineering,* Vol. 97, No. 2 (February), pp. 167–170.

STHANUSUBRAMONIAN, T., J. RINDERLE, and S. FINGER. 1991. "Transformation From Functional Specifications to Physical Form," Proceeding of the 1991 NSF Design and Manufacturing Systems Conference, University of Texas at Austin, Texas, 1991, pp. 1163–1168. Washington, DC: National Science Foundation.

SUH, N. P. 1990. *The Principles of Design.* Oxford, UK: Oxford University Press.

TURNER, J. 1991. "Research in Automated Tolerancing at RPI," Proceeding of the 1991 NSF Design and Manufacturing Systems Conference, University of Texas at Austin, TX, pp. 625–631. Washington, DC: National Science Foundation.

ULLMAN, D. G. 1991. *The Mechanical Design Process,* New York: McGraw-Hill Publishing Company (August).

ULRICH, K. T., and S. D. EPPINGER. 1994. *Product Design and Development,* New York: McGraw-Hill.

WOLTER, J., and P. CHANDRASEKARAN. 1991. "Knowledge Representation for Functional Design of Mechanical Assemblies." Proceedings of the 1991 NSF Design and Manufacturing Systems Conference. University of Texas at Austin, TX, pp. 633–639, Washington DC: National Science Foundation.

WOO, T. C. 1987. "Automatic Disassembly." Proceeding of the SAE International Computer Graphics Conference. Detroit, MI, pp. 163–168.

TEST PROBLEMS: PRODUCT REALIZATION TAXONOMY

9.1. Describe what roles interactions play in the product design and development (PDD) process?

9.2. Explain the "track and loop methodology" for product realization? What are the benefits of the loop methodology concept?

9.3. What are the fundamental issues of track and loop methodology? Discuss 2-T Loops and 3-T Loops. What is the relationship between the number of intergroup interactions and tracks?

9.4. What are the factors that affect the taxonomy scope and contents? How do you relate those effects with the CE aims discussed in Chapter 4?

9.5. How could customer satisfaction be achieved or ensured through product realization taxonomy (PRT)?

9.6. How do you determine the level of details in a product or a process decomposition? How does this relate to the level of taxonomy distributions?

9.7. What are the benefits of product realization taxonomy? In CE, how do interactions play a major role in improving the overall production?

9.8. How can PRT be integrated to the QFD product development cycle? How do QFD, Taguchi, and SPC fit together with VOC? Give examples.

9.9. Can you elaborate on why a circular shape in Figure 9.13 was used? Why were the CE teams placed in the center?

9.10. Why is product/process development a complex activity? Provide some example of system decomposition using Figure 8.6 as a guide.

9.11. What are the differences between CE taxonomy and SE? Compare and contrast the taxonomy of CE with that of SE.

9.12. Describe the parallel tracks involved in the taxonomy development procedure.

9.13. What are the major benefits of taxonomy? Describe and label the transformation steps and the corresponding specifications in making a bookshelf.

9.14. What are the loops that form 3-T interactions? Describe the loops. What could be a drawback of using 3-T loops for PDD process?

9.15. Why is it often necessary to make design revisions? Why, during an early stage, does one commonly make improper designs or insufficient interactions?

9.16. How does product realization taxonomy ensure that customers' needs and demands will be met? List the merits/benefits of applying PRT.

9.17. How does program timing relate to PRT? How can the various levels of prototypes be reduced under PRT? List the criteria that determine the choice for alternate sequences.

9.18. How would you implement an edge in "superior delivery and service" through PRT?

9.19. Discuss three representations of taxonomy descriptions. How is frame description used?

9.20. What are the various loops involved during the product management cycle? Describe parallel tracks.

9.21. How do you exploit the process track in PRT? List typical items for an automotive engine part that are found in a process track.

9.22. What are the major activities of the product engineering process? Describe the major steps in the product engineering loop.

9.23. Explain the subloops of a design synthesis loop. Define a V-process of the design synthesis loop.

9.24. What is behind the concept of the design loop? Which tracks interact in this case? What are the steps in the conceptual design phase?

9.25. What is the main point that PRT contributes to CE? Can we shorten the product development cycle?

9.26. Larger numbers of group interactions lead to more balanced output, but what are some of the disadvantages.

9.27. In Figure 9.19 aggregation follows decomposition, how could you design something when this process is reversed?

9.28. Describe the various prototyping techniques. What is the meaning of "soft" prototype? How does it differ from "hard" prototype?

9.29. How would you use an information flow diagram to analyze the control aspects of a manufacturing or an assembly track?

9.30. Compare the production and manufacturing tracks. Describe the information flow in a production track.

9.31. How can unforeseen situations that pop up during customer usage of the product be resolved during product realization taxonomy, so that warranty costs can be minimized?

9.32. What differentiates the effectiveness of simulation or analyses in solving problems? Does it matter if problems are at system-level or component-level?

INDEX